ADVENTITIOUS ROOT FORMATION IN CUTTINGS

ISBN 0-931146-10-0
Printed in Hong Kong

DIOSCORIDES PRESS
9999 SW Wilshire
Portland, Oregon 97225

Library of Congress Cataloging-in-Publication Data

Adventitious root formation in cuttings / edited by Tim D. Davis,
 Bruce E. Haissig, Narendra Sankhla.
      p.   cm. -- (Advances in plant sciences series ; v. 2)
   Includes bibliographies and index.
   ISBN 0-931146-10-0
   1. Plant cuttings--Rooting.   I. Davis, Tim D. II. Haissig, Bruce
E. III. Sankhla, Narendra. IV. Series.
 SB123.75.A38 1988
 635'.0435--dc19   88-3711
                                                          CIP

ADVENTITIOUS ROOT FORMATION IN CUTTINGS

edited by

Tim D. Davis
Associate Professor of Horticulture
Brigham Young University, Provo, Utah

Bruce E. Haissig
Supervisory Plant Physiologist
USDA-Forest Service, Rhinelander, Wisconsin

Narendra Sankhla
Professor and Head
Department of Botany, University of Jodhpur, India

Advances in Plant Sciences Series
VOLUME 2
Theodore R. Dudley, Ph.D., General Editor

DIOSCORIDES PRESS
Portland, Oregon

Contents

Contributors 6

Foreword 8

Preface 9

CHAPTER 1
**Donor Plant Maturation and Adventitious
Root Formation**   Wesley P. Hackett................................... 11

CHAPTER 2
**Etiolation and Banding Effects on Adventitious
Root Formation**   Brian K. Maynard and Nina L. Bassuk................. 29

CHAPTER 3
Genetic Effects on Adventitious Rooting
Bruce E. Haissig and Don E. Riemenschnieder.......................... 47

CHAPTER 4
Mineral Nutrition and Adventitious Rooting
Frank A. Blazich.. 61

CHAPTER 5
**Relations Between Carbohydrates and Adventitious
Root Formation**   Bjarke Veierskov 70

CHAPTER 6
Photosynthesis During Adventitious Rooting
Tim D. Davis 79

CHAPTER 7
Enzyme Activities During Adventitious Rooting
Nikhil C. Bhattacharya 88

CHAPTER 8
Water Relations and Adventitious Rooting
K. Loach 102

CHAPTER 9
Auxin Metabolism During Adventitious Rooting
Thomas Gaspar and Michel Hofinger 117

CHAPTER 10
**Chemicals and Formulations Used to Promote
Adventitious Rooting**   Frank A. Blazich............................. 132

CHAPTER 11
Effect of Ethylene on Rooting
Kenneth W. Mudge...150

CHAPTER 12
**Influence of Gibberellins on Adventitious Root
Formation**   Jürgen Hansen 162

CHAPTER 13
**Effect of Shoot Growth Retardants and Inhibitors on
Adventitious Rooting**   Tim D. Davis and Narendra Sankhla 174

CHAPTER 14
Cytokinins and Adventitious Root Formation
J. Van Staden and A. R. Harty 185

CHAPTER 15
Polyamines and Adventitious Root Formation
Narendra Sankhla and Abha Upadhyaya...............................202

CHAPTER 16
**Stock Plant Environment and Subsequent Adventitious
Rooting**   Roar Moe and Arne Skytt Andersen..........................214

CHAPTER 17
Storage of Unrooted Cuttings   Volker Behrens 235

CHAPTER 18
**Controlling Environmental Conditions to
Improve Adventitious Rooting**   K. Loach 248

CHAPTER 19
**Bioassay, Immunoassay, and Verification of
Adventitious Root Promoting Substances**
Charles W. Heuser 274

CHAPTER 20
Agrobacterium rhizogenes: **A Root Inducing
Bacterium**   Gary A. Strobel and Avi Nachmias 284

CHAPTER 21
Adventitious Rooting of Tissue Cultured Plants
Brent H. McCown..289

CHAPTER 22
**Future Directions in Adventitious Rooting
Reseach**   Bruce E. Haissig 303

Index 311

Contributors

Arne Skytt Andersen
Horticultural Institute
Royal Veterinary and Agricultural University
Rolighedsvej 23, DK-1958 Frederiksberg C
Denmark

Nina L. Bassuk
The Urban Horticulture Institute
Department of Floriculture and Ornamental
    Horticulture
Cornell University
Ithaca, New York 14853 USA

Volker Behrens
Institut für Obstbau and Baumschule
Universität Hannover
Am Steinberg 3
3203 Sarstedt
West Germany

Nikhil C. Bhattacharya
Department of Agricultural Sciences and GWC
    Experiment Station
Tuskegee University
Tuskegee, Alabama 36088 USA

Frank A. Blazich
Department of Horticultural Science
North Carolina State University
Raleigh, North Carolina 27695-7609 USA

Tim D. Davis
Department of Agronomy and Horticulture
271 WIDB
Brigham Young University
Provo, Utah 84602 USA

Thomas Gaspar
Institut de Botanique B22
Université de Liège-Sart Tilman
B-4000 Liège
Belgium

Wesley P. Hackett
Department of Horticultural Science and
    Landscape Architecture
University of Minnesota
St. Paul, Minnesota 55108 USA

Bruce E. Haissig
USDA-Forest Service
North Central Forest Experiment Station
Forestry Sciences Laboratory
P.O. Box 898
Rhinelander, Wisconsin 54501 USA

Jürgen Hansen
Institute of Glasshouse Crops
Kirstinebjergvej 10
DK-5792 Aarslev
Denmark

H. T. Hartmann
Department of Pomology
University of California
Davis, California 95616 USA

A. R. Harty
UN/CSIR Research Unit for Plant Growth and
    Development
Department of Botany
University of Natal
Pietermaritzburg 3200
Republic of South Africa

Charles W. Heuser
Department of Horticulture
101 Tyson Building
The Pennsylvania State University
University Park, Pennsylvania 16802 USA

Michel Hofinger
Institut de Botanique B-22
Université de Liège-Sart Tilman
B-4000 Liège
Belgium

K. Loach
Institute for Horticultural Research-
    Littlehampton
Worthing Road, Littlehampton, West Sussex
BN17 6LP
England

Brian K. Maynard
Department of Floriculture and Ornamental
    Horticulture
Cornell University
Ithaca, New York 14853 USA

Brent H. McCown
Department of Horticulture
University of Wisconsin—Madison
Madison, Wisconsin 53706 USA

Roar Moe
Department of Horticulture
Agricultural University of Norway
P.O. Box 22
N-1432 Aas-NLH
Norway

Kenneth W. Mudge
Department of Floriculture and Ornamental
    Horticulture
Cornell University
Ithaca, New York 14853 USA

Avi Nachmias
Agricultural Research Organization
Gilat Regional Experiment Station
Mobile Post Negev
Israel

Don E. Riemenschneider
USDA-Forest Service
North Central Forest Experiment Station
Forestry Sciences Laboratory
P.O. Box 898
Rhinelander, Wisconsin 54501 USA

Narendra Sankhla
Department of Botany
University of Jodhpur
Jodhpur, India 342001

Gary A. Strobel
Department of Plant Pathology
Montana State University
Bozeman, Montana 59717 USA

J. van Staden
UN/CSIR Research Unit for Plant Growth and
    Development
Department of Botany
University of Natal
Pietermaritzburg 3200
Republic of South Africa

Abha Upadhyaya
Department of Botany
University of Jodhpur
Jodhpur, India 342001

Bjarke Veierskov
Department of Plant Physiology
Royal Veterinary and Agricultural University
Thorvaldsenvej 40
1871 Frederiksberg C
Denmark

Foreword

Due to the tremendous numbers of plants, especially ornamentals, that are propagated by cuttings, there is great interest among plant propagators and nursery people in the rooting of cuttings. Plant physiologists and plant anatomists, too, are intrigued by the activities that must occur in plant tissue that can lead to the formation of adventitious roots and thus the formation of a new plant.

Rooting of hardwood cuttings must be the most ancient form of asexual propagation. Early man may have been astonished to find his pointed spear, if made of certain kinds of wood, starting to grow when stuck deeply into the ground. Observant people in those early days would soon have found that woody shoots broken from such plants as the fig, olive, grape, mulberry, and pomegranate would start to grow, permitting them to establish a sort of orchard or vineyard.

Later, after glass had been invented and closed high humidity enclosures admitting light were possible, the use of leafy cuttings was feasible, thereby greatly increasing the scope of plants that could be rooted by keeping the cuttings alive long enough for roots to form. In more recent years glass enclosures were largely replaced by the cheaper and more flexible polyethylene covered structures. Then in the most modern era of micropropagation, cuttings could be but tiny shoots forming under sterile conditions by the action of cytokinins and subsequently rooted, either under sterile or non-sterile conditions, by the action of auxins.

The discovery in the mid-1930s that auxins would stimulate adventitious root formation in cuttings became a major milestone in propagation history. And for 50 years no other chemical has been found to equal auxin in this ability.

The use of mist to prolong the life of leafy cuttings until roots had a chance to form, developed in the 1940s and 1950s, was soon found to be so effective that in all commercial and experimental nurseries mist propagation became a standard practice. About this time, and with the advent of mist propagation it was found that the kind of rooting media used could be a most important factor in obtaining good root formation.

The scope of plants that can be rooted, and rooted faster and better, by cuttings has been increased tremendously by research efforts which led to the use of auxins, mist, and sterile culture. Nonetheless, much is still unknown about the intricate details of adventitious root formation and how this process is affected by the many environmental factors surrounding cuttings during rootings, as well as the plants from which the cuttings are taken.

The scientific and trade literature abounds with articles relating to studies dealing in some manner with adventitious rooting of cuttings. This book attempts to bring much of this information together as well as adding new information that has not been widely publicized.

Hudson T. Hartmann   April 16, 1986
Department of Pomology
University of California
Davis, California 95616

Preface

Adventitious root formation has been of interest to plant scientists for many years. Horticulturists and foresters have been interested in rooting because it is essential to the multiplication of many economically important plants. Most ornamental shrubs, many commercial greenhouse crops, and numerous fruit and forest tree crops are propagated by cuttings. In addition, propagation by cuttings is the only practical means of preserving unique characteristics of some plants. Because of the wide commercial basis, many plant physiologists and biochemists have been interested in the control mechanism(s) concerned with the initiation and development of new roots. The process has been particularly intriguing because it is presumably under hormonal control, and changes in gene expression must be involved. Unfortunately, much of the fundamental biology of adventitious root formation is still poorly understood. Nevertheless, basic research has contributed significantly to practical propagation, most notably through the discovery, characterization, and improvement of auxins.

Because rooting has been of interest for such a long time, literature on the subject is voluminous. The purpose of this book is to bring together, review, and interpret the research that has been done. We hope to foster use of available knowledge in further research and development of commercial applications. This would be an enormous task for one individual, hence we chose to have each chapter written by an experienced worker in the appropriate area of rooting research. We are particularly pleased with the wide array of distinguished authors who have contributed chapters to the book. The authors come from ten countries and represent a range of disciplines including horticulture, forestry, plant physiology and pathology, biochemistry, genetics, and molecular biology. We hope that the problems often associated with multi-author books have been minimized and are more than compensated for by the overall comprehensive and balanced scope of the book and nature of its contents. The authors have been encouraged to strongly document their statements and contentions by citing appropriate literature where possible. The reference sections at the end of the chapters will be useful to readers in finding specific literature dealing with adventitious root formation. In addition to reviewing and interpreting the literature, we have asked each author to identify areas of future research that are needed regarding their particular subject, in order to foster studies to achieve further fundamental knowledge regarding rooting.

We envisioned that this book would be of particular interest to those conducting research, and research and development on vegetative propagation, including botanists, horticulturists, foresters, and agronomists. Consequently, we have instructed the authors to write at the highly technical level typical of most plant science reference books. The book will also be of value as a textbook for specialized graduate-level courses or short courses. Although the book has not been specifically aimed at industry, it may be of interest to nurserymen who rely heavily on adventitious root formation for successful cutting propagation.

The editors are completely responsible for the general topics covered in the book as well as for the selection of contributors. Each chapter has been thoroughly reviewed by at least two of the editors. In some cases, outside reviews were sought when we felt that the scope of the chapter was beyond our expertise. The precise content of each chapter, within broad editorial guidelines, is the responsibility of the individual author. Final versions of chapters were submitted during the summer of 1987.

The book has been divided into five sections: Development, Physiology and Biochemistry, Growth Regulators, Environmental Considerations, and Future Outlook. One topic related to adventitious rooting that has not been covered in detail in this volume is the anatomical aspects of root initiation and development. Anatomical aspects have been thoroughly discussed in excellent reviews published during recent years [e.g. Haissig, 1974, *N. Z. J. For. Sci.* 4: 299–310; Lovell and White, 1986, In *New Root Formation in Plants and Cuttings* (M. B. Jackson, ed), pp. 111–140, Martinus Nijhoff Publishers, Dordrecht/Boston/Lancaster. ISBN 90-247-3260-3].

We express sincere appreciation to Dioscorides Press for recognizing the importance of this subject and for their efforts in publishing this book. In particular, we thank Mr. Richard Abel for his thorough editorial guidance and support. The editorial comments of Dr. Ted Dudley, General Editor for Dioscorides Press, were also much appreciated. We also thank our families who were very understanding of the long hours required to complete the book, and who participated directly in its preparation. Specifically, Patti Davis helped with indexing and Karin Haissig reviewed foreign language references for accuracy. Last, but certainly not least, we thank the distinguished contributors to the book. We greatly appreciate their cooperation and willingness to conform to our editorial guidance.

Editing this book has been a tremendous education; we hope that the reader will learn from the product.

Tim D. Davis
Bruce E. Haissig
Narendra Sankhla
August 26, 1987

CHAPTER 1

Donor Plant Maturation and Adventitious Root Formation

Wesley P. Hackett

Department of Horticultural Science and Landscape Architecture
University of Minnesota
St. Paul, MN 55108

INTRODUCTION. 11
EXAMPLES OF CHANGE IN ROOTING POTENTIAL IN
   RELATION TO MATURATION 12
     Hedera helix L.. 13
     Eucalyptus spp.. 14
     Pinus radiata D. Don. 14
     Ficus pumila L.. 15
     Malus domestica Borkh.. 15
PROCEDURES TO OBTAIN CUTTINGS WITH HIGH ROOTING
   POTENTIAL 16
     Severe Pruning and Hedging. 16
     Grafting Mature Scions on Juvenile Rootstocks 17
     Initiation of Adventitious Buds 17
     In Vitro Propagation 17
FACTORS ASSOCIATED WITH REDUCED ROOTING
   CAPACITY DURING MATURATION 18
     Auxin 18
     Other Plant Growth Substances. 18
     Rooting Cofactors and Promoters 20
     Rooting Inhibitors 21
     Non-Mobile or Non-Extractable Factors .22
CONCLUSION 24
REFERENCES 25

Additional key words: cutting, juvenility, phase change, rooting bioassay, rooting cofactors, rooting inhibitors, rooting potential, rooting promoters.
Abbreviations: ABA, abscisic acid; ECGLC, electron capture gas liquid chromatography; GA, gibberellin; GA_3, gibberellin A_3; IAA, indole-3-acetic acid; IBA, indole-3-butyric acid; NAA, naphthaleneacetic acid; PPO, polyphenol oxidase.

INTRODUCTION

In the development of all woody plants from seed there is a so-called juvenile phase, lasting up to 30–40 yr in certain forest tree species, during which flowering does not occur and cannot be induced by the normal flower initiating treatment or condition. In time, however, the ability to flower is achieved and maintained under natural conditions. At this stage, the tree is considered to have attained the sexually mature condition. This transition from the juvenile to the mature phase has been referred to as phase change by Brink (1962), ontogenetic aging by Fortanier and Jonkers (1976), or meristem aging by Seelinger (1924) and Oleson (1978). Associated with this transition are progressive changes in many morphological, developmental, and physiological characteristics (Hackett 1985). Changes in such characteristics during development are not consistent from species to species; most change gradually during the period preceding the mature phase, and usually no distinct change in any one characteristic is apparent at the time the ability to flower is attained. For this reason it is unclear whether these associated characteristics are directly or causally related to reproductive maturation. For a more detailed discussion see Hackett (1985).

It is difficult to distinguish between characteristics associated with maturation and those such as loss of vegetative vigor and type of branching associated with physiological aging because in both cases aging is involved. Wareing (1959) suggested that the greater stability of maturation-related characteristics during vegetative propagation can be used to distinguish them from physiological aging characteristics which disappear quickly after vegetative propagation. However, most experimentation was not planned to have appropriate controls to make it possible to distinguish between characteristics related to maturation and physiological aging. This means that experimental results purported to involve maturation characteristics must be considered with some skepticism unless appropriate controls were included (Hackett 1985).

One of the characteristics which has been observed to change in many species with developmental age is potential for adventitious root initiation. It has frequently been observed that rooting ability of cuttings from many woody plant species, particularly tree species, declines with increasing age of seedling-derived mother plants. This inverse relationship between ontogenetic age and rooting was reported as early as 1929 by Gardner (1929), and has since been observed repeatedly. High adventitious rooting potential, therefore, is considered a juvenile characteristic. The loss of rooting potential with maturation is particularly severe in many long lived tree species and limits the success or efficiency in clonally propagating desirable mature individuals (after the expenditure of much time and effort in the evaluation and selection process). In vitro microcutting propagation as well as conventional cuttage propagation is limited by this decreased potential for rooting (Mullins 1985).

Where it has been possible to analyze changes in rooting potential associated with ontogenetic development, it has been shown that the upper and peripheral parts of a plant are the first to exhibit reduced rooting potential. For example, in Eucalyptus grandis seedlings, it has been demonstrated that the cotyledonary node has a very high rooting potential but by the 15th node rooting capacity is almost completely lost (Paton et al. 1970). In Olea europaea L. (olive, Porlingis and Therios 1976) and Picea abies (L.) Karst. (Roulund 1973), it has also been reported that cuttings taken from shoots formed in the basal region or lower portion of the crown of trees have a higher capacity to root than those from shoots in the more distal, upper portions of the same plant. These findings are not surprising in light of many studies showing that other characteristics such as phyllotaxis, leaf shape, leaf retention, thorniness, and pigmentation, which are associated with juvenility, are maintained in the basal portions of mature plants of many species (Hackett 1985).

The purposes of this chapter are to document the nature of the loss of rooting potential during maturation using specific examples; to describe some procedures for maintaining high rooting potential in tissues of juvenile plants and for increasing rooting potential in mature plants; and to review and evaluate evidence for the involvement of physiological and morphogenetic factors in the loss of rooting potential. The reader should be aware that it is difficult to distinguish between the phenomena of maturation and physiological aging, and, therefore, some of the results of investigations involving loss or gain in rooting potential may be related to physiological age rather than maturation.

EXAMPLES OF CHANGE IN ROOTING POTENTIAL IN RELATION TO MATURATION

Several species and genera have been studied extensively because of their economic importance or because of unique characteristics that make it easier to study the influence of ontogenetic age on adventitious rooting. These are discussed below as examples.

Hedera helix L.

This species is a classical example of dimorphism whose characteristics are summarized in Table 1. Observation indicates that it takes 10 or more yr from seed germination before the first mature flowering branches are formed (Hackett and Srinivasan 1985). Wareing and Frydman (1976) observed that leaves at various heights on vines attached to trees show a change in shape and become intermediate in character between fully juvenile and fully mature types. They considered this to be a zone of transition between juvenile and mature regions of the vine. Characteristics at the base of these vines were fully juvenile and at the top were fully mature.

Table 1. Characteristics of juvenile and mature forms of Hedera helix.

Juvenile	Mature
5-lobed palmate leaves	Entire, ovate leaves
Distichous (1,2) phyllotaxis	Spiral (2,3) phyllotaxis
Anthocyanin stem pigmentation	No anthocyanin pigmentation
Stem aerial roots	No aerial roots
Plagiotropic growth habit	Orthotropic growth habit
Absence of flowering	Presence of flowering

The juvenile and mature forms can be maintained as separate plants through use of cuttage propagation or they can co-exist on the same plant. However, under high temperature conditions, grafting a mature scion shoot onto a leafy juvenile rootstock will induce juvenile characteristics in the shoot developing from the originally mature scion (Doorenbos 1954). Treatment of a mature shoot with GA_3 will also induce development of juvenile characteristics (Rogler and Hackett 1975).

Aerial adventitious roots routinely form on internodes of juvenile stems but very infrequently, or not at all, on internodes of mature stems (Table 1). Hess (1964) reported 100% rooting success for juvenile and only 16% rooting success for mature stem cuttings with or without auxin treatment. No rooting data are available on cuttings from so-called transitional forms (Wareing and Frydman 1976) which display varying degrees of juvenile and mature characteristics.

Girouard (1967ab) found that the anatomical process of root initiation is quite different in juvenile and mature cuttings. Six to 10 d after juvenile stem cuttings were made, he observed that cells of the outermost edge of the phloem, probably ray parenchyma, started to increase in size. This was followed immediately by anticlinal divisions which resulted in formation of root initials. Root initials originated in the median and lateral part of the phloem, and emerged after 10–14 d. Several days after cuttings of mature stems were made, cells below the suberized cut surface began to divide in the pith, secondary xylem, phloem, vascular cambium, and cortex. This resulted in formation of large amounts of what Girouard called wound wood and callus at the base of the cutting. Roots began to initiate by 2–4 weeks in phloem ray parenchyma and in callus at the base of the cutting. Roots emerged after 3–5 weeks.

Mokhtari (1985, M.S. Thesis, Univ. of Minnesota, St. Paul, USA) found that the rooting potential of detached juvenile and mature leaves was very similar to that of stem cuttings. Auxin treatment enhanced root initiation in juvenile petioles as compared to controls, and initiation occurred as a result of localized division of parenchyma cells at the interface of the phloem and cortex near ducts associated with the vascular bundles. Auxin treatment of mature petioles induced extensive random cell divisions in the cortex, pith, and phloem but seldom induced root initiation, and when it did, root primordia originated from auxin induced callus.

Geneve (1985, Ph.D. Thesis, Univ. of Minnesota, St. Paul, USA) studied root initiation in de-bladed juvenile and mature petioles in vitro using a complete liquid nutrient medium including sucrose as a carbon source. He, likewise, found that the rooting potential of juvenile and mature petioles was very similar to that of stem cuttings. However, de-bladed juvenile petioles initiated roots only if auxin was in the culture medium, presumably because the lamina is a source of auxin. De-bladed mature petioles also responded to auxin, but instead of forming root initials, as the juvenile ones did, they formed callus. Without auxin, mature petioles formed little or no callus. In both juvenile and mature, de-bladed petioles, the first cell divisions in response to auxin were observed at day 6 in epithelial cells of ducts that are adjacent to the vascular bundles. Beyond day 6, the response to auxin in juvenile and mature de-bladed petioles was very different. By day 9 in the juvenile petioles, localized division in the cortical parenchyma associated with the vascular bundle was observed. By day 12 well defined root primordia were apparent. In the mature petioles, by day 9, divisions had occurred throughout the cortical parenchyma. These divisions lacked orientation and did not usually result in an organized root meristem. When a root initial did form in mature petioles, it lacked vascular connection to the petiolar vascular bundle, and appeared to arise from divisions in the cortically derived callus.

Geneve (1985, previously cited Ph.D. Thesis) concluded that in vitro cultured, de-bladed petioles of the two phases have several important advantages as an experimental bioassay system to analyze the morphogenetic, physiological, and biochemical basis of root initiation potential. These advantages are: 1) variable rooting potential in tissue having similar physiological age, and identical anatomical organization and genetic makeup; 2) relatively simple, uniform, fully differentiated tissue systems in which endogenous sources of plant growth substances are minimized; 3) easy in vitro manipulation of environment, and for the precise provision of nutrients and plant growth substances as pulses or continuously fed from proximal or distal ends; 4) relatively rapid, specific, and morphogenetically distinct responses to auxin; and 5) tissues readily available in large amounts on a year-round basis.

Eucalyptus Spp.

All species of this genus root easily from stem tissue provided that leafy cuttings are taken from very young seedlings (Pryor and Willing 1963). However, there is an ontogenetic loss of capacity to form adventitious roots. The steepness of the gradient for ontogenetic loss of rooting capacity varies with species. For example, rooting ability was completely lost by the 4th node in E. viminalis Labill and E. pauciflora, and by the 15th node in E. grandis (Paton 1984). In E. camaldulensis Dehnhardt, 40–50% rooting was obtained in cuttings from the 100th node, and in E. deglupta 100% rooting was obtained in cuttings from the 100th and even higher nodes (Paton 1984). According to Davidson (1974), E. deglupta cuttings from trees up to 1-yr-old rooted easily; those from 5-yr-old trees did not. In trees of E. robusta Smith, aerial roots originated on shoots 30 feet or higher above the ground (Pryor and Willing 1963). Thus, in this genus, the influence of ontogenetic age on rooting capacity varies from species to species. The loss of rooting capacity with ontogenetic age is not strongly correlated with other developmental phase related characteristics such as leaf shape or flowering ability (Jacobs 1955, Paton et al. 1970).

Cuttings from epicormic shoots arising from accessory buds in the secondary phloem at the bases of mature trees of several species often root easily, whereas similar shoots arising above node 20 do not root (Paton and Willing 1974). This indicates that retention of rooting ability (and other juvenile characteristics) is related to position on the tree rather than chronological age (Hartney 1980). That such rooting ability can be maintained is indicated by the fact that stem cuttings from low-pruned hedges derived from cuttings of basal epicormic shoots of E. grandis have continued to produce cuttings with no decline in rooting ability after more than 7 yr (Hartney 1980).

Many species of Eucalyptus develop lignotubers (swelling in the axils of the cotyledons and first few nodes) that contain numerous buds and meristematic tissue (Chattaway 1958). In mature trees of some species the lignotuber may develop into a massive structure, while in others it remains small and becomes embedded as the stem grows in diameter. Shoots developing from lignotubers have juvenile leaf morphology and often retain the ability to form roots even when taken from lignotubers on mature trees (Hartney 1980).

Pinus radiata D. Don

Jacobs (1939), Mirov (1944), Fielding (1954, 1969), and Thulin and Faulds (1968) have demonstrated that the ease with which cuttings of this species can be rooted declines with age of the parent stock plant. Ease of rooting is reflected not only in frequency of rooting, but also in rooting time, and site of root initiation. In IBA treated hypocotyl cuttings from 20-d-old seedlings, Smith and Thorpe

(1975) found root primordia microscopically 10–13 d after excision from the parent plant. Root formation occurred in nearly 100% of the cuttings treated with an optimal IBA concentration. In these hypocotyl cuttings, root initiation had its origin in a single cell from which a meristemoid and then a root primordium subsequently developed. The events leading to root initiation occurred on the margin of resin ducts or in parenchyma external to these cells.

In contrast, Cameron and Thomson (1969) found that non-auxin treated cuttings from 5-yr-old trees showed a 70% rooting success 100 d after first callusing. They reported that adventitious roots did not arise directly by simple formation of root meristems in callus, but instead were preceded by the lateral extension of callus xylem that was centripetally located with reference to callus meristems. They concluded that in cuttings from 5-yr-old (or older) trees callus formation must precede root initiation.

Fielding (1954) reported data from experiments in which success in rooting fell from 88% for cuttings from 3-yr-old trees to 68% at age 5 yr, and 11% at age 26 yr. Libby and Conkle (1966) report smaller differences than this for cuttings treated with auxin, but only compared materials from trees up to 17 yr old. In a report of additional results 15 yr later, Fielding (1969) again found a reduction in rooting percentage in cuttings taken from 3-yr-old trees (71%) in comparison to those taken from 6-yr-old trees (45%). However, Thulin and Faulds (1968) report a very high rooting percentage in cuttings taken from 9-yr-old trees (100%) and a relatively small reduction even in those taken from 20-yr-old trees (70%). They noted, however, that cuttings from trees more than 20 yr old were very difficult to root. Although there are some inconsistencies in these reports, it appears that there is a substantial loss of rooting potential with increased tree age. Fielding (1969) speculated that the inconsistency in reduction in rooting percentage with age reported by various investigators could have been due to different environmental conditions of the mother trees. Another possible explanation is genotype effects.

With 4-yr-old trees, Fielding (1969) found that cuttings from branches in the lower crown tended to root better (76% vs. 63%) than those from the upper crown. However, in earlier experiments, Fielding (1954) was unable to detect differences in the rooting ability of cuttings from the upper and lower branches of the crown of trees over 20 yr old. In this latter case, the rooting percentage was low for cuttings from both positions.

Fielding (1954) reported that hedged trees of seedling origin remained physiologically juvenile in that cuttings from them rooted more readily than cuttings from unhedged trees of the same age. With clones derived from trees ranging in age from 2–17 yr, Libby et al. (1972) concluded that hedging arrested the decline in rooting percentage, as assessed by a comparison of the rooting of pairs of hedged and tree form clones over a 5 yr period.

Ficus pumila L.

Ficus pumila is a woody, ornamental vine that exhibits dimorphism (Condit 1969) and differences in rooting ability between juvenile and mature forms. The juvenile form has nodal aerial roots while the mature form does not. Using leaf-bud cuttings, Davies and Joiner (1980) found that mature cuttings required higher exogenous auxin concentration (3000 mg·l⁻¹ vs. 1000 mg·l⁻¹ IBA) to obtain maximum rooting percentage and root number. When root initiation was calculated as a daily rate, IBA treated mature cuttings rooted slower than IBA treated juvenile cuttings. However, by day 20 after sticking, rooting equaled juvenile cuttings not treated with IBA. They suggested that these results give evidence that endogenous auxin levels act as a possible limiting factor in rooting in mature cuttings.

Davies et al. (1982) concluded that the developmental pattern of root initiation was the same in juvenile and mature cuttings, but the process was slower in mature than juvenile cuttings. The process of root initiation involved dedifferentiation of cells in the phloem ray parenchyma. The first anticlinal divisions in phloem ray parenchyma cells that resulted in root primordia were observed 4 d after IBA treatment in juvenile cuttings and 6 d after treatment in mature cuttings.

Malus domestica Borkh.

Gardner (1929) reported that softwood cuttings made from the tops of apple seedlings during the first year of growth could be rooted easily, whereas cuttings from 2–4-yr-old seedlings rooted poorly. This research was done before the advent of rooting hormones (IBA and NAA) and, therefore, these results may under-estimate rooting potential. However, Wilkinson (1973) reported that hardwood cuttings from trees of 64 cultivars of fruiting age gave a mean rooting frequency of only 9%. Even at present, it is still difficult to propagate many apple cultivars from cuttings.

Stoutemyer (1937) demonstrated that two distinct developmental phases exist in apple trees. The juvenile condition was recognized by the thinness of the leaves, reduced leaf and stem pubescence, abundant anthocyanin production in stems and leaves, and, in some species, entire leaf form, compared to deeply lobed leaves in the mature form. These juvenile characteristics were found not only in shoots of young seedling trees but also in shoots developing from adventitious buds on roots, and trunk sphaeroblasts of severely pruned, fruiting trees. Cuttings from shoots with juvenile phase characteristics rooted readily while those with mature characteristics did not.

Robinson and Schwabe (1977) confirmed Stoutemyer's findings that cuttings from shoots developing from adventitious buds on root pieces of mature trees were easy to root. With such IBA treated softwood cuttings under mist they reported rooting percentage of up to 90%. They also noted juvenile morphological characteristics of shoots developing from root adventitious buds but, in addition, found that trees developing from cuttings of root-derived shoots flowered only 15 months after rooting. This early flowering contrasts with time to flowering for seedlings, which has taken from 6–12 yr (Visser 1964). This raises the question of whether root-derived shoots are truly juvenile. It also raises the possibility that juvenile behavior, as regards rooting ability, may be separable from juvenile behavior as regards flowering ability. A similar separation of rooting ability from flowering ability has been found with *in vitro* derived microshoots of apple cultivars that have shown increased rooting potential with increasing number of subcultures (Mullins 1985). Plantlets derived from such rooted microshoots flowered within 2.5 yr of removal from the culture flask.

PROCEDURES TO OBTAIN CUTTINGS WITH HIGH ROOTING POTENTIAL

As has been illustrated in the previous section, adventitious rooting potential of many woody species, particularly tree species, may decrease during ontogenetic aging and maturation. By the time individual seedlings have been tested and selected for desirable characteristics, they may have lost much of their potential to root and, therefore, cannot be clonally propagated by conventional cuttings or by *in vitro* microcuttings, at least not with commercially acceptable efficiency.

Thus, I will discuss some methods that have been used successfully with some species to arrest or reverse (rejuvenate) the maturation process in order to obtain large numbers of cuttings with high rooting potential. Some of the methods cited are controversial because it is not clear whether the effect of the method and the characteristics observed are related to reversal of physiological age (invigoration) or maturation (rejuvenation). A classification, evaluation, and more detailed discussion of these procedures appeared in a recent review (Hackett 1985).

Severe Pruning and Hedging

Buds near the base of seedling plants or asexually propagated clones can be forced into growth by severe pruning to form hedges or stool beds. This method has been used successfully to obtain shoots with increased rooting potential in apple (Hatcher 1959), *Pinus radiata* (Libby et al. 1972), and *Eucalyptus* spp. (Martin and Quillet 1974), and is the basis of their commercial clonal propagation. The basis of the increased rooting potential has been interpreted in several ways. Libby et al. (1972) indicated for *Pinus radiata* that maturation had been arrested by hedging at the ontogenetic stage of the nodes from which hedge shoots were formed. However, their results could also be interpreted in terms of a rejuvenation of tissue from a relatively more mature state because plants resulting from rooted cuttings from hedged plants of *Pinus radiata* have a growth habit similar to seedlings (Libby et al. 1972). Based on their work with apple root stock clones, Garner and Hatcher (1962) considered hedged plants to be physiologically invigorated, not rejuvenated, because 1-yr shoots on stool beds can flower even though they have high rooting potential.

Grafting Mature Scions on Juvenile Rootstocks

As detailed in later sections, there are several reports indicating that grafting mature scions on juvenile rootstocks causes an induction of juvenile characteristics in general, or an induction of increased rooting potential by shoots that develop from mature scions. With the exception of *H. helix,* successful enhancement of rooting requires repeated grafting in serial fashion at short intervals on successive young juvenile seedlings (Muzik and Cruzado 1958, Franclet 1979). In *Eucalyptus* spp., where serial grafting has been used successfully on a commercial scale, Martin and Quillet (1974) and Franclet (1979) observed changes in foliar characteristics and growth rate as well as rooting potential. Chaperon (1979) stressed the importance of high vigor of the juvenile rootstock and proximity of the scion to the seedling rootstock for successful rejuvenation. Paton (1984) also found that proximity of the bud to the juvenile rootstock was important, and stressed the importance of time required for grafting influences to occur. Paton (1984) has interpreted increased rooting potential of mature scions grafted on juvenile seedlings in terms of a reduction of G compound concentrations from a rooting inhibitory level to a promotive level. Paton's interpretations were based on physiochemical estimates of G compounds in tissues, and rooting bioassays with mung bean. The nature of G compounds is discussed in a later section.

Initiation of Adventitious Buds

As previously noted, Stoutemyer (1937) observed that shoots developing adventitiously on roots had juvenile morphological characteristics and high rooting potential, whereas cuttings from the mature mother tree failed to root. Similar observations have been reported for root suckers of *Populus tremula* L., *Populus tremuloides* Michx. (*P. graeca* Loud.), and *Ulmus* spp. (Heybrock and Visser 1976). Wellensiek (1952) used adventitious shoots from sphaeroblasts of a number of species to obtain shoots with juvenile characteristics and high rooting potential. Similar observations have been made with shoots from sphaeroblasts on apple (Stoutemyer 1937) and olive (Baldini and Scaramuzzi 1957) clones. As would be expected, Garner and Hatcher (1962) concluded that the rooting potential of shoots of adventitious origin decreases in shoots of succeeding vegetative generations.

Robinson and Schwabe (1977) concluded that adventitious shoots from root cuttings of 'Lord Lamborne' apple clones had sufficiently high rooting potential to provide a commercial means for propagating own-rooted trees for high density plantings. As noted earlier, they found that although adventitious shoots of root origin had high rooting potential and juvenile morphological characteristics, plantlets derived from them flowered in 18 months from the time of rooting. This observation that flowering potential (reproductive maturation) can be separated from rooting potential and other juvenile characteristics in "rejuvenation" procedures is very important horticulturally.

In Vitro Propagation

Relatively recent studies indicate that the phase-related characteristics, including rooting potential of excised buds from mature plants, can be modified as a result of *in vitro* culture (see Chapter 21 by McCown). With *Sequoia sempervirens* Endl., Boulay (1979) found that when the primary explant was of plagiotropic orientation, the number of orthotropic shoots produced increased with the number of subcultures (orthotropic shoot formation was enhanced by inclusion of activated charcoal in the medium). He also reported that there was increasing root formation after a variable number of subcultures and suggested that conditions in the culture produced ontogenetic rejuvenation. Franclet (1979) indicated that *in vitro*-produced plantlets were impossible to distinguish from seedlings when planted in the nursery.

Microcuttings derived from shoot tips of some apple cultivars have an increased rooting potential after about 10 subcultures at monthly intervals (Mullins 1985). Both Zimmerman (1981) and Sriskandarajah et al. (1982) reported some slight changes to juvenile morphology with increasing number of subcultures, but little loss of ability to flower, the ultimate mature characteristic. Changes in biochemical and anatomical characteristics during subcultures *in vitro* will be described in the next section. Shen and Mullins (1983) and Mullins (1985) reported similar increases in rooting potential of prune and pear microshoots with increasing numbers of subcultures. Mullins (1985) further reported that the culture conditions required for this increased rooting potential were different for apple and prune.

Similar findings have been reported for several *Eucalyptus* spp. Several investigators have shown that the rooting potential of microshoots derived from shoot tips of mature trees increases with successive transfers on multiplication medium containing cytokinin. Gupta et al. (1981) were probably the first to report the increase in rooting potential with number of subcultures of *E. citriodora*. Boulay (1979) noted that explants taken from suckers and sprouts on stumps, or from mature scions after grafting and regrafting on juvenile seedlings, improved the speed with which rooting potential *in vitro* is increased.

By recovery of existing juvenile material at the base of mature plants or by use of one or more rejuvenation methods, it has been possible with some species to establish stock block plantings that yield cuttings of higher rooting potential and less intraclonal variability than cuttings taken from the crown of a mature plant. In theory, cuttings from such stock blocks should have, in addition to high rooting potential, the desirable juvenile characteristics of vegetative vigor, orthotropic growth habit, and branching pattern. But their potential to flower may be reduced, an undesirable side effect for some purposes. Plants in stock blocks should be severely pruned to increase cutting production and maintain juvenile characteristics. Periodically, it might be necessary to re-rejuvenate stock plants by the original techniques to establish a new stock block.

In vitro rejuvenation may offer a means to easily study how maturation influences rooting. Mullins (1985) has used this approach effectively by comparing the physiological basis of rooting potential in mature and rejuvenated microshoots of apple.

FACTORS ASSOCIATED WITH REDUCED ROOTING CAPACITY DURING MATURATION

Auxin

Auxin is essential for initiation of adventitious roots on cuttings (Haissig 1974). Even with easy-to-root cuttings from the juvenile phase, auxin treatment may increase the rate of root initiation and number of roots formed. Auxin treatment often has much less effect or no effect on difficult-to-root cuttings from mature phase plants of the same species (Hess 1959, Porlingis and Therios 1976). This latter observation suggests that *(endogenous) auxin level is not the factor limiting rooting in mature plants.*

Endogenous auxin levels have been measured in juvenile and mature tissues of several species. Hess (1964) reported that the endogenous auxin content, as measured by the avena coleoptile straight growth test, was virtually the same in juvenile and mature *H. helix* stem tissue. However, more auxin and other growth promoters measured by bioassay were found in juvenile than mature cuttings of *Castanea sativa* Mill. (Vieitez and Vieitez 1976). Using physicochemical methods, Caruso et al. (1978) measured 2.9 μg IAA $\cdot$ g^{-1} fresh wt. of shoot tips of Douglas-fir seedlings, while actively growing shoot tips from a 40-yr-old tree measured 1.6 μg $\cdot$ g^{-1} fresh wt. However, they concluded that factors in addition to auxin influence rooting of cuttings because treatment with auxin did not improve the rooting of cuttings from mature trees. In *Sequoia sempervirens,* Fouret et al. (1986) found much higher levels of IAA in *in vitro* derived, microshoot stem tissue of seedling and 50-yr-old trees (592 and 425 ng $\cdot$ g^{-1} dry wt., resp.) than from 500-yr-old trees (244 ng $\cdot$ g^{-1} dry wt.). Microshoots from the seedling and 50-yr-old trees formed roots at percentages of 75 and 82, respectively, while those from the 500-yr-old tree did not root. Rooting of microshoots from the 500-yr-old tree could not be increased to that of the 50-yr-old tree by treatment with auxin (or other additives) at any concentration (Bekkaoui et al. 1984). In comparative studies with *in vitro* derived non-rooting (mature) and rooting (rejuvenated) microcuttings of Jonathan and Granny Smith apple cultivars, Mullins (1987) detected no differences in free IAA; non-rooting microcuttings had little or no rooting response to auxin treatments (Mullins 1985). These results indicate that endogenous auxin levels do not limit rooting in mature tissues. However, if other limiting factors were sequentially removed, auxin might sometimes be the limiting factor.

Other Plant Growth Substances

Several naturally occurring plant growth substances, when applied exogenously, promote or inhibit adventitious rooting, depending on species, maturation state, and other factors (see Chapter 13 by Davis and Sankhla, Chapter 12 by Hansen, and Chapter 11 by Mudge). Some recent research has been done on endogenous ethylene, ABA, and gibberellins in relation to rooting ability of juvenile (or rejuvenated) and mature cuttings. Geneve (1985, Ph.D. Thesis, Univ. of Minnesota, St. Paul, USA) used in vitro cultured, de-bladed petioles to analyze the possible involvement of endogenous ethylene in the different rooting potentials of juvenile and mature *H. helix*. He found significant differences in the time course of ethylene evolution from petioles of the two forms treated with the optimal concentration of NAA for rooting of the juvenile petioles. Ethylene evolution from juvenile petioles subsided from a maximum level at 24 h after treatment to near control (non-treated) petiole levels after 6 d. Ethylene evolution in mature petioles was similar to that of juvenile petioles after 24 h. In contrast to juvenile petioles, evolution in mature petioles continued to increase slowly over the 14 d experimental period. The correlation of the timing of subsidence of ethylene evolution in juvenile petioles with the period when root initials were being formed suggested that root initiation potential might be related to ethylene synthesis or action. Geneve used several methods including treatments with ethylene or ethylene precursors, inhibitors of ethylene synthesis or action, and ethylene scrubbers to test the hypothesis that ethylene metabolism or action was causally involved in the difference in rooting potential of petioles of the two phases. Although elevated ethylene levels from exogenous treatments inhibited root initial outgrowth in auxin treated juvenile petioles, he found no evidence that reduction of ethylene levels or interference with ethylene action would promote rooting in mature petioles. He also found that ethylene treatment of either juvenile or mature petioles not treated with auxin had no effect on rooting or cell division. He, therefore, rejected the hypothesis that ethylene metabolism or action was causally involved in the observed differences in rooting potential, and also concluded that auxin promotion of rooting in juvenile petioles was not mediated through induction of ethylene synthesis.

Fouret et al. (1986) studied endogenous abscisic acid levels in *in vitro* derived microshoot stem tissue from *S. sempervirens* trees of different ages. As stated earlier, microshoots from trees of these different ages have different rooting potential. They found that the ABA level in microshoots from the 500-yr-old tree, as determined by ECGLC, was 1.5 times as high (1416 $ng \cdot g^{-1}$ dry wt.) as in microshoots from the 50-yr-old tree (983 $ng \cdot g^{-1}$ dry wt.) or seedling (983 $ng \cdot g^{-1}$ dry wt.) trees. Similarly, Mullins (1985) reported lower ABA levels in rooting (rejuvenated) than non-rooting (mature) microcuttings of Jonathan and Granny Smith apple cultivars. He and his coworkers also showed that exogenous ABA inhibited rooting in the rooting type microcuttings. Raviv and coworkers (1986b) found that declining rooting potential of avocado (*Persea americana* Mill.) with increasing age (up to 1 yr) was correlated with a 5-fold increase in ABA levels in leaves as determined by ECGLC. Spraying of seedlings with ABA did not inhibit their growth or inhibit rooting of cuttings taken from them. In contrast, Hillman et al. (1974) showed that the leaves of the juvenile form of *H. helix* contained five times the quantity of ABA as the mature form on a fresh wt. basis. Chin and coworkers (1969) reported a promotive effect of ABA on rooting in juvenile *H. helix* cuttings.

Takeno et al. (1983) used the dwarf rice assay to determine GA levels in *in vitro* derived non-rooting (mature) and rooting (rejuvenated) microcuttings of Jonathan apple. They found that the GA level of non-rooting microcuttings was 40 ng GA_3 equivalents $\cdot g^{-1}$ dry wt., but 14 ng in rooting type microshoots. Mullins (1987) reported that exogenously applied GA_3 inhibited rooting of the rooting type microcuttings. Using the barley endosperm bioassay, Raviv and coworkers (1986b) determined the level of gibberellin-like substances in the leaves of 1-month- and 1-yr-old avocado seedlings. They found that the level of gibberellin-like substances decreased more than 10-fold with increasing age (up to 1 yr). Application of GA_3 to seedlings promoted shoot elongation but did not enhance rooting of cuttings therefrom. Similarly, Frydman and Wareing (1973) reported higher levels of gibberellin-like substances in juvenile *H. helix* shoot tips than in those of mature plants. Leaves of the two forms had about equal quantities.

These reports of levels of endogenous ABA and gibberellin-like substances in mature and juvenile or re-juvenated tissues show no consistent relations between endogenous level of either plant growth substance and loss of rooting potential. In one case, application of ABA inhibited rooting of rejuvenated tissue, in another it promoted rooting of juvenile tissue, and in a third it had no effect. Likewise, there was no consistent effect of GA_3 on rooting of juvenile or mature tissues. Only in apple tissue did there appear to be a consistent relation between endogenous levels of ABA and GA_3, effects of exogenous application of these plant growth substances on rooting, and rooting potential of juvenile and mature tissues.

Rooting Cofactors and Promoters

Rooting cofactors were first isolated from *H. helix* by Hess (1959). Based on paper chromatographic separation and activity in the mung bean [*Vigna radiata* (L.) Wilczek] rooting assay (see Chapter 19 by Heuser), he suggested that there were four groups of cofactor compounds, and numbered them 1, 2, 3, and 4. He called them rooting cofactors because they were most active in the rooting bioassay in the presence of IAA. He and his coworkers concluded that easy-to-root juvenile cuttings of *H. helix* contained higher amounts of rooting cofactors than difficult-to-root mature cuttings, based on activity in the mung bean rooting assay (Hess 1964).

Hess and his coworkers were not successful in purifying and unequivocally identifying any of the cofactors. Part of the rooting activity in cofactor 3 has been attributed to the phenolic compound isochlorogenic acid (Hess 1965). Tests by Hess (1962) with other phenolic compounds showed that catechol had the highest specific activity and that the highest activity occurred when the phenolic hydroxyl groups were *ortho* to each other with a free *para* position. Girouard (1969) found that juvenile phase *H. helix* had a higher phenolic content than the mature phase. Hess (1963) reported that cofactor 4 was a mixture of oxygenated terpenes that are lipid-like. Heuser and Hess (1972) purified three of the substances in cofactor 4, but were unable to identify them because the pure substances were unstable.

Based on the work cited above, Hess (1963) postulated that the presence of greater amounts of these rooting cofactors in the juvenile than the mature form of *H. helix* might account for the higher rooting ability of juvenile cuttings as compared to mature ones. Subsequently, considerable research has been done on the relation of rooting promoter content to ease of rooting. Evidence for root promoting activity has been found in extracts from many woody species using the mung bean rooting assay with and without auxin treatment (Heuser 1976). Investigators reported that the amount of activity present in tissues was not necessarily related to plant age (Zimmerman 1963, Quamme and Nelson 1965, Vasquez and Gesto 1982) or rooting ability (Lanphear and Meahl 1963, Richards 1964, Al Barazi and Schwabe 1985). *However, in none of this work did anyone test for rooting promoter activity using cuttings of the species from which the promoters were extracted.* As a result, no one has demonstrated that rooting cofactors promote rooting of difficult-to-root cuttings. This situation was a consequence of the lack of a convenient rooting bioassay based on both easy- and difficult-to-root tissue.

Hackett (1970) reported the development of an *in vitro* bioassay for root initiation using excised shoot apices of the juvenile and mature form of *H. helix* (see Chapter 19 by Heuser). Following Hess's procedures for extraction and separation of the rooting cofactors, Hackett used the *in vitro Hedera* shoot apex system to bioassay for root initiating activity. He found ample evidence for methanol extractable factors from juvenile and mature phase tissue that promoted rooting in easy-to-root, juvenile ivy shoot tips, but no evidence that these factors promoted rooting of difficult-to-root, mature shoot tips. These results bring into question the physiological significance of such cofactors in explaining the higher rooting capacity of the juvenile as compared to the mature form of *H. helix*. Because most other evaluations of root promoting activity in extracts were made on easy-to-root cuttings of mung bean, the role, if any, of the rooting promoters in phase related changes in rooting potential is obscure.

However, recently Raviv and coworkers (1986a) have isolated and identified four rooting promoters with an acetylenic moiety (1, 2, 4-trihyroxy-n-hepta-deca-16yn is most active) that accumulate faster (based on a mung bean bioassay without auxin) in the bases of juvenile cuttings of avocado during the course of rooting than in mature cuttings and difficult-to-root cuttings (Raviv et al. 1986b). No correlation was found between plant age and concentration of these promoters in the leaves, as assayed with mung bean cuttings without auxin (M. Raviv. 1981, Ph.D. Thesis. Hebrew University, Jerusalem, Israel). However, these non-auxinic root promoters were found to promote rooting of cuttings from both juvenile and mature (1-yr-old) avocado plants (Raviv and Reuveni 1984).

Rooting Inhibitors

The best evidence for the involvement of rooting inhibitors as the physiological basis for the relation between rooting potential, ontogenetic age, and position has been obtained by Australian researchers using mainly *Eucalyptus grandis* and *E. deglupta*. Using physicochemical methods to extract and purify endogenous substances, and cuttings of seedling *E. grandis* as a rooting assay, Nichols and coworkers (1972) found that the concentration of rooting inhibitors increased in successively older leaves of *E. grandis*. This increase in concentration of inhibitors correlated with decreased rooting ability of cuttings taken from successively higher internodes (Paton et al. 1970, Paton and Willing 1974). Three of these inhibitors have been identified as closely related, fused, bicyclic compounds with a peroxide link (Sterns 1971), and are structurally related to the β-triketones. Each of these compounds at 10^{-4} mol·l^{-1} inhibited the rooting of cuttings from seedlings of *E. grandis* and *E. deglupta*. The natural concentration in mature leaves of *E. grandis* is about 2×10^{-3} mol·l^{-1}, which is more than sufficient to inhibit adventitious rooting (Nichols et al. 1972). These inhibitors have been isolated and synthesized, and shown not to be artifacts of extraction (Crow et al. 1976). In addition, another inhibitor, grandinol, was found in mature leaves of *E. grandis*. Grandinol is presumed to be derived from phloroglucinol, as are the other three inhibitors (Crow et al. 1977).

Dhawan et al. (1979) and Paton (1984) have reported that the so-called G inhibitors (G is for *grandis,* the species from which they were first isolated) actually promoted rooting of mung bean cuttings optimally at 10^{-5} mol·l^{-1}, and inhibited at 10^{-4} mol·l^{-1} and higher, compared to a water control. Dhawan et al. (1979), and Paton and Willing (1974) have also shown that these G compounds at 0.1–0.5 mg·g^{-1} of inert powder can replace auxin in promoting rooting of cuttings of *E. grandis* seedlings, *Lagerstroemia indica* L., and *Rhododendron* spp. In addition, Dhawan and coworkers (1979) have demonstrated that the G compounds at 5×10^{-6} mol·l^{-1} double the growth of *Avena* coleoptile sections as compared to a water control, which suggested that they have auxin-like activity.

A survey of chromatographed methanolic extracts of the mature form of nine *Eucalyptus* spp. and the juvenile form of four of these species, using the mung bean bioassay, revealed that the G inhibitors were present in six and absent in three mature forms, but absent in all of the juvenile forms (Paton and Willing 1974). According to Paton and Willing (1974), *E. deglupta,* an easily rooted species, did not have any G inhibitor in either the mature or juvenile forms. However, Davidson (1974) found that the decline in rooting ability of cuttings from older trees of *E. deglupta* correlated with an increase in concentration of rooting inhibitors in tissue extracts.

To be regarded as a true endogenous rooting inhibitor, a substance should fulfill the following criteria: 1) the active substance must be identified; 2) a correlation must be established between endogenous levels of identified substances and rooting capacity; and 3) the extracted substance must be effective when applied to the species from which it was extracted. The G inhibitors found in *E. grandis* and other species appear to fulfill the above criteria. From evidence such as that cited earlier on tissue concentrations, Paton (1984) concluded that the range in G concentrations from promotive (10^{-5} mol·l^{-1}) to inhibitory (10^{-3} mol·l^{-1}) in the mung bean bioassay is associated reasonably well with ontogenetic loss of rooting capacity of seedlings of *E. grandis* between the cotyledons and about node 10.

Juvenile tissue grafted in close proximity to mature tissue maintains its capacity for adventitious rooting (Paton and Willing 1974). As pointed out earlier, epicormic shoots arising from accessory buds at the base of mature trees (stump sprouts) often root easily, whereas similar shoots arising above node 20 do not root. These observations suggest that there may be a physiological barrier to the transport of G inhibitors in the basal phloem (Paton and Willing 1974). However, Paton and coworkers (1981) found that the epicormics from the *basal* accessory buds of *mature* scions grafted on seedlings exhibited juvenility as reflected in both increased rooting capacity (30–50%) and lowered G inhibitor content (0.28 mg·g^{-1}). In contrast, the shoots derived from distal buds on the mature scions maintained their original mature characteristics of non-rooting and a high G inhibitor content (2.3 mg·g^{-1}). The epicormics at the base of mature scions thus exhibit a level of juvenility approaching that of stump sprouts. Paton and coworkers (1981) pointed out that there is an important difference because stump sprouts *retain* a level of juvenility that is related to their origin within the first few seedlings nodes whereas the basal epicormics of mature scions *regain*, not retain, their juvenility. They concluded that the basal epicormics on mature scions *regain* their juvenility as reflected in rooting ability as a result of their close proximity to the seedling roots for six months or longer. Because there is no evidence (Paton and Willing 1974) that root extracts degrade inhibitor G, or that roots act as a sink for high levels of G from mature leaves, they concluded that the time requirements for regaining rooting potential involves reduced synthesis of G inhibitors in basal shoot tissue of the previously mature scion. Paton (1984) concluded that the problem of rooting mature stem tissue is one of reducing G content to promotive levels, as previously discussed.

Inhibitors, as reflected by the growth response of *Avena* coleoptile sections, have been measured in extracts from juvenile and mature tissues. Hess (1963) found roughly equivalent levels of growth inhibition in extracts of juvenile and mature *H. helix* whether plants were growing or dormant. In contrast, mature chestnut (*Castanea* spp.) cuttings contained more growth inhibitors than did cuttings from young (juvenile) seedlings (Vieitez and Vieitez 1976). The significance of these results are difficult to assess because the bioassays used did not involve adventitious root initiation on the plant studied. However, using the mung bean bioassay, Hess's (1963) work with paper chromatographed extracts of juvenile and mature *H. helix* gave no evidence for rooting inhibitor fractions at the dilutions he used.

Using reciprocally approach-grafted, detached juvenile and mature leaves of *H. helix* as cuttings (Fig. 1), Mokhtari (1985, M.S. Thesis, Univ. of Minnesota, St. Paul, USA) found no evidence for inhibition of rooting in juvenile petioles by mature lamina grafted thereon. He concluded that the difference in rooting potential of juvenile and mature leaf petioles was not due to the production of a translocatable rooting inhibitor in mature leaf lamina. Similarly, Mullins (1985) found no evidence for a transmissible rooting inhibitor in an *in vitro* reciprocal grafting study with non-rooting (mature) and rooting (rejuvenated) microcuttings of Jonathan apples.

Non-Mobile or Non-Extractable Factors

The possibility that non-mobile or non-extractable factors are involved in the difference in rooting ability of juvenile and mature tissues has been raised by several authors (Bouillenne and Bouillenne-Walrand 1955, Hackett 1970, Haissig 1974, Heuser 1976). This possibility is based on three types of experimental evidence: 1) differences in anatomical characteristics of juvenile and mature stems; 2) inability to influence rooting by reciprocal grafting of juvenile and mature cuttings; and 3) inability to demonstrate that extracts from easy-to-root juvenile or environmentally modified mature tissues can promote rooting of difficult-to-root mature tissues.

Stoutemyer's (1937) study was the first to investigate the anatomical characteristics of juvenile and mature stems. He found that mature phase stems of apple contained more pericycle fibers than the juvenile form. Similarly, Hatcher and Garner (1955), and Beakbane (1961) reported that juvenile apple stems were relatively free from fibers and sclereids in the primary phloem as compared to difficult-to-root mature shoots of the same clones. Both Gooden (1965) and Girouard (1967ab) studied the anatomy in juvenile and mature *H. helix* stems. Girouard showed that the phloem fibro-bundle caps were thicker in mature stems than in juvenile stems. Goodin observed that there was greater overall lignification in the mature stem and that, in addition, phloem fibro-bundle caps were very rare in juvenile stems. He found that stems with transitional morphology (intermediate between juvenile and mature) possessed a few fibers that became more numerous as the maturing plant increased in age. Davies and coworkers (1982) studied anatomical characteristics of juvenile and mature stems of *Ficus pumila*, and concluded that the continuous ring of perivascular sclerenchyma of the mature had more cell layers than the juvenile.

The researchers cited above all concluded that the greater amount of sclerenchyma in mature tissue than in juvenile tissue could not account for the difference in rooting ability on the basis of mechanical obstruction to rooting. As Beakbane (1961) pointed out, mechanical restriction by the fiber sheath could not be the cause of failure of some mature plants to root because root initials do not form readily within the sheath of these plants. This conclusion does not rule out a biochemical relationship between fiber formation, lignification, and ease of rooting in juvenile and mature plants (Goodin 1965).

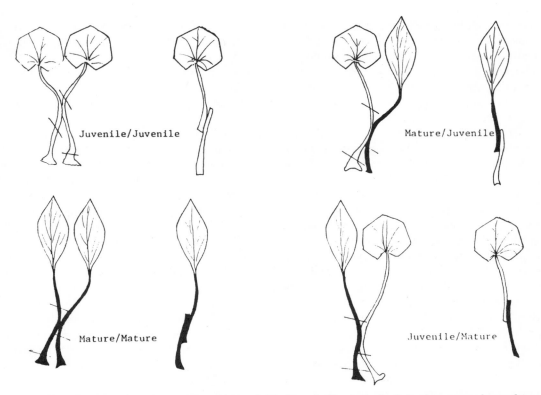

Figure 1. Schematic representation of *Hedera helix* detached leaf approach grafting procedure used to obtain cuttings with reciprocal combinations of juvenile and mature petioles with juvenile and mature lamina.

Few grafting experiments have been done with *cuttings* of juvenile and mature plants. Mokhtari (1985, M.S. Thesis, Univ. of Minnesota, St. Paul, USA) used detached reciprocally approach-grafted juvenile and mature leaves of *H. helix* (Fig. 1) to obtain evidence regarding the possible translocation of substances involved in control of root initiation and/or acquisition of rooting potential. Approach graft unions were anatomically complete before appropriate lamina or petioles were excised to give all four combinations of juvenile and mature petioles (receptors) with juvenile and mature lamina (donors). Auxin treated juvenile petioles with either juvenile or mature lamina were maximally rooted in 24 d with a mean of 12 roots per petiole. At 24 d, auxin treated mature petioles with either juvenile or mature lamina had not rooted. However, 21 d later 100% of mature petioles with juvenile lamina had rooted with a mean of 8 roots per petiole, while mature petioles with mature lamina had only callus formation. Microscopic observation showed that root initials in juvenile petioles formed directly from cortical parenchyma, whereas those on mature petioles with juvenile lamina formed only in callus.

These data strongly suggest that root initiation is a function of rooting potential of cells in the petiole and that, at least initially, root initiation is affected little by the type of lamina. That is, there is no evidence of a rooting promoter being translocated from juvenile lamina, and no evidence of a rooting inhibitor being translocated from mature lamina. However, the results also indicate that with time, newly formed (and apparently dividing) cells in callus at the base of mature petioles with juvenile lamina have an increased rooting potential as compared with cells in callus at the base of mature petioles with mature lamina. This increase in rooting potential of dividing cells in response to a juvenile lamina may be similar to rejuvenation in terms of morphological and physiological characteristics of mature *H. helix* shoot scions when grafted onto leafy, juvenile rootstocks (Doorenbos 1954). The case with mature *Hedera* scions is similar to other cases such as *Hevea brasiliensis* L. (Muzik and Cruzado 1958) and *Eucalyptus* spp. (Martin and Quillet 1974) in which grafting to juvenile stocks influences rooting potential only after considerable time has passed and growth of the scion has occurred. Mullins (1985) similarly found no evidence of transmissible promoters or inhibitors of rooting based on reciprocal grafts or co-culturing of *in vitro* derived non-rooting (mature) and rooting (rejuvenated) microcuttings of Jonathan apple.

The overall conclusion from the discussion above is that rooting potential of juvenile and mature tissue is a characteristic of cells at the site of root initiation; it is not related to translocated rooting promoters or inhibitors. However, as discussed in the previous section the rooting potential of cells can be changed.

Hackett (1970) showed that juvenile shoot tips of *H. helix in vitro* responded to NAA or IAA plus catechol treatment by forming more roots while mature apices did not respond to these factors. In addition, juvenile apices responded to both mature and juvenile, fractionated methanol extracts whereas mature apices did not. As noted in the section on rooting cofactors, these results indicate that extractable cofactors are not limiting root initiation in mature shoot tips. It was further shown that a reduction in light intensity, or growth in darkness, brings about a qualitative change in rooting response of mature shoot tips to IAA plus catechol: so while mature shoot tips fail to respond to auxin plus catechol in high light, they do respond when cultured in low light or darkness. Furthermore, the light controlled factor was not methanol extractable from etiolated tissue as evidenced by rooting bioassays using light cultured juvenile and mature apices. Taken together with results in the reciprocal grafting experiments, these results provide strong evidence for a non-mobile component controlling root initiation potential in *H. helix*.

Similarly, Mullins (1985) found that Jonathan apple microcuttings from recently established *in vitro* cultures responded little to auxin and/or phenolics while microcuttings from the same source, subcultured 10 times at monthly intervals, rooted at a 95% frequency with nearly 10 roots per microcutting in response to auxin only. In addition to their change in rooting potential with increasing number of subcultures, the microcuttings had decreased vascularization and increased accumulation of anthocyanin when subjected to auxin-rich medium. These results, likewise, provide strong evidence for a non-mobile component of root initiation potential in apple as was demonstrated in *H. helix*.

What might be the nature of a non-mobile component? It has been suggested that only certain parenchyma cells, called target cells, are involved in root organogenesis (Nougarède and Rondet 1982). Perhaps only certain cells are competent to respond to auxin by forming root initials. This could be based on an enzyme or enzyme system required to carry out the auxin requiring process (Haissig 1974): poor rooting could be due to lack of the enzyme(s) or compartmentalization of the enzyme(s) separate from substrates. For example, Al Barazi and Schwabe (1984) presented evidence that PPO levels decreased with the age of *Pistacia vera* L. plants, hence corresponding to loss in rooting potential. Most of the PPO activity was located in tissues where root primordia were expected to develop. Another possibility is that cells lack binding sites or receptors (Katekar 1979) required for auxin action in root morphogenesis.

CONCLUSION

The loss of adventitious rooting potential with ontogenetic age or maturation severely limits the success or efficiency in clonally propagating selected, desirable mature individuals of many species, especially tree species. The reasons for loss of rooting potential with maturation are very poorly understood. In some species the anatomical origin of root primordia is very different in easy-to-root juvenile and difficult-to-root mature tissue. Usually, treatment of cuttings with auxins enhances root initiation in those from juvenile plants while such treatment enhances cell division but not root initiation in difficult-to-root cuttings from mature plants. This indicates that rooting of mature cuttings is not limited by endogenous auxin.

Several procedures including hedging of stock plants, serial grafting of mature scions on seedling rootstocks, serial subculturing of mature explants *in vitro,* and use of cuttings of adventitious origin have been used successfully in some species to maintain or enhance rooting potential. Hedging is thought to arrest maturation at the ontogenetic stage of the nodes from which hedge shoots are formed. The other methods mentioned are thought to reverse the maturation process resulting in a more juvenile state for rooting potential but not necessarily for other characteristics such as flowering.

Endogenous levels of plant growth substances, rooting cofactors and promoters, and rooting inhibitors have been investigated in juvenile and mature tissues of various species. The best evidence for a causal relationship between the concentration of specific, extractable compounds and loss of rooting potential with age is for *Eucalyptus grandis* (and some other *Eucalyptus* spp.). The so-called G compounds that have been identified and chemically synthesized, occur at high concentration in mature tissues and low concentration in juvenile tissues, and are inhibitory to rooting of *E. grandis* cuttings from seedling plants when applied at high concentration, and promotive of rooting when applied at low concentration. In other species or genera there is little or no convincing evidence that extractable or translocatable factors are physiologically important in explaining loss of rooting potential with maturation. Many of the investigations of the involvement of extractable rooting cofactors in loss of rooting potential are suspect because of the use of easy-to-root tissue (most often of other species) as the sole bioassay for rooting activity. Indirect evidence from grafting and co-culturing experiments suggest that non-mobile factors may be important in explaining loss of rooting potential in some species.

An important requirement for investigation and understanding of the physiological basis of loss of rooting potential is an experimental rooting system that involves both easy-to-root juvenile and difficult-to-root mature tissue. Two systems appear to fulfill this requirement and are experimentally convenient: 1) *in vitro* subcultured microcuttings derived from mature shoot apices of apple and other species; and 2) *in vitro* cultured de-laminated leaf petioles from juvenile and mature *H. helix*. These systems provide a means for testing the physiological significance of putative endogenous root promoting and inhibiting substances by assaying with easy- and difficult-to-root tissue from the same genotype from which extracts were made. They may also provide a means for investigating the genetic basis for control of rooting.

REFERENCES

Al Barazi, Z. and W. W. Schwabe. 1984. The possible involvement of polyphenol oxidase and the auxin oxidase system in root formation and development in cuttings of *Pistacia vera*. *J. Hortic. Sci.* 59:453–461.

————— ————— 1985. Studies on possible internal factors involved in determining ease of rooting in cuttings of *Pistacia vera* and *Prunus avium*, cvs Colt and F 12/1. *J. Hortic. Sci.* 60:439–445.

Baldini, E. and F. Scaramuzzi. 1957. Indagini et asservazioni sugli sferoblasti delli olivo. *Ann. Sper. Agr.* 11:723–740.

Beakbane, A. B. 1961. Structure of the plant stem in relation to adventitious rooting. *Nature* 192:954–955.

Bekkaoui, F., Y. Arnaud, C. Larrieu and E. Miginiac. 1984. Etude comparative de la rhizogénèse *in vitro* du *Sequoia sempervirens* chez deux clones d'age differents. *Annales Assoc. Forêt-Cellulose 1983*, Paris, France. pp. 5–25.

Bouillenne, R. and M. Bouillenne-Walrand. 1955. Auxines et bouturage *Proc. 14th Int. Hortic. Cong.* 1:231–238.

Boulay, M. 1979. Multiplication et clonage rapide du *Sequoia sempervirens* par la culture *in vitro*. In *Micropropagation d'Arbres Forestiers*. Assoc. Forêt-Cellulose. Nangis, France. pp. 49–56.

Brink, R. A. 1962. Phase change in higher plants and somatic cell heredity. *Quart. Rev. Biol.* 37:1–22.

Cameron, R. J. and G. V. Thomson. 1969. The vegetative propagation of *Pinus radiata*: root initiation in cuttings. *Bot. Gaz.* 130:242–251.

Caruso, J., R. G. Smith, L. M. Smith, T.-Y. Cheng and G. D. Daves, Jr. 1978. Determination of indole-3-acetic acid in Douglas fir using a deuterated analog and selected ion monitoring. Comparison of microquantities in seedling and adult tree. *Plant Physiol.* 62:841–845.

Chaperon, H. 1979. Maturation et bouturage des arbres forestiers. In *Micropropagation d'Arbres Forestiers. Assoc. Forêt-Cellulose*. Nangis, France. pp. 19–31.

Chattaway, M. M. 1958. Bud development and lignotuber formation in eucalyptus. *Aust. J. Bot.*

6:103–115.

Condit, I. J. 1969. *Ficus: The Exotic Species*. Univ. of California Div. of Agric. Sci., Berkeley, CA, USA.

Chin, T.-Y., M. Meyer, Jr. and L. Beevers. 1969. Abscisic acid stimulated rooting of stem cuttings. *Planta* 88:192–196.

Crow, W. D., T. Osawa, K. M. Platz and D. S. Sutherland. 1976. Root inhibitors in *Eucalyptus grandis* II. Synthesis of the inhibitors and origin of the peroxide linkage. *Aust. J. Chem.* 29:2525–2531.

——— ——— D. M. Paton and R. R. Willing. 1977. Structure of grandinol: a novel root inhibitor from *Eucalyptus grandis*. *Tetrahedron Lett.* 12:1073–1074.

Davidson, J. 1974. Reproduction of *Eucalyptus deglupta* by cuttings. *N. Z. J. For. Sci.* 4:191–203.

Dhawan, A. K., D. M. Paton and R. R. Willing. 1979. Occurrence and bioassay responses of G, a plant growth regulator in *Eucalyptus* and other Myrtaceae. *Planta* 146:419–422.

Doorenbos, J. 1954. Rejuvenation of *Hedera helix* in graft combinations. *Proc. K. Ned. Akad. Wet. Amst.*, Ser. C. 57:99–102.

Fielding, J. M. 1954. *Methods of Raising Monterey Pine from Cuttings in the Open Nursery*. For. and Timber Bur. Bull. 32, Canberra, Aust.

——— 1969. *Factors Affecting the Rooting and Growth of Pinus radiata Cuttings in the Open Nursery*. For. and Timber Bur. Bull. 45, Canberra, Aust.

Fortanier, E. J. and H. Jonkers. 1976. Juvenility and maturity of plants as influenced by their ontogenetic and physiological aging. *Acta Hortic.* 56:37–44.

Fouret, Y., Y. Arnaud, R. Maldiney, B. Sotta and E. Miginiac. 1986. Relation entre rhizogénèse teneur en auxine et acide abscissique chez trois clones de *Sequoia sempervirens* Endl. issus d' arbres d'age different. *C.R. Acad. Sci.* Paris. 4:135–138.

Franclet, A. 1979. Rajeunissement des arbres adults en vue de leur propagation végétative. In *Micropropagation d'Arbres Forestiers*. Assoc. Forêt-Cellulose. Nangis, France. pp. 3–18.

Frydman, V. M. and P. F. Wareing. 1973. Phase change in *Hedera helix* L. I. Gibberellin-like substances in the two growth phases. *J. Exp. Bot.* 24:1131–1138.

Gardner, F. E. 1929. The relationship between the age and the rooting of cuttings. *Proc. Amer. Soc. Hortic. Sci.* 26:101–104.

Garner, R. J. and E. S. J. Hatcher. 1962. Regenerating in relation to vegetative vigor and flowering. In *Proc. 16th Int. Hortic. Cong.*, Brussels, Belgium. pp. 105–111.

Girouard, R. M. 1967a. Initiation and development of adventitious roots in stem cuttings of *Hedera helix*. Anatomical studies of the juvenile growth phase. *Can. J. Bot.* 45:1877–1881.

——— 1967b. Initiation and development of adventitious roots in stem cuttings of *Hedera helix*. Anatomical studies of the mature growth phase. *Can. J. Bot.* 45:1883–1886.

——— 1969. Physiological and biochemical studies of adventitious root formation. Extractable rooting cofactors from *Hedera helix*. *Can. J. Bot.* 47:687–699.

Goodin, J. R. 1965. Anatomical changes associated with juvenile-to-mature phase transition in *Hedera*. *Nature* 208:504–505.

Gupta, P. K., A. F. Mascarenhas and V. Jagganattran. 1981. Tissue culture of forest trees: clonal propagation of mature trees of *Eucalyptus citriodora* by tissue culture. *Plant Sci. Lett.* 20:195–201.

Hackett, W. P. 1970. The influence of auxin, catechol and methanolic extracts on root initiation in aseptically cultured shoot apices of the juvenile and mature forms of *Hedera helix*. *J. Amer. Soc. Hortic. Sci.* 95:398–402.

——— 1985. Juvenility, maturation and rejuvenation in woody plants. *Hortic. Rev.* 7:109–155.

——— C. Srinivasan. 1985. *Hedera helix* and *H. canariensis*. In *Handbook of Flowering* (A. H. Halevy, ed). CRC Press, Boca Raton, FL, USA. Vol. III. pp. 89–97. ISBN 0-8493-3913-8.

Haissig, B. E. 1974. Influences of auxins and auxin synergists on adventitious root primordium initiation and development. *N. Z. J. For. Sci.* 4:311–323.

Hartney, V. J. 1980. Vegetative propagation of eucalyptus. *Aust. For. Res.* 10:191–211.

Hatcher, E. J. S. 1959. The propagation of rootstocks from stem cuttings. *Ann. Appl. Biol.* 47:635–639.

_____ R. J. Garner. 1955. The production of sphaeroblast shoots of apple for cuttings. *Rep. E. Malling Res. Sta. for 1954.* pp. 73–75.

Hess, C. E. 1959. A study of plant growth substances in easy and difficult to root cuttings. *Proc. Int. Plant Prop. Soc.* 9:39–43.

_____ 1962. Characterization of rooting cofactors extracted from *Hedera helix* L. and *Hibiscus rosa-sinensis* L. *Proc. 16th Int. Hortic. Cong.* 4:382–388.

_____ 1964. Naturally occurring substances which stimulate root initiation. In *Rég. Nat. Croiss. Végét., 5th Int. Conf. on Plant Growth Sub.* (J. P. Nitsch, ed). pp. 517–527. C.N.R.S. Gif-sur-Yvette, France.

_____ 1965. Rooting cofactors—identification and functions. *Proc. Int. Plant Prop. Soc.* 15:181–186.

Heuser, C. W. 1976. Juvenility and rooting cofactors. *Acta Hortic.* 56:251–259.

_____ C. E. Hess. 1972. Isolation of three lipid root-initiating substances from juvenile *Hedera helix* shoot tissue. *J. Amer. Soc. Hortic. Sci.* 97:571–574.

Heybroek, H. H. and T. Visser. 1976. Juvenility in fruit growing and forestry. *Acta Hortic.* 56:71–80.

Hillman, J. R., I. Young and B. A. Knight. 1974. Abscisic acid in leaves of *Hedera helix* L. *Planta* 119:263–266.

Jacobs, M. R. 1939. *The Vegetative Reproduction of Forest Trees. Experiments with Cuttings of Pinus radiata Don.* Commonwealth For. Bul. 25. Canberra, Aust.

_____ 1955. *Growth Habits of the Eucalyptus.* Commonwealth Government Printer. Canberra, Aust.

Katekar, G. F. 1979. Auxins: on the nature of the receptor site and molecular requirements for auxin activity. *Phytochemistry* 18:223–233.

Libby, W. J., Jr. and M. T. Conkle. 1966. Effects of auxin treatment, tree age, tree vigor and cold storage on rooting young Monterey pine. *For. Sci.* 12:484–502.

_____ A. G. Brown and J. M. Fielding. 1972. Effects of hedging radiata pine on production, rooting and early growth of cuttings. *N. Z. J. For. Sci.* 2:263–283.

Lanphear, F. O. and R. P. Meahl. 1963. Influence of endogenous rooting cofactors and environment on seasonal fluctuation in root initiation of selected evergreen cuttings. *Proc. Amer. Soc. Hortic. Sci.* 83:811–818.

Martin, B., and G. Quillet. 1974. Bouturage des arbres forestiers en Congo. Résultats des essais effectués à Pointe-Noire de 1969 à 1973. Rajeunissement des arbres plus et constitution du parc à bois. *Bois et Forêt des Tropiques* 157:21–40.

Mirov, N. J. 1944. Experiments in rooting pines in California. *J. For.* 42:199–204.

Mullins, M. G. 1985. Regulation of adventitious root formation in microcuttings. *Acta Hortic.* 166:53–61.

_____ 1987. Propagation and genetic improvement of temperate fruits: the role of tissue culture. In *Plant Tissue and Cell Culture* (C. E. Green, D. A. Somers, W. P. Hackett and D. D. Biesboer, eds). Alan R. Liss, Inc., NY, USA. pp. 395–406. ISBN 0-8451-1802-1.

Muzik, T. J. and H. J. Cruzado. 1958. Transmission of juvenile rooting ability from seedlings to adults of *Hevea brasiliensis. Nature* 181:1288.

Nichols, W., W. D. Crow and D. M. Paton. 1972. Chemistry and physiology of rooting inhibitors in adult tissue of *Eucalyptus grandis.* In *Plant Growth Substances 1970* (D. J. Carr, ed). Springer-Verlag, Berlin, W. Germany.

Nougarède, A. and P. Rondet. 1982. Rhizogénèse adventive dans l'épicotyle du Pois: initiation et structuration de la racine. *Can. J. Bot.* 60:261–280.

Olesen, P. O. 1978. On cyclophysis and topophysis. *Silvae Genetica* 27:173–178.

Paton, D. M. 1984. Vegetative propagation of adult eucalyptus. In *Colloque International sur les Eucalyptus Résistants au Froid.* Sept., 1983. CSIRO-AFOCEL. Bordeaux, France. pp. 570–586.

_____ R. R. Willing. 1974. Inhibitor transport and ontogenetic age in *E. grandis.* In *Plant Growth Substances 1973.* Hirokawa Pub., Tokyo, Japan. pp. 126–132.

_____ _____ L. D. Pryor. 1981. Root-shoot gradients in Eucalyptus ontogeny. *Ann Bot.* 47:835–838.

_____ _____ W. Nicholls and L. D. Pryor. 1970. Rooting of stem cuttings of Eucalyptus: A rooting inhibitor in adult tissues. *Aust. J. Bot.* 18:175–183.

Porlingis, I. C. and I. Therios. 1976. Rooting response of juvenile and adult leafy olive cuttings to various factors. *J. Hortic. Sci.* 51:31–39.

Pryor, L. D. and Willing, R. R. 1963. The vegetative propagation of Eucalyptus—an account of progress. *Aust. For.* 27:52–62.

Quamme, H. A. and S. Nelson. 1965. Root promoting substances in the juvenile phase of *Malus robusta* 5. *Can. J. Plant Sci.* 4:509–511.

Raviv, M. and O. Reuveni. 1984. Endogenous content of a rooting promoter extracted from leaves of difficult and easy-to-root avocado cuttings. *J. Amer. Soc. Hortic. Sci.* 109:284–287.

_____ D. Becker and Y. Sahali. 1986a. The chemical identification of root promoters extracted from avocado tissues. *Plant Growth Regul.* 4:371–374.

_____ O. Reuveni and E. E. Goldschmidt. 1986b. The physiological basis for loss of rootability with age of avocado seedlings. *Tree Physiol.* 3:112–115.

Richards, M. 1964. Root formation on cuttings of *Camellia reticulata* var. 'Capt. Rawes'. *Nature* 204:601–602.

Robinson, J. C. and W. W. Schwabe. 1977. Studies on the regeneration of apple cultivars from root cuttings. I. Propagation aspects. *J. Hortic. Sci.* 52:205–220.

Rogler, C. E. and W. P. Hackett. 1975. Phase change in *Hedera helix*: Induction of the mature to juvenile phase change by gibberllin A_3. *Physiol. Plant.* 34:141–147.

Roulund, H. 1973. The effect of cyclophysis and topophysis on rooting ability of Norway spruce cuttings. *For. Tree Improv. Arbor.* (Hørsholm) 5:21–41.

Seelinger, R. 1924. Topophysis und Zyklophysis pflanzlicher Organe und ihre Bedeutung für die Pflanzenkultur. *Angew. Bot.* 6:191–200.

Shen, X. S. and M. G. Mullins. 1983. *In vitro* propagation of pear *Pyrus communis* L. cvs Packham's Triumph, William's 'Bon Chretien and Beurre' Bosc. *Scientia Hortic.* 23:51–57.

Smith, D. R. and T. A. Thorpe. 1975. Root initiation in cuttings of *Pinus radiata* seedlings. I. Developmental sequence. *J. Exp. Bot.* 26:184–192.

Sriskandarajah, D., M. G. Mullins and Y. Nair. 1982. Induction of adventitious rooting *in vitro* in difficult-to-root cultivars of apple. *Plant Sci. Lett.* 24:1–9.

Sterns, M. 1971. Crystal and molecular structure of a root inhibitor from *Eucalyptus grandis*, 4-ethyl-1-hydroxy-4,8,8,10,10-pentamethyl-7,9-dioxo-2,3-dioxabicyclo (4.4.0) decene-5. *J. Crystl. Mol. Struct.* 1:373–381.

Stoutemyer, V. T. 1937. Regeneration in various types of applewood. *Iowa Agric. Exp. Bull.* 220. pp. 307–352.

Takeno, K. J., S. Taylor, S. Sriskandarajah, R. P. Pharis and M. G. Mullins. 1983. Endogenous gibberellin and cytokinin-like substances in cultured shoot tissues of apple, *Malus pumila* cv Jonathan, in relation to adventitious root formation. *Plant Growth Regul.* 1:261–268.

Thulin, I. J. and T. Faulds. 1968. The use of cuttings in the breeding and afforestation in *Pinus radiata*. *N. Z. J. For. Sci.* 13:66–77.

Vasquez, A. and D. V. Gesto. 1982. Juvenility and endogenous rooting substances in *Castanea sativa* Mill. *Biol. Plant.* 24:48–52.

Vieitez, E. and A. M. Vieitez. 1976. Juvenility factors related to the rootability of chestnut cuttings. *Acta Hortic.* 56:269–274.

Visser, T. 1964. Juvenile phase and growth of apple and pear seedlings. *Euphytica* 13:119–129.

Wareing, P. F. 1959. Problems of juvenility and flowering in trees. *J. Linn. Soc.* (London). 44:1402–1406.

_____ V. M. Frydman. 1976. General aspects of phase change with special reference to *Hedera helix* L. *Acta Hortic.* 56:57–68.

Wellensiek, S. J. 1952. Rejuvenation of woody plants by formation of sphaeroblasts. *Proc. K. Ned. Akad. Wet. Amst.*, Ser. C. 55:567–573.

Wilkinson, E. H. 1973. Growing apple cultivars on their own roots. In *Fruit Present and Future*. Royal Hortic. Soc. London. 2:81–76.

Zimmerman, R. H. 1963. Rooting cofactors in some southern pines. *Proc. Int. Plant Prop. Soc.* 13:71–74.

_____ 1981. Micropropagation of fruit plants. *Acta Hortic.* 120:217–222.

CHAPTER 2

Etiolation and Banding Effects on Adventitious Root Formation

Brian K. Maynard and Nina L. Bassuk

The Urban Horticulture Institute
Department of Floriculture and Ornamental Horticulture
Cornell University
Ithaca, NY 14853

INTRODUCTION 30
   Definitions 30
   Historical Perspectives 31
   Practical Aspects of Etiolation 33
      Timing 33
      Environment 33
      Levels of shade 33
   Practical Aspects of Banding 34
      Location 34
      Width of band 35
      Banding variations 35
   Differential Responses to Etiolation and Banding 35
FACTORS ASSOCIATED WITH ENHANCED ROOTING OF
  ETIOLATED TISSUE 37
   Anatomical Factors 37
      Presence of sclerified tissues 37
      Transmission of the etiolation stimulus 38
      Other anatomical factors 38
   Physiological Factors 39
      Endogenous activity of indole-3-acetic acid 39
      Transport of auxin 40
      Tissue sensitivity to auxin 40
      Meristematic activity 40
      Auxin cofactors 41
CONCLUSION 42
ACKNOWLEDGMENTS 43
REFERENCES 43

Additional key words: blanching, plant propagation, light exclusion, rooting.
Abbreviations: ELISA, enzyme-linked immuno sorbance assay; IAA, indole-3-acetic acid; IAA-O, indole-3-acetic acid oxidase (EC 1.11.1.8); IBA, indole-3-butyric acid; PAL, phenylalanine ammonia-lyase (EC 4.3.1.5); PPO, polyphenol oxidase (EC 1.10.3.1).

INTRODUCTION

Definitions

Although etiolation is defined as the total exclusion of light, etiolation as it is used by the plant propagator also refers to forcing new shoot growth under conditions of heavy shade. This new growth is then used as the cutting propagule. Banding, localized light exclusion, is a pretreatment, adjunct to etiolation, which involves excluding light from that portion of the stem that is to become the cutting base. An adhesive band may either be applied to etiolated shoots whose tops are subsequently allowed to turn green in the light, or applied to light-grown developing shoots which are still in the softwood stage. In the latter case the band is said to "blanch" the underlying tissues. The responses to etiolation and banding are similar: the tissues shielded from light are characteristically chlorotic and in a soft or succulent condition. The beneficial effects of etiolation and banding have been reviewed a number of times (Frolich 1961, Herman and Hess 1963, Kawase 1965, Ryan 1969, Delargy and Wright 1978, and Harrison-Murray 1982). The exclusion of light during stock plant growth has been firmly established as a successful approach to propagation with cuttings (see Table 1 for a list of studies involving etiolation, shading, or banding). Other vegetative plant propagation methods, including mound layering (stooling) and air layering, also result in the exclusion of light from the area to be rooted for a time before excising propagules (Ryan 1969). The standard practice of placing cuttings in an opaque medium also facilitates rooting by excluding light (Frolich 1961).

Table 1. Reports of improved cutting propagation by use of stock plant etiolation, shading, and banding pretreatments.

Treatment	Plant Species	Reference
Banding/ Blanching:	Acer platanoides L.	Howard (1981, 1982, 1983)
	Tilia cordata Mill.	Howard (1981, 1982, 1983)
	Tilia europaea L.	Howard (1981, 1982, 1983)
	Pinus elliottii Engelm. 'elliotti'	Hare (1978)
	Platanus occidentalis L.	Hare (1976)
	Rhododendron cvs.	Davis and Potter (1983)
	Rubus idaeus L.	Howard (1983)
Etiolation:	Artocarpus heterophyllus Lam.	Mukherjee and Chatterjee (1979)
	Bryophyllum tubiflorum Harvey	Nanda et al. (1967)
	Camphora officinarum Ness.	Blackie et al. (1926), Reid (1923)
	Clematis spp.	Smith (1924)
	Corylus maxima Mill.	Rowell (1982)
	Cotinus coggygria Scop.	Rowell (1982)
	Polygonum baldschuanicum Regel.	Rowell (1982)
	Syringa vulgaris L. cvs.	Rowell (1982)
	Malus sylvestris Mill.	Doud and Carlson (1977) Anderson (1982) Harrison-Murray et al. (1982) Harrison-Murray (1982) Howard et al. (1985)
	Mangifera indica L.	Mukherjee et al. (1967), Bid and Mukherjee (1972)
	Persea americana Mill.	Frolich (1951), (1961), Frolich and Platt (1972), Ernst and Holtzhausen (1978), Raviv (1981), Mohammed and Sorhaindo (1984)

Treatment	Plant Species	Reference
	Prunus domestica L.	Knight and Witt (1937)
	Rubus idaeus L. 'Meeker'	Doss et al. (1980)
	Tilia tomentosa Moench. 'Szeleste'	Schmidt (1982)
Etiolation plus Banding:	*Acer* spp.	Maynard and Bassuk (1986)
	Betula papyrifera Marsh.	Maynard and Bassuk (1986)
	Carpinus betulus L.	Maynard and Bassuk (1986)
	Castanea mollissima Bl.	Maynard and Bassuk (1986)
	Corylus americana Marsh. 'Rush'	Maynard and Bassuk (1986)
	Pinus spp.	Maynard and Bassuk (1986)
	Quercus spp.	Maynard and Bassuk (1986)
	Carpinus betulus L.	Bassuk et al. (1985)
	Pinus strobus L.	Bassuk et al. (1985)
	Hibiscus rosa-sinensis L. cvs.	Hermann and Hess (1963)
	Malus sylvestris Mill.	Gardner (1937), Delargy and Wright (1978, 1979), Howard (1979–1984)
	Persea americana Mill.	Reuveni and Goren (1982)
	Pistachia vera L.	Al Barazi and Schwabe (1984)
	Syringa vulgaris L.	Gardner (1937)
	Syringa vulgaris L. cvs.	Miske and Bassuk (1985)
	Tilia cordata L.	Howard (1983)
	T. europaea L.	Howard (1983)
Shading:	*Crassula argentea* Thunb.	Paterson and Rost (1979)
	Euonymus japonicus L.	Ooishi et al. (1982)
	Hibiscus rosa-sinensis L.	Johnson and Hamilton (1977)
	Picea sitchensis (Bong.) Carr.	van den Driessche (1985)
	Rhododendron spp.	Johnson and Roberts (1971)
	Rosa spp.	Khosh-kui and Sink (1982)
	Schefflera arboricola Merrill	Rauch (1981)

Historical Perspectives

The earliest report of etiolation as a cutting propagation tool is that of Reid (1923), who credits L. B. Stewart, Royal Botanic Garden, Edinburgh, with the discovery that etiolation accelerates the rooting of cuttings. Ryan (1969) cites the 1894 work of J. Sachs as the first use of light exclusion to promote adventitious rooting. Sachs studied the inhibiting effect of light on the rooting of *Cactus specious* (sic), *Tropaeolum majus* L., and *Hebe speciosa* Cockayne & Allen.

In the 1920's black paper coverings were used to produce etiolated cuttings of *Camphora* (Reid 1923, Blackie et al. 1926) and *Clematis* (Smith 1924); neither species had previously been known to root adventitiously. In 1937, Gardner viewed etiolation as a method of considerable practicality. His work with apple and other species was the first mention of using etiolation and banding together to facilitate the rooting of difficult-to-root species. By wrapping black insulation tape close to the growing point, cuttings were blanched while they were still attached to the tree and rooting was increased by 70%. Gardner (1937) also built light-tight boxes for pre-etiolation of apple shoots. Cuttings from stock plant shoots banded for up to one year rooted successfully at a 98% rate. The response to etiolation plus banding varied among apple varieties, although in all instances it increased rooting percentage. Gardner also was the first to mention the appearance of root primordia, up to 1.2 cm in length, produced under banded stems still on the stock plant. The procedure developed by Gardner, although modified in several useful ways, had remained essentially unchanged since that time (Fig. 1).

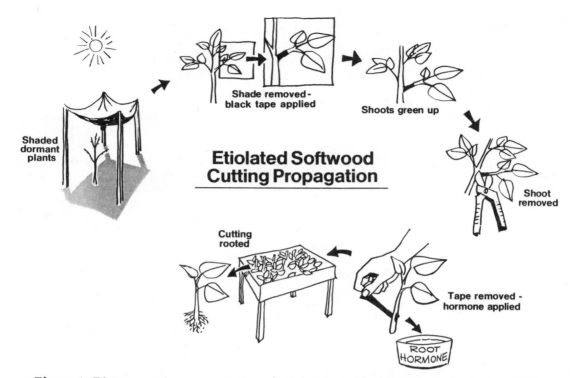

Figure 1. Diagrammatic representation of etiolation and banding as stock plant pretreatments to softwood cutting propagation. Shoots are forced in darkness until they reach a suitable length for banding. Bands remain in place for about 4 weeks. [Bassuk et al. (1985)]

Etiolation is usually achieved by covering the stock plant with black cloth or black polyethylene, fastened over a structure which encloses the plant. Banding employs black tape, paper, tubing, aluminum foil (Biran and Halevy 1973, Hare 1976, Davis and Potter 1983), or even black paste (Reuveni and Goren 1982). Most recently a reusable adhesive fabric, known as Velcro, has been added to the list of useful banding materials (Maynard and Bassuk 1986).

Knight and Witt (1937) compared etiolation under black cloth with stooling (mounding earth over coppiced stock plants to produce etiolated growth) as pretreatments in the propagation of plum cuttings. They found that etiolated cuttings rooted better than stooled cuttings, but were also more susceptible to *Botrytis* infections.

Interest in the practical uses of etiolation and the physiological and/or anatomical basis for the etiolation effect has increased steadily since Frolich (1961) used etiolation to propagate avocado (*Persea americana* Mill.). Frolich's technique made use of stooling following etiolation. The procedure called for shoot extension in a dark chamber to a length of 7.5 cm before placement into a tar-paper collar filled with vermiculite to exclude light (Frolich 1951, 1961, Frolich and Platt 1972). The technique has been modified by Ernst and Holtzhausen (1978) to obtain etiolated cutting wood by forcing excised hardwood or semi-hardwood stem sections of avocado. A further modification to increase the production of etiolated shoots by applying gibberellic acid (GA_3) and 6-benzylamino purine was developed by Mohammed and Sorhaindo (1984). Schmidt (1982) compared etiolation, banding, and stooling in the propagation of *Tilia tomentosa* Moench 'Szeleste' and found that the promotion of rooting was due to the etiolation component, and not the other component influences of stooling (moisture, temperature).

Perhaps the greatest contribution to the development of a practical system for etiolation and banding has come from research at East Malling Research Station, U.K. Researchers there have provided great insight into the applicability of these pretreatments to a wide range of plant materials. Refinements in the technique developed at East Malling include the use of heavy shade instead of complete darkness, ventilation of the shading structure, greening shoots for specified periods of time before taking cuttings, and comparisons of etiolation with banding as an additional pretreatment.

Pretreatments which have been used successfully with etiolation include hedging to induce shoot vigor (Mukherjee et al. 1967, Delargy and Wright 1978), girdling (Delargy and Wright 1978), and defoliation of the cutting prior to sticking (Howard 1984).

Practical Aspects of Etiolation

Timing

The first study on the importance of synchronizing etiolation with the development of shoots was undertaken by Reid (1923), who compared etiolating shoots of *Camphora officinarum* Nees. for periods of 14 vs. 28 d. The shoots treated for 28 d were noticeably more etiolated, though the 14 d treatment alone promoted callusing (from 0 to 80%, without any mention of rooting). Smith (1924) observed that shoots of *Clematis* spp. etiolated for as long as three weeks were weakened and tended to die off in the rooting environment. Gardner (1937) etiolated apple trees just before bud break, and banded shoots for up to 1 yr before taking cuttings, noting a tremendous increase in rooting percentage. The rooting response of etiolated shoots followed a pattern which Gardner generally observed in woody cuttings of apple—good through August, then poor until bud break the next spring. Evidence suggests that etiolation need only be applied for a length of time sufficient to yield shoots which are usable for banding or taking cuttings (Delargy and Wright 1978, Howard 1979). Etiolation has been applied up to three weeks after bud break with no decrease in its effect on rooting percentage (Howard 1982). Miske and Bassuk (1985) working with *Syringa vulgaris* L. cultivars concluded that etiolation extended the period of time over which lilac cuttings can be taken with acceptable rooting results.

The absence of chlorophyll, which is characteristic of etiolation, appears to have no bearing on rooting success (Ryan 1969, Howard 1981). In general, greening occurs rapidly (days) while rooting percentages decline very slowly (months) upon exposure to light (Harrison-Murray and Howard 1982). It has been shown that in some cases 3–6 weeks of greening after etiolation caused no decrease in rooting (Howard 1980, Schmidt 1982). Moreover, up to nine months of light exposure following etiolation only decreased rooting percentage to a level which was still 40% above that of non-etiolated shoots (Howard 1982, 1983).

Our recent work with *Carpinus betulus* L. has shown that after eight weeks of greening, etiolated shoots rooted at 83% as compared to 37% rooting of light grown shoots. After 16 weeks of greening, the percentage rooting of etiolated shoots fell to 18% vs. 8% rooting of light grown shoots. By comparison shoots which were banded after etiolation rooted at 69% after 16 weeks greening of the shoot distal to the band.

Environment

Questions about the etiolation effect on rooting often are directed to associated environmental changes which occur during the etiolation process. Howard (1983) demonstrated that ventilating or using reflective covers to reduce the temperature under an etiolation structure could partially reverse the enhancement of rooting by etiolation. Work completed the following year (Howard 1984), however, indicated that root number, not rooting percentage, was affected. Temperatures under black plastic averaged 5 °C above uncovered controls, whereas temperatures under clear plastic increased 13 °C without corresponding increases in rooting (Howard 1984). Humidity, with minimal ventilation, did not increase under plastic coverings. An interesting observation was that clear polyethylene reduced light transmission by 30% and increased the rooting percentage of M.9 apple shoots by 10% vs. uncovered control shoots (Howard 1984).

Levels of shade

A very important contribution to making etiolation a more practical technique grew out of the findings that complete exclusion of light during the growth of the shoot is not required to stimulate rooting. The use of 50% Saran shading to "etiolate" stock plants of three *Dahlia* cultivars improved rooting percentage from 7% to 75% (Biran and Halevy 1973), while 96% shade promoted the rooting of *Tilia tomentosa* Moench 'Szeleste' to the same extent as 100% shade (Schmidt 1982). Scientists at the East Malling Research Station, U.K., tested a range of shading and found no decrease in rooting percentage with up to a 20% transmission of light; even 70% shade promoted rooting to levels 25% greater than light grown controls (Howard 1982, 1983, 1984). Thirty-eight percent shading of sitka spruce [*Picea sitchensis* (Bong.) Carr.] increased rooting percentage between 10 and 25%, depending upon the clone tested (van den Driessche 1985). The benefit of shading in this instance was attributed only in part to a concomitant reduction of water stress (determined with a pressure bomb) in the shaded stock plants.

The primary benefit of using partial vs. complete shade is that shoots grown in a low light level adjust more easily to full irradiance, are more resistant to *Botrytis* infection (Howard 1983, 1984), and resist scorching, which a premature exposure to full sunlight sometimes produces in etiolated shoots lacking protective pigmentation (Gardner 1937, Howard 1984). In terms of production scheduling, cuttings may be taken up to two weeks earlier if recovery from etiolation is not needed (Howard 1983).

The alternative to shading has been to acclimate etiolated shoots to higher irradiance by uncovering them in stages over about a week before removing the shade enclosure entirely (Howard 1980).

The observation that any reduction of irradiance can increase rooting warrants rethinking regarding a number of reports dealing with the growth of stock plants under irradiation of varying intensity. In tests of stock plant irradiances in the range of $7-68 \ W \cdot m^{-2}$, growing stock plants under lower irradiance improved rooting in cuttings of *Pisum sativum* L. 'Alaska' (Hansen and Eriksen 1974, Andersen et al. 1975), *Hedera helix* L. (Poulsen and Andersen 1980), *Malus* (Christensen et al. 1980), *Forsythia* × *intermedia* Zabel 'Lynwood' (Loach and Gay 1979), and *Vaccinium corymbosum* L. (Waxman 1965). In only a few instances has higher irradiance resulted in increased rooting, and then only root number was increased in response to higher carbohydrate levels; rooting percentages were 100% in each case (Fisher and Hansen 1975, Eliasson 1978—*Pisum*, Borowski et al. 1980—*Chrysanthemum*).

Practical Aspects of Banding

The importance of using opaque banding material to maintain an etiolated section of stem, or in the case of blanching to exclude light from a section of stem grown in the light, has been emphasized in research done since Gardner's initial report in 1937 (Biran and Halevy 1973, Howard 1978–1984, Delargy and Wright 1978, 1979, Harrison-Murray 1982, Schmidt 1982). Consistent increases in rooting percentage have been noted, from 45% in cuttings of *Dahlia* (Biran and Halevy 1973) to greater than 70% in *Malus sylvestris* Mill. 'Bramley's Seedling' (Delargy and Wright 1979).

Location

Gardner (1937) experimented with banding of developing apple shoots close to the growing tip, or 2.5, 5.0, and 7.5 cm behind the tip. He left these bands on as the shoot matured, taking cuttings the next spring. Banding very close to the developing tip gave the greatest response, while rooting percentage declined on those stems banded further from the shoot tip (Gardner 1937). These results indicated that light must be excluded early in the ontogeny of the shoot, before the cells derived from the shoot meristem had differentiated substantially. Gardner envisioned etiolation as an effective alternative to banding closely to the easily damaged shoot tip, in which the exclusion of light could influence the growth and differentiation of the shoot from the time of bud break. Banding could then be applied further from the shoot tip, to maintain etiolation without endangering the growth of the apical meristem.

Howard (1982) observed that blanching apple stems immediately below the apex was nearly as effective as etiolation, in contrast to basal blanching that was relatively ineffective, which again suggested that undifferentiated tissue was the most responsive to the exclusion of light. Frolich's method of maintaining the etiolation effect was similar because the frequent addition of vermiculite to the tube surrounding the elongating shoots kept light from the stem immediately below the apex

(Frolich 1951, 1961). Herman found that blanching was less effective than etiolation in enhancing the rooting percentage of *Phaseolus vulgaris* L. and *Hibiscus rosa-sinensis* L. 'Brilliant'; however, these results might reflect the placement of the band on a physiologically older part of the stem (D. E. Herman. 1967, Ph.D. Thesis, Purdue Univ., Bloomington, Indiana, USA).

Width of band

Very few reports of work with banding have indicated the effect of band width on root production. Delargy and Wright (1979) working with *Malus sylvestris* Mill. 'Bramley's Seedling' compared 2.5, 5, and 7.5 cm bands, and showed that 5 and 7.5 cm bands performed equally well, producing a rooting percent response 15% better than the 2.5 cm band and 70% better than the unbanded control. Howard et al. (1985) studied the effect on rooting of applying up to three adhesive tapes to cause local blanching of M.9 and MM.106 apple stems. Slight increases in rooting percentages were noted when more bands were added, with similar increases in root number per rooted cutting. A recent experiment with *Carpinus betulus* L. 'Fastigiata', using multiple bands of Velcro to maintain etiolation, showed that rooting percentages of 53, 63, and 87% were obtained using 2.5, 5, and 7.5 cm band widths, respectively (Maynard and Bassuk unpublished).

Banding variations

The same questions relating to the precise nature of the etiolation effect on rooting can be asked about the banding effect. Although several studies have shown that clear plastic bands fail to promote rooting entirely, while opaque bands work quite well (Krul 1969, Delargy and Wright 1979, Kawase and Matsui 1980), few comparisons of clear banding vs. no banding, or measurements of the environmental conditions extant under black bands vs. clear bands have been reported. In a comparison of banding apple shoots with black or clear polyethylene vs. an unbanded control, the clear polyethylene consistently resulted in rooting percentages and root numbers closer to those of the control and significantly lower than the responses obtained using opaque bands (Howard et al. 1985). Davis and Potter (1982), using aluminum foil to blanch 11 rhododendron cultivars, found that only two of the 11 cultivars showed any increases in rooting percentages. These disappointing results might reflect a need for prior etiolation in the propagation of rhododendrons, or perhaps a decrease in effectiveness when a reflective banding material is used. However, Biran and Halevy (1973) shaded stock plants of three *Dahlia* cultivars before banding with aluminum foil and achieved marked improvements in rooting percentage.

Two methods exist at present for applying root promoting substances as a part of the banding process. Hare (1976, 1977, 1978) has experimented successfully with girdling stems of *Quercus nigra* L., *Pinus elliottii* Engelm., and *Platanus occidentalis* L. followed by covering the girdle with a slurry of auxin contained within a wrapping of saran film and aluminum foil. The exclusion of light by the aluminum foil purportedly contributed to the induction of roots in this method. Maynard and Bassuk (1986) have developed a method to apply root promoting substances using Velcro bands. In this method the bands are coated with a talc preparation of IBA before placement on the shoot to be propagated. Dramatic increases in rooting percentage were obtained in trials with more than 20 difficult-to-root woody ornamental species (Table 2).

Differential Responses to Etiolation and Banding

The majority of studies comparing the relative effectiveness of etiolation and banding have been made with respect to the cutting propagation of apple. Beginning with the research of Gardner (1937) through that of Delargy and Wright (1978, 1979) and Howard's group (1979–1982), it has repeatedly been shown that for this species etiolation is about twice as effective as banding in promoting rooting percentage. However, the combination of the two pretreatments always exceeded the response obtained with either treatment alone.

The range of variation in rooting percentage among six *Syringa* cultivars subjected to etiolation, banding, or a combination of the two treatments is displayed in Fig. 2. Some cultivars responded more to one treatment than the other, though the combination of etiolation and banding gave the highest rooting response in five of seven cultivar trials. Averaging the lilac cultivar responses within

Table 2. Effect of etiolation and banding (Velcro coated with 0.8% indole-3-butyric acid in talc) stock plant pretreatments on the percent rooting of 12 woody ornamental plant species.

| Plant Species | Stock Plant Pretreatment | | | | Rooting Time (weeks) |
| | Light Grown | | Etiolated | | |
	No Band	Banded	No Band	Banded	
Acer griseum Pax. 1-yr-old seedlings	7 a[1]	12 ab	14 ab	34 b	4
A. griseum Pax. 30-yr-old trees	0	0	0	5	4
A. saccharum Marsh. 1-yr-old seedlings	47	64	65	86[2]	2
Betula papyrifera Marsh. 1-yr-old seedlings	51 a	65 ab	71 b	100 c	2
Carpinus betulus L. 1-yr-old seedlings	0 a	63 b	5 a	94 c	2
C. betulus L. 10-yr-old hedge	19 a	65 b	96 c	92 c	2
C. betulus L. 30-yr-old hedge	14 a	52 bc	37 ab	72 c	2
Castanea mollissima Bl. 4-yr-old seedlings	0	0	44[2]	100[2]	2
Corylus americana Marsh. 'Rush' 20-yr-old hedge	4 a	83 b	0 a	87 b	4
Pinus mugo Turra. 3-yr-old seedlings	41	64[2]	—[3]	—[3]	12
Pinus strobus L. 3-yr-old seedlings	29 a	79 c	58 ab	83 c	12
Quercus coccinea Muenchh. 1-yr-old seedlings	0	0	0	46[2]	4
Q. palustris Muenchh. 1-yr-old seedlings	31 a	24 a	50 a	44 a	4
Q. robur L. 1-yr-old seedlings	36 a	70 b	53 b	58 b	4
Q. robur L. 30-yr-old hedge	0 a	9 a	27 b	36 b	4
Q. rubra L. 2-yr-old seedlings	37 a	50 a	29 a	35 a	4

[1] — letters indicate separation within rows by Waller-Duncan means separation (P=0.05), based on three replications of 10 cuttings.

[2] — significantly different from the no band, light grown control, probability of a greater Chi Square value < 0.05, based on 10–18 cuttings per treatment without replication.

[3] — the indicated treatment was not applied to this species.

treatment reveals that the rooting of banded shoots with or without etiolation (54% and 47%, resp.) was greater than that of etiolated, unbanded shoots (33%), which in turn was greater than untreated shoots (24%).

The interspecific differences in the rooting responsiveness between different woody species to etiolation, blanching, or etiolation followed by banding is strikingly apparent in the study of Maynard and Bassuk (1986) with a wide variety of ornamental species (Table 2). From these results it is evident that certain species respond to etiolation alone while others respond to blanching, however several

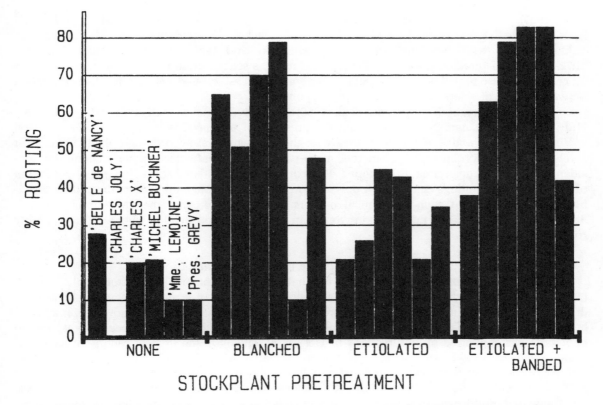

Figure 2. Effect of etiolation and banding pretreatments on the rooting response of six lilac cultivars. Four-year-old containerized stock plants were forced in greenhouse conditions under either 16 h photoperiod or 99% shade (ETIOLATED). Bands were applied after two weeks to either light grown (BANDED) or etiolated (ETIOLATED + BANDED) shoots, and left in place for four weeks prior to taking cuttings. Light grown, unbanded shoots (NONE) served as the control treatment. Rooting was assessed after five weeks in a misted propagation bed.

species would appear to require etiolation followed by banding as stock plant pretreatments for an optimum rooting response. If these differential responses are proved to be species related we would have an interesting handle for elucidating the factors through which etiolation and banding act, i.e. whether different anatomical or physiological changes are occurring between species.

FACTORS ASSOCIATED WITH ENHANCED ROOTING OF ETIOLATED TISSUE

Anatomical Factors

The most apparent changes associated with the etiolation of shoots are those related to the anatomy of the rooting zone in the resulting cuttings. The effects of etiolation include a lack of chlorophyll (as seen in the characteristic chlorotic appearance), increases in internodal length, increased succulence, and decreased mechanical strength of stem tissues. A number of researchers have suggested a role for these changes in the stimulation of rooting by etiolation (for reviews see Stoutemyer 1961 and Hess 1969).

Presence of sclerified tissues

As early as the 1920s, gross anatomical changes associated with etiolation were being correlated with improved rooting. Reid (1923) found that etiolated shoots of *Camphora officinarum* Nees. showed no traces of lignification [a possible mechanical barrier to adventitious rooting (Beakbane

1961, 1969)] before the sixth stem node, vs. light grown stems which were lignified up to the third node. Etiolated stems also had thinner cell walls, and lacked the continuous fiber sheath found in light grown tissues (also see Smith 1924—clematis, Blackie et al. 1926—camphor, Bid and Mukherjee 1972—mango). Doud and Carlson (1977) found much less sclerification in etiolated stems of Malus clones, which also rooted better than light grown stems, and established a strong negative correlation between degree of sclerification and rooting response.

On the contrary, both Christensen et al. (1980) and Raviv (M. Raviv. 1981, Ph.D. Thesis, Hebrew Univ. of Jerusalem, Israel) concluded that the presence of a fiber sheath had no direct effect on rooting. Hartmann and Kester (1983) agree with these researchers; even though plants with lower levels of lignification and fiber often root more easily, these anatomical changes cannot be considered the primary determinants in rooting. For example, roots initiated within the thick fiber sheath of carnation stem cuttings merely grow down before emerging through the base of the cutting. Beakbane (1961) noted that such mechanical barriers to rooting were not the only possibilities for decreased rooting, because initials often would not form even centripetal to fiber sheaths in poorly rooted cuttings.

Transmission of the etiolation stimulus

One of the most often cited arguments for anatomical involvement derives from Frolich's work (1961) which showed that in avocado there was no transmission of a rooting stimulus either up or down the stem from an etiolated section bordered by light grown tissue. This, however, may have been due to a localized chemical effect, not necessarily an anatomical one. The localized response to etiolation was supported by observations of Doss et al. (1980) on the in vitro rooting of shoots of Rubus idaeus L. 'Meeker' which had been etiolated or blanched. While the length of the etiolated section was varied, roots consistently initiated only in the etiolated sections. Delargy and Wright (1978) applied bands at positions on the stem directly adjacent to and distally removed from the cutting base and found that rooting occurred twice as well when the rooting zone, the cutting base in this instance, was the same section of stem as that from which light had been excluded.

Shapiro (1958) used Lombardy poplar as a model system to investigate the effect of irradiation on the growth of dormant root primordia which are found in the phloem of 2-yr-old stems as unorganized groups of meristematic cells. Shapiro noted that the banding of stems induced root primordia growth in the shaded areas, but did not promote root growth in adjacent irradiated areas. Thus, Shapiro hypothesized that there were immobile factors affecting root growth which were influenced by light.

Evidence for a transmissible etiolation effect was presented by Kawase and Matsui (1980) working with hypocotyls of Phaseolus vulgaris L. wrapped in black or clear plastic bands. In this rather elegant system, roots formed only under black plastic, not under clear plastic. In an experiment using multiple opaque bands placed along the stem, they found that the presence of a band stimulated the rooting in the band immediately below, though not under a band basally removed by > 4 cm. This suggested that there was a transmission of a root promoting stimulus. The short distance over which the influence acted may indicate that the movement of substances from the banded area was caused by diffusion instead of active transport. Howard (1983) reported that etiolation immediately distal to a basal opaque band increased root number under the band from 4.9 to 21.7 roots per cutting; this suggests that substances produced in etiolated tissues might move basipetally and influence rooting in other areas shaded from light. More research is needed to resolve this particular matter.

Other anatomical factors

Other anatomical factors which have been studied include the effects of etiolation on internode length. The longer nodes which are characteristic of etiolated stems have been correlated with higher root numbers in pea (Veierskov 1978). Two factors which deserve more consideration in the future include: 1) the effect of etiolation on the proportion of undifferentiated parenchyma in the pericycle and phloem ray tissues of the stem, both of which are sites of root initiation (Snyder 1962); and 2) irradiation effects on the suberization of stem tissues, which has recently been negatively correlated with the rooting success of 19 Australian woody plant species (Williams et al. 1984). Suberization may act as an anatomical barrier to root emergence.

Etiolation and banding increase the proportion of stem tissues occupied by undifferentiated parenchyma (Herman and Hess 1963, Schmidt 1982). Undifferentiated parenchyma is thought to be an intermediate in the initiation of adventitious roots (Hartmann and Kester 1983) and could be associated with the increase in herbaceous character caused by etiolation and which is correlated with improved rooting (Frolich 1961, Bid and Mukherjee 1972, Biran and Halevy 1973, Christensen et al. 1980). In a study of the anatomical changes affecting root initiation in cuttings of *Agathis australis* D. Don, White and Lovell (1984) concluded that the absence of initiation sites in stem tissues is more likely to play a decisive role in the rooting of cuttings than is the presence of sclerenchymatous tissue. However, they noted that there are many correlations of the presence of sclerenchymatous tissue with reduced rooting; consequently, they suggested that the parent plant be grown under conditions which are not conducive to the production of sclerenchymatous tissues, i.e. exclusion of light from the developing shoots.

Physiological Factors

The physiological changes associated with etiolation related to root initiation are discussed here in the broad categories of: 1) auxin metabolism and transport; 2) tissue sensitivity to auxin; and 3) the presence and activity of secondary metabolites.

Endogenous activity of indole-3-acetic acid

A number of researchers have examined the differences between endogenous auxin contents in etiolated and light grown tissues. Two groups in particular used the *Avena* mesocotyl straight growth bioassay to measure auxin contents in tissues of *Phaseolus* spp. and found that auxin activity was greater in etiolated stem tissues. Herman and Hess (1963) tentatively identified the putative auxin in *Phaseolus vulgaris* L. as IAA by its mobility in paper chromatography. Kawase (1965) described an auxin activity in etiolated mung bean (*Phaseolus aureus* Roxb.) cuttings which decreased by 80% after 24 h irradiation, but by only 56% after 48 h in the dark; thus providing partial evidence for the alteration of endogenous auxin content by light. However, in a 10 d rooting period at least 7 d of light exposure had to follow etiolation for a decrease in root number per rooted cutting. Other studies using similar assay techniques have found no difference in auxin activity between light- and dark-grown tissues of *Phaseolus, Hibiscus* (Herman and Hess 1963), or *Pisum sativum* L. 'Alaska' (Galston and Hand 1949, Scott and Briggs 1963).

More recently Tillberg (1974) showed that there was higher IAA content in shoot tissue of light grown than etiolated *Phaseolus vulgaris* L. using the *Avena* bioassay. However, etiolated shoots allowed to green in the light did not exhibit a concomitant increase in IAA content. In the most recent study employing bioassays to measure auxin levels, Kawase and Matsui (1980) concluded that etiolation did not affect the level of endogenous auxin in hypocotyls of *Phaseolus vulgaris* L.

These conflicting reports reflect, in part, the confusion which often surrounds studies employing bioassays to quantify endogenous contents of auxins, auxin cofactors, or other plant growth regulators.

Weigel et al. (1984) used an ELISA to examine the levels of both free and ester bound forms of IAA in stock plants of *Chrysanthemum morifolium* Ramat. grown at either low (4.5 $W \cdot m^{-2}$) or high (40 $W \cdot m^{-2}$) irradiance (see Chapter 19 by Heuser for more information on ELISA). Both forms of IAA increased sharply when cuttings were made; this response was interpreted as an increase in auxin synthesis induced by the excision of the shoot and the interruption of a putative basipetal flux of IAA. However, the increase in IAA was delayed in light grown tissues and final total levels of the hormone after 12 d of irradiance were greater in low irradiance than in high irradiance cuttings. These results support the studies of Herman and Hess (1963) and Kawase (1965). Root numbers per rooted cutting were also favored under the lower irradiance and rooting delayed under high irradiance, contrary to results obtained with *Chrysanthemum* by Fischer and Hansen (1974)—cv. Improved Mefo, and Borowski et al. (1980)—cv. Horim, who applied three and four weeks of varying irradiance pretreatments, respectively.

Transport of auxin

It is commonly assumed that IAA is transported basipetally in plant stems (Audus 1972). The uptake and transport of auxin differs markedly between light and dark grown tissues. Two studies done with *Pisum sativum* L. 'Alaska' epicotyls showed that irradiation increased the rate at which auxin was taken up from solution (Galston and Hand 1949, Scott and Briggs 1963). Auxin diffused more slowly from dark grown epicotyls (Scott and Briggs 1963) which suggested that auxin was transported more efficiently in light grown than in dark grown epicotyls. Thimann and Wardlaw (1963) confirmed that irradiation stimulated the uptake of ^{14}C-IAA from solution by pea stems, although they found no differences in the transport of labelled auxin. In a study of *Zea mays* L. coleoptiles, Naqvi and Gordon (1967) found that irradiation decreased the amount of diffusible auxin and reduced the basipetal transport of apically applied ^{14}C-IAA, although the velocity of transport was unaffected. It appears, therefore, that irradiation acts to increase the uptake and retention of auxin. Perhaps this phenomenon is related to a higher metabolic rate which develops in irradiated tissues. The possibility has also been raised that temperature differences associated with differing irradiances might affect the uptake of exogenously applied auxins (Baadsmand and Andersen 1984). Finally, increased transpiration under higher irradiance may also account for differences in the uptake of materials from solution. Shaded cuttings would lose less water from the leaves and hence not take up as much exogenous solution into the transpiration stream.

Baadsmand and Andersen (1984) examined the effect of high and low irradiance (38 and 16 $W \cdot m^{-2}$) on the transport velocity and accumulation of 2-[^{14}C]-IAA in cuttings of *Pisum sativum* L. They identified and quantified IAA and its conjugates by thin layer chromatography followed by liquid scintillation counting. In tissues grown under either high or low irradiance, the transport of IAA proceeded at ~ 1 $cm \cdot h^{-1}$. However, the accumulation of IAA at the cutting base was greater under higher irradiance. Under lower irradiance, IAA was more broadly distributed along the stem which correlated with the production of more scattered and numerous roots. Similar rooting patterns, perhaps related to differing intensities of auxin polarity, were observed by Knight and Witt (1937), Shapiro (1958), and Eliasson (1980). Baadsmand and Andersen (1984) proposed that these rooting patterns resulted from a form of apical dominance exerted by the early initiation of a few roots under the influence of high auxin levels which then inhibited subsequent root initiation. In etiolated stems, more evenly distributed levels of auxin (weak polarity) might result in the formation of more root initials before any one root primordium might exert dominance over the others.

Tissue sensitivity to auxin

The process of root initiation is auxin-sensitive; the process of root elongation which follows is either auxin-insensitive or inhibited by exogenous auxins (Mohammed and Eriksen 1974, Mitsuhashi-Kato et al. 1978). The auxin-sensitive phase in cuttings of *Malus pumila* Mill. lasts for 6 d. During this time root initiation is very sensitive to incident light (James 1983). In several studies, the rooting of woody plants in tissue culture required darkness during the auxin-sensitive phase for root initiation to occur (Pierik and Steegmans 1975—*Rhododendron*, Druart et al. 1982, Zimmerman 1984—*Malus*).

Irradiation has been observed to depress the growth response of *Avena sativa* L. internodes to applied auxin (Galston and Hand 1949, Naqvi and Gordon 1967). Furthermore, irradiation greatly decreased the efficiency of IAA as a root promotor without changing the endogenous auxin content, leading Galston and Hand (1949) to wonder if irradiation was decreasing the sensitivity of tissues to auxin.

Eliasson (1980) rooted pea stem cuttings in solution culture at either high or low irradiance (30 or 6 $W \cdot m^{-2}$). Applying high irradiance to the cutting base decreased root number; however, this reduced rooting was reversed by the addition of exogenous auxin. Furthermore, illuminated cuttings formed roots only at the base, in contrast to low-irradiance cuttings which rooted all along the stem. This result suggested to Eliasson that irradiation acted to decrease the sensitivity of stem tissues to auxin such that in well-illuminated cuttings a large basal buildup of auxin was needed to stimulate root initiation, while the more sensitive low-irradiance tissues responded to the lower auxin levels found distal to the base. This offers an intriguing alternative to the apical dominance hypothesis of Baadsmand and Andersen (1984) discussed earlier.

Kawase and Matsui (1980) developed an interesting system for etiolating hypocotyl sections of *Phaseolus vulgaris* L. with a wrap of black plastic. In this system, they found that the stimulation of root initiation by etiolation required the presence of the shoot apex; however, a lanolin paste containing auxin could be substituted for the shoot apex. Hypocotyls banded with clear plastic produced only a meager number of roots even when very high concentrations of auxin were applied to the decapitated shoot apex. The endogenous auxin activity of light and dark grown seedlings, as determined in the *Avena* coleoptile elongation bioassay, did not differ. This evidence suggests that irradiation strongly influences the response of the hypocotyl to auxin produced in the shoot apex.

Delargy and Wright (1979) utilized both clear and opaque plastic bands as a banding pretreatment on apple shoots and applied auxins at the time the cuttings were taken. Excluding light with an opaque band enhanced rooting percentage much more than adding auxin alone, although auxin plus light exclusion produced the best response. Again it appears that etiolation either increases auxin activity or the stem tissue responsiveness to exogenous auxin. Although the differential sensitivity of stem tissues to applied auxin is an attractive hypothesis, it would seem impossible to demonstrate this theory unequivocally. Ryan (1969) also made the observation that in some plants, such as *Philodendron* or juvenile forms of *Hedera helix* L., rooting shows no sensitivity to irradiance.

Meristematic activity

As early as 1918, a correlation had been established between the C/N ratio of a tissue and the ability of that tissue to form roots (Kraus and Kraybill 1918). Smith (1924) suggested that etiolation, by decreasing endogenous starch levels, established a C/N ratio favoring root initiation through the maintenance of the vitality of those meristematic cells involved in root initiation.

Root initiation is thought to involve cellular dedifferentiation followed by a redifferentiation to form root initials (Hartmann and Kester 1983). By their nature, meristematic cells are undifferentiated and therefore might be more sensitive to the rooting stimulus, i.e. one step closer to becoming root initials than differentiated cells. Reid (1923) attributed this to an anatomical difference between etiolated and non-etiolated tissues; etiolated tissue remained more undifferentiated and was more readily available for meristematic activation (also Herman and Hess 1963).

Irradiation induced a boron requirement for rooting in *Phaseolus aureus* Roxb. (Jarvis and Ali 1984). In tissues already limited in boron, irradiation might create a relative deficiency, thus reducing the tissue's ability to respond to rooting stimuli. Darkness during the rooting period partially overcame the requirement for an exogenous source of boron. Jarvis and Ali (1985) hypothesized that irradiation might act to make boron unavailable by modifications in boron metabolism, i.e. through the synthesis of *ortho*-dihydroxy phenols with *cis*-hydroxy groups capable of complexing with boron.

Ironically, despite all we know about auxin metabolism and tissue sensitivity, simply adding more exogenous auxin fails to induce rooting in many instances. Given the evidence that etiolation alters the sensitivity of stem tissues to both endogenous and exogenous auxins, and with a minimal knowledge of the optimal auxin concentration required for rooting, we should be able to combine etiolation and auxin treatments to increase the propagation success of most difficult-to-root species.

Auxin cofactors

Hess (1961) proposed that cofactors are present in tissues and interact with auxin to promote rooting. Although one study indicated that etiolation acts to increase the titer of endogenous rooting cofactors (Punjabi and Basu 1976), other studies have found no significant correlations between etiolation and rooting cofactor content (Herman and Hess 1963). The activity of rooting cofactors has been attributed to the formation, via reactions catalyzed by PPO, of phenol-auxin conjugates which are more active in stimulating rooting than auxin alone (Bassuk et al. 1981).

It is of great interest that there may be interactions occurring between irradiation, phenolic acids, and enzymes involved in auxin and phenolic acid metabolism which could act to establish levels of auxins or rooting cofactors which are capable of inducing root initiation. The relationship of auxin responsiveness to the presence of certain phenolic acids has attracted some attention. Phenolics have been implicated as auxin cofactors or "protectors," as well as inhibitors and stimulators of the

IAA-O. Phenolic substances are also precursors in the synthesis of lignin and suberin which have been implicated as possible anatomical constraints to adventitious rooting (Shapiro 1958, Herman and Hess 1963, Beakbane 1969, Williams et al. 1984).

Working with cotyledons of *Pharbitis nil* Choisy, Konishi and Galston (1964) determined that etiolation favored the endogenous titer of IAA-O inhibitors, while irradiation decreased the level of these substances. Zenk and Muller (1963) showed that certain classes of phenolic acids had differential effects on the activity of IAA-O *in vitro*. *Para*-coumaric acid acted as an IAA-O cofactor, stimulating the destruction of IAA, while chlorogenic acid inhibited IAA-O action. It is now generally assumed that monophenols (e.g. *p*-coumaric acid) are IAA-O cofactors, while polyphenols (e.g. chlorogenic acid, phloroglucinol, caffeic acid, catechol) inhibit IAA destruction and stimulate IAA associated growth phenomena. Zimmerman (1984) has shown that phloroglucinol was effective in promoting rooting in light grown apple cultures, but it was less effective in dark grown cultures; this suggests that phloroglucinol might be replacing lost IAA-O inhibitors in light grown tissues. Leopold and Kreideman (1975) offered an explanation of this behavior in terms of irradiance effects on phenolic acid synthesis. Irradiation promotes the synthesis of *o*-coumaric acid, a monophenolic cofactor of IAA-O, by stimulating the activity of PAL. This, in turn, increases the activity of IAA-O and leads to the oxidation of endogenous IAA. However, other research has shown that *o*-coumaric acid and other IAA-O cofactors are also active as root promotors (Gorter 1969). These findings conceivably relate to evidence that *p*- and *o*-coumaric acids inhibit PAL (Hodgins 1971), thus altering the synthesis of other endogenous phenolics required as IAA-O cofactors. For further information regarding IAA metabolism see Chapter 9 by Gaspar and Hofinger.

Herman found that phenolic content was generally lower in etiolated tissues of mung bean and *Hibiscus rosa-sinensis* L. (D. E. Herman. 1967, Ph.D. Thesis, Purdue Univ., Bloomington, Indiana, USA). Herman proposed that phenolics play a role in the lignification of stem tissues and suggested that phenolic acid biosynthesis was blocked by etiolation, thus creating anatomical changes which affect rooting. In the context of more recent information, these results could indicate that phenolic levels were lowered in favor of the dihydroxy phenols, causing an inhibition of IAA-O and concomitant increases in auxin activity. Druart and coworkers (1982) examined the effect of darkness on the total complement of phenolic substances and peroxidase activity in cultured apple shoots. The two were inversely correlated; dark treatments resulted in increased phenolic acid content and decreased peroxidase activity.

Using 2–3 mm shoot apices of the adult and juvenile forms of *Hedera helix,* Hackett (1970) developed a bioassay in which root number per rooted cutting was observed over the period of one month. He found that an irradiation-induced decline in the responsiveness of juvenile apices to applied IAA was prevented by the addition of catechol, an *o*-diphenol which is known to inhibit the activity of IAA-O (Leopold and Kreideman 1975). Catechol did not stimulate rooting as much in the darkness; this suggests that the activity of catechol is specifically related to the decreased ability of roots to initiate in the light. The rooting response of adult *Hedera helix* presents an interesting system. Adult shoot apices root only at low irradiance or in the dark, and fail to respond in the light to the addition of auxin, catechol, or juvenile tissue extracts (Hackett 1970). If light acts by destroying auxin via IAA-O then the addition of enough auxin should yield rooting comparable to that obtained by etiolation. Since this expected result fails to occur in the *Hedera helix* system, some other factors must be involved in addition to those governing auxin levels.

Al Barazi and Schwabe (1984) examined PPO and IAA-O activities in cuttings of pistachio (*Pistacia vera* L.) in relation to etiolation and banding. Etiolation increased PPO activity by 100%, while banding alone increased activity 25% over that of the light controls. IAA-O activity was also slightly lower in both etiolated and banded shoots, as compared to the control.

CONCLUSION

The physiological and anatomical consequences of light and darkness on root initiation in cuttings are quite complex. They range from changes in stem anatomy which are correlated with ease of rooting to changes in the light control of auxin metabolism, factors affecting the activity of IAA-O, and in the activity of rooting cofactor substances. The changes induced by etiolation may be viewed as alterations in meristematic activity, hormone action, and the ability of tissues to support the growth and development of root primordia.

To gain a meaningful comprehension of the etiolation effect more comprehensive studies on the anatomical and physiological changes occurring in response to the exclusion of light are needed. A major work of this type was that by Herman and Hess (1963) with *Phaseolus vulgaris* L. and three *Hibiscus rosa-sinensis* L. cultivars. Future examinations of light effects on auxin metabolism must make use of the most up-to-date methods of auxin detection, e.g. gas chromatography combined with mass spectrometry. The issue of tissue sensitivity to auxin must be addressed in terms of our present knowledge of auxin transport and action at the cellular and molecular levels. For example, does etiolation affect the number or activity of membrane-bound auxin receptors?

Ryan (1969) stated that "with our present knowledge of the rooting response to exclusion of light, the propagator can continue to make use of the etiolation effect on otherwise difficult-to-propagate plants without knowing why it is so effective." And yet, for etiolation and banding pretreatments to gain wider usage refinements in the present state of the art must be discovered. Etiolation of shoots must be adapted to the scale of the stock plant and the skill of the propagator. Banding materials should be easy to apply, inexpensive, and effective in promoting rooting. The ability to apply growth promoting substances via the band might result in banding someday becoming even easier, more effective, and therefore more feasible than etiolation. Studies should also be conducted on the growth and performance of cuttings rooted from etiolated and/or banded shoots, permitting an assessment of the effect of stock plant pretreatments on the post propagation vigor and survival of rooted cuttings.

ACKNOWLEDGMENTS

The authors wish to thank Kenneth W. Mudge and Henry E. deVries, II for their assistance in reviewing this manuscript.

REFERENCES

Al Barazi, Z. and W. W. Schwabe. 1984. The possible involvement of polyphenol-oxidase and the auxin-oxidase system in root formation and development in cuttings of *Pistacia vera*. *J. Hortic. Sci.* 59:453–461.

Andersen, A. S., E. N. Eriksen, J. Hansen and B. Veierskov. 1975. Stock plant conditions and root initiation on cuttings. *Acta Hortic.* 54:33–37.

Anderson, W. C. 1981. Etiolation as an aid to rooting. *Proc. Int. Plant Prop. Soc.* 31:138–141.

Audus, L. J. 1972. In *Plant Growth Substances, Vol. 1: Chemistry and Physiology*. Barnes & Noble Books, NY, USA. ISBN 0-249-44085-7.

Baadsmand, S. and A. S. Andersen. 1984. Transport and accumulation of indole-3-acetic acid in pea cuttings under two levels of irradiance. *Physiol. Plant.* 61:107–113.

Bassuk, N. L., L. D. Hunter and B. H. Howard. 1981. The apparent involvement of polyphenol oxidase and phloridzin in the production of apple rooting cofactors. *J. Hortic. Sci.* 56:313–322.

_____ D. M. Miske and B. K. Maynard. 1985. Stock plant etiolation for improved rooting of cuttings. *Proc. Int. Plant Prop. Soc.* 35:543–550.

Beakbane, A. B. 1961. Structure of the plant stem in relation to adventitious rooting. *Nature* 192:954–955.

_____ 1969. Relationships between structure and adventitious rooting. *Proc. Int. Plant Prop. Soc.* 19:192–202.

Bid, N. N. and S. K. Mukherjee. 1972. Studies into the effects of forced shoot etiolation and different media on the rootage of *Mangifera indica* L. cuttings. *Acta Hortic.* 24:77–81.

Biran, I. and A. H. Halevy. 1973. Stock plant shading and rooting of *Dahlia* cuttings. *Scientia Hortic.* 1:125–131.

Blackie, J. J., R. T. D. Graham and L. B. Stewart. 1926. Propagation of camphor. *Kew Bull.* pp. 380–381.

Borowski, E., P. Hagen and R. Moe. 1981. Stock plant irradiation and rooting of *Chrysanthemum* cuttings in light or dark. *Scientia Hortic.* 15:245–253.

Christensen, M. V., E. N. Eriksen and A. S. Andersen. 1980. Interaction of stock plant irradiance and auxin in the propagation of apple rootstocks by cuttings. *Scientia Hortic.* 12:11–17.

Davis, T. D. and J. R. Potter. 1983. Effect of localized etiolation of stock plants on the rooting of *Rhododendron* cuttings. *J. Environ. Hortic.* 1:96–98.

Delargy, J. A. and C. E. Wright. 1978. Root formation in cuttings of apple (cv. Bramley's seedlings) in relation to ringbarking and to etiolation. *New Phytol.* 81:117–127.

———— ———— 1979. Root formation in cuttings of apple in relation to auxin application and to etiolation. *New Phytol.* 82:341–347.

Doss, R. P., L. C. Torre and B. H. Barritt. 1980. An investigation of the influence of etiolation on the rooting of red raspberry (*Rubus idaeus* L., cv. Meeker) root-shoots. *Acta Hortic.* 112:77–80.

Doud, S. L. and R. F. Carlson. 1977. Effects of etiolation, stem anatomy, and starch reserves on root initiation of layered *Malus* clones. *J. Amer. Soc. Hortic. Sci.* 102:487–491.

Druart, P., C. Keevers, P. Boxus and T. Gaspar. 1982. *In vitro* promotion of root formation by apple shoots through darkness effect on endogenous phenols and peroxidases. *Z. Pflanzenphysiol.* 108:429–436.

Eliasson, L. 1978. Effects of nutrients and light on growth and root formation in *Pisum sativum* cuttings. *Physiol. Plant.* 43:13–18.

———— 1980. Interaction of light and auxin in regulation of rooting in pea stem cuttings. *Physiol. Plant.* 48:78–82.

Ernst, A. A. and L. C. Holtzhausen. 1978. New promising technique for rooting difficult-to-root avocado (*Persea americana* Mill.) cuttings. *Citrus and Subtropical Fruit J.* pp. 6–10.

Fisher, P. and J. Hansen. 1977. Rooting of *Chrysanthemum* cuttings. Influence of irradiance during stock plant growth and of decapitation and disbudding of cuttings. *Scientia Hortic.* 7:171–178.

Frolich, E. F. 1951. Rooting Guatemalan avocado cuttings. *Calif. Avocado Soc. Yearbook* 34:136–138.

———— 1961. Etiolation and the rooting of cuttings. *Proc. Int. Plant Prop. Soc.* 11:277–283.

———— R. G. Platt. 1972. Use of the etiolation technique in rooting avocado cuttings. *Calif. Avocado Soc. Yearbook* 55:97–109.

Galston, A. W. and M. E. Hand. 1949. Studies on the physiology of light action. I. Auxin and the light inhibition of growth. *Amer. J. Bot.* 36:85–94.

Gardner, F. E. 1937. Etiolation as a method of rooting apple variety stem cuttings. *Proc. Amer. Soc. Hortic. Sci.* 34:323–329.

Gorter, C. J. 1969. Auxin synergists in the rooting of cuttings. *Physiol. Plant.* 15:88–95.

Hackett, W. P. 1970. The influence of auxin, catechol, and methanolic tissue extracts on root initiation in aseptically cultured shoot apices of the juvenile and adult forms of *Hedera helix. J. Amer. Soc. Hortic. Sci.* 95:398–402.

Hansen, J. and E. N. Eriksen. 1974. Root formation of pea cuttings in relation to the irradiance of the stock plants. *Physiol. Plant.* 32:170–173.

Hare, R. C. 1976. *Girdling and Applying Chemicals Promote Rapid Rooting of Sycamore Cuttings.* USDA-Forest Service, Res. Note No. SO-202. South. For. Exp. Sta., New Orleans, LA, USA.

———— 1977. Rooting of cuttings from mature water oak. *Southern J. Appl. For.* 1:24–25.

———— 1978. Effect of shoot girdling and season on rooting of slash pine cuttings. *Can. J. For. Res.* 8:14–16.

Harrison-Murray, R. S. 1982. Etiolation of stock plants for improved rooting of cuttings. I. Opportunities suggested by work with apple. *Proc. Int. Plant Prop. Soc.* 31:386–392.

———— B. H. Howard. 1982. Effects of prior etiolation on adventitious rooting of apple cuttings. *Int. Soc. Hortic. Sci. 21st Int. Hortic. Cong.,* Hamburg, W. Germany. Vol. 2 Abst. No. 1281.

Hartmann, H. T. and D. E. Kester. 1983. *Plant Propagation: Principles and Practices.* Prentice-Hall, Englewood Cliffs, NJ, USA. 4th ed. ISBN 0-13-681007-1.

Herman, D. E. and C. E. Hess. 1963. The effect of etiolation upon the rooting of cuttings. *Proc. Int. Plant Prop. Soc.* 13:42–62.

Hess, C. E. 1961. The physiology of root initiation in easy- and difficult-to-root cuttings. *Hormolog*

3:3–6.

_____ 1969. Internal and external factors regulating root initiation. In *Root Growth* (W. J. Whittington, ed). Butterworths, London, England. pp. 42–53.

Hodgins, D. S. 1971. Yeast phenylalanine ammonia-lyase. *J. Biol. Chem.* 246:2977–2985.

Howard, B. H. 1978. *Plant Propagation*. Rep. E. Malling Res. Sta. for 1977. pp. 67–73.

_____ 1979. *Plant Propagation*. Rep. E. Malling Res. Sta. for 1978. pp 71–82.

_____ 1980. *Plant Propagation*. Rep. E. Malling Res. Sta. for 1979. pp. 67–84.

_____ 1981. *Plant Propagation*. Rep. E. Malling Res. Sta. for 1980. pp. 59–72.

_____ 1982. *Plant Propagation*. Rep. E. Malling Res. Sta. for 1981. pp. 57–72.

_____ 1983. *Plant Propagation*. Rep. E. Malling Res. Sta. for 1982. pp. 59–75.

_____ 1984. *Plant Propagation*. Rep. E. Malling Res. Sta. for 1983. pp. 131–132.

_____ R. S. Harrison-Murray and S. B. Arjyal. 1985. Responses of apple summer cuttings to severity of stockplant pruning and to stem blanching. *J. Hortic. Sci.* 60:145–152.

James, D. J. 1983. Adventitious root formation 'in vitro' in apple rootstocks (*Malus pumila*) I. Factors affecting the length of the auxin-sensitive phase in M.9. *Physiol. Plant.* 57:149–153.

Jarvis, B. C. and A. H. N. Ali. 1984. Irradiance and adventitious root formation in stem cuttings of *Phaseolus aureus* Roxb. *New Phytol.* 97:31–36.

Johnson, C. R. and D. F. Hamilton. 1977. Rooting of *Hibiscus rosa-sinensis* L. cuttings as influenced by light intensity and ethephon. *HortScience.* 12:39–40.

_____ A. N. Roberts. 1971. The effect of shading *Rhododendron* stock plants on flowering and rooting. *J. Amer. Soc. Hortic. Sci.* 96:166–168.

Kawase, M. 1965. Etiolation and rooting in cuttings. *Physiol. Plant.* 18:1066–1076.

_____ H. Matsui. 1980. Role of auxin in root primordium formation in etiolated 'Red Kidney' bean stems. *J. Amer. Soc. Hortic. Sci.* 105:898–902.

Khosh-Khui, M. and K. C. Sink. 1982. Rooting enhancement of *Rosa hybrida* for tissue culture propagation. *Scientia Hortic.* 17:371–376.

Knight, R. C. and A. W. Witt. 1937. The propagation of fruit tree stocks by stem cuttings. *J. Pomol.* 6:47–60.

Konishi, M. and A. W. Galston. 1964. Light-induced changes in phenolic inhibitors of indoleacetic acid oxidase in cotyledons of *Pharbitis nil. Phytochemistry* 3:559–568.

Kraus, E. J. and H. R. Kraybill. 1918. *Vegetation and Reproduction with Special Reference to the Tomato.* Oreg. Agric. Exp. Sta. Bull. 149.

Krul, W. R. 1968. Increased root initiation in pinto bean hypocotyls with 2,4-dinitrophenol. *Plant Physiol.* 43:439–441.

Leopold, A. C. and P. E. Kriedemann. 1975. *Plant Growth and Development.* McGraw-Hill, NY, USA. 2nd ed. ISBN 0-07-037200-4.

Loach, K. and A. P. Gay. 1979. The light requirement for propagating hardy ornamental species from leafy cuttings. *Scientia Hortic.* 10:217–230.

Maynard, B. and N. Bassuk. 1986. Etiolation as a tool for rooting cuttings of difficult-to-root woody plants. *Proc. Int. Plant Prop. Soc.* 36:488–495.

Miske, D. M. and N. L. Bassuk. 1985. Propagation of hybrid lilacs using stock plant etiolation. *J. Environ. Hortic.* 3:111–114.

Mitsuhashi-Kato, M., H. Shibaoka and M. Shimokoriyama. 1978. The nature of the dual effect of auxin on root formation in Azukia cuttings. *Plant Cell Physiol.* 19:1535–1542.

Mohammed, S. and E. N. Eriksen. 1974. Root formation in pea cuttings. IV. Further studies on the influence of indole-3-acetic acid at different developmental stages. *Physiol. Plant.* 32:94–96.

_____ C. A. Sorhaindo. 1984. Production and rooting of etiolated cuttings of West Indian and hybrid avocado. *Tropical Agric.* 61:200–204.

Mukherjee, S. K. and B. K. Chatterjee. 1979. Effects of forcing, etiolation and indole butyric acid on rooting of cuttings of *Artocarpus heterophyllus* Lam. *Scientia Hortic.* 10:295–300.

_____ P. K. Majumder, N. N. Bid and A. M. Goswami. 1967. Standardization of rootstocks of mango (*Mangifera indica* L.) II. Studies on the effects of source, invigoration and etiolation on the rooting of mango cuttings. *J. Hortic. Sci.* 42:83–87.

Nanda, K. K., A. N. Purohit and A. Bala. 1967. Effect of photoperiod, auxins and gibberellic acid on rooting of stem cuttings of Bryophyllum tubiflorum. Physiol. Plant. 20:1096–1102.

Naqvi, S. M. and S. A. Gordon. 1967. Auxin transport in Zea mays coleoptiles. II. Influence of light on the transport of indoleacetic acid-2-^{14}C. Plant Physiol. 42:138–143.

Ooishi, A., Y. Shiobara, H. Machida and T. Hosoi. 1982. Role of light in rooting of softwood cuttings of Euonymus japonicus Thunb. J. Jap. Soc. Hortic. Sci. 50:511–515.

Paterson, K. E. and T. L. Rost. 1979. Effects of light and hormones on regeneration of Crassula argentea from leaves. Amer. J. Bot. 66:463–470.

Pierik, R. L. M. and H. H. M. Steegmans. 1975. Analysis of adventitious root formation in isolated stem explants of Rhododendron. Scientia Hortic. 3:1–20.

Poulsen, A. and A. S. Andersen. 1980. Propagation of Hedera helix: Influence of irradiance to stock plants, length of internode and topophysis of cutting. Physiol. Plant. 49:359–365.

Punjabi, B. and R. N. Basu. 1976. The effect of etiolation of shoot on the levels of endogenous rooting factors. Indian Biol. 8:41–44.

Rauch, F. D. 1981. The influence of shading and mist on rooting of selected foliage plants. Plant Propagator 27:8–9.

Reid, O. 1923. The propagation of camphor by stem cuttings. Trans. Bot. Soc. Edin. 28:184–188.

Reuveni, O. and M. Goren. 1982. Production and rooting of base-etiolated Avocado cuttings. In Int. Soc. for Hortic. Sci. 21st Int. Hortic. Cong. Vol. I. Abst. No. 1348.

Rowell, D. J. 1982. Etiolation of stock plants for the improved rooting of cuttings. II. Initial experiences with hardy ornamental nursery stock. Proc. Int. Plant Prop. Soc. 31:392–397.

Ryan, G. F. 1969. Etiolation as an aid in propagation. Proc. Int. Plant Prop. Soc. 19:69–74.

Schmidt, G. 1982. Different methods of etiolation for increasing the rooting of the Silver Lime, Tilia tomentosa, softwood cuttings. In Proc. Int. Soc. for Hortic. Sci. 21st Int. Hortic. Cong. Vol. 2. Abst. No. 1785.

Scott, T. K. and W. R. Briggs. 1963. Recovery of native and applied auxin from the dark grown 'Alaska' pea seedling. Amer. J. Bot. 50:652–657.

Shapiro, S. 1958. The role of light in the growth of root primordia in the stem of the lombardy poplar. In The Physiology of Forest Trees (K. V. Thimann, ed). Ronald Press, NY, USA. pp. 445–465.

Smith, E. P. 1924. The anatomy and propagation of clematis. Trans. Bot. Soc. Edin. 29:17–26.

Snyder, W. E. 1962. Plant anatomy as related to the rooting of cuttings. Proc. Int. Plant Prop. Soc. 12:43–47.

Stoutemyer, V. T. 1961. Light and Propagation. Proc. Int. Plant Prop. Soc. 11:252–260.

Thimann, K. V. and I. F. Wardlaw. 1963. The effect of light on the uptake and transport of indoleacetic acid in the green stem of pea. Physiol. Plant. 16:368–377.

Tillberg, E. 1974. Levels of indole-3yl-acetic acid and acid inhibitors in green and etiolated bean seedlings. (Phaseolus vulgaris). Physiol. Plant. 31:106–111.

van den Driessche, R. 1985. The influence of cutting treatments with indole-3-butyric acid and boron, stock plant moisture stress, and shading on rooting in sitka spruce. Can. J. For. Res. 15:740–742.

Veierskov, B. 1978. A relationship between length of basis and adventitious root formation in pea cuttings. Physiol. Plant. 42:146–150.

Waxman, S. 1965. Propagation of blueberries under fluorescent light at various intensities. Proc. Int. Plant Prop. Soc. 15:154–158.

Weigel, U., W. Horn and B. Hock. 1984. Endogenous auxin levels in terminal stem cuttings of Chrysanthemum morifolium during adventitious rooting. Physiol. Plant. 61:422–428.

White, J. and P. H. Lovell. 1984. Anatomical changes which occur in cuttings of Agathis australis (D. Don) Lindl. 2. The initiation of root primordia and early root development. Ann. Bot. 54:633–645.

Williams, R. R., A. M. Taji and J. A. Bolton. 1984. Suberization and adventitious rooting in Australian plants. Aust. J. Bot. 32:363–366.

Zenk, M. H. and G. Muller. 1963. In vivo destruction of exogenously applied indolyl-3-acetic acid as influenced by naturally occurring phenolic acids. Nature 200:761–763.

Zimmerman, R. H. 1984. Rooting apple cultivars in vitro: Interactions among light, temperature, phloroglucinol and auxin. Plant Cell Tissue Organ Cult. 3:301–311.

CHAPTER 3

Genetic Effects on Adventitious Rooting

Bruce E. Haissig and Don E. Riemenschneider

USDA-Forest Service
North Central Forest Experiment Station
Forestry Sciences Laboratory
P.O. Box 898, Rhinelander, WI 54501

INTRODUCTION. 47
TERMINOLOGY 48
GENETIC EFFECTS ON VEGETATIVE ORGAN
   REGENERATION. 49
     Organogenesis and Somatic Embryogenesis *In Vitro* 49
     Rooting of Cuttings 51
EXPERIMENTAL GUIDELINES. 52
     Environmental Effects 53
     Genetic Models. 54
       Half-sib families. 54
       Full-sib families 54
       Factorial mating. 56
     Biochemical and Physiological Measurements. 56
CONCLUSION 57
ACKNOWLEDGMENTS 57
REFERENCES 57

Additional key words: tissue culture, microculture, vegetative propagation, cutting, plant breeding, C-effects.

INTRODUCTION

Adventitious rooting of cuttings may be directly or indirectly controlled by genetics. However, genetic aspects of rooting by cuttings have not been discussed much in the literature or ever reviewed. Genetic effects on rooting of cuttings have been little studied and considered unimportant. However, we have recently noted more interest in genetic effects on adventitious rooting. This interest has been catalyzed partially by the growing number of genetically based studies of vegetative regeneration *in vitro* and practical propagation successes that have arisen from them. Little literature is available about genetic effects on only rooting of cuttings. Yet, much literature exists about genetic effects on vegetative regeneration by various tissue and organ explants propagated *in vitro*, which may be useful in designing similar experiments concerning rooting of cuttings.

Therefore, the present review provides an introduction to genetically based vegetative regeneration research in general and specifically to studies concerning rooting of cuttings. We have written this review to encourage further genetically based experimentation in the hope that such experiments will lead to an improved fundamental understanding of rooting by cuttings.

TERMINOLOGY

In the following presentation, *Microculture* is *in vitro* culture of plant protoplasts, cells, tissues, organs, or parts of organs (i.e. *tissue culture*). *Rooting* means adventitious rooting, without reference to state of anatomical or biochemical differentiation, when it refers to cuttings (Haissig 1974, 1986). *Rooting* means any type of rooting when it refers to microcultures because often the type of rooting [adventitious, lateral, basal (seminal), and radicle initiation] is uncertain.

Clone refers to a genetically unique individual. *Clonal differences* are differences between individuals of differing genotype and are synonymous with genetic or genotypic differences in the absence of *C-effects*. *C-effects* (Lerner 1958) are nonrandom parental or environmental effects (e.g. age and physiological condition of the ortet) that cause ramets to resemble each other more strongly than would be expected according to only their genetic relation. A *ramet* is a member of a group of genetically identical individuals that are vegetatively descended from a clone (termed the *ortet*). *Family* refers to a group of clones having at least one sexual parent in common.

The nucleus and the cytoplasm are the two general origins of heritable traits (characters). *Nuclear inheritance* involves transmission of hereditary traits via chromosomal genes; *Cytoplasmic inheritance* involves nonnuclear genes. There are two main types of nuclear and, probably, cytoplasmic inheritance (Fig. 1). The first type, *qualitative inheritance,* includes traits that are controlled by only a few genes and discontinuously distributed, in clearly defined classes. The second type, *quantitative inheritance,* includes traits that are controlled by many genes and continuously, or metrically, distributed. A calculated value, termed *heritability,* estimates the degree to which a quantitative trait is genetically controlled. Heritabilities can include all (*broad sense*) or only additive genetic traits (*narrow sense*). Genes are said to act additively in the absence of intralocus (*dominance*) and interlocus (*epistasis*) interactions. Traits that are mostly influenced by additive gene action may be

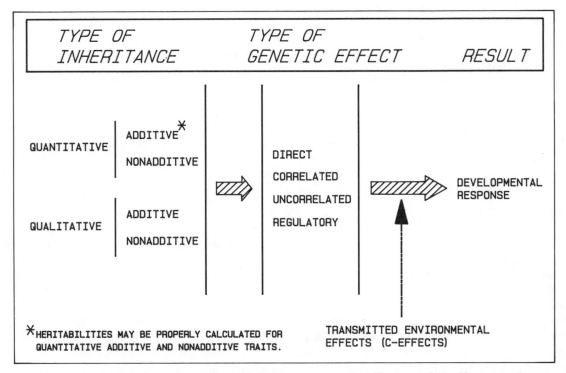

Figure 1. Relations between types of inheritance, genetic effects, and C-effects, as they influence plant developmental responses. Genes may originate in the nucleus or cytoplasm, or both, which complicates genetic relations in plant developmental responses. Plant developmental responses are also conditioned by environmental influences other than C-effects.

improved by using simple breeding schemes such as mass selection. Improvement of traits that are influenced by non-additive gene action (dominance and epistasis) requires complex breeding procedures. Detailed discussions of gene action and selective breeding methods can be found in standard texts (Falconer 1960, Allard 1960, Fehr 1987ab).

The terminology of inheritance *per se* does not always make it easier to understand genetically based experiments. First, it may be unclear whether inheritance is cytoplasmic or nuclear, or both. Cytoplasmic inheritance can only be distinguished if the hereditary cytoplasmic traits are unequally contributed by the parents and, therefore, distinguishable by reciprocal cross. Second, cytoplasmic inheritance is more poorly understood than nuclear inheritance, which means, for example, that its descriptive terminology is unclear. Third, quantitative and qualitative inheritance may be difficult to distinguish experimentally. This distinction may be difficult because a trait that is controlled by only a few genes can have a continuous distribution if genetic influences are much smaller than environmental influences. Fourth, heritabilities may have been calculated without adequate supporting evidence that only quantitative traits were involved. Finally, there is no standard terminology for describing gene actions that influence plant developmental and regenerative processes (what we term *genetic effects*).

For the purposes of this chapter, we have defined four genetic effects: *direct, correlated, uncorrelated,* and *regulatory*. Although our categories are somewhat arbitrary, they are useful because they bridge the area of discussion between all types of inheritance and all developmental processes (Fig. 1). *Direct effects* relate to genes whose expression is specifically required for rooting. *Correlated effects* relate to genes whose expression controls some developmental process other than rooting but where the other process positively or negatively influences rooting. *Uncorrelated effects* relate to genes that control developmental processes but whose expression never influences rooting. *Regulatory effects* determine expression of genes of any category. They influence all plant developmental processes, including (sometimes) rooting.

Apparent examples of direct, correlated, and uncorrelated effects can be found in the microculture or rooting literature. Vivipary mutation and ortet vigor (related to potassium uptake), which are genetically based, have appeared to be uncorrelated genetic effects with respect to somatic embryogenesis in maize (*Zea mays* L.) microcultures. Possession of these traits by ortets did not influence regeneration ability of their tissues (Duncan et al. 1985). Maturation of the ortet appears to be a correlated effect with respect to rooting of tree cuttings. For example, rooting potential of cuttings has often declined with increasing age of ortet (Haissig 1983) but other developmental characteristics such as propensity for flowering and wood formation also change dramatically during maturation. Direct effects have been demonstrated for bud initiation by tobacco (*Nicotiana* spp.) microcultures, in which budding has been related to specific ortet genotype (Ogura and Tsuji 1977). Direct effects were also described in a two gene model for bud initiation by alfalfa (*Medicago sativa* L.) in microculture (Reisch and Bingham 1980). However, it has not been determined whether apparent direct effects, such as the previous examples, relate to structural or regulatory genes. The potential influences of regulatory effects are obscure, but they are important because current theory suggests that lack of essential genes or their inadequate expression may preclude organ regeneration.

GENETIC EFFECTS ON VEGETATIVE ORGAN REGENERATION

Organogenesis and Somatic Embryogenesis *In Vitro*

Propagating plants from protoplast, cell, callus, and some types of organ cultures is developmentally more complex than propagating by conventional cuttings. Such *in vitro* cultures often are initially devoid of bud and root primordia. Thus, to form a plantlet, both buds and roots must be regenerated, either separately or in the form of a somatic embryo. Plantlet formation often occurs only after callus has formed. Previous research with microcultures, like that with cuttings, primarily has studied nongenetic effects on vegetative regeneration (Thorpe 1980, 1982). However, genetic effects on many aspects of microculture have been observed (refs. in Keyes et al. 1980) and the literature is rapidly expanding.

The microculture literature concerning genetic effects on organogenesis and somatic embryogenesis is sometimes confusing with regard to type of vegetative regeneration. Some commonly studied species are known to regenerate by both organogenesis and somatic embryogenesis (e.g. Green 1981). However, most microculture studies have not included supporting, detailed microscopic examination of the cultures. Thus, types of simultaneously occurring regeneration may have been confused. In addition, good regenerability of cultures may be associated with formation of a morphologically distinct type of callus (Rines and McCoy 1981, Hibberd 1984). Formation of such callus sometimes has been reported in lieu of the actual type of vegetative regeneration (e.g. Duncan et al. 1985). Such reporting further confuses interpretation because callusing, apart from plantlet regeneration, has been shown to be genetically controlled in, for example, maize (Tabata and Motoyoshi 1965) and alfalfa (Keyes and Bingham 1979). Thus, the microculture literature cannot be unerringly interpreted.

Failures to obtain high frequency plantlet regeneration with certain herbaceous crop plants have led to studies of the genetic control of vegetative organ regeneration in microculture, mostly using embryo culture or embryo derived callus cultures. Such studies have examined somatic embryogenesis, adventitious budding, the generic plantlet formation, and rooting. Rooting has been investigated less frequently than the others because shoot formation is usually the limiting factor for propagating herbaceous crop species in microculture.

As an example of past research, recurrent selection of tetraploid alfalfa was used to obtain varieties with markedly enhanced regeneration frequency from callus cultures (Bingham et al. 1975). Subsequent research established that budding ability in diploid alfalfa callus cultures was genetically controlled (Reisch and Bingham 1980, summarized by McCoy and Walker 1983). Similarly, research with red clover (*Trifolium pratense* L.) callus cultures indicated that the propensities for rooting and somatic embryogenesis were inherited (Keyes et al. 1980). The authors suggested that selection and breeding would effectively enhance regeneration frequency of plants from callus (Keyes et al. 1980). Differences in rooting ability *in vitro* between *Lycopersicon* spp. suggested that "regeneration genes" might be introduced into tomato (*L. esculentum* Mill.) from wild-types (Locy 1983). Marked genotypic differences also have been found, for example, in:

—bud and embryo formation by cultures of wheat [*Triticum aestivum* L. (Maddock et al. 1983, Mathias and Simpson 1986)], sunflower [*Helianthus annuus* L. (Paterson and Everett 1985)], and barleys (Jørgensen et al. 1986). There was a pronounced genotype-medium (coconut milk) interaction with wheat cultures that influenced shoot regeneration (Mathias and Simpson 1986).

—embryo formation by cultures of potato [*Solanum tuberosum* L. (Jacobsen and Sopory 1978)], *Citrus* spp. (Moore 1985), maize (Duncan et al. 1985, Fahey et al. 1986, Hodges et al. 1986), wheat (He et al. 1986), and several clovers [*Trifolium* spp. (Pederson 1986)].

—budding and rooting by alfalfa [cv. 'Regen S' (Walker et al. 1978) and *Petunia hybrida* Vilm. callus cultures (Mitchell et al. 1980, Skvirsky et al. 1984).

—budding by tobacco (Ogura and Tsuji 1977), pea [*Pisum sativum* L. (Malmberg 1979)], wheat (Sears and Deckard 1982, Mathias and Fukui 1986), Indian mustard [*Brassica juncea* Coss (Fazekas et al. 1986)], and beets [*Beta vulgaris* L. (Saunders and Shin 1986)]. In wheat, enhanced budding ability has been associated either with gain of promotive or loss of inhibitory gene(s) on specific chromosomes (Mathias and Fukui 1986, Anon. 1987).

—plantlet formation by cultures of oat [*Avena* spp. (Cummings et al. 1976, Rines and McCoy 1981)], rice [*Oryza sativa* L. spp. *indica* (Reddy et al. 1985)] and barley [*Hordeum vulgare* L. (Hanzel et al. 1985)].

Genotypically based potentials for *in vitro* regeneration ability, such as described above, have been positively and negatively modified by the environment to which the ortet was exposed (e.g. Gharyal and Maheshwari 1983, Maddock et al. 1983) and by the nature of the culture medium (e.g. Reddy et al. 1985). Modifying culture media has been shown to alleviate, sometimes greatly, genotypic restrictions on microculture regeneration. For example, improved environmental conditions have resulted in embryogenesis in all *Triticum* spp. tested (R. W. Zobel, personal communication). However, in comprehensive tests of many genotypes, modifying culture conditions has not always overcome apparent genotypic restrictions on regeneration. Such findings can be interpreted in two major ways: Either there are culturally insurmountable genotypic restrictions on microculture regeneration, depending upon species, or cultural conditions have not been adequately studied. Indeed, some aspects of culture, such as influences of CO_2 and ethylene (Zobel 1978, 1986b), may not always have been adequately considered.

One recent comprehensive study with maize (Duncan et al. 1985) demonstrated how an efficient experimental design can properly probe for genetic effects on microculture regeneration and interactions with culture environment and ortet genomic characteristics. Similar experimental designs and results can be gleaned individually from previous literature, but the catholic nature of the Duncan et al. (1985) study makes it particularly suitable for detailed analysis here. Duncan et al. tested the hypotheses that lack of success in regenerating maize indicated that: 1) there was a genetic limitation; 2) media and methods were insufficient; or 3) existing surveys of available genotypes had not properly assessed potential for regenerability. The experimental material was immature embryos of 101 (Test 1) or 218 additional (Test 2) selfed inbred lines and germplasm stocks. The authors used morphological type of callus formed (which in maize is related to potential plantlet formation by somatic embryogenesis) and plantlet formation as the data bases. In Test 1, only 49% of the genotypes produced callus. Plants were regenerated from 38 (92%) of the genotypes that produced apparently regenerable callus. Plantlet regeneration was not related to callus growth rate, vivipary mutation, or vigor ratings of the ortets. F_1 hybrid embryos produced regenerable callus only if at least one parent produced that type of callus. In Test 2, with improved media, 199 (91%) of the additional lines or stocks produced apparently regenerable callus (plantlet regeneration was not attempted). Of 23 lines that did not produce regenerable callus in the first trial, 83% did so on the better media. However, the number of lines producing callus was less than for a general survey of genotypes. Results of Tests 1 and 2 supported the authors' three hypotheses. Both tests established a genetic basis for developing regenerable callus and somatic embryogenesis by maize embryos. Test 2 indicated that regenerability could be improved by proper formulation of culture media, but that such external factors did not correct for very deficient genomic regeneration characteristics. Finally, both tests indicated that a large sample size may be required to properly select for and study a genetically based regeneration trait such as somatic embryogenesis. Even larger sample sizes might have been required if inbred lines had not been used because of possible masking of recessive traits.

However, the Duncan et al. (1985) study may not have been completely definitive. First, only limited cultural conditions were tested; other conditions *might* have overcome genetic restrictions of embryogenesis. Second, only 20% of recessive genes in maize may be expressed even in isogenic lines (Zobel 1983). Finally, differences in the environmental (cultural) history of ortets and source tissues may not have been fully considered (R. W. Zobel, personal communication).

Evidence is much more limited for the genetic control of vegetative regeneration of woody species in microculture. For example, observations with Norway spruce [*Picea abies* (L.) Karst.] cultures indicated genotypic effects on adventitious budding (von Arnold 1984). Genotypic effects on microculture regeneration ability of poplars have also been suggested (Ahuja and Muhs 1982, Ahuja 1983). However, such genetic effects have not been directly studied.

Rooting of Cuttings

The possible role of genetics in predisposing conventional propagules to root has not received nearly as much attention as the role of genetics in controlling root system development (Zobel 1975, 1986a). Nevertheless, substantial evidence exists that rooting by cuttings is genetically controlled (Haissig 1986). For example, there is consistently different rooting ability within and between genera (Rauter 1971, Hardwick 1979, Locy 1983), with *Populus* spp. a good example (Zsuffa 1976 and refs. therein). Similarly, certain cultivars of horticultural species have consistently been easy- or difficult-to-root. Preformed root primordia form only in certain species (Haissig 1974). Rooting has sometimes been observed to be positively related to activities of specific enzymes or with specific types of intermediary metabolism (Haissig 1986). Finally, mutation may lead to loss of rooting ability (Zobel 1986a).

Genotypic differences have been observed in rooting of cuttings from agronomic plants such as alfalfa (Georgiev and Vassileva 1982). In one study with alfalfa (Reisch and Bingham 1981), diploid, tetraploid, aneuploid, and hexaploid plants were regenerated from cell suspension cultures derived from a diploid clone. Rooting of cuttings from the original diploid clone was 100%. Rooting of cuttings from seven new diploid subclones ranged from 19–100%; rooting of cuttings from 11 new tetraploid, aneuploid, and hexaploid subclones ranged from 56–100%. Thus, genetic changes that occurred during regeneration of the subclones from cell suspensions apparently influenced rooting.

Clonal differences in rooting ability of tree cuttings were noted years ago. In 1929, Zimmerman and Hitchcock reported differences in rooting for clones of holly cuttings within but not between different geographic origins (Massachusetts, 40–82% rooted; New Jersey, 64–79% rooted). Snow (1939) found statistically significant differences between the percentages of rooted red maple (*Acer rubrum* L.) cuttings that originated from 20 separate clones of stump sprout origin. Clonal variation in rooting ability of cuttings has also been found for other angiospermous forest trees such as sugar maple [*Acer saccharum* Marsh. (Dunn and Townsend 1954)], eastern cottonwood [*Populus deltoides* Bartr. ex Marsh. (Wilcox and Farmer 1968, Ying and Bagley 1977)], and American sycamore [*Platanus occidentalis* L. (Cunningham 1986)], and for numerous tree or shrub horticultural species.

With regard to gymnosperms, Deuber (1940) noted marked differences between rooting percentages of Norway spruce cuttings originating from 38 clones of 26- or 40-yr-old trees. Broad sense heritability for rooting of untreated Norway spruce cuttings from 30-yr-old trees (15 clones) was estimated at about 0.7 for cuttings collected in either August or November (Mergen 1960). Mergen (1960) concluded that rooting percentages and volume of roots produced by Norway spruce cuttings were genetically controlled. Clonal differences in rooting ability also have been noted for other gymnosperms such as eastern white pine [*Pinus strobus* L. (Kiang and Garrett 1975)], shortleaf pine [*P. echinata* Mill. (Fancher and Tauer 1981)], several spruces [*Picea* spp. (Rauter 1971)], and western hemlock [*Tsuga heterophylla* (Raf.) Sarg. (Brix and Barker 1975, Foster et al. 1984)].

Results of all older studies may not be definitive if confounding C-effects were not taken into account (Wilcox and Farmer 1968, Foster et al. 1984). However, in a properly controlled test, vegetative propagating ability of eastern cottonwood cuttings differed markedly between clones and was found to be highly heritable (Wilcox and Farmer 1968, Zsuffa 1976). More recently, direct evidence has been obtained for genetic control of rooting in western hemlock cuttings (Foster et al. 1984). In that study, C-effects were properly evaluated and five rooting traits of western hemlock cuttings were found to be highly heritable.

In summary, current evidence suggests that the capacities of microcultures to initiate vegetative organs and somatic embryos, and of cuttings to initiate roots, are influenced by genetics but not necessarily the same genes. Further, these capacities can be enhanced in a population through selection and breeding. The thrust of most research has been toward producing regenerable strains of herbaceous crop plants. Past research has not determined whether the genetic effects were direct, correlated, or regulatory, nor how they were biochemically and physiologically manifested. Therefore, much remains to be studied.

EXPERIMENTAL GUIDELINES

The literature cited above clearly demonstrates that genotype is an important factor in vegetative regeneration in microculture and from conventional cuttings. However, little is known of the possible genetic modes of action (e.g. direct vs. correlated effects, additive vs. non-additive gene action) or genetic variation in physiological and biochemical mechanisms that mediate gene expression. Nevertheless, based on past research, we have formulated some guidelines for further studies of genetic effects. These guidelines concern control of potentially confounding environmental effects, types of genetic models (including genetic structure of the experimental population), and measurement of physiological and biochemical effects related to gene expression. Each of the foregoing is discussed in detail below.

Environmental Effects

In a definitive experiment, factors causing differential physiological preconditioning of parent ortets are either eliminated or included in the experimental design. These nongenetic factors have been termed *C-effects* and are present when "the environments of members of a family are more alike than those of a group of individuals picked at random from a given population" (Lerner 1958). Extending this terminology from sexual to vegetative offspring (Libby and Jung 1962) has caused confusion. Burdon and Shelbourne (1974) divided the vegetative analogue of C-effects into two classes, M and m. M are the general effects of the parent ortet that are transmitted to its ramets (e.g. age, sex, physiological status of ortet); m are effects that relate to characteristics of individual propagules (e.g. cutting diameter, crown position, crown aspect). Because M-effects from unreplicated parental ortets are impossible to separate from genetic effects, the two are irrevocably confounded unless M-effects are eliminated before the test begins. Serial vegetative propagation may cause rejuvenation in some species (Libby et al. 1971). However, physiological "equilibration" between ramets descended from ortets of different ages remains an untestable and, therefore, invalid assumption. Thus, studies should begin either with seed or with a population of ortets of known, equal age in order to eliminate M-effects. Compared to M-effects, m-effects are less difficult to control and may be included as design variables, selected for uniformity, or randomized by treatment. Control by experimental design or selection for uniformity must be applied systematically throughout a test. Selection for uniformity may limit the scope of inference because, for example, results may only apply to cuttings of a certain diameter taken from a specific crown position. This limitation also applies to control of M-effects by selection for uniform ortet age. Randomization of m-effects among treatments is valid for avoiding experimental bias, but the resulting reduction in experimental precision may preclude its use in many studies.

The first explicit attempt to separate C-effects from genetic effects in the study of rooting was made by Wilcox and Farmer (1968). In their study, 49 two-yr-old eastern cottonwood clones formed the experimental population. Six cuttings were taken from each ortet and planted in a replicated design with cutting diameter sorted by replication. These plants were the "primary ramets." In Test 1, two cuttings were taken from four primary ramets per clone and root production in two soil types was determined. Genotype × environment interactions were great enough to warrant separate analyses of each soil type. Heritabilities for number of roots, root length, and root weight were estimated at 0.56, 0.52, and 0.58 (clay soil) and 0.44, 0.33, and 0.36 (sandy soil), respectively. In Test 2, 10 clones were used to evaluate potential C-effects among primary ramets. Four cuttings were taken from each primary ramet and root production was determined in "standard" potting mixture. Variation attributable to primary ramets (C-effects) was significant for root number and root weight (11 and 9% of total variation, respectively) but was only associated with three of 10 clones. In addition, a potential bias was demonstrated in estimating broad sense heritabilities for root number (0.39, biased vs. 0.31, unbiased) and root weight (0.38, biased vs. 0.32, unbiased).

In the foregoing tests, Wilcox and Farmer (1968) provided an impeccable model for controlling environmental preconditioning in genetic experiments. Potentially confounded effects among parental ortets were minimized by using plants of equal age on a uniform site. C-effects introduced by experimental conditions were controlled by including this factor as an experimental design variable.

Foster et al. (1984) conducted a similar study using western hemlock. Rooting was determined for cuttings taken directly from large trees (Test 1) and after serial vegetative propagation of parental ortets (Test 2). C-effects among primary ramets were included as an experimental design variable. Vertical crown position effects were pronounced as demonstrated by greater rooting of cuttings taken from lower than upper crown regions. Broad sense heritabilies for number of roots and root length were 0.92 and 0.87, respectively. Variation due to C-effects was significant but comprised only from 2–6% of total variation. The primary source of bias noted by the authors was the potential influence of age differences between parental ortets, which ranged in age from 25–68 yr. This type of bias probably results in overestimation of genetic variation and expected response to selection.

Genetic Models

Sophistication of the genetic model is an important consideration in experimental design of rooting studies. Most research has used clonal populations (Wilcox and Farmer 1968, Foster et al. 1984). Such tests are logistically straightforward but provide little insight into possible genetic mechanisms beyond accumulating evidence that some kind of genetic control exists. Studies of clonal populations allow estimation of broad sense heritability, that is, the ratio of total genetic variance to phenotypic variance. Estimates of broad sense heritability for rooting are useful for predicting genetic response to clonal selection but they neither provide guidelines for improvement through selective breeding nor promote a deeper understanding of physiological and biochemical aspects of gene expression. Future experiments should concentrate on populations derived from more sophisticated mating designs (e.g. Keyes et al. 1980) that allow estimates of additive and dominance effects as well as total genetic variance.

Half-sib families

Few studies of genetic effects on rooting in trees have incorporated family structure in the experimental population. Of these studies, most have used half-sib families (i.e. one common parent) obtained from open-pollinated ortets. Theoretical considerations allow the partitioning of total additive genetic variance into between- and within-family components (Falconer 1960). Variance between half-sib families can be equated to 25% of the additive genetic variance present in the population (Falconer 1960) and thus used as an estimator for additive genetic variance. Additive genetic variance is important because it determines whether genetic improvement can be achieved through simple selective breeding methods (i.e. mass selection). Open-pollinated families may not be strictly half-sib because of self-pollination or multiple pollinations by the same male parent. The assumption of obtaining half-sib families from open-pollinated ortets is probably acceptable only when ortets are selected from large, random mating populations.

A recent study of rooting of cuttings from a population of American sycamore open-pollinated families indicated that root number and root weight differed between clones within families but not between families (Cunningham 1986). Cunningham's (1986) results equally supported the two mutually exclusive conclusions that: 1) genetic control of rooting was primarily determined by genes with non-additive effects; or 2) precision of the experiment (20 families, four trees per family) was insufficient to detect interfamilial variation in rooting. Earlier work by Ying and Bagley (1977) found significant differences between open-pollinated families of eastern cottonwood and differences between individuals (clones) within families. Compared to Cunningham (1986), Ying and Bagley (1977) studied more families (48 in greenhouse and 47 in field tests) from more diverse geographic origins. Broad generalizations regarding the importance of additive genetic effects on rooting presently cannot be formulated because of limited experimental results. Rooting was found to be partially controlled by additive effects in eastern cottonwood (Ying and Bagley 1977) and thus simple selective breeding methods may prove effective in that species. On the contrary, results from American sycamore (Cunningham 1986) indicated that clonal selection or more complex breeding methods may be required to achieve genetic improvement in rooting. Additional experiments with greater precision will be required before substantive conclusions can be reached.

Full-sib families

Full-sib families consist of clones having common parents. Variance between full-sib, as between half-sib, families has a genetic expectation and can be used for estimation. Variance between full-sib families can be equated to 50% of the additive genetic variance in the population plus 25% of the dominance variance and variance due to several higher order interactions. Thus, it is impossible to obtain unconfounded estimates of a single kind of genetic effect from full-sib populations: Narrow-sense heritability estimates are biased upwards by the presence of dominance variance. The primary advantage of full-sib, compared to half-sib, families is that more genetic variance is distributed between families. Thus, interfamilial differences are easier to detect in experiments with full-sib families.

For the purposes of this example, we conducted an experiment to evaluate rooting by eastern white pine seedling cuttings from each of 23 full-sib families. The objectives were to determine whether or not families differed significantly in rooting and whether genetic effects were correlated or direct. Seeds from 23 single-pair matings were obtained from the USDA-Forest Service, Oconto River Seed Orchard, Lakewood, WI. Seedlings were grown from seed in a growth chamber (18 h d, 22 °C d, 20 °C nights) under fluorescent (Sylvania F96T12/CWX/VHO) and incandescent (Sylvania 1950L/P25/8) light (ca 350 μmol·m^{-2}·s^{-1}). After 95 d of stock plant growth, cuttings were made by severing the hypocotyl just above the root collar. Then cuttings were placed in a greenhouse rooting bench with bottom heat and intermittent mist (5 s mist at 3 min intervals). The experimental design was two replications of 20-cutting plots. One-half of the cuttings from each plot was harvested after 30 d in the mist bench and the remainder after 45 d. Number of roots per cutting was counted. In addition, fresh weights were determined for roots plus the basal 1 cm of stem, next adjacent (second) 1 cm of stem, and remainder of the aerial portion of each cutting. Data from each harvest date were analyzed separately using analyses of variance and covariance. Relations between traits and harvest dates were evaluated using family mean correlation analysis. Intertrait family mean correlations measure trait relatedness, and include genetic and environmental components.

The average number of rooted cuttings for all families was 15% (after 30 d) and 36% (after 45 d). Families differed for all measured traits at both harvest dates (P < 0.05). Heritabilities for fresh weight of roots plus basal 1 cm stem were 0.36 (30 d) and 0.28 (45 d). Analysis of covariance indicated that families differed (P < 0.01) in number of roots per cutting and fresh weight of roots plus basal stem, after effects due to upper stem and plant fresh weight were statistically removed. Family mean correlations between traits at each harvest date were significant only between immediately adjacent plant parts (Fig. 2). Furthermore, the fresh weights of roots plus basal stem and second 1 cm of stem at day 45 were not related to the fresh weight of the aerial portion of the cutting at 30 d.

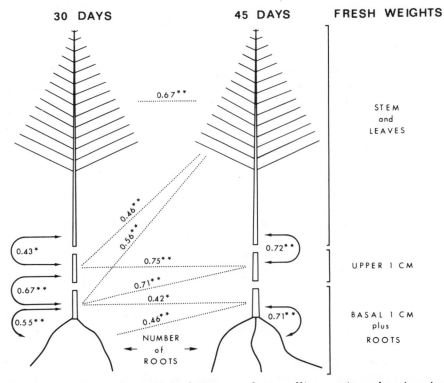

Figure 2. Diagrammatic representation of *Pinus strobus* seedling cuttings showing significant family mean correlation coefficients for fresh weight relations between various parts of the cuttings after 30 d (15% rooted) and 45 d (36% rooted) in the rooting bench. The cuttings originated from 95-d-old plants of 23 full-sib families. See text for details of experimental design and statistical analyses (* denotes P < 0.05, ** denotes P < 0.01).

In our test, strong correlations between rooting and stem plus leaf biomass would have supported the hypothesis that correlated effects were important in rooting of eastern white pine. Under those circumstances, we would have concluded that the genes or gene blocks that affected growth of the aerial portion of the plant were also involved in development of adventitious roots. However, this pattern of intertrait correlations was not observed. Absence of significant intertrait correlations between remote plant parts, as we observed in this test, indicated that genetic effects on rooting were primarily direct. Results indicated that rooting was not related to the amount of biomass that cuttings accumulated prior to day 30. Instead, accumulation of biomass at day 45 was related to rooting at day 30, which would have been expected because the previous rooting increased biomass and, most probably, enhanced aerial growth of the cuttings. Aerial growth would have been enhanced due to increased absorption of water and nutrients from the rooting medium, and as a response to cytokinin(s) synthesized in the roots. Overall, the experiment demonstrated that ancillary measurements as simple as the biomass of different plant parts can be used to provide insights into genetic control of rooting in full-sib progenies.

Factorial mating

Sophisticated mating designs can be used to obtain unbiased estimates of additive and dominance genetic variance. Such designs have not been used in investigations of genetic effects on rooting. They have, however, been used in the study of vegetative regeneration of some agronomic crops in microculture. Keyes et al. (1980) used three sets of a 4 × 4 factorial mating scheme to estimate additive and dominance genetic variance for callus growth and rooting by red clover in microculture. Additive genetic variance was significant for most traits regardless of media formulations. Dominance variance was significant for rooting characteristics in subcultures two and three but the expression of dominance depended on the medium used. Phenotypic and additive genetic correlations between two media formulations were positive and highly significant. Reciprocal effects tended to be significant for most traits studied. Heritability for rooting during three subcultures ranged from 0.30–0.80.

Three important conclusions can be formulated for red clover from Keyes et al. (1980) study: 1) regeneration of roots was genetically controlled; 2) additive genetic effects were stable across media but expression of intra-allelic interactions were medium dependent; and 3) simple selective breeding schemes such as recurrent selection would be effective in modifying potential for rooting. Of those three conclusions only the first could have been made if the investigators had not used the factorial mating design. The factorial design led to a more basic understanding of gene action and to guidelines for applying selective breeding to improve rooting.

Biochemical and Physiological Measurements

A basic understanding of genetic control will not be forthcoming until physiological and biochemical factors become routine response variables in genetic experiments. It is clear from previous experiments that genes that promote rooting, even though present throughout a plant's life cycle, are not expressed at all stages of development (Deuber 1940 and refs. therein). Identifying controlling genes, and elucidating their regulation and expression will require detailed measurements of biochemical and physiological processes involved in rooting as these become better known. Such processes as the synthesis and activity of key enzymes, synthesis and partitioning of carbohydrates, and regulation of plant water status may prove to be particularly fruitful areas of investigation (Haissig 1986). Veierskov et al. (1982), for example, compared relations between rooting and carbohydrate accumulation in wild-type pea and a mutant pea that was deficient in photosystem II activity.

Methods of increasing experimental precision will be required because measuring physiological and biochemical processes can be costly and time consuming. Studying clonal populations is one obvious method but, as previously stated, clonal populations provide only limited genetic inference compared to sexually derived populations. Studying inbred lines may prove to be more advantageous. Inbreeding increases and redistributes genetic variance, increasing it between lines and reducing it within lines (Allard 1960). Thus, by using inbreds, genetic differences could be measured more precisely or, conversely, fewer measurements would achieve precision equal to that attainable with outcrossed families. Alternatively, rapid progress may be made by using suitable isogenic mutant lines, if such lines are available or can be produced for rooting tests.

CONCLUSION

Genetic effects are among the least studied of factors that control or modify rooting by cuttings. In other areas of rooting research, experimentation has closely followed initial observations of directly and indirectly related phenomena. In comparison, observations of strong genetic effects on rooting by cuttings were made at least 50 years ago but have been studied little in the interim. As a consequence, little is known about the various possible genetic effects on rooting of cuttings. Future research should control potentially confounding environmental effects, use sophisticated genetic models, and investigate physiological and biochemical aspects of gene expression. Such research will aid a basic understanding of genetic effects on rooting. Unambiguous separation of direct, correlated, and regulatory genetic effects will be more difficult and may require application of new methods such as use of transposable elements. Transposable elements are capable of limiting or suspending the function of individual genes (Freeling 1984, Doring and Starlinger 1986) and thus provide the equivalent of isogenic or single-point mutated lines (cf. Zobel 1986a). Lines in which the ability to produce roots has been genetically disabled could be compared with their untransposed, normal counterparts. The location and structure of controlling genes might then be determined through restriction enzyme digests and molecular probes. Such procedures would be most effective if genes with large effects, i.e. qualitative genes, were important in rooting. Related physiological and biochemical pathways also might be detected based on their absence or changed activities in disabled lines. Tomato would be the best initial test species because it is genetically well characterized and has a low amount of genetic duplication; pea may also prove useful (Zobel 1986a and refs. therein).

ACKNOWLEDGEMENTS

We thank Drs. E. T. Bingham, T. D. Davis, W. P. Hackett, and R. W. Zobel for critically reviewing this chapter; Mr. K. M. Hansen for conducting the rooting test with eastern white pine; Mr. E. Bauer for preparing Figure 2; and Ms. Sandra Haissig for clerical assistance.

REFERENCES

Anon. 1987. Wheat regeneration chromosomes identified. *Agricell Rept.*, February. p. 12.

Ahuja, M. R. 1983. Somatic cell differentiation and rapid clonal propagation of aspen. *Silvae Genetica* 32:3–4.

_____ H. J. Muhs. 1982. Control of growth and differentiation in tissues and protoplast derived callus in different genotypes of aspen. In *Proc. 5th Int. Cong. Plant Tissue and Cell Culture. Plant Tissue Culture 1982*. pp. 177–178.

Allard, R. W. 1960. *Principles of Plant Breeding*. John Wiley and Sons, NY, USA. 485 pp.

Bingham, E. T., L. V. Hurley, D. M. Kaatz and J. W. Saunders. 1975. Breeding alfalfa which regenerates from callus tissue in culture. *Crop Sci.* 15:719–721.

Brix, H. and H. Barker. 1975. *Rooting Studies of Western Hemlock Cuttings*. Canadian For. Serv., Pacific For. Res. Centre, Victoria, British Columbia, Canada. Report BC-X-131. 14 pp.

Burdon, R. D. and Shelbourne, C. J. A. 1974. The use of vegetative propagules for obtaining genetic information. *N. Z. J. For. Sci.* 4:418–425.

Cummings, D. P., C. E. Green and D. D. Stuthman. 1976. Callus induction and plant regeneration in oats. *Crop Sci.* 16:465–470.

Cunningham, M. W. 1986. Genetic variation in rooting ability of American sycamore cuttings. In

58   Bruce E. Haissig and Don E. Riemenschneider

    Proc. TAPPI Res. and Dev. Conf. TAPPI Press, Atlanta, GA, USA. pp. 1–6.

Deuber, C. G. 1940. Vegetative propagation of conifers. *Trans. Conn. Acad. Arts and Sci.* 34:1–83.

Doring, H.-P. and P. Starlinger. 1986. Molecular genetics of transposable elements. *Ann. Rev. Genet.* 20:175–200.

Duncan, D. R., M. E. Williams, B. E. Zehr and J. M. Widholm. 1985. The production of callus capable of plant regeneration from immature embryos of numerous *Zea mays* genotypes. *Planta* 165:322–332.

Dunn, S. and R. J. Townsend. 1954. Propagation of sugar maple by vegetative cuttings. *J. For.* 52:678–679.

Fahey, J. W., J. N. Reed, T. L. Readdy and G. M. Pace. 1986. Somatic embryogenesis from three commercially important inbreds of *Zea mays*. *Plant Cell Repts.* 5:35–38.

Falconer, D. S. 1960. *Introduction to Quantitative Genetics.* The Ronald Press, NY, USA. 365 pp.

Fancher, G. A. and C. G. Tauer. 1981. Natural variation in rooting ability of western provenances of shortleaf pine. In *Proc. 16th Southern For. Tree Improv. Conf.* pp. 189–199.

Fazekas, G. A., P. A. Sedmach and M. V. Palmer. 1986. Genetic and environmental effects on *in vitro* shoot regeneration from cotyledon explants of *Brassica juncea*. *Plant Cell Tissue Organ Cult.* 6:177–180.

Fehr, W. R. 1987a. *Principles of Cultivar Development. Vol. 1: Theory and Technique.* Macmillan Pub., NY, USA. ISBN 0-02-949920-8.

_____ 1987b. *Principles of Cultivar Development. Vol. 2: Crop Species.* Macmillan Pub., NY, USA. ISBN 0-02-949181-9.

Foster, G. S., R. K. Campbell and W. T. Adams. 1984. Heritability, gain, and C effects in rooting of western hemlock cuttings. *Can. J. For. Res.* 14:628–638.

Freeling, M. 1984. Plant transposable elements and insertion sequences. *Ann. Rev. Plant Physiol.* 35:277–298.

Georgiev, Z. and B. Vassileva. 1982. Effect of the genotype, the stem zone and the phase of development on the rooting of alfalfa cuttings. *Plant Sci.* (Sofia) 19:70–75.

Gharyal, P. K. and S. C. Maheshwari. 1983. Genetic and physiological influences on differentiation in tissue cultures of a legume, *Lathyrus sativus*. *Theor. Appl. Genet.* 66:123–126.

Green, C. E. 1981. Tissue culture in grasses and cereals. In *Genetic Engineering for Crop Improvement* (K. O. Rachie and J. M. Lyman, eds). The Rockefeller Foundation, NY, USA. pp. 107–122.

Haissig, B. E. 1974. Origins of adventitious roots. *N. Z. J. For. Sci.* 4:299–310.

_____ 1983. N-phenyl indolyl-3-butyramide and phenyl indole-3-thiolobutyrate enhance adventitious root primordium development. *Physiol. Plant.* 57:435–440.

_____ 1986. Metabolic processes in adventitious rooting. In *New Root Formation in Plants and Cuttings.* (M. B. Jackson, ed). Martinus Nijhoff Pub., Dordrecht/Boston/Lancaster. pp. 141–189. ISBN 90-247-3260-3.

Hanzel, J. J., J. P. Miller, M. A. Brinkman and E. Fendos. 1985. Genotype and media effects on callus formation and regeneration in barley. *Crop Sci.* 25:27–31.

Hardwick, R. C. 1979. Leaf abscission in varieties of *Phaseolus vulgaris* (L.) Merrill—a correlation with propensity to produce adventitious roots. *J. Exp. Bot.* 30:795–804.

He, D. G., G. Tanner and K. J. Scott. 1986. Somatic embryogenesis and morphogenesis in callus derived from the epiblast of immature embryos of wheat (*Triticum aestivum*). *Plant Sci.* 45:119–124.

Hibberd, K. A. 1984. Induction, selection, and characterization of mutants in maize cell cultures. *Cell Cult. and Somatic Cell Genet. of Plants* 1:571–576.

Hodges, T. K., K. K. Kamo, C. W. Imbrie and M. R. Becwar. 1986. Genotype specificity of somatic embryogenesis and regeneration in maize. *Bio/Technology* 4:219–223.

Jacobsen, E. and S. K. Sopory. 1978. The influence and possible recombination of genotypes on the production of microspore embryoids in anther cultures of *Solanum tuberosum* and dihaploid hybrids. *Theor. Appl. Genet.* 52:119–123.

Jørgensen, R. B., C. J. Jensen, B. Andersen and R. Von Bothmer. 1986. High capacity of plant regeneration from callus of interspecific hybrids with cultivated barley (*Hordeum vulgare L.*).

Plant Cell Tissue Organ Cult. 6:199–207.

Keyes, G. J. and E. T. Bingham. 1979. Heterosis and ploidy effects on the growth of alfalfa callus. *Crop Sci.* 19:473–476.

_____ G. B. Collins, and N. L. Taylor. 1980. Genetic variation in tissue cultures of red clover. *Theor. Appl. Genet.* 58:265–271.

Kiang, Y. T. and P. W. Garrett. 1975. Successful rooting of eastern white pine cuttings from a 17-year-old provenance planting. In *Proc. 22nd Northeastern Tree Improv. Conf.* State Univ. of New York, College Environmental Sci. and For., Syracuse, NY, USA. pp. 24–34.

Lerner, I. M. 1958. *The Genetic Basis of Selection.* John Wiley and Sons, NY, USA. 298 pp.

Libby, W. J. and E. Jung. 1962. Variance associated with cloning. *Heredity* 17:533–540.

_____ A. G. Brown and J. M. Fielding. 1971. Effects of hedging radiata pine on production, rooting, and early growth of cuttings. *N. Z. J. For. Sci.* 2:263–283.

Locy, R. D. 1983. Callus formation and organogenesis by explants of six *Lycopersicon* species. *Can. J. Bot.* 61:1072–1079.

Maddock, S. F., V. A. Lancaster, R. Risiott and J. Franklin. 1983. Plant regeneration from cultured immature embryos and inflorescences of 25 cultivars of wheat (*Triticum aestivum*). *J. Exp. Bot.* 34:915–926.

Malmberg, R. L. 1979. Regeneration of whole plants from callus culture of diverse genetic lines of *Pisum sativum* L. *Planta* 146:243–244.

Mathias, R. J. and K. Fukui. 1986. The effect of specific chromosome and cytoplasm substitutions on the tissue culture response of wheat (*Triticum aestivum*). *Theor. Appl. Genet.* 71:797–800.

_____ E. S. Simpson. 1986. The interaction of genotype and culture medium on the tissue culture responses of wheat (*Triticum aestivum* L. em. thell) callus. *Plant Cell Tissue Organ Cult.* 7:31–37.

McCoy, T. and K. Walker. 1983. Alfalfa. *Handbook of Plant Cell Culture* 3:171–92.

Mergen, F. 1960. Variation and heritability of physiological and morphological traits in Norway spruce. In *Proc. 5th World For. Cong.* 2:755–757.

Mitchell, A. Z., M. R. Hanson, R. C. Skvirsky and F. M. Ausubel. 1980. Anther culture of *Petunia:* Genotypes with high frequency of callus, root, or plantlet formation. *Z. Pflanzenphysiol.* 100:131–146.

Moore, G. A. 1985. Factors affecting *in vitro* embryogenesis from undeveloped ovules of mature *Citrus* fruit. *J. Amer. Soc. Hortic. Sci.* 110:66–70.

Ogura, H. and S. Tsuji. 1977. Differential responses of *Nicotiana tabacum* L. and its putative progenitors to de- and redifferentiation. *Z. Pflanzenphysiol.* 83:419–426.

Paterson, K. E. and N. P. Everett. 1985. Regeneration of *Helianthus annuus* inbred plants from callus. *Plant Sci.* 42:125–132.

Pederson, G. A. 1986. *In vitro* culture and somatic embryogenesis of four *Trifolium* species. *Plant Sci.* 45:101–104.

Rauter, M. 1971. Rooting of *Picea* cuttings in plastic tubes. *Can. J. For. Res.* 1:125–129.

Reddy, V. S., S. Leelavathi and S. K. Sen. 1985. Influence of genotype and culture medium on microspore callus induction and green plant regeneration in anthers of *Oryza sativa*. *Physiol. Plant.* 63:309–314.

Reisch, B. and E. T. Bingham. 1980. The genetic control of bud formation from callus cultures of diploid alfalfa. *Plant Sci. Lett.* 20:71–77.

_____ _____ 1981. Plants from ethionine-resistant alfalfa tissue cultures: Variations in growth and morphological characteristics. *Crop Sci.* 21:783–788.

Rines, H. W. and T. J. McCoy. 1981. Tissue culture initiation and plant regeneration in hexaploid species of oats. *Crop Sci.* 21:837–842.

Saunders, J. W. and K. Shin. 1986. Germplasm and physiologic effects on induction of high-frequency hormone autonomous callus and subsequent shoot regeneration in sugarbeet. *Crop Sci.* 26:1240–1244.

Sears, R. G. and E. L. Deckard. 1982. Tissue culture variability in wheat: Callus induction and plant regeneration. *Crop Sci.* 22:546–550.

Skvirsky, R. C., M. R. Hanson, and F. M. Ausubel. 1984. Intraspecific genetic variation in cytokinin-controlled shoot morphogenesis from tissue explants of *Petunia hybrida*. *Plant Sci. Lett.* 35:237–246.

Snow, A. G., Jr. 1939. *Clonal Variation in Rooting Response of Red Maple Cuttings*. USDA-Forest Service, Northeastern For. Exp. Sta. Tech. Note No. 29. 2 pp.

Tabata, M. and F. Motoyoshi. 1965. Heredity control of callus formation in maize endosperm cultured *in vitro*. *Jap. J. Genet.* 40:343–355.

Thorpe, T. A. 1980. Organogenesis *in vitro*: structural, physiological, and biochemical aspects. In *Int. Rev. Cytol. Suppl. 11A* (I.K. Vasil, ed). Academic Press, NY, USA. pp. 71–111.

_____ 1982. Callus organization and *de novo* formation of shoots, roots and embryos *in vitro*. In *Application of Plant Cell and Tissue Culture to Agriculture & Industry* (D. T. Tomes, B. E. Ellis, P. M. Harney, K. J. Kasha and R. L. Peterson, eds). Plant Cell Cult. Centre, Univ. of Guelph, Ontario, Canada. pp. 115–138.

Veierskov, B., A. S. Andersen, B. M. Stummann and K. W. Henningsen. 1982. Dynamics of extractable carbohydrates in *Pisum sativum*. II. Carbohydrate content and photosynthesis of pea cuttings in relation to irradiance and stock plant temperature and genotype. *Physiol. Plant.* 55:174–178.

von Arnold, S. 1984. Importance of genotype on the potential for *in vitro* adventitious bud production of *Picea abies*. *For. Sci.* 30:314–318.

Walker, K. A., P. C. Yu, S. J. Sato and E. G. Jaworski. 1978. The hormonal control of organ formation in callus of *Medicago sativa* L. cultured *in vitro*. *Amer. J. Bot.* 65:654–659.

Wilcox, J. R. and R. E. Farmer, Jr. 1968. Heritability and C effects in early root growth of eastern cottonwood cuttings. *Heredity* 23:239–245.

Ying, Ch.-Ch. and W. T. Bagley. 1977. Variation in rooting capability of *Populus deltoides*. *Silvae Genetica* 26:204–206.

Zimmerman, P. W. and A. E. Hitchcock. 1929. Vegetative propagation of holly. *Amer. J. Bot.* 16:570–577.

Zobel, R. W. 1975. The genetics of root development. In *The Development and Function of Roots* (J. G. Torrey and D. T. Clarkson, eds). Third Cabot Symp., Academic Press, NY, USA. pp. 261–275.

_____ 1983. Genic duplication: A significant constraint to molecular and cellular genetic manipulation in plants. *Comments Mol. Cell. Biophys.* 1:355–364.

_____ 1986a. Rhizogenetics (root genetics) of vegetable crops. *HortScience* 21:956–959.

_____ 1986b. Gaseous compounds of soybean tissue cultures: Carbon dioxide and ethylene evolution. *Environ. Exp. Bot.* 27:223–226.

_____ L. W. Roberts. 1978. Effects of low concentrations of ethylene on cell division and cytodifferentiation in lettuce pith explants. *Can. J. Bot.* 56:987–990.

Zsuffa, L. 1976. Vegetative propagation of cottonwood by rooting cuttings. In *Proc. Symp. on Eastern Cottonwood and Related Species*. Louisiana State Univ., Dept. Continuing Education, Baton Rouge, LA, USA. pp. 99–108.

CHAPTER 4

Mineral Nutrition and Adventitious Rooting

Frank A. Blazich

Department of Horticultural Science
North Carolina State University
Raleigh, NC 27695-7609

INTRODUCTION. 61
EFFECTS OF INDIVIDUAL NUTRIENTS ON ROOTING 62
    Root Initiation. 62
    Root Growth and Development 63
    Root Inducing Factors and Other Considerations 65
LEACHING OF NUTRIENTS DURING MIST PROPAGATION. 65
CONCLUSION 67
REFERENCES 67

Additional key words: auxin, cuttings, leaching, mist, nutrient mobilization, propagation.
Abbreviations: B, boron; Ca, calcium: Cl, chlorine; Fe, iron; IAA, indole-3-acetic acid; K, potassium; Mg, magnesium; Mn, manganese; Na, sodium; N, nitrogen; P, phosphorus; S, sulfur; Zn, zinc.

INTRODUCTION

Mineral nutrition is one of many factors which influence adventitious rooting in cuttings. In fact, a discussion of factors influencing the phenomenon would not be complete without including the relationship between adventitious rooting and the mineral nutrient status of the cutting material and the stock plant. In addition to being of fundamental scientific interest, the subject of mineral nutrition and adventitious rooting is of practical importance to the propagator who must make decisions as to fertilization of stock plants. Because fertility can have a pronounced effect on plant growth and development, and since plant developmental history can profoundly affect subsequent rooting of cuttings taken from that plant (see Chapter 16 by Moe and Andersen), such decisions are of considerable importance to successful propagation by cuttings.

Although adventitious rooting and mineral nutrition are intimately related, the subject is difficult to deal with because root formation on stem cuttings is a multi-stage process and few studies have distinguished between mineral effects at the various stages (Hartmann and Kester 1983). However, the various stages can be reduced to two general stages consisting of: 1) root initiation; and 2) root growth and development. Therefore, when considering the influence of various mineral nutrients on adventitious rooting one must consider the role of a particular nutrient in each stage of the process.

The following review discusses mineral nutrition as it directly influences root initiation, and root growth and development. Other influences of mineral nutrition which are worthy of consideration are also presented, such as leaching of mineral nutrients during mist propagation. Lack of information regarding various aspects of the subject are mentioned to encourage research directed towards furthering understanding of the phenomenon and to improve the rooting of cuttings.

EFFECTS OF INDIVIDUAL NUTRIENTS ON ROOTING

Root Initiation

Root initiation involves dedifferentiation of specific cells, leading to the formation of root meristems (Hartmann and Kester 1983). Initiation is dependent on the presence of auxin whether endogenous or artificially applied (Gautheret 1969, Haissig 1972). Excluding the mobilization of mineral nutrients into the base of a cutting during root initiation due to auxin application, one could argue that any nutrient involved in the multitude of metabolic processes associated with dedifferentiation and root meristem formation is essential for root initiation. Based on the previous statement one could undoubtedly prepare a long list of mineral nutrients essential for root initiation which would likely include most of the nutrients required by plants for growth in general. For example, Ca is required for cell elongation and cell division (Burström 1968), which immediately suggests importance of this element in root initiation. However, for various reasons, researchers have neglected to identify those nutrients essential for root initiation based on indirect evidence and basically have studied the relative importance of particular nutrients on adventitious rooting in general. As a consequence, most of these studies have not provided a clear picture of the importance of specific nutrients essential to root initiation. The few studies which have examined the mobilization or redistribution of mineral nutrients within cuttings during rooting (mobilization studies) have clearly pointed to the role mineral nutrients play in root initiation.

Some mobilization studies have examined the redistribution of particular mineral nutrients following treatment of cuttings with exogenous auxin. Such studies have usually included control (non-auxin treated) cuttings. The appropriate controls such as nontreated cuttings are very important because if omitted the question is posed as to whether the initial movement of nutrients into the base of a cutting during root initiation is a direct result of auxin application or the indirect result of auxin-induced growth stimulation and subsequent movement of various substances to these areas of growth (Booth et al. 1962, Blazich et al. 1983). Two studies whose design included such controls demonstrated that N was mobilized during root initiation in stem cuttings of kidney bean [*Phaseolus vulgaris* L. cv. Calapproved (Stuart 1938)] and plum [*Prunus* L. cv. Marianna 2624 (Strydom and Hartmann 1960)], and redistribution was accelerated by auxin treatment. On the other hand, N was not mobilized nor was any redistribution of P, K, Ca, and Mg detected during root initiation in stem cuttings of Japanese holly [*Ilex crenata* Thunb. cv. Convexa (Blazich and Wright 1979, Blazich et al. 1983)]. However, mobilization of P but not N, K, and Ca (Good and Tukey 1967) was reported during root initiation in cuttings of chrysanthemum (*Chrysanthemum morifolium* Ramat. cv. Indianapolis White). Conflicting reports as to mobilization of particular nutrient elements during root initiation may somehow be related to species differences.

An argument could be made that mobilization of a mineral nutrient into the base of a cutting during root initiation is a strong indication that the nutrient is important in this stage of the rooting process. However, lack of mobilization should not rule out need for a particular nutrient since some such as Ca are considered immobile (Mengel and Kirkby 1982).

Because there are limited data from mobilization studies showing movement of particular mineral nutrients, specifically N (Stuart 1938, Strydom and Hartmann 1960) and P (Good and Tukey 1967), into the base of a cutting during root initiation, one might argue that root initiation is primarily influenced by the initial tissue nutrient contents within the base of a cutting. If such is the case then one needs to examine data from studies which have generally attempted to correlate rooting with particular stock plant fertility levels (fertility studies). Such studies have been conducted by growing stock plants under varying levels of fertility followed by rooting of cuttings taken from them. If one, therefore, subscribes to the concept that mineral nutrients are not mobilized during root initiation then fertility studies reporting such data as percent rooting and root numbers should be useful in elucidating the importance of a particular nutrient in root initiation. Fertility treatments showing increased percent rooting, increased root numbers, or similar data could be interpreted as reflecting greater root initiation. On the other hand, caution should be exercised in use of such data because root initiation takes place on a cellular level so these data are but indirect measures of root initiation. Unless anatomical or similar studies are conducted one does not necessarily have a true measure of root initiation.

In considering the previous arguments, two fertility studies examined the influence of stock plant fertility levels on rooting. Stock plants of geranium [*Pelargonium* × *hortorum* Bailey cv. Ricard (Haun and Cornell 1951)] and azalea [*Rhododendron* L. cvs. Hexe, Gardenia Supreme, and Vervaeneana (Preston et al. 1953)] were grown at low, medium, and high levels of N, P, and K utilizing a factorial arrangement of treatments. The geranium study (Haun and Cornell 1951) showed that N had the greatest influence on root initiation. The low and medium levels of N resulted in higher rooting percentages than the high level. Overall, the N treatments had a greater influence on the rooting response of the cuttings than the P or K treatments, although a positive response for P and K was also noted for certain treatment combinations. Enhanced rooting percentage at low N levels was previously reported in another fertility study utilizing stock plants of grape [*Vitis vinifera* L. cv. Waltham Cross (Pearse 1943)]. The azalea study (Preston et al. 1953) reported results similar to the geranium study (Haun and Cornell 1951). Varied levels of P and K had no influence on rooting as reflected by rootball diameter. On the other hand, N had the greatest effect on rooting. However, the level producing the greatest response was related to the growth stage of the cutting at the time of excision.

The importance of N in root initiation is clearly supported in light of findings of mobilization and fertility studies, and the need for N in nucleic acid and protein synthesis. However, the promotive influence of N in root initiation may also be manifested by the manner in which it relates to carbohydrate content and metabolism (Hartmann and Kester 1983; also see Chapter 5 by Veierskov).

A discussion of the role of various mineral nutrients in root initiation would not be complete without mention of Zn and Mn, both of which influence endogenous auxin levels. The former being required for production (Tsui 1948, Salami and Kenefick 1970) of the auxin precursor tryptophan (Thimann 1935, Goodwin and Mercer 1983) and the latter acting as an activator of IAA oxidase which destroys native auxin (Thomaszewski and Thimann 1966). The relationship, however, between Zn, tryptophan, and IAA is not simple because some data suggest that Zn is required for the synthesis of IAA from tryptophan (Takaki and Kushizaki 1970).

Because tryptophan is a precursor of auxin, one would assume poor rooting to result from cuttings taken from Zn deficient stock plants. Unfortunately, critical Zn tissue levels that might influence rooting have yet to be determined. One study demonstrated that Zn fertilization of stock plants of grape (*Vitis* L. cv. Chasselas × Berlandieri 41 B) increased the percentage of rooted cuttings and improved the quality of the resulting plants which were scions of *Vitis* L. cv. Muscat-Hambourg initially grafted onto unrooted cuttings of *Vitis* L. cv. Chasselas × Berlandieri 41 B (Samish and Spiegel 1958). Increased rooting was also associated with increased levels of tryptophan.

Similar to the case of Zn there is only a single study which demonstrates the influence of Mn on root initiation. This investigation showed that high Mn levels were found in leaf tissue of cuttings taken from difficult-to-root cultivars of avocado (*Persea americana* Mill.) while an opposite relationship was noted for cuttings of easy-to-root cultivars (Reuveni and Raviv 1981).

Based on previous research it is apparent that with the exception of N, the importance of various mineral nutrients in root initiation has not been clearly defined. If one concurs with the limited findings that mineral nutrients are generally not mobilized during root initiation, one must conclude that within limits the nutritional status of cuttings or stock plants is more important in terms of root growth and development than root initiation. There are, however, some nutrients such as Ca which are considered immobile (Mengel and Kirkby 1982). Although a nutrient may not be mobilized into the base of a cutting during root initiation, one cannot conclude that it is unimportant. The absence of redistribution during root initiation may partially explain the reason that the role of particular mineral nutrients in root initiation is so uncertain.

Root Growth and Development

Unlike root initiation in which the importance of various mineral nutrients has not been clearly demonstrated, the situation with respect to root growth and development is much clearer, although many questions remain. Greater knowledge of this stage exists because studies undertaken to investigate the influence of mineral nutrition on adventitious rooting have by design resulted in data which provide a better understanding of the need of various mineral nutrients in root growth and development as contrasted to root initiation. As in the case of root initiation, studies providing information on root growth and development can be grouped into two basic kinds of experiments: mobilization studies and fertility studies. In addition, a third approach has been used consisting of treating stem cuttings with solutions of various mineral nutrients.

Mobilization studies have been particularly useful since once roots are initiated on the basal portion of a cutting the developing roots create sinks for various mineral nutrients. Movement of nutrients to the base of a cutting following root initiation is an indication that these nutrients are needed for root growth and development. Although nutrients are mobilized to the base of a cutting to support root growth and development, one study showed that a critical number of roots must be initiated before mobilization occurs (Blazich et al. 1983). The same experiment also showed that despite extensive root initiation in the base of a cutting, certain nutrients such as Ca and Mg are sometimes mobilized from the base to support growth in the upper portion of the cutting.

Despite evidence provided by fertility studies of the relative need for various nutrients during root growth and development, inconsistencies between individual fertility studies have emerged as well as cases where findings from fertility studies do not support mobilization studies. The only nutrient for which differences do not appear is N. The importance of N has been borne out by both fertility (Haun and Cornell 1951, Preston et al. 1953) and mobilization studies (Good and Tukey 1967, Blazich et al. 1983). The role of P, K, Ca, and Mg has been conflicting and in some cases inconclusive results have occurred.

Stimulation of root growth and development by P and K was demonstrated in a study involving stock plant fertilization of geranium [*Pelargonium* × *hortorum* Bailey cv. Ricard (Haun and Cornell 1951)], while a similar study utilizing stock plants of azalea [*Rhododendron* L. cvs. Hexe, Gardenia Supreme, and Vervaeneana (Preston et al. 1953)] indicated that varying levels of P and K had no significance. Mobilization of P into the bases of cuttings of chrysanthemum [*Chrysanthemum morifolium* Ramat. cv. Indianapolis White (Good and Tukey 1967)] and Japanese holly [*Ilex crenata* Thunb. cv. Convexa (Blazich et al. 1983)] has been reported. The former investigation noted no movement of K while the latter did.

Fertility and mobilization studies have generally not demonstrated the need for Ca, undoubtedly due to the immobile nature of this nutrient (Mengel and Kirkby 1982). Another factor which may explain why mobilization studies have failed to show a need for Ca during root growth and development is that Ca is transported poorly in the phloem (Rios and Pearson 1964, Marschner and Richter 1974). However, importance of this nutrient in promoting root growth and development is without question whether one is dealing with nonadventitious or adventitious roots. Interruption of the Ca supply to nonadventitious roots immediately results in reduced growth followed by browning and subsequent death of root tips (Mengel and Kirkby 1982). The inability to sustain growth of adventitious roots of cuttings of pea (*Pisum sativum* L. cv. Weibull's Marma) in the absence of Ca was also reported (Eliasson 1978). The fact that Ca is immobile and has not been detected during mobilization experiments studying adventitious rooting but is obviously of critical importance to root growth and development, points to a major weakness of attempting to rely solely on mobilization experiments to study the relationship between mineral nutrition and root growth and development.

Although considered immobile, redistribution of Ca was reported during rooting of cuttings of Japanese holly [*Ilex crenata* Thunb. cv. Convexa (Blazich et al. 1983)]. Calcium was not mobilized to support root growth and development, but was apparently redistributed to support tissue development on the upper portions of the cuttings.

Studies investigating the need for Mg during root growth and development are extremely limited. Based on findings that Mg deficiency inhibits protein synthesis (Mengel and Kirkby 1982), one must suspect its importance for rooting. Surprisingly, a mobilization study failed to demonstrate the need for Mg although the same experiment showed a redistribution of Mg to support shoot development on the upper stem (Blazich et al. 1983).

Boron is often mentioned in discussions involving adventitious rooting. An early investigation reported that it stimulated rooting of cuttings by promoting root growth and development rather than initiation (Hemberg 1951). A later study substantiated these earlier findings and concluded that in cuttings of mung bean [*Vigna radiata* (L.) R. Wilcz., syn. *Phaseolus aureus* Roxb.)] rooting was initiated by auxin but growth from early stages of development ("pre-primordial") was dependent on the presence of B (Middleton et al. 1978). The obligatory role of B in root growth and development appears to be certain but the manner by which it affects metabolic processes influencing rooting (Haissig 1986) as well as at the whole plant level (Lewis 1980) remains speculative. Research has suggested that B influences rooting by regulating endogenous auxin levels through enhancement of IAA oxidase activity. (Jarvis et al. 1983, 1984). It has also been demonstrated that Ca can enhance the rooting response of B treated cuttings (Jarvis and Yasmin 1985).

Conflicting outcomes of fertility and mobilization studies regarding the need for various mineral nutrients demonstrate that further study is needed. Inconsistencies associated with fertility studies suggest that the relative proportion of one mineral nutrient to another may explain why one study has shown a stimulatory effect attributable to a particular nutrient while another study has not (Haun and Cornell 1951).

Root Inducing Factors and Other Considerations

Mineral nutrition as it directly influences root initiation, and root growth and development has been considered, but there are other influences which are worthy of mention. For example, it has long been known that the presence of buds and leaves on cuttings can influence rooting (Hartmann and Kester 1983). Various root inducing factors (rooting cofactors or auxin synergists) are produced in buds and leaves which interact with auxin to promote rooting (Hartmann and Kester 1983). Logic suggests that the nutritional status of the stock plant or cutting influences the production of such substances but no research has been reported to confirm or refute these relationships. The same is true of relationships between mineral nutrition, carbohydrate production, and rooting.

A practical consideration which deserves study is the influence of mineral nutrition on post-propagation growth and vigor which has unfortunately not been investigated to any great extent. A fertility study which considered survival following rooting of cuttings of azalea (*Rhododendron* L.) found that stock plant fertility influenced post-rooting survival (Preston et al. 1953). This phenomenon was affected by stock plant N levels, while various levels of P and K had no effect. Data indicated that N levels affecting survival were influenced by the growth stage of the cuttings when taken for rooting.

LEACHING OF NUTRIENTS DURING MIST PROPAGATION

Development of mist propagation during the 1940s, and subsequent modification and widespread adoption in the next two decades for rooting cuttings, revolutionized the art and science of plant propagation (Synder 1965). This innovation made it possible to root cuttings of various species which had previously been virtually impossible. The promotive influence of mist results from maintaining turgidity during the rooting period. Prior to the development of mist, desiccation of cuttings during rooting was a major problem which misting helped alleviate.

As techniques for using mist were modified and refined, researchers initiated experiments to study the physiological status of cuttings during mist propagation. These investigations, which continue at present, were also aimed at elucidating how mist actually promoted rooting. Early studies suggested that mineral nutrients were leached from cuttings during mist propagation (Evans 1951, Sharpe 1955). These findings were later confirmed by more detailed studies which demonstrated that mineral nutrients such as N, P, K, Ca, and Mg are leached from cuttings while under mist (Good and Tukey 1966, Blazich et al. 1983). The extent of leaching may be dependent on the growth stage of the cutting material with leafy hardwood cuttings apparently more susceptible than softwood or herbaceous cuttings (Good and Tukey 1966). Greater leaching from hardwood cuttings has been attributed to an increased proportion of the nutrients being in an exchangeable form, whereas in young, growing tissues nutrients are quickly metabolized within cells and cell walls which are difficult to leach (Good and Tukey 1966). In addition to tissue maturity, there are also other factors which influence this phenomenon such as relative leachability of a particular nutrient (Tukey et al. 1958). Experiments classifying the leaching of nutrients as easily, moderately, and leached with difficulty resulted in the following (Tukey et al. 1958): Na, Mn (easily); Ca, Mg, S, K (moderately); Fe, Zn, P, Cl (with difficulty).

Leaching of mineral nutrients during mist propagation partially explains why leaves on cuttings often show signs of mineral nutrient deficiency. These symptoms intensify as root growth and development proceed because nutrients are mobilized from the older tissues including leaves to support the newly developing roots and, if occurring, new shoot development on the upper stem (Good and Tukey 1966, Blazich et al. 1983). Thus, leaching and mineral nutrient mobilization contribute to foliar nutrient deficiency symptoms.

When studying the mineral nutrient status of cuttings during mist propagation, caution must be exercised as to experimental protocol and expression of the nutrient data. Cuttings must be rooted under conditions such that there is no opportunity to absorb any additional nutrients. This restraint immediately rules out use of tap water or a rooting medium containing any mineral nutrients. If either is used nutrients lost by leaching could not be detected due to reabsorption from mist or the rooting medium. Another important consideration involves expression of data. Because cuttings often show an increase in dry weight during mist propagation, expressing nutrient content as percent dry weight can lead to false conclusions (Good and Tukey 1966, 1967, Blazich and Wright 1979, Blazich et al. 1983). Such erroneous conclusions can arise because an increase in dry weight can result in dilution of nutrients in the cuttings. Hence, even if no nutrients are lost by leaching, data expressed as percent dry weight will make it appear that nutrients were lost. To avoid misleading results of expressing nutrient data as percent dry weight, one must express nutrient content in terms of total per cutting or percent of total if measuring the mineral nutrient content of various tissues such as leaves.

Loss of mineral nutrients during mist propagation is often cited as a problem. This conclusion has lead to research on the use of nutrient mist to replenish mineral nutrients lost by leaching. Although the concept of foliar nutrition appears in principle to have merit in overcoming so called adverse effects of nutrient leaching, results with nutrient mist have been mixed. Rooting cuttings of various herbaceous, softwood, and hardwood species under nutrient mist resulted in increased root quality and greater top growth than propagation under a water mist (Wott and Tukey 1967). On the other hand, nutrient mist propagation had a deleterious effect on cuttings of several azalea cultivars (*Rhododendron* L. cvs. Gloria, Prize, Solitaire, Whitewater, Dorothy Gish, Skyliner, and Kingfisher) resulting in injured foliage and inhibition of rooting (Keever and Tukey 1979). Nutrient mist also caused sanitation problems by encouraging growth of algae (Wott and Tukey 1967, Coorts and Sorensen 1968).

As an alternative to nutrient mist, controlled release fertilizers either top-dressed or incorporated into the rooting medium have been used during mist propagation. Despite enhancement of overall root quality and subsequent cutting growth (Johnson and Hamilton 1977, Ward and Whitcomb 1979) percent rooting was unaffected (Johnson and Hamilton 1977). The inability to stimulate percent rooting either with slow-release (Johnson and Hamilton 1977) or water soluble (Booze-Daniels et al. 1984) fertilizer suggests that supplemental nutrition during rooting has little or no promotive influence on root initiation. These findings may also be related to the inability of unrooted cuttings to absorb nutrients. A study with cuttings of chrysanthemum (*Chrysanthemum morifolium* Ramat. cv. Giant No. 4 Indianapolis White) demonstrated that only small quantities of P were absorbed from the rooting medium even though P levels were high (Wott and Tukey 1969). This was in contrast to absorption of large quantities by the foliage applied through nutrient mist. Inconclusive findings as to the role of supplemental nutrition in stimulating root initiation reinforces the importance of initial nutrient levels when taking cuttings.

Although supplemental nutrition during mist propagation has been the main approach to counter nutrient leaching, one study attempted to prevent leaching by treating cuttings with an antitranspirant (Baggott and Joiner 1974). Results, however, were totally unsatisfactory.

Leaching of mineral nutrients during mist propagation is an established fact. The extent to which it prevents or retards rooting and affects post propagation growth and vigor has not been clearly established. There is no doubt that if leaching is excessive, resulting from the growth stage of the cutting (Good and Tukey 1966) or other factors (Tukey et al. 1958), it may be critical. However, a more overriding consideration may involve the nutritional status of the stock plants such that if cuttings are taken from stock plants growing under poor or marginal fertility, leaching may influence rooting. On the other hand, any deleterious influence of leaching may be offset by cuttings taken from stock plants growing under optimum fertility levels. This conclusion poses the question of what are optimum levels of fertility for rooting, which have yet to be determined.

Despite all the criticisms directed at mist in connection with mineral nutrient leaching during propagation, it is still one of the most powerful tools ever developed for rooting cuttings.

CONCLUSION

Despite the recognized significance of the relationship between mineral nutrition and adventitious rooting, the importance of various mineral nutrients in the process is not clearly understood. Understanding has been slow due to the nature of rooting which actually consists of several stages but which can be reduced to two consisting of: 1) root initiation; and 2) root growth and development. Therefore, when considering the influence of a particular mineral nutrient on adventitious rooting, one must consider the role or importance of that nutrient in each stage of the process.

One could argue that any nutrient involved in the multitude of metabolic processes associated with dedifferentiation and root meristem formation is essential for root initiation. Such a line of reasoning immediately points to the need for N based on its role in nucleic acid and protein synthesis. In addition, direct and indirect experimental evidence suggests that P, Ca, Mn, and Zn are also important for root initiation. The situation in root growth and development is clearer thanks to studies specifically directed at this stage of the rooting process which are easier to design and conduct. Although some conflicting reports have emerged, sufficient experimental evidence has accumulated to confirm the need for N, P, K, Ca, and B during root growth and development. In addition to directly influencing root initiation, and root growth and development, nutrition may also be of significance by influencing the production of various root inducing factors and affecting post propagation growth and vigor. This statement coupled with the need to determine optimum tissue nutrient levels for both root initiation and subsequent growth and development, point to potential areas for research which to the present writing have been virtually unexplored.

There are limited data to demonstrate mobilization of mineral nutrients during root initiation which is quite different from mobilization during root growth and development. This suggests that within limits the nutritional status of a stock plant or a cutting taken from that plant has a greater impact on root growth and development than root initiation. This observation also strongly suggests that root initiation as influenced by mineral nutrition is highly dependent on initial nutrient levels within that portion of the cutting where roots are to form.

Leaching of N, P, K, Ca, and Mg during mist propagation is an established fact and has lead to various approaches to replenish such losses. Results have generally been mixed. When a promotive influence has been observed the response has consisted primarily of enhancement of overall root quality and subsequent cutting growth. The results demonstrate that supplemental nutrition has little or no positive influence on root initiation. Despite leaching of mineral nutrients from cuttings during mist propagation, the extent to which it prevents or retards rooting and influences post propagation growth and vigor are unknown and deserving of study.

REFERENCES

Baggott, A. J., Jr. and J. N. Joiner. 1974. Effects of shade, mist and antitranspirant on rooting and nutrient leaching of *Ligustrum japonicum* and *Chrysanthemum morifolium* cuttings. *Proc. Florida State Hortic. Soc.* 87:474–477.

Blazich, F. A. and R. D. Wright. 1979. Non-mobilization of nutrients during rooting of *Ilex crenata* Thunb. cv. Convexa stem cuttings. *HortScience* 14:242.

68   Frank A. Blazich

_____ _____ H. E. Schaffer. 1983. Mineral nutrient status of 'Convexa' holly cuttings during intermittent mist propagation as influenced by exogenous auxin application. *J. Amer. Soc. Hortic. Sci.* 108:425–429.

Booth, A., J. Moorby, C. R. Davies, H. Jones and P. F. Wareing. 1962. Effects of indolyl-3-acetic acid on the movement of nutrients within plants. *Nature* 194:204–205.

Booze-Daniels, J. N., R. D. Wright and R. E. Lyons. 1984. Effect of timed fertilizer application on *Ilex crenata* 'Helleri' cuttings during propagation. *J. Environ. Hortic.* 2:43–45.

Burström, H. G. 1968. Calcium and plant growth. *Biol. Rev.* 43:287–316.

Coorts, G. D. and C. C. Sorenson. 1968. Organisms found growing under nutrient mist propagation. *HortScience* 3:189–190.

Eliasson, L. 1978. Effects of nutrients and light on growth and root formation in *Pisum sativum* cuttings. *Physiol. Plant* 43:13–18.

Evans, H. 1951. Investigations on the propagation of cacao. *Tropic. Agric.* 28:147–203.

Gautheret, R. J. 1969. Investigations on the root formation in the tissues of *Helianthus tuberosus* cultured *in vitro*. *Amer. J. Bot.* 56:702–717.

Good, G. L. and H. B. Tukey, Jr. 1966. Leaching of metabolites from cuttings propagated under intermittent mist. *Proc. Amer. Soc. Hortic. Sci.* 89:727–733.

_____ _____ 1967. Redistribution of mineral nutrients in *Chrysanthemum morifolium* during propagation. *Proc. Amer. Soc. Hortic. Sci.* 90:384–388.

Goodwin, T. W. and E. I. Mercer. 1983. *Introduction to Plant Biochemistry*. Pergamon Press, NY, USA. 2nd ed. pp. 574–576. ISBN 0-08-024921-3.

Haissig, B. E. 1972. Meristematic activity during adventitious root primordium development. Influences of endogenous auxin and applied gibberellic acid. *Plant Physiol.* 49:886–892.

_____ 1986. Metabolic processes in adventitious rooting of cuttings. In *New Root Formation in Plants and Cuttings* (M. B. Jackson, ed). Martinus Nijhoff Publishers, Dordrecht/Boston/Lancaster. pp. 150–152. ISBN 90-247-3260-3.

Hartmann, H. T. and D. E. Kester. 1983. *Plant Propagation: Principles and Practices*. Prentice-Hall, Englewood Cliffs, NJ, USA. 4th ed. pp. 234–297. ISBN 0-13-681007-1.

Haun, J. R. and P. W. Cornell. 1951. Rooting response of geranium (*Pelargonium hortorum,* Bailey var. Ricard) cuttings as influenced by nitrogen, phosphorus, and potassium nutrition of the stock plant. *Proc. Amer. Soc. Hortic. Sci.* 58:317–323.

Hemberg, T. 1951. Rooting experiments with hypocotyls of *Phaseolus vulgaris* L. *Physiol. Plant.* 11:1–9.

Jarvis, B. C., A. H. N. Ali and A. I. Shaheed. 1983. Auxin and boron in relation to the rooting response and ageing of mung bean cuttings. *New Phytol.* 95:509–518.

_____ S. Yasmin. 1985. The influence of calcium on adventitious root development in mung bean cuttings. *Biochem. Physiol. Pflanzen* 180:697–701.

_____ _____ , A. H. N. Ali and R. Hunt. 1984. The interaction between auxin and boron in adventitious root development. *New Phytol.* 97:197–204.

Johnson, C. R. and D. F. Hamilton. 1977. Effects of media and controlled-release fertilizers on rooting and leaf nutrient composition of *Juniperus conferta* and *Ligustrum japonicum* cuttings. *J. Amer. Soc. Hortic. Sci.* 102:320–322.

Keever, J. G. and H. B. Tukey, Jr. 1979. Effect of intermittent nutrient mist on the propagation of azaleas. *HortScience* 14:755–756.

Lewis, D. H. 1980. Boron, lignification and the origin of vascular plants—a unified hypothesis. *New Phytol.* 84:209–229.

Marschner, von H. and Ch. Richter. 1974. Calcium translocation in roots of maize and bean seedlings. *Plant and Soil* 40:193–210.

Mengel, K. and E. A. Kirkby. 1982. *Principles of Plant Nutrition*. Int. Potash Inst., Bern, Switzerland. 3rd ed. pp. 444–450.

Middleton, W., B. C. Jarvis and A. Booth. 1978. The boron requirement for root development in stem cuttings of *Phaseolus aureus* Roxb. *New Phytol.* 81:287–297.

Pearse, H. L. 1943. The effect of nutrition and phytohormones on the rooting of vine cuttings. *Ann.*

Bot. 7:123–132.

Preston, W. H., Jr., J. B. Shanks and P. W. Cornell. 1953. Influence of mineral nutrition on production, rooting and survival of cuttings of azaleas. *Proc. Amer. Soc. Hortic. Sci.* 61:499–507.

Reuveni, O. and M. Raviv. 1981. Importance of leaf retention to rooting of avocado cuttings. *J. Amer. Soc. Hortic. Sci.* 106:127–130.

Rios, M. A. and R. W. Pearson. 1964. The effect of some chemical environmental factors on cotton root behavior. *Soil Sci. Soc. Amer. Proc.* 28:232–235.

Salami, A. U. and D. G. Kenefick. 1970. Stimulation of growth in zinc-deficient corn seedlings by the addition of tryptophan. *Crop Sci.* 10:291–294.

Samish, R. M. and P. Spiegel. 1958. The influence of nutrition of the mother vine on the rooting of cuttings. *Ktavim* (Records of the Agric. Res. Sta., State of Israel) 8:93–100.

Sharpe, R. H. 1955. Mist propagation studies with emphasis on mineral content of foliage. *Proc. Florida State Hortic. Soc.* 68:345–347.

Strydom, D. K. and H. T. Hartmann. 1960. Effect of indolebutyric acid on respiration and nitrogen metabolism in Marianna 2624 plum softwood stem cuttings. *Proc. Amer. Soc. Hortic. Sci.* 76:124–133.

Stuart, N. W. 1938. Nitrogen and carbohydrate metabolism of kidney bean cuttings as affected by treatment with indoleacetic acid. *Bot. Gaz.* 100:298–311.

Synder (sic, Snyder), W. E. 1965. A history of mist propagation. *Proc. Int. Plant Prop. Soc.* 15:63–67.

Takaki, H. and M. Kushizaki. 1970. Accumulation of free tryptophan and tryptamine in zinc deficient maize seedlings. *Plant Cell Physiol.* 11:793–804.

Thimann, K. V. 1935. On the plant growth hormone produced by *Rhizopus sinuis. J. Biol. Chem.* 109:279–291.

Thomaszewski, M. and K. V. Thimann. 1966. Interactions of phenolic acids, metallic ions, and chelating agents on auxin-induced growth. *Plant Physiol.* 41:1443–1454.

Tsui, C. 1948. The role of zinc in auxin synthesis in the tomato plant. *Amer. J. Bot.* 35:172–179.

Tukey, H. B., Jr., H. B. Tukey and S. H. Wittwer. 1958. Loss of nutrients by foliar leaching as determined by radioisotopes. *Proc. Amer. Soc. Hortic. Sci.* 71:496–506.

Ward, J. D. and C. E. Whitcomb. 1979. Nutrition of Japanese holly during propagation and production. *J. Amer. Soc. Hortic. Sci.* 104:523–526.

Wott, J. A. and H. B. Tukey, Jr. 1969. Absorption of phosphorus by *Chrysanthemum morifolium* cuttings propagated under nutrient mist. *J. Amer. Soc. Hortic. Sci.* 94:382–384.

_____ _____ 1967. Influence of nutrient mist on the propagation of cuttings. *Proc. Amer. Soc. Hortic. Sci.* 90:454–461.

CHAPTER 5

Relations Between Carbohydrates and Adventitious Root Formation

Bjarke Veierskov

Department of Plant Physiology
Royal Veterinary and Agricultural University
Thorvaldsenvej 40, 1871 Frederiksberg C
Denmark

INTRODUCTION: 70 YEARS OF CONTROVERSY..............................70
STOCK PLANT CARBOHYDRATE STATUS 71
CARBON/NITROGEN RATIO 73
CARBOHYDRATE ACCUMULATION DURING ROOTING........................73
EFFECT OF EXOGENOUSLY SUPPLIED CARBOHYDRATES.....................74
POSSIBLE ROLES OF CARBOHYDRATES IN ROOTING 75
CONCLUSION 76
REFERENCES...76

Additional key words: carbon/nitrogen ratio, energy, phosphorylated carbohydrates.
Abbreviation: C/N, carbon/nitrogen ratio; DCMU, N'-(3,4-dichlorophenyl)N,N-dimethylurea.

INTRODUCTION: 70 YEARS OF CONTROVERSY

The relationship between carbohydrates and adventitious root formation has remained controversial for 70 years. Some of the earliest work dealing with the influence of carbohydrates on rooting of cuttings was performed by Kraus and Kraybill (1918), and later by Bouillene and Went (1933) and Went and Thimann (1937). Kraus and Kraybill (1918) reported that stem segments from tomato plants with a high C/N ratio formed more adventitious roots than segments with a low C/N ratio. Their observations led to the formulation of a hypothesis that high C/N ratios in cuttings are conducive to the formation of adventitious roots whereas low ratios reduce rooting capacity. This hypothesis was fairly widely accepted and the rooting ability of cuttings has since been frequently discussed in relation to carbohydrate content. Surprisingly, however, few studies have critically evaluated the role of carbohydrates in rooting and the relationship between carbohydrates and rooting remains somewhat obscure.

Several general approaches have been used to investigate the influence of carbohydrates on the rooting of cuttings: 1) use of exogenously supplied carbohydrates to stock plants or cuttings; 2) determination of endogenous carbohydrate content of stock plants or cuttings in relation to rooting; 3) alteration of the environment under which stock plants or cuttings are maintained in order to alter endogenous carbohydrate content; and 4) study of the influence of carbohydrates in cultivars which differ in their rooting ability.

Each of these experimental approaches has limitations. The carbohydrate pools in plant tissue consist of an array of compounds, of which supplying only one or a few in relatively high concentrations could easily cause unwanted gross perturbations in metabolism. Many structural forms of carbohydrates exist, and endogenous determination is usually only performed on a fraction of these forms without consideration of what role individual carbohydrates may play in rooting compared to other unmeasured compounds. Also carbohydrate determinations in cuttings have largely been made in stem bases (i.e. organ-level measurements) although rooting involves only a few obscure primordial cells. Such analytical techniques may completely mask carbohydrate status in the primordial cells. When attempting to alter the endogenous carbohydrate content via environmental manipulation, stock plants or cuttings are often altered in ways that affect many physiological parameters in addition to carbohydrate content (e.g. auxin, water status, state of differentiation). The use of easy- and difficult-to-root cultivars to study the role of carbohydrates in rooting assumes that most of the genotypic variation between the cultivars is related to rooting. This is true if most of the cultivars' genomes are homologous, an assumption which has often been made without validation (see Chapter 3 by Haissig and Riemenschneider).

Another general limitation of past studies dealing with carbohydrates and rooting has been that analyses have not been keyed with the anatomical events linked to the various stages of rooting. It is possible that carbohydrates play different roles in the different stages which may confound our understanding of the process.

When discussing the relationship between carbohydrates and rooting one should distinguish between cuttings which can and cannot photosynthesize (i.e. leafy vs. non-leafy cuttings). It is possible that the carbohydrate requirement for rooting and the role of carbohydrates may differ between these two general types of cuttings. These distinctions have often been ignored in the past.

Working hypotheses are generally altered together with enlarged biochemical understanding. It is thus important to reevaluate the literature related to carbohydrates and rooting on the basis of our present understanding of metabolic processes. The aim of this chapter is to summarize the main conclusions of past work and thereby elucidate possible roles of carbohydrates in the rooting process.

STOCK PLANT CARBOHYDRATE STATUS

It is well documented that stock plant growth conditions are of utmost importance for the rooting of cuttings (reviewed by Andersen 1986; also see Chapter 16 by Moe and Andersen). Carbohydrate content is but one parameter which reflects stock plant developmental history and may thereby show a coincidental correlation to the rooting ability of cuttings without having any regulatory role in rooting. Although carbohydrate status of the stock plant may influence rooting, high carbohydrate content has not always been associated with high rooting potential. For example, when light was used to alter stock plant carbohydrate content, increased stock plant irradiance caused an increased content of carbohydrates but diminished root number, i.e. a negative correlation between carbohydrates and rooting was observed (Hansen et al. 1978, Veierskov et al. 1982a). Similarly by using pea seedlings grown from seeds harvested in different years it was possible to obtain stock plants varying in total carbohydrate content and a high negative correlation (−0.95) was found between initial content of total carbohydrates and number of roots produced (Veierskov et al. 1982a). Nanda and Anand (1970) and Okoro and Grace (1976) found that the failure of Populus tremula L. to root was not caused by insufficient carbohydrate reserves; it was also shown that starch content was not related to rooting.

In other studies, carbohydrate content has been positively correlated with rooting. For instance, by using temperature sensitive photosynthetic pea mutants it was possible to alter the carbohydrate content in the stock plants without changing the light intensity (Veierskov et al. 1982b). Photosynthetic electron transport in these mutants was blocked if plants were grown at a temperature higher than 25 °C. Data in Fig. 1 show a positive correlation between carbohydrate content and root number in cuttings from different mutants of pea plants. The slope of the correlation curves were similar, whereas the number of roots formed at a given content of carbohydrate depended on the variety used. This demonstrates that the level of rooting is determined by factors other than total carbohydrate content. However, within the level of carbohydrate found in a given plant a correlation

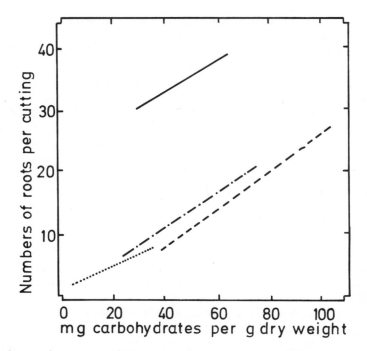

Figure 1. Relations between carbohydrate content and root number in four varieties of pea cuttings after 13 d rooting at 38 W · m⁻². The carbohydrate content of the cuttings was controlled by stock plant temperature. The following pea plants were used: (———) Dark Skinned Perfection, (-------) wild type from chlorina-5657, (· · · · ·) chlorina-5657, (-·-·-·-) chlorotica-887. [For further details on growing conditions see Veierskov et al. (1982b)]

may be observed between content of total carbohydrate and root number. This may explain some of the discrepancies observed with regard to a correlation between total carbohydrate content and number of roots formed. It also should be noted that the correlation was only observed if cuttings were rooted under 38 W·m⁻² and not at a lower level of irradiance (16 W·m⁻²) (Veierskov et al. 1982b).

It has been documented that the initial carbohydrate content must be sufficient to supply the cutting with energy reserves throughout the rooting period for optimum rooting to occur under photosynthesis-limited conditions. This has been demonstrated by preventing net CO_2 fixation during the rooting period by using low irradiance during propagation, use of defoliated cuttings, or the use of genetic mutants (Howard and Sykes 1966, Nanda and Jain 1972, Breen and Muraoka 1974, Loach and Whalley 1978, Veierskov et al. 1982ab).

That suboptimal rooting may be observed because the initial carbohydrate content is too low has also been found in work with leafless hardwood cuttings (Vieitez et al. 1980) and etiolated stock plants (e.g. Nanda et al. 1971, Pal and Nanda 1981). Hence, the carbohydrate content of the stock plant apparently has some effect on the rooting ability of cuttings, at least under some conditions. Unfortunately, the optimum carbohydrate content for rooting has not been defined under any conditions. This is partly because altering the carbohydrate status of stock plants is very difficult without interactions with other important developmental parameters. For instance, altering the level of irradiance to stock plants grown in growth chambers causes an increase in leaf temperature, and altered auxin transport and accumulation in the subsequent cuttings (Baadsmand and Andersen 1984). Therefore, when trying to manipulate the carbohydrate content in stock plants one may easily cause alterations in other parameters of importance to rooting.

Although stock plant carbohydrate content and rooting may sometimes be positively correlated, one should be cautious about concluding that carbohydrates have a regulatory role in rooting. A positive correlation between carbohydrate content and rooting may, for example, only reveal that the supply of current photosynthate is insufficient to support optimal rooting.

CARBON/NITROGEN RATIO

According to the C/N ratio hypothesis of Kraus and Kraybill (1918), the ratio between carbohydrate and nitrogen contents should be related to the rooting ability of cuttings. In the experiments performed by Kraus and Kraybill (1918), stem segments containing varying C/N ratios were obtained by controlling nitrogen fertilization. Stems with high C/N ratios (those with low N fertilization) rooted much better than those with low C/N ratios. That stock plants with low C/N ratios produce cuttings with low rooting potential has also been shown by Reid (1924) and Knoblauch (1979). It is important to understand, however, that these plants likely varied in many biochemical and physiological aspects in addition to C/N ratio because applied N stimulates vegetative growth.

In pea stock plants originating from different seed lots, Veierskov et al. (1982a) obtained cuttings where the C/N ratio varied from 0.9–1.6, but no significant correlation to root number was observed. Because total nitrogen includes inorganic and organic nitrogen with different turnover rates, total nitrogen might not be related to the rooting process, but rather reflect the stock plant's general nutritional status and vigor. Hence, C/N ratio is not likely to always be an adequate indicator of rooting ability.

When the C/N ratio is altered within a plant due to nitrogen fertilization, the carbohydrate pool is also modified. If nitrogen is limiting, the photosynthetically fixed CO_2 will first go into the carbohydrate pool, but this cannot be metabolized further because most organic compounds contain nitrogen. In this case, the plant transports the carbohydrates as sucrose to long term storage locations such as the stem. The triosephosphates formed in photosynthesis may either be converted into starch or sucrose. Recent experiments have determined that fructose 2,6-bisphosphate is an important sugar phosphate which regulates enzymes that influence carbohydrate formation and pools (Stitt et al. 1987). It is of interest to note that the precursor for sucrose formation is fructose-6-phosphate, whereas starch formation depends on glucose-1-phosphate, and that sugar phosphates may be involved in the initiation process. Also, fructose 2,6-bisphosphate regulates catabolic activities between the pentose phosphate and glycolytic-citric acid pathways, a change which also may be related to root initiation (Haissig 1982b).

Measurements in cuttings have shown that whereas the content of amino acids changed little during rooting, a large accumulation of carbohydrates occurred (Hansen et al. 1978, Davis and Potter 1981, Haissig 1982a, Veierskov and Andersen 1982). Cuttings are normally unable to increase the content of nitrogen after excision, at least until new roots are formed, but they may alter the different pool levels (e.g. inorganic nitrogen, amino acids, proteins). Hence, rooting may be related to specific pools rather than to total contents (Haissig 1986). Determinations in rooted cuttings revealed that proline was the most abundant amino acid in newly rooted cuttings whereas in intact plants asparagine was predominant (Suzuki 1982). Haissig (1974) also suggested that rooting ability was reflected in the amino acid composition in such a way that poor rooters contained a relatively high content of arginine, histidine, and especially α-aminobutyric acid. Determinations of amino acid contents in pine (*Pinus* spp.) cuttings revealed that only minor changes were observed in the base of the cuttings (Haissig 1982a) whereas an increased translocation of amino acids was noted from the needles to the upper stem.

CARBOHYDRATE ACCUMULATION DURING ROOTING

A decrease in the carbohydrate content in cuttings may sometimes be observed during the first few days of the rooting period (Haissig 1982a, Veierskov et al. 1982a). This decrease may be due to closure of stomata (Gay and Loach 1977, Orton 1979) which diminishes the cutting's ability to fix CO_2. However, a net accumulation of carbohydrates normally occurs until roots emerge from leafy cuttings propagated in the light (Fig. 2; Lovell et al. 1974, Hansen et al. 1978, Davis and Potter 1981, Veierskov et al. 1982a, Spellenberg 1985). The changes in carbohydrate content during rooting should be discussed in relation to the various phases of rooting. The first phase in rooting is a lag phase which may vary considerably in length (Eriksen 1973, White and Lovell 1984). The lag phase is

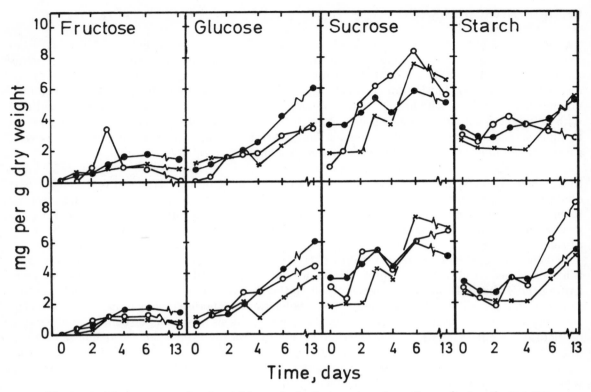

Figure 2. The content of extractable carbohydrates at various times during the rooting period in the three parts of pea cuttings: (●) top, (○) middle, (×) base section. The stock plants were grown at 16 W·m⁻² (upper figures) or 38 W·m⁻² (lower figures). [For further description of growing conditions see Veierskov et al. (1982b)]

followed by the initiation phase after which the time to root emergence appears to be relatively constant (White and Lovell 1984). In pea cuttings, a decrease in carbohydrate content occurred during the lag phase (Eriksen 1973, Veierskov et al. 1982a) and may, therefore, affect the subsequent initiation process.

Rapid accumulation of carbohydrates has been observed in cuttings if auxin or increased rooting temperature (from 15–25 °C) were used to hasten rooting (Altman and Wareing 1975, Haissig 1982a, Veierskov and Andersen 1982). If the speed of rooting was hastened by controlling stock plant irradiance, however, delayed carbohydrate accumulation and rooting was observed (Hansen et al. 1978). This apparent discrepancy might be explained by differences in experimental protocol. In the work of Hansen et al. (1978) carbohydrates were analyzed in the whole cutting whereas Haissig (1982a) and Veierskov and Andersen (1982) performed their measurements on various parts of the cuttings. Accumulation of sugars and starch in cuttings begins in the leaves, at a later point in the stem, and lastly in the base of the cutting (Haissig 1982a, Veierskov et al. 1982).

EFFECT OF EXOGENOUSLY SUPPLIED CARBOHYDRATES

A number of investigators have attempted to study the role of carbohydrates in rooting by observing responses of cuttings to exogenously supplied carbohydrates (Oliemann-van der Meer et al. 1970, Lovell et al. 1972, Nanda and Jain 1972, Lovell et al. 1974, Nanda et al. 1974, Eliasson 1978, Loach and Whalley 1978). Unfortunately there have been several problems associated with this experimental approach. First, cuttings take up and translocate exogenous sugars from solutions primarily via the xylem, unlike endogenous carbohydrates which are primarily transported in the phloem (Haissig 1986). Second, applied sugars can exert strong osmotic effects apart from their role as a carbon source. Third, sugars have often been applied to cuttings at concentrations that greatly exceed physiological concentrations and, in some cases, maybe even at toxic levels. Hence the assumption that applied carbohydrates act in the same manner as endogenous ones is likely often to be invalid.

That endogenous carbohydrate content may be a limiting factor during the rooting process was shown by Eliasson (1978) who doubled the root number in pea cuttings by supplying a 1% sucrose solution to the rooting medium. Loach and Whalley (1978) increased rooting percentage by 33% by supplying a 2% sucrose solution to cuttings of various woody species, but the effect was observed only if the cuttings were rooted under a low level of irradiance. This is in accordance with the findings that exogenous carbohydrates could reverse inhibition of rooting caused by DCMU, a photosynthesis inhibitor, in *Raphanus sativus* L. cuttings (Lovell et al. 1972).

Relatively high concentrations of carbohydrates supplied in the rooting medium generally have suppressed rooting. This effect has been more pronounced on primordia development than on primordia initiation (Lovell et al. 1973). The concentration used when supplying sugars has generally been between 0.5 and 2% (w:v) depending on type of sugar used (Lovell et al. 1972, Moore et al. 1972, Nanda and Jain 1972). If sucrose is supplied at 2% (w:v), the concentration is about 60 mmol $\cdot$ l^{-1} and the osmotic potential is about -0.15 MPa. Estimates of carbohydrate uptake by cuttings treated with such a sugar solution indicated that the uptake may well cause non-physiological levels in the plants (Howard and Sykes 1966, Eliasson 1978, Veierskov et al. 1982a). Because the transport of exogenously supplied carbohydrates occurs in the xylem, a strong influence on the water balance in the cuttings might also occur.

POSSIBLE ROLES OF CARBOHYDRATES IN ROOTING

Do carbohydrates then have a role in controlling root formation in cuttings? One function of course is to serve as an energy source and to deliver the needed carbon skeletons for new metabolic products (Van't Hoff 1968). The amount of carbohydrate required to fulfill this function has not been critically defined. Although the supply of carbohydrates during the rooting period may regulate the number of roots to be supported and their subsequent growth, this is not necessarily a controlling mechanism.

Are carbohydrates able to take part in the dedifferentiation or initiation processes? It has been found that reducing sugars such as glucose-6-phosphate or glucose are able to glycosylate DNA and thereby alter transcription. Reducing sugars, especially the phosphorylated ones, are known to react non-enzymatically with nuclear proteins, and cause modification in about 10% of the protein (Bojanovic et al. 1970, Monnier and Cerami 1983). This non-enzymatic browning of proteins (the Maillard process) causes the formation of highly reactive products which may be mutagenic (Aeschbacher et al. 1981). The products formed by the reaction of proteins or nucleic acids with reducing sugars has been suggested to be the cause of tumorigenesis (Bucala et al. 1984). Although these findings were in procaryotic and mammalian cells, they might also occur in plant tissue. Reducing sugars do not start to accumulate in the base of most cuttings until several days after excision from the stock plant (Haissig 1982a, Veierskov 1982b) and in the pea cutting system this is after root initiation (Eriksen 1973). Because phosphorylated sugars most readily glycosylate proteins and nucleic acids (Stevens et al. 1977), these may be candidates for regulating root initiation. The content and metabolism of phosphorylated carbohydrates has not, however, been determined in cuttings. Studies on excised tobacco leaves have revealed that detachment caused the adenylate energy charge to increase and reducing sugars to become phosphorylated (Macnicol 1973), with a 400% increase in fructosebisphosphate within the first 48 h.

The phosphorylated sugars may also play another very important role in regulating root initiation. It was suggested by Gibbs and Beevers (1955) that in mature tissues the pentose phosphate pathway dominated respiratory metabolism whereas the glycolytic pathway dominated in meristematic tissues. The possible significance of this change in intermediary metabolism with respect to root initiation has been discussed by Haissig (1974, 1982b). It seems likely that during root initiation the respiratory pathway changes from the pentose phosphate path to the glycolytic path. This may enable the new meristematic tissue to receive carbon skeletons which are needed for root development. It is then of interest that the sugar phosphate, fructose 2,6-bisphosphate, is able to control the respiratory activity between the two pathways (see Huber 1986 for review). Haissig (1982b) investigated the participation of the two pathways during rooting and found that enzyme activities did indeed change during rooting suggesting that there was an interrelated operation of the two pathways. The exact role of the two pathways in rooting is difficult to ascertain, however, because all cells in the rooting zone were used for determination of enzyme activities although only a few of these are associated with the root initials. Hence, it is likely that we are only obtaining a picture of the general metabolic changes in the rooting zone, but not necessarily a knowledge of what is specifically happening in those cells which will become meristematic. The development of microanalytical techniques may aid in addressing this problem in the future.

Carbohydrates may also play other roles in the formation of adventitious roots. For instance, it is possible that carbohydrate accumulation in the rooting zone may influence osmoregulation, cellular solvent capacity, and other physicochemical phenomena (Haissig 1986). Accumulation of osmotically active solutes such as sugars may have a profound effect on metabolic processes which are involved in rooting. This possibility has not been evaluated to any great extent in cuttings.

CONCLUSION

Past research has revealed that there can be a relationship between carbohydrate content and the rooting of cuttings. However, determination of total carbohydrate content has not been adequate for accurately predicting rooting. This is probably because the carbohydrate status of the stock plant and cuttings is determined by environmental and developmental parameters which also alter other factors known to influence rooting (e.g. auxin, state of differentiation). Hence the influence of carbohydrates on rooting may often be masked by other overriding factors. Also, carbohydrate analyses have primarily been made on organ-level tissues which may obscure what is happening in primordial cells. Microanalytical techniques for measuring carbohydrate content and metabolism at the cellular level would greatly enhance our understanding of the role of carbohydrates, particularly if such measurements are keyed to anatomical stages of rooting.

Carbohydrates serve as an energy source and yield the carbon skeletons needed for the production of new tissues. This means that without some threshold level of carbohydrates, growth and development will cease. Therefore, if stock plants are depleted of carbohydrates and subsequent cuttings are rooted under conditions where net photosynthesis cannot occur, the energy charge will be too low to support rooting. Under these conditions any source of supplemental carbon will be beneficial to the rooting process.

Based on the results from experiments without any direct connection to the rooting of cuttings, we may deduce that it is also possible that carbohydrates play a role in the root initiation process. Phosphorylated carbohydrates are of special interest in this regard because they may play a key role in regulating carbohydrate metabolism. Further work should therefore be aimed toward specific components of pools such as phosphorylated carbohydrates and amino acids. The translocation and metabolism of such compounds in relation to rooting needs further study.

REFERENCES

Aesbacher, H. U., C. Chappuis, M. Manganel and R. Aeschbacher. 1981. Investigation of maillard products in bacterial mutagenicity test systems. *Proc. Food Nutr. Sci.* 5:279–293.

Altman, A. and P. F. Wareing. 1975. The effect of IAA on sugar accumulation and basipetal transport of ^{14}C-labelled assimilates in relation to root formation in *Phaseolus vulgaris* cuttings. *Physiol. Plant.* 33:32–38.

Andersen, A. S. 1986. Stock plant conditions. In *New Root Formation in Plants and Cuttings* (M. B. Jackson, ed). Martinus Nijhoff Pub., Dordrecht/Boston/Lancaster. pp. 223–255. ISBN 90-247-3260-3.

Baadsmand, S. and A. S. Andersen. 1984. Transport and accumulation of indole-3-acetic acid in pea cuttings under two levels of irradiance. *Physiol. Plant.* 61:107–113.

Bojanovic, J. J., A. D. Jevtovic, D. Pantic, S. M. Dugandzic and D. Javonivic. 1970. Thymus nucleoproteins. Thymus histones in young and adult rats. *Gerontologia* 16:304–312.

Bouillene, R. and F. Went. 1933. Recherches experimentales sur la néoformation des racines dans les plantules et les boutures des pes supérieures. *Ann. Jard. Bot. Buitenzorg* 43:25–202.

Breen, P. J. and T. Muraoka. 1974. Effect of leaves on carbohydrate content and movement of ^{14}C-assimilate in plum cuttings. *J. Amer. Soc. Hortic. Sci.* USA. 99:326–332.

Bucala, R., P. Model and A. Cerami. 1984. Modification of DNA by reducing sugars: A possible mechanism for nucleic acid aging and age-related dysfunction in gene expression. *Proc. Nat. Acad. Sci.* USA 81:105–109.

Davis, T. D. and J. R. Potter. 1981. Current photosynthate as a limiting factor in adventitious root formation in leafy pea cuttings. *J. Amer. Soc. Hortic. Sci.* 106:278–282.

Eliasson, L. 1978. Effects of nutrients and light on growth and root formation in *Pisum sativum* cuttings. *Physiol. Plant.* 43:13–18.

Eriksen, E. N. 1973. Root formation in pea cuttings. I. Effects of decapitation and disbudding at different developmental stages. *Physiol. Plant.* 28:503–506.

Gay, A. P. and K. Loach. 1977. Leaf conductance changes in leafy cuttings of *Cornus* and *Rhododendron* during propagation. *J. Hortic. Sci.* 52:509–516.

Gibbs, M. and H. Beevers. 1955. Glucose dissimilation in the higher plant. Effect of age of tissue. *Plant Physiol.* 30:343–347.

Haissig, B. E. 1974. Metabolism during adventitious root primordium initiation and development. *N. Z. J. For. Sci.* 4:324–327.

_____ 1982a. Carbohydrate and amino acid concentrations during adventitious root primordium development in *Pinus banksiana* Lamb. cuttings. *For. Sci.* 28:813–821.

_____ 1982b. Activity of some glycolytic and pentose phosphate pathways enzymes during the development of adventitious roots. *Physiol. Plant.* 55:261–272.

_____ 1986. Metabolic processes in adventitious rooting of cuttings. In *New Root Formation in Plants and Cuttings* (M. B. Jackson ed). Martinus Nijhoff Pub., Dordrecht/Boston/Lancaster. pp. 141–191. ISBN 90-247-3260-3.

Hansen, J., L. H. Strömquist and A. Ericsson. 1978. Influence of the irradiance on carbohydrate content and rooting of cuttings on pine seedlings (*Pinus sylvestris* L.). *Plant Physiol.* 61:975–979.

Howard, B. H. and J. T. Sykes. 1966. Regeneration of the hop plant (*Humulus lupulus* L.) from softwood cuttings. II. Modification of the carbohydrate resources within the cutting. *J. Hortic. Sci.* 41:155–163.

Huber, S. C. 1986. Fructose 2,6-bisphosphate as a regulatory metabolite in plants. *Ann. Rev. Plant Physiol.* 37:233–246.

Knoblauch, F. 1976. *Nutrition of Hypericum hysan in Containers for Production of Cuttings and Plants to be Sold* (in Danish). *Statens Forsogsvirksomhed i Plantkultur Medd.* 1275.

Kraus, E. J. and H. R. Kraybill. 1918. *Vegetation and Reproduction With Special Reference to Tomato.* Oregon Agric. Coll. Exp. Sta. Bull. 149.

Loach, K. and D. N. Whalley 1978. Water and carbohydrate relationships during the rooting of cuttings. *Acta Hortic.* 79:161–168.

Lovell, P. H., A. Illsley and K. G. Moore. 1972. The effects of light intensity and sucrose on root formation, photosynthetic ability, and senescence in detached cotyledons of *Sinapis alba* L. and *Raphanus sativus* L. *Ann. Bot.* 36:123–134.

_____ _____ _____ 1973. The effect of sucrose on primordium development and on protein and RNA levels in detached cotyledons of *Sinapis alba* L. *Ann. Bot.* 37:805–816.

_____ _____ _____ 1974. Endogenous sugar levels and their effects on root formation and petiole yellowing of detached mustard cotyledons. *Physiol. Plant.* 31:321–326.

Macnicol, P. K. 1973. Metabolic regulation in senescing tobacco leaf. II. Change in glycolytic metabolite levels in the detached leaf. *Plant Physiol.* 51:798–801.

Monnier, V. M. and A. Cerami. 1983. Nonenzymic glycosylation and browning of proteins in diabetes. *Clin. Endocrinol. Med.* 11:431–452.

Moore, K. G., A. Cobb and P. H. Lovell. 1972. Effects of sucrose on rooting and senescence in detached *Raphanus sativus* L. cotyledons. *J. Exp. Bot.* 23:65–74.

Nanda, K. K. and V. K. Anand. 1970. Seasonal changes in auxin effects on rooting of stem cuttings of *Populus nigra* and its relationship with mobilization of starch. *Physiol. Plant.* 23:99–107.

_____ M. K. Jain. 1972. Utilization of sugars and starch as carbon sources in the rooting of etiolated stem segments of *Populus nigra. New Phytol.* 71:825–828.

_____ M. K. Jain and S. Malhotra. 1971. Effects of glucose and auxins in rooting etiolated stem segments of *Populus nigra. Physiol. Plant.* 24:387–391.

_____ N. C. Bhattacharya, and V. K. Kochhar. 1974. Biochemical basis of adventitious root formation on etiolated stem segments. *N. Z. J. For. Sci.* 4:347–358.

Okoro, O. O. and J. Grace. 1976. The physiology of rooting *Populus* cuttings. I. Carbohydrates and photosynthesis. *Physiol. Plant.* 36:133–138.

Olieman-van der Meer, A. W., R. L. M. Pierek and S. Roest. 1970. Effects of sugar, auxin, and light on adventitious root formation in isolated stem explants of *Phaseolus* and *Rhododendron. Medd. Rijks. Landbo. Gent.* 1970, pp. 511–518.

Orton, P. J. 1979. The influence of water stress and abscisic acid on the root development of *Chrysanthemum morifolium* cuttings during propagation. *J. Hortic. Sci.* 54:171–180.

Pal, M. and K. K. Nanda. 1981. Rooting of etiolated stem segments of *Populus robusta*—interaction of temperature, catechol, and sucrose in the presence of IAA. *Physiol. Plant.* 53:540–542.

Reid, M. E. 1924. Relation of kind of food reserves to regeneration in tomato plants. *Bot. Gaz.* 77:103–110.

Spellenberg, B. 1985. Verbesserung des Vermehrungserfolges bei schwer vermehrbaren Laubgehölzen. I. Der Einfluss des Vermehrungsklimas auf Inhaltstoffe und weiteres Wachstum der bewurzelten Stecklinge. *Gartenbauwissenschaft* 50:71–77.

Stevens, V. J., H. Vlassara, A. Abati and A. Cerami. 1977. Nonenzymic glycosylation of hemoglobin. *J. Biol. Chem.* 252:2998–3002.

Stitt, M. R., Gerhardt, I. Wilke and H. W. Heldt. 1987. The contribution of fructose 2,6-bisphosphate to the regulation of sucrose synthesis during photosynthesis. *Physiol. Plant.* 69:377–386.

Suzuki, T. 1982. Changes in total nitrogen and free amino acids in stem cuttings of mulberry (*Morus alba* L.). *J. Exp. Bot.* 132:21–28.

Van't Hoff, J. 1968. Control of cell progression through the mitotic cycle by carbohydrate provision. *J. Cell Biol.* 37:773–778.

Veierskov, B. and A. S. Andersen. 1982. Dynamics of extractable carbohydrates in *Pisum sativum.* III. The effect of IAA and temperature on content and translocation of carbohydrates in pea cuttings during rooting. *Physiol. Plant.* 55:179–182.

_____ _____ E. N. Eriksen. 1982a. Dynamics of extractable carbohydrates in *Pisum sativum.* I. Carbohydrate and nitrogen content in pea plants and cuttings grown at two different irradiances. *Physiol. Plant.* 55:167–173.

_____ _____ B. M. Stummann and K. W. Henningsen. 1982b. Dynamics of extractable carbohydrates in *Pisum sativum.* II. Carbohydrate content and photosynthesis of pea cuttings in relation to irradiance and stock plant temperature. *Physiol. Plant.* 55:174–178.

Vieitez, A. M., A. Ballester, M. T. Garcia and E. Vieitez. 1980. Starch depletion and anatomical changes during the rooting of *Castanea sativa* Mill. cuttings. *Scientia Hortic.* 13:261–266.

Went, F. W. and K. V. Thimann. 1937. *Phytohormones.* MacMillan Co., NY, USA.

White, J. and P. H. Lovell. 1984. The anatomy of root initiation in cuttings of *Griselinia littoralis* and *Griselinia lucida. Ann. Bot.* 54:7–20.

CHAPTER 6

Photosynthesis During Adventitious Rooting

Tim D. Davis

Department of Agronomy and Horticulture
Brigham Young University
Provo, UT 84602

INTRODUCTION...79
RATES AND PATTERNS OF PHOTOSYNTHESIS IN CUTTINGS................80
FACTORS INFLUENCING PHOTOSYNTHESIS IN CUTTINGS...................81
DOES CURRENT PHOTOSYNTHESIS INFLUENCE
   ADVENTITIOUS ROOT FORMATION? 82
     Current Photosynthesis May Influence Rooting.........................82
     Contrary Evidence 83
     Rationalizing the Conflicts..83
POSSIBLE ROLES OF PHOTOSYNTHESIS IN ROOT
   FORMATION 84
     Carbohydrates 84
     Auxins..84
     Other Factors...84
CONCLUSION 84
ACKNOWLEDGMENTS 85
REFERENCES..85

Additional key words: CO_2 fixation, current photosynthate, environment, light.
Abbreviations: IAA, indole-3-acetic acid; IBA, indole-3-butyric acid; PAR, photosynthetically active radiation; P_n, net photosynthesis.

INTRODUCTION

There has been a general feeling that leafy cuttings (defined as cuttings which possess leaves, as opposed to leafless hardwood cuttings) should be rooted in an environment that is conducive to photosynthesis. For example, Scott and Marston (1967) stated: "It is suggested that photosynthesis proceeds faster under mist . . . thereby encouraging both growth and regeneration." Hess and Snyder (1955) similarly ascribed the improved rooting of cuttings under mist propagation to high rates of photosynthesis. More recently, Eliasson and Brunes (1980) stated that ". . . success in rooting depends on an optimum light treatment providing appropriate energy for photosynthesis but being as little detrimental to the rooting process as possible." Implicit in such statements is the assumption that photosynthesis by cuttings positively influences rooting. However, there is little scientific evidence to support or reject this assumption. Although considerable research has been focused on carbohydrate metabolism of cuttings, surprisingly little work has been done on photosynthesis even though it exerts considerable influence on carbohydrate pools.

The purposes of this review are to describe patterns of photosynthesis in cuttings during rooting and to examine whether or not such photosynthetic activity (hereafter referred to as "current photosynthesis" to distinguish it from photosynthesis prior to excision from the stock plant) influences adventitious root formation. In addition, possible roles of current photosynthesis in rooting will be discussed.

RATES AND PATTERNS OF PHOTOSYNTHESIS IN CUTTINGS

Perhaps some of the reluctance to study photosynthesis by cuttings has been due to the difficulty in measuring CO_2 exchange. Rates of P_n by cuttings are low and hence rather sensitive instrumentation is required to accurately monitor CO_2 fixation. With the advent of modern gas analysis and environmental control systems, however, this limitation should no longer be as serious.

There are relatively few studies which have actually measured rates of P_n during the rooting period. Fig. 1 illustrates the pattern of P_n exhibited by leafy pea cuttings under moderate PAR (280 $\mu mol \cdot m^{-2} \cdot s^{-1}$) after excision from the stock plant. Similar patterns of P_n have been observed with leafy cuttings of other species and rates are considerably lower than those for intact plants (Cameron and Rook 1974, Okoro and Grace 1976, Machida 1977, Eliasson and Brunes 1980). Net photosynthesis drops rapidly upon excision, particularly during the first 24–48 h of the rooting period. This drop in P_n is presumably due primarily to stomatal closure (Gay and Loach 1977, Eliasson and Brunes 1980) although the precise contributions of stomatal versus non-stomatal limitations to P_n during this period have not been studied in detail. Water stress can also directly influence P_n via non-stomatal effects and the stomatal limitation to P_n under water stress may often be overestimated (Farquhar and Sharkey 1982).

Relatively low rates of P_n are exhibited until adventitous roots begin to emerge from the cutting. Low P_n rates prior to root emergence have been attributed to a number of factors in addition to water stress. For instance, it has been suggested that photo-oxidative damage to the chloroplast, due to light

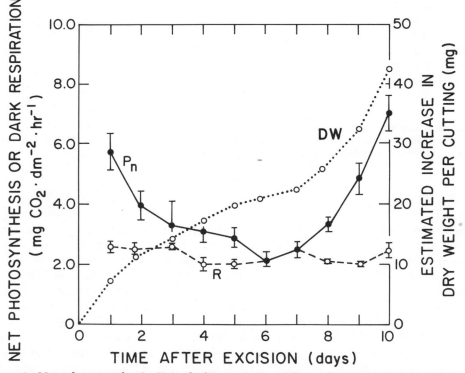

Figure 1. Net photosynthesis (P_n), dark respiration (R), and estimated increase in dry weight (DW) by 'Alaska' pea cuttings during the rooting period. Bars indicate standard error of the mean. The PAR during the P_n measurements was 280 $\mu mol \cdot m^{-2} \cdot s^{-1}$. Estimated increase in DW was calculated daily based upon the leaf area and unit leaf area photosynthetic and respiratory rates. [From Davis and Potter (1981)]

absorption when stomata are closed, may contribute to decreased P_n of cuttings (Eliasson and Brunes 1980). This hypothesis has not been tested, however. Humphries and Thorne (1964) attributed reduced P_n in bean leaf-petiole cuttings to reduced assimilate transport within the cutting. Reduced P_n of cuttings under mist propagation has been attributed to the inundation of intercellular spaces with water which presumably would result in impeded CO_2 diffusion (Okoro and Grace 1976).

Net photosynthesis gradually increases upon the emergence of roots (Fig. 1), probably at least in part due to the alleviation of water stress and the re-opening of stomata. Other factors, however, may also be responsible for the increase in P_n after root emergence. Humphries and Thorne (1964) suggested that P_n of bean leaf-petiole cuttings was influenced by the sink activity of the root system. P_n declined when roots were removed or when their growth was chemically restricted. It has also been suggested that P_n in cuttings increases after rooting because roots supply leaves with cytokinins which may increase the activity and/or amount of carboxylating enzymes (Okoro and Grace 1976). This possibility has not been critically evaluated.

Using the patterns of CO_2 exchange shown in Fig. 1, it is possible to estimate dry matter accumulation during the course of the rooting period. The estimated daily increase in dry weight of leafy pea cuttings was greatest later in the rooting period after roots emerged. Nevertheless, considerable dry matter accumulated during the first 6 d of the rooting period and cutting dry weight nearly doubled during the course of the entire 10 d rooting period. Hence, even low rates of P_n can significantly influence the carbon balance and dry matter accumulation within a cutting.

FACTORS INFLUENCING PHOTOSYNTHESIS IN CUTTINGS

Net photosynthesis of most intact C_3 species is saturated at about one-half of full sunlight (ca. 1000–1200 $\mu mol \cdot m^{-2} \cdot s^{-1}$). Light saturation of P_n by unrooted leafy cuttings is generally at a much lower PAR level, however. For cuttings of *Camellia japonica* L., *Forsythia suspensa* Vahl., and *Chrysanthemum morifolium* Ramat., P_n was saturated at about 500 $\mu mol \cdot m^{-2} \cdot s^{-1}$ or about one-quarter of full sunlight (Machida et al. 1977). Above this PAR level, P_n declined considerably probably due to water stress. Similarly, P_n of pea cuttings increased up to about 350 $\mu mol \cdot m^{-2} \cdot s^{-1}$ and declined above about 450 $\mu mol \cdot m^{-2} \cdot s^{-1}$ (Davis and Potter 1981). Finally, rhododendron cuttings rooted under either 0 or 95% shade had similar rates of P_n throughout an 18 week rooting period (Davis and Potter 1987). These data suggest that only moderate PAR is needed to saturate P_n of unrooted cuttings.

Most photosynthesis-light response curves have been determined by short-term (1–2 h) exposure of leaves to various PAR levels. Data obtained from such experiments may overestimate the light saturation point of unrooted leafy cuttings in the long-term. Such prolonged exposure to relatively high PAR may result in a slowly developing water deficit within the cutting which might not be detected in a short-term exposure. This phenomenon has been observed with leafy pea cuttings which were light-saturated at about 350 $\mu mol \cdot m^{-2} \cdot s^{-1}$ for one to two h, but began to wilt after about 3 h at such PAR levels (Davis and Potter 1981). Similarly, Loach and Gay (1979) reported that dry weight gain (presumably a rough indicator of photosynthetic activity) of *Forsythia* × *intermedia* Zabel. and *Weigela florida* A. DC. cuttings was initially very rapid under relatively high PAR but declined after prolonged exposure.

It is likely that P_n by cuttings is a result of a complex interaction between PAR, relative humidity, stomatal opening, and water potential. It may be that a light regime of relatively low PAR during the early stages of rooting (before roots emerge) and higher PAR after roots emerge would result in maximized P_n. More detailed work on how PAR levels during rooting influence P_n in the long-term would seem worthwhile.

Indirect evidence indicates that P_n by cuttings can be influenced by the ambient CO_2 concentration during rooting. Under CO_2-enriched conditions, dry weight of cuttings from a variety of species was increased, suggesting that P_n was increased (Molnar and Cumming 1968, Davis and Potter 1983). The increased P_n is probably due to increased carboxylation efficiency but might also be due to improved water status due to partial stomatal closure induced by high CO_2 (Raschke 1975). When the ambient CO_2 concentration during rooting was maintained at the compensation point (ca. 100 $\mu l \cdot l^{-1}$), the dry weight and carbohydrate content of pea cuttings was reduced considerably compared to cuttings rooted at 320 $\mu l \cdot l^{-1}$ (Davis and Potter 1981), again illustrating that ambient CO_2 can influence dry matter accumulation.

Various exogenously applied compounds may influence P_n rates of cuttings. For example, cuttings are often treated with auxins to stimulate rooting and, although little work has been done on the effects of auxins on P_n by cuttings, sprays of IAA have been reported to increase P_n of detached bean (*Phaseolus vulgaris* L.) leaves by promoting the dark reactions of photosynthesis (Turner and Bidwell 1965). Also, treatment of detached *Chrysanthemum* spp. leaves with IBA reportedly reduced transpiration and increased water-retaining capacity which could influence P_n (Mochvan 1979).

The application of sucrose to cotyledon cuttings reduced their ability to photosynthesize (Lovell et al. 1972). This was not considered to be an osmotic effect because osmotica such as mannitol and polyethylene glycol had no effect on the cuttings, but rather appeared to be associated with enhanced chlorophyll loss and leaf yellowing (Moore et al. 1972). These results indicate that supplying exogenous carbohydrates to cuttings is not necessarily the same as increasing cutting P_n, a false assumption which has often been made when studying the influence of carbohydrates and photosynthesis on rooting.

Certainly other factors such as temperature and mineral nutrition can influence the photosynthetic rate of cuttings, but these have not been studied in any detail. The number of leaves per cutting may also have an effect on the rate of P_n on a leaf area basis. For instance, photosynthesis decreased faster after excision in cuttings with three or four leaves compared to one or two leaves (Machida et al. 1977). The greater transpirational surface on cuttings with more leaves may have resulted in water stress sooner.

DOES CURRENT PHOTOSYNTHESIS INFLUENCE ADVENTITIOUS ROOT FORMATION?

Photosynthesis by cuttings is not an absolute requirement for root formation. This statement is supported by the observation that leafy cuttings from a number of species can form at least some roots when placed in the dark (van Overbeek et al. 1946, Davis and Potter 1981). The fact that many plants can be propagated from non-photosynthetic leafless hardwood cuttings further supports this contention. These observations do not preclude the possibility that photosynthesis by cuttings can in some manner quantitatively influence root formation under circumstances where photosynthesis occurs.

From an experimental point of view it is difficult to test the hypothesis that current photosynthesis influences root formation on leafy cuttings. To test such a hypothesis, ideally one should utilize an experimental system where photosynthetic rate can be varied independently of all other factors which might affect rooting. Unfortunately such a protocol is impossible to achieve. For example, one can alter PAR levels during rooting to vary photosynthesis. Such a treatment, however, may also directly influence auxin content (Scott and Briggs 1963, Heide 1968) and water relations (Loach and Gay 1979), both of which influence rooting. Likewise, altering ambient CO_2 content may affect P_n, but also water relations (Paez et al. 1983, Morison and Gifford 1984), ethylene activity (Dhawan et al. 1981), auxin content (Avery et al. 1937, Hillman and Galston 1961, Vardar 1968), and senescence (Widholm and Ogren 1969). Given these limitations, the evidence supporting the hypothesis that current photosynthesis can influence root formation follows.

Current Photosynthesis May Influence Rooting

Several studies wherein ambient CO_2 was varied during propagation suggest that current photosynthesis may influence rooting. Hardh (1967) and Molnar and Cumming (1968) reported that CO_2 enrichment increased both the dry weight and number of roots per cutting for several species. Although critically controlled experiments were not conducted, both Hare (1974) and Sorenson (1978) likewise suggested that CO_2 enhanced the rooting of cuttings. More recently, however, two studies (Lin and Molnar 1980, Davis and Potter 1983) indicated that rooting of some species or cultivars did not respond to ambient CO_2. Hence although rooting of most species tested has been promoted by elevated CO_2, there have been exceptions.

A number of studies using exogenously applied carbohydrates support the view that current photosynthesis can influence rooting, although the assumption that treating cuttings with carbohydrates is the same as increasing photosynthesis is most likely invalid (Haissig 1986). Under low PAR, where photosynthesis was presumably relatively low, the application of sugars to cuttings has been reported to increase rooting (Howard and Sykes 1966, Lovell et al. 1972, Eliasson 1978, Loach and Whalley 1978). When photosynthesis of cotyledon cuttings was reduced by DCMU [N'-(3,4-dichlorophenyl)N,N dimethylurea; a photosystem II inhibitor], rooting was inhibited but this inhibition could be overcome with the application of sucrose (Lovell et al. 1972).

Hess and Snyder (1955) reported that a combination of environmental conditions (relative humidity, light, and temperature) during propagation, which resulted in relatively high rates of P_n, also resulted in enhanced rooting compared to conditions where P_n was low. High carbohydrate contents in the cuttings accompanied the improved rooting.

Numerous studies have been conducted wherein light intensity during the rooting period was varied. In some of these studies (Foster 1963, Howard 1965, Eliasson 1978, Davis and Potter 1981) rooting increased with increasing light intensity which suggested that rooting was enhanced by photosynthesis. Increasing light intensity has not always promoted rooting and the diverse effects of light on cuttings (auxin, water status, anatomy, etc.) make such experiments very difficult to interpret.

Experiments wherein P_n by leafy pea cuttings was reduced by several different treatments indicate that current photosynthesis can limit root formation (Davis and Potter 1981). When P_n was reduced to the compensation point by reducing the PAR, reducing ambient CO_2 levels, or by blocking CO_2 diffusion with a wax-forming antitranspirant, the number of roots per cutting was reduced by about 50% compared to untreated controls. Similarly, rooting was reduced by about 50% when cuttings were rooted under a short photoperiod which maintained a carbon balance similar to that of the treatments which reduced P_n to the compensation point. Each of the above treatments substantially reduced the carbohydrate content in the base of the cuttings which was measured just prior to root emergence. Taken together, these data suggest that at least 50% of the rooting was associated with current photosynthesis.

Contrary Evidence

Current photosynthesis appears to be of little importance to root formation on some leafy cuttings. For example, leafy cuttings of *Hibiscus rosa-sinensis* L. reportedly rooted equally well in darkness and moderate light suggesting that photosynthesis had no effect on rooting (van Overbeek et al. 1946). Breen and Muraoka (1974), using $^{14}CO_2$, reported that current photosynthate remained in the upper portion of leafy plum cuttings until after roots formed. Assuming that the products of photosynthesis must be at the site of root formation to affect rooting, these data suggested that the cuttings contained sufficient carbohydrate reserves to satisfy the demands of rooting. Stomatal conductance on *Cornus alba* L. cuttings remained very low until after roots emerged, suggesting that photosynthesis before root formation was minimal (Gay and Loach 1977). The authors concluded that root formation in *C. alba* depended upon reserves rather than current photosynthate. Actual rates of P_n during rooting were not measured, however, and even low rates of P_n can contribute significantly to the carbon budget of a given plant part (Okoro and Grace 1976, Flinn et al. 1977).

Rationalizing the Conflicts

From the previous discussion, it appears that root formation on many types of leafy cuttings is influenced by current photosynthesis while in other leafy cuttings current photosynthesis is of little or no apparent importance. Influences of photosynthesis on rooting may depend upon several characteristics of the cutting. For example, some cuttings may have sufficient carbohydrate reserves to satisfy the demands of rooting. However, if the rooting period is quite long, even cuttings with substantial reserves such as *Picea abies* L. (Karst.) may require additional photosynthate (Strömquist and Eliasson 1979). The amount of shoot growth on cuttings during the rooting period may also determine whether current photosynthesis influences root formation. Some cuttings continue to elongate and produce new leaves during the rooting period. Such shoot growth may compete with the base of the cutting for assimilates. For instance, actively growing shoots on *Populus tremula* L. cuttings influenced the growth of the roots (Eliasson 1971). Hence, under conditions of rapid shoot growth the supply of current photosynthate may limit rooting. In this regard, the application of growth retardants which restrict shoot growth during rooting have been found to sometimes promote root formation (Krishnamoorthy 1972, Davis et al. 1985, also see Chapter 13 by Davis and Sankhla).

POSSIBLE ROLES OF PHOTOSYNTHESIS IN ROOT FORMATION

Carbohydrates

It has generally been assumed that photosynthesis during rooting provides carbohydrates to the base of the cutting. Indeed, carbohydrates often do accumulate in the base of cuttings during the rooting period (Altman and Wareing 1975, Hansen et al. 1978, Davis and Potter 1981, Haissig 1982, 1984, Veierskov et al. 1982). With leafy pea cuttings the amount of carbohydrate accumulated was positively correlated with photosynthetic activity after excision (Davis and Potter 1981). It has long been thought that the carbohydrate content of cuttings is important to rooting, although the precise role in the process is not completely understood (Haissig 1986 and Chapter 5 by Veierskov).

Auxins

Auxins are well-known promoters of adventitious root formation. Auxin transport and/or synthesis may be reduced under conditions of low photosynthetic activity (Hillman and Galston 1961, Scott and Briggs 1963, Heide 1968, Vardar 1968). Thus photosynthesis by cuttings may indirectly influence rooting by affecting auxin supply to the base of the cutting. Although this hypothesis has not been critically tested, preliminary evidence by Kumpula and Potter (1984) tend to support it. They found that both rooting and basipetal transport of ^{14}C-labelled IAA in leafy pea cuttings was reduced when photosynthesis was reduced by lowering the PAR level or by blocking CO_2 exchange with a wax-forming antitranspirant.

Other Factors

The rate of P_n by cuttings may influence the basipetal translocation of other factors within the cutting as it has generally been observed that plants with high rates of P_n have high translocation rates (Salisbury and Ross 1985). A number of factors other than IAA may be basipetally transported during rooting. For example, rooting cofactors (see Chapter 1 by Hackett) may originate in leaves and buds and may be transported to the base of the cutting where they promote rooting.

It is also possible that photosynthesis influences the formation of a noncarbohydrate, non-auxin component which may be involved in rooting. For example, considerable phenolic biosynthesis occurs in the chloroplast and hence may be related to photosynthetic activity. In this regard, a number of phenolic compounds have been found to act as rooting cofactors.

CONCLUSION

By the foregoing synopsis, I conclude that current photosynthesis may influence root formation on most types of leafy cuttings. In some cases the effects of photosynthesis may be minor and masked by other, possibly overriding, factors such as auxin or water status. Nevertheless, from a practical view, the past recommendations that photosynthesis during rooting should be maximized seem sound. It must be understood, however, that P_n of unrooted cuttings is saturated at relatively low PAR levels and that high PAR will not likely maximize P_n and may only serve to desiccate the cuttings.

Future work should focus on determining the role of photosynthesis in rooting. The following questions need to be addressed: What does photosynthetic activity contribute to rooting? Is it the process or the product that is important? How does photosynthesis relate to carbohydrate metabolism and accumulation in the cutting? More detailed measurements of photosynthesis during rooting are also needed. It would be of interest to know how environmental parameters such as light, temperature, and relative humidity influence photosynthesis over the course of the entire rooting period. Such information might suggest ways to maximize photosynthesis by manipulation of the rooting environment.

ACKNOWLEDGMENTS

Appreciation is expressed to Drs. Bruce E. Haissig (Forestry Sciences Lab., Rhinelander, Wisconsin) and R. D. Horrocks (Brigham Young University, Provo, Utah) who critically reviewed this manuscript. Gratitude is also expressed to Dr. John R. Potter (USDA Horticultural Crops Lab., Corvallis, Oregon) whose research philosophies have been of great value to the author.

REFERENCES

Altman, A. and P. F. Wareing. 1975. The effect of IAA on sugar accumulation and basipetal transport of ^{14}C-labelled assimilates in relation to root formation in *Phaseolus vulgaris* cuttings. *Physiol. Plant.* 33: 32–38.

Avery, G. S. Jr., P. R. Burkholder and H. B. Creighton. 1937. Growth hormone in terminal shoots of *Nicotiana* in relation to light. *Amer. J. Bot.* 24: 666–673.

Breen, P. J. and T. Muraoka. 1974. Effect of leaves on carbohydrate content and movement of ^{14}C-assimilate in plum cuttings. *J. Amer. Soc. Hortic. Sci.* 99: 326–332.

Cameron, R. J. and D. A. Rook. 1974. Rooting stem cuttings of radiata pine: environmental and physiological aspects. *N. Z. J. For. Sci.* 4: 291–298.

Davis, T. D. and J. R. Potter. 1981. Current photosynthate as a limiting factor in adventitious root formation on leafy pea cuttings. *J. Amer. Soc. Hortic. Sci.* 106: 278–282.

_____ _____ 1983. High CO_2 applied to cuttings: effects on rooting and subsequent growth in ornamental species. *HortScience* 18: 194–196.

_____ _____ 1987. Physiological response of rhododendron cuttings to different light levels during rooting. *J. Amer. Soc. Hortic. Sci.* 112:256–259.

_____ N. Sankhla, R. H. Walser and A. Upadhyaya. 1985. Promotion of adventitious root formation on cuttings by paclobutrazol. *HortScience* 20: 883–884.

Dhawan, K. R., P. K. Bassi and M. S. Spencer. 1981. Effects of carbon dioxide on ethylene production and action in intact sunflower plants. *Plant Physiol.* 68: 831–834.

Eliasson, L. 1971. Adverse effect of shoot growth on root growth in rooted cuttings of aspen. *Physiol. Plant.* 25: 268–272.

_____ 1978. Effects of nutrients and light on growth and root formation in *Pisum sativum* cuttings. *Physiol. Plant.* 43: 13–18.

_____ L. Brunes. 1980. Light effects on root formation in aspen willow cuttings. *Physiol. Plant.* 48: 261–265.

Farquhar, G. D. and T. D. Sharkey. 1982. Stomatal conductance and photosynthesis. *Ann. Rev. Plant Physiol.* 33: 317–345.

Flinn, A. M., C. A. Atkins and J. S. Pate. 1977. Significance of photosynthetic and respiratory exchanges in the carbon economy of developing pea fruit. *Plant Physiol.* 60: 412–418.

Foster, R. E. 1963. Aeration, light, and type of cutting for vegetative propagation of muskmelon (*Cucumis melo* L.) *Proc. Amer. Soc. Hortic. Sci.* 83: 596–598.

Gay, A. P. and K. Loach. 1977. Leaf conductance changes on leafy cuttings of *Cornus* and *Rhododendron* during propagation. *J. Hortic. Sci.* 52: 509–516.

Haissig, B. E. 1982. Carbohydrate and amino acid concentrations during adventitious root primordium development in *Pinus banksiana* Lamb. cuttings. *For. Sci.* 28: 813–821.

_____ 1984. Carbohydrate accumulation and partitioning in *Pinus banksiana* seedlings and seedling cuttings. *Physiol. Plant.* 61: 13–19.

86   Tim D. Davis

_____ 1986. Metabolic processes in adventitious rooting of cuttings. In *New Root Formation on Plants and Cuttings* (M. B. Jackson, ed). Martinus-Nijhoff, Dordrecht/Boston/Lancaster. pp. 141–189. ISBN 90-247-3260-3.

Hansen, J., L. Strömquist and A. Ericsson. 1978. Influence of the irradiance on carbohydrate content and rooting of cuttings of pine seedlings (*Pinus sylvestris* L.) *Plant Physiol.* 61: 975–979.

Hardh, J. E. 1967. Trials with carbon dioxide, light, and growth substances on forest tree plants. *Acta For. Fennica* 81: 1–10.

Hare, R. C. 1974. Chemical and environmental treatments promoting rooting of pine cuttings. *Can. J. For. Res.* 4: 101–106.

Heide, O. M. 1968. Auxin level and regeneration of *Begonia* leaves. *Planta* 81: 153–159.

Hess, C. E. and W. E. Snyder. 1955. A physiological comparison of the use of mist with other propagation procedures used in rooting cuttings. *Rept. 14th Int. Hortic. Cong.* pp. 1133–1139.

Hillman, W. S. and A. W. Galston. 1961. The effect of external factors on auxin content. In *Handbuch der Pflanzenphysiologie* (W. Ruhland, ed) Vol. 14. Springer Verlag, Berlin, W. Germany. pp. 683–702.

Howard, B. H. 1965. Regeneration of the hop plant (*Humulus lupulus* L.) from softwood cuttings. I. The cutting and its rooting environment. *J. Hortic. Sci.* 10: 181–191.

_____ J. T. Sykes. 1966. Regeneration of the hop plant (*Humulus lupulus* L.) from softwood cuttings. II. Modification of the carbohydrate resources within the cutting. *J. Hortic. Sci.* 41: 155–163.

Humphries, E. C. and G. N. Thorne. 1964. The effect of root formation on photosynthesis of detached leaves. *Ann. Bot.* 28: 391–400.

Krishnamoorthy, H. N. 1972. Effect of growth retardants and abscisic acid on the rooting of hypocotyl cuttings of muskmelon (*Cucumis melo* 'Kutana') *Biochem. Physiol. Pflanzen* 163: 513–517.

Kumpula, C. L. and J. R. Potter. 1984. Photosynthesis linked auxin transport during adventitious root formation in leafy pea cuttings. *Plant Physiol.* (suppl.) 75: 43. (Abst.).

Lin, W. C. and J. M. Molnar. 1980. Carbonated mist and high intensity supplementary lighting for propagation of selected woody ornamentals. *Proc. Int. Plant Prop. Soc.* 30: 104–109.

Loach, K. and A. P. Gay. 1979. The light requirement for propagating hardy ornamental species from leafy cuttings. *Scientia Hortic.* 10: 217–230.

_____ D. N. Whalley. 1978. Water and carbohydrate relationships during the rooting of cuttings. *Acta Hortic.* 79: 161–168.

Lovell, P. H., A. Illsley and K. G. Moore. 1972. The effects of light intensity and sucrose on root formation, photosynthetic ability, and senescence in detached cotyledons of *Sinapis alba* L. and *Raphanus sativus* L. *Ann. Bot.* 36: 123–134.

Machida, H., A. Ooshi, T. Hosoi, H. Komatsu, and F. Kamota. 1977. Studies on photosynthesis in cuttings during propagation. I. Changes in the rate of apparent photosynthesis in the cuttings of several plants after planting. *J. Jap. Soc. Hortic. Sci.* 46: 274–282.

Mochvan. L. T. 1979. Water exchange in isolated leaves, and growth regulators. *Fiziol. Biok. Kullt. Rast.* 11: 258–261.

Molnar, J. M. and W. A. Cumming. 1968. Effect of carbon dioxide on propagation of softwood conifer and herbaceous cuttings. *Can. J. Plant Sci.* 48: 595–599.

Moore, K. G., A. Cobb and P. H. Lovell. 1972. Effects of sucrose on rooting and senescence in detached *Raphanus sativus* L. cotyledons. *J. Exp. Bot.* 23: 65–74.

Morison, J. I. L. and R. M. Gifford. 1984. Plant growth and water use with limited water supply in high CO_2 concentrations. I. Leaf area, water use, and transpiration. *Aust. J. Plant Physiol.* 11: 361–374.

Okoro, O. O. and J. Grace. 1976. The physiology of rooting *Populus* cuttings. I. Carbohydrates and photosynthesis. *Physiol. Plant.* 36: 133–138.

Paez, A., H. Hellmers, and B. R. Strain. 1983. CO_2 enrichment, drought stress, and growth of Alaska pea plants (*Pisum sativum*). *Physiol. Plant.* 58: 161–165.

Raschke, K. 1975. Stomatal action. *Ann. Rev. Plant Physiol.* 26: 309–340.

Salisbury, F. B. and C. W. Ross. 1985. *Plant Physiology.* Wadsworth, Belmont. p. 226. ISBN 0-534-00562-4.

Scott, M. A. and M. E. Marston. 1967. Effects of mist and basal temperature on the regeneration of *Saintpaulia ionantha* Wendl. from leaf cuttings. *Hortic. Res.* 7: 50–60.

Scott, T. K. and W. R. Briggs. 1963. Recovery of native and applied auxin from the dark-grown 'Alaska' pea seedling. *Amer. J. Bot.* 50: 652–657.

Sorenson, S. E. 1978. CO_2 as an aid to propagation. *Proc. Int. Plant Prop. Soc.* 28: 175–176.

Stromquist, L. and L. Eliasson. 1979. Light inhibition of rooting in Norway spruce (*Picea abies*) cuttings. *Can. J. Bot.* 57: 1314–1316.

Turner, W. B. and R. G. S. Bidwell. 1965. Rates of photosynthesis in attached and detached bean leaves, and the effect of spraying with indoleacetic acid solution. *Plant Physiol.* 40: 446–451.

van Overbeek, J., S. A. Gordon and L. E. Gregory. 1946. An analysis of the function of the leaf in the process of root formation in cuttings. *Amer. J. Bot.* 33: 100–107.

Vardar, Y. 1968. Agents modifying longitudinal transport of auxin. In *Transport of Plant Hormones* (Y. Vardar, ed). North-Holland Pub., Amsterdam, The Netherlands. pp. 156–191.

Veierskov, B., A. S. Andersen, B. M. Stumman and K. W. Henningsen. 1982. Dynamics of extractable carbohydrates in *Pisum sativum.* II. Carbohydrate content and photosynthesis of pea cuttings in relation to irradiance and stock plant temperature and genotype. *Physiol. Plant.* 55: 174–178.

Widholm, J. M. and W. L. Ogren. 1969. Photorespiratory-induced senescence of plants under conditions of low carbon dioxide. *Proc. Nat. Acad. Sci. USA* 63: 668–675.

CHAPTER 7

Enzyme Activities During Adventitious Rooting

Nikhil C. Bhattacharya

Department of Agricultural Sciences and GWC Experiment Station
Tuskegee University
Tuskegee, AL 36088

INTRODUCTION 88
RESPIRATORY ENZYMES 89
OXIDASES 89
    Peroxidases 89
    IAA Oxidase 90
    Polyphenol Oxidase 92
HYDROLYTIC ENZYMES 93
    Amylases 93
    Invertase (Sucrase) 94
    Phosphatases 95
    Nucleases 95
GLUTAMATE DEHYDROGENASE 96
CONCLUSION 96
ACKNOWLEDGMENTS 97
REFERENCES 97

Additional key words: isoenzymes, sucrose, auxin, phenols, purine, pyrimidine, metabolism, root regeneration, vegetative propagation.
Abbreviations: AP, acid phosphatase (EC 3.23.2); AC, actinomycin-D; ATP, adenosine triphosphate; ALK, alkaline phosphatase (EC 3.13.1); Amy, amylase (EC 3.2.1.1); CY, cycloheximide; COx, cytochrome oxidase (EC 1.9.3.1); DNase, deoxyribonuclease (EC 3.1.4.5); L-DOPA oxidase, o-diphenol oxidase (EC 1.14.18.1); GDH, glutamate dehydrogenase (EC 1.4.1.2); IAA-O, IAA oxidase (EC 1.11.1.8); INV, invertase (EC 3.2.1.26); MDH, malate dehydrogenase (EC 1.1.1.37); PO, peroxidase (EC 1.11.1.7); PPO, polyphenol oxidase (EC 1.10.3.1); RNase, ribonuclease (EC 3.1.4.22); SD, succinic dehydrogenase (EC 1.3.99.1).

INTRODUCTION

It has been observed that rooting ability of stem cuttings may vary considerably with the annual growth cycle of a plant species. Effectiveness of applied auxin on rooting may also vary with season. It is well established that rooting responses of cuttings are determined by the interactions between internal and environmental factors. Based on the earlier reports on auxins in relation to adventitious rooting, it seems that auxins are associated with division and elongation of meristematic cells, differentiation of root primordia, and also mobilization of reserve food materials to the site of rooting.

Enzyme activities in the rooting zone of cuttings may provide one of the easy, fast, and reliable means of assessing cellular differentiation into roots. According to Haissig (1986), the physiology of

Dedicated to the memory of Professor K. K. Nanda.

rooting in woody cuttings is not clearly understood because of the slowness and irregularity of primordium development. Cytological-histological methods have been used to estimate the physiological status of cells that were forming primordia (Molnar and LaCroix 1972ab) but it is still unclear whether these techniques provide adequate information about the earliest stages of rooting. During the past several years, concerted attempts have been made by a number of investigators (Nanda 1975, 1979, Bhattacharya 1978, Bhattacharya and Kumar 1980, Haissig 1986, Upadhyaya et al. 1986) to envisage the macromolecular changes during initiation and development of adventitious roots in different plant species in order to better understand the underlying physiology and biochemistry. In the present review, I have described those enzymes and isoenzymes (isozymes) which have been associated with adventitious rooting in stem or hypocotyl cuttings of various plant species.

RESPIRATORY ENZYMES

The requirement of oxygen during rooting has been adequately demonstrated by using uncouplers or inhibitors of oxidative phosphorylation (Wirth 1960, Turetskaya and Kof 1965, Krul 1968, Nanda et al. 1978), which usually inhibit rooting. Nanda et al. (1978), however, reported stimulation of rooting in the presence of uncouplers of oxidative phosphorylation, through regulation of endogenous ATP levels.

SD and COx exhibited high activities with the onset of primordium initiation in *Hydrangea macrophylla* (Thunb.) Ser and their activities continued to increase with time (Molnar and LaCroix 1972a). Avers (1958) also detected activity of SD in meristematic cells of roots and root caps, and high SD activity has been reported in the epidermis of onion (*Allium* spp.) root (DeJong 1967). Recently, NAD-dependent MDH activity has been shown to increase dramatically in paclobutrazol treated cuttings in which rooting was two times greater than in untreated controls (Upadhyaya et al. 1986). In that study, MDH activity was correlated with rooting.

OXIDASES

Peroxidases
Several plant POs have been shown to catalyze oxidation of various phenols and amines by peroxide, and also aerobic oxidation of considerably different substances without H_2O_2. Such multiple catalytic properties of a single enzyme species are inconsistent with the general concept of high enzyme specificity, i.e. a strictly limited enzymic action on one substrate and its close relatives. The IAA-O extracted from *Phaseolus mungo* L., *Impatiens balsamina* L., and other plant species appears to show allosteric behavior with two sites, one representing an oxidase (site I) and the other a PO (site II). Involvement of PO in lignification, apart from its mediation of auxin levels in cuttings during rooting, has been reviewed by Haissig (1986).

Transient activity of PO concomitant with tissue differentiation has been reported by many investigators (Siegel and Galston 1967, Galston and Davis 1969, Gordon and Alldridge 1971, Chandra et al. 1971, Nanda et al. 1972, 1973d, 1974, Dhaliwal et al. 1974, Gurumurti and Nanda 1974, Frenkel and Hess 1974, Bhattacharya et al. 1975b, 1976a, Bhattacharya and Nanda 1979, Chibbar et al. 1980). Molnar and LaCroix (1972b) showed that PO was the first enzyme whose activity increased during the initiation and development of roots in *Hydrangea* spp. cuttings. Gaspar (1980) correlated an induction phase in rooting with a rise in total PO activity of the whole cutting. In the latter test, root primordium initiation took place following the peak activity, during a decrease of the basic iso-PO. Druart et al. (1982) defined adventitious rooting in two distinct phases: an "inductive" phase and an "initiative" phase during which primordia were initiated. The initiative phase started with the onset of planting of test organs in a medium suitable for rooting. In one of these experiments, the investigators used axillary cuttings from apple meristem tips cultured either in light (L) or darkness (D) during the inductive phase. Each of these types of cultures were later maintained only in light or darkness during the initiative phase. The percentage of rooted cultures increased in the order: LL, DL, and DD. It was concluded that darkness in the initiative phase produced the best rooting (Fabijan et al. 1981). The DD treatment resulted in the highest PO activity and lowest endogenous phenolic levels in the explants at the beginning of the initiative phase, compared to treatments in light. DD treatment also resulted in a significant increase in PO activity and the highest endogenous phenolic content in explants later in the initiative phase (Legrand et al. 1976). Bhattacharya and Kumar (1980) also determined the changes in phenolic content during the initiative (Phase I) and development phase (Phase II) of roots in dark grown cuttings of Phaseolus mungo. Their results indicated that phenolic content in the cuttings increased during Phase I followed by a decrease in Phase II.

In subsequent rooting trials with cultures of Cynara scolymus L. in the presence of NAA and vitamin D_2, a positive correlation between PO activity and rooting was established by Moncousin and Gasper (1983). Gaspar and associates suggested that PO may be useful as a biochemical marker to predict the success of vegetative propagation in vitro and in vivo (Thorpe et al. 1978, Gaspar et al. 1985). PO activity was also investigated by Bhattacharya and coworkers during rooting stem and hypocotyl cuttings of Populus nigra L. (Bhattacharya et al. 1975a), Salix tetrasperma Roxb. (Bhattacharya et al. 1978a), Impatiens balsamina (Dhaliwal et al. 1974), and Phaseolus mungo (Gurumurti and Nanda 1974, Bhattacharya et al. 1976a, Bhattacharya and Nanda 1978). Those studies used various cultural conditions and IAA, morphactin, or nitrogenous bases as treatments, in addition to no treatment. In most of those studies, a decrease in PO activity was correlated with the availability of a high concentration of auxin for initiation of roots; a subsequent increase in PO activity and lowered concentration of IAA was correlated with root development. These observations support Nanda's (1970, USDA, PL480 Project A7-FS-11) postulate that a high concentration of auxin is needed for root initiation and a low concentration for subsequent elongation. Nanda's contention is further supported by the findings of Chibbar et al. (1980), who studied changes in PO in maleic hydrazide treated hypocotyl cuttings of Phaseolus mungo during adventitious rooting. The latter investigation conclusively demonstrated that high PO activity was a prerequisite to root development, rather than initiation.

Synchronous changes in isoenzyme patterns of PO have been positively correlated with initiation and development of rooting in several plant species (Nanda et al. 1973a, Dhaliwal et al. 1974, Bhattacharya et al. 1975b, 1978a, Thorpe et al. 1978, Moncousin and Gaspar 1983). Chandra et al. (1971) found that at least four new isoenzymes of PO were associated with rooting of Phaseolus mungo cuttings; isoenzymes were completely inhibited in puromycin treated cuttings which did not root. Treatment of Phaseolus mungo cuttings with maleic hydrazide at different times confirmed that different isoenzymes developed during the two stages of rooting. Maleic hydrazide inhibits division of cells but not elongation (Chibbar et al. 1980).

IAA Oxidase

Metabolism of endogenous IAA has been associated with the activity of IAA-O, which may regulate the concentration of auxin in cuttings during rooting. Bhattacharya et al. (1978a) investigated the activity of IAA-O during rooting in etiolated stem segments of Salix tetrasperma by using inhibitors of nucleic acid and protein synthesis. Their results confirmed earlier observations (Saito and Ogasawara 1960) that free IAA levels declined in cuttings of Salix gracilistyla Mqi. prior to root primordium initiation. The decline may have occurred because IAA oxidase increased as a result of IAA induced PO activity or because of enhanced PPO activity (Chandra et al. 1971, Haissig unpublished). Decreased activity of PO, IAA-O, and PPO prior to primordium initiation in coumarin treated hypocotyl cuttings of Impatiens balsamina has been reported by Dhawan and Nanda (1982). In contrast to these findings, Dhaliwal et al. (1974) found that a high activity of IAA-O was associated with rooting in a medium containing different concentrations of glucose and IAA. Furthermore, by comparing isoenzyme profiles of IAA-O and PO in CY treated hypocotyl cuttings of Impatiens balsamina (Fig. 1), it was demonstrated that certain isoenzymes of IAA-O corresponded with isoenzyme profiles of PO (Dhaliwal et al. 1974). The exact physiological role of PO in rooting has yet to be elucidated, but it appears that some POs act as IAA-Os.

Indeed, Meudt (1971) postulated that oxidation of IAA was essential for its activity as a growth hormone. This postulate was supported by the research of Bhattacharya et al. (1977) in which a low

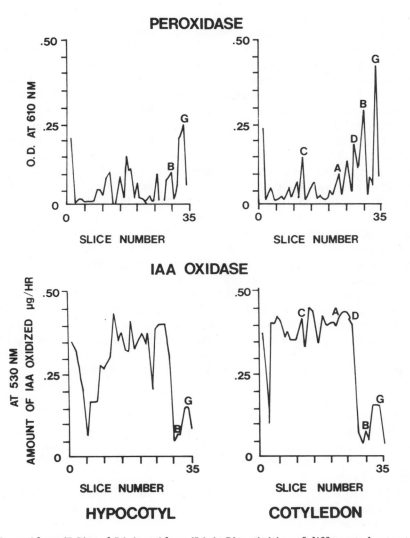

Figure 1. Peroxidase (PO) and IAA-oxidase (IAA-O) activities of different eluates taken at 24 h from hypocotylar and cotyledonary extracts of *Impatiens balsamina* cuttings cultured in cycloheximide. Ten peaks in the activity of both PO and IAA-O were observed in hypocotyls. In cotyledons, the number of peaks was 10 and 13 for PO and IAA-O, respectively. The total activity of *in vitro* IAA-O both in the cotyledons and the hypocotyls was, however, much higher (5-fold) than that of PO. Peaks A, B, C, D, and G of PO activity in the cotyledons, and B and G in the hypocotyls coincided with the respective peaks of IAA-O activity. (From Dhaliwal et al. (1974)]

concentration of AC plus IAA enhanced rooting in hypocotyl cuttings of *Phaseolus mungo*. This increased rooting was associated with increased IAA-O activity during the first 20–40 h in AC plus IAA (where root initials formed), which resulted in IAA oxidation. Temporal changes in IAA-O activity were also investigated in etiolated hypocotyl cuttings of *Phaseolus mungo* with and without longitudinal incisions (Bhattacharya and Kumar 1980). In this species, high IAA-O activity in sucrose plus IAA was also associated with pronounced rooting, compared to untreated controls. Chibbar et al. (1979) and Brunner (1978) also found increased PO and IAA-O activity during rooting of bean (*Phaseolus vulgaris* L.) cuttings. Rooting was also associated with enhanced PO and IAA-O activities in *Tradescantia albiflora* L. (Pingel 1976). In *Tradescantia*, phenolic promoters and inhibitors of IAA-O stimulated rooting, with promoters being most effective (Pingel 1976).

Bansal and Nanda (1981) investigated IAA-O activities in easy- and difficult-to-root species. Cuttings of *Salix tetrasperma, Populus robusta* L., *Hibiscus rosa-sinensis* L., and *Eucalyptus citriodora* Hook rooted 100%, 100%, 30%, and 0%, respectively. IAA oxidase activity was highest in those cuttings where rooting was most pronounced and least in those with few to no roots. Moreover, protein extracted from stem cuttings of *Eucalyptus* spp. was found to inhibit IAA-O and PO activity *in vitro* in a concentration-dependent manner (Bansal and Nanda 1981). The differential behavior of the IAA-O system has been studied extensively in a number of plant species by Nanda and associates by incubating varying concentrations of enzyme and substrate (Gurumurti et al. 1973, Nanda et al. 1975, Chibbar et al. 1979). In one of these investigations (Nanda et al. 1975), the activity curve of IAA-O showed a sigmoidal relation with the higher concentration of enzyme. This type of catalytic behavior was not found in the acetone precipitated proteins extracted from *Ipomoea fistulosa* L., where the activity curve was hyperbolic. In the latter species, rooting response varied markedly, as compared to *Salix tetrasperma, Hibiscus rosa-sinensis,* and *Phaseolus mungo.* The foregoing results suggested an allosteric nature of IAA-O. The authors postulated that IAA-O consists of two sites: one representing the oxidase (Site I) and the other PO (Site II). Nanda et al. (1975) concluded that Site I was concerned with production of active IAA oxidation products, while Site II merely detoxified the supraoptimal concentration of IAA for rooting (Nanda et al. 1975). Whereas Galston and Davies (1969) suggested that IAA-O has a detoxifying role, others have suggested that the physiological responses elicited by IAA are due to its oxidation products (Meudt 1967, 1971, Hoyle 1974, Tuli and Moyed 1979, Ockerse et al. 1970). In contrast, Haissig (1974b) suggested that the available evidence (Tomaszweski 1959, Stenlid 1963, Tomaszweski and Thimann 1966) tends to refute the IAA oxidation growth control theory as it relates to rooting. The production of new isoenzymes of IAA-O that may have been related to rooting of cuttings of *Phaseolus aureus, Populus nigra,* and *Salix tetrasperma* has also been noted but not thoroughly studied (Nanda et al. 1973b, Frenkel and Hess 1974, Bhattacharya et al. 1978a).

Polyphenol Oxidase

The significance of certain cofactors in enhancing rooting in easy-, difficult-, and obstinate-to-root species is well documented (Fadl and Hartmann 1967, Nanda 1975). Identification of these factors as phenolics and auxin conjugates has been reported by Hess and associates (cf. Haissig 1974c), but their physiological roles in rooting are unknown. The synergistic effect of auxins and phenols, and the probable involvement of oxidases and phenolases in causing synergistically enhanced rooting have been reviewed (Haissig 1986). According to Haissig (1974c), difficult-to-root tissues may not respond to applied rooting cofactors because they lack necessary active enzymes or substrates, or both. Several investigators have indirectly shown that phenolic substances undergo oxidative transformation before becoming active (Poapst and Durkee 1967, Poapst et al. 1970, Jankiewiez et al. 1973). PPO may oxidize endogenous or applied phenolics which then conjugate with oxidation products of auxin to form cofactors necessary for rooting.

Seasonal changes in rooting response of M.26 apple cuttings appeared to be associated with PPO activity and phenolic contents (Bassuk et al. 1981). For example, the product of a reaction between PPO and phloridzin *in vitro* promoted rooting of *Phaseolus mungo,* as did extracted endogenous cofactors which were separated by chromatography (Bassuk et al. 1981). Increased PPO activity during oxidative metabolism of phenolics was observed by Kaminski (1959), who was able to identify potential root primordia by histochemical localization of PPO activity. In rooting etiolated hypocotyl cuttings of *Phaseolus mungo,* the activities of phenolases (L-DOPA oxidase, monophenolase, and PPO) remained low during the initial period of rooting, which corresponded with a high concentration of endogenous phenolics. However, an inverse relation between the activities of phenolases and phenolic contents was found at later stages of rooting in this species, compared to earlier stages (Bhattacharya and Kumar 1980). However, it is not clear whether phenolics or IAA are being utilized as such or are first being oxidized. A similar trend in PPO activity was also reported during rooting of etiolated segments of *Populus nigra* (Bhattacharya et al. 1976b). Promotion of rooting by coumarin treatment in hypocotyl cuttings of *Impatiens balsamina* was associated with earlier cell division that preceded primordium initiation and with an enhanced PPO activity, all compared to control cuttings where rooting was insignificant. In marked contrast to earlier reports, a significant increase in production of roots after paclobutrazol treatment did not seem to be directly mediated through PPO activity (Upadhyaya et al. 1986). Frenkel and Hess (1974) found that in *Phaseolus aureus* cuttings, propagated in the presence of catechol and IBA, one electrophoretic band represented IAA-O, PO, and PPO activities during rooting. The kinetic properties of these enzymes and their mutual relations have been discussed (Haissig 1986).

HYDROLYTIC ENZYMES

Amylases

The need for high levels of carbohydrates at the initial stage of rooting was demonstrated by Borthwick et al. (1937). One might expect that starch, in starch storing species, might be a sugar supply (even if there were a net accumulation of starch during rooting). Thus, high Amy activity during rooting might sometimes be expected. It has been shown that there is a close correlation between the disappearance of starch and increased activity of Amy during rooting (Nanda et al. 1967, 1969, 1970, Nanda and Anand 1970, Molnar and LaCroix 1972b). Seasonal changes in rooting response of *Populus nigra* and its relation with the fluctuation of Amy activity was investigated by Nanda and Anand (1970). They reported that Amy activity was high during April–August (when rooting was profuse) but decreased in October (when rooting decreased) and was nil in December (when rooting was poor). The poor rooting in winter was, therefore, ascribed to the low temperature that depressed Amy activity (Nanda and Anand 1970).

Nanda and Jain (1972) reported that etiolated stem segments of *Populus nigra* produced roots when cultured in a medium containing starch as the carbon source, which suggested that starch was hydrolyzed to simple sugars by enzymes that leached from cuttings into the liquid rooting medium. In etiolated stem segments of *Salix tetrasperma,* rooting occurred most profusely in the segments cultured in glucose+IAA but not in water. Both CY and AC treatments suppressed rooting completely when these were added to glucose+IAA. The most interesting point that emerged from this investigation was that Amy activity did not change much in glucose+IAA at 24 h and increased sharply at 48–96 h, compared to its marked inhibition in glucose+IAA+CY/+AC. Based on these results, Bhattacharya et al. (1978a) suggested that Amy activity was somehow associated with the release of free sugars from reserve carbohydrates in the cuttings. The changes in Amy activity, as observed in rooting of etiolated stem segments of *Salix tetrasperma,* were similar in etiolated stem segments of *Populus nigra* during rooting (Bhattacharya et al. 1976b). Bhattacharya and Nanda (1978) investigated Amy activity in *Phaseolus mungo* hypocotyl cuttings, with intact apex and leaves, cultured in combinations of purine or pyrimidine, auxin, and sucrose. In the foregoing test, an increase in Amy activity during 12–24 h suggested the hydrolysis of starch to glucose after sucrose was depleted in the rooting medium. Thus, in the absence of free sugars, starch hydrolysis may have to meet the greater demand for carbohydrates during adventitious rooting, as proposed by Borthwick et al. (1937).

According to Upadhyaya et al. (1986), Amy activity was not directly related to rooting of *Phaseolus vulgaris* L. cuttings, because paclobutrazol treatment, which enhanced rooting, did not increase basal starch hydrolysis. Bhattacharya et al. (1977) reported that light grown cuttings of *Phaseolus mungo* rooted in water, IAA, and (most profusely) AC supplied with IAA. In the latter test, the paradoxical effect of AC stimulation of rooting in the presence of IAA was associated with an increase in Amy activity, which suggested the breakdown of starch into soluble sugars. Inhibition of rooting by AC treatment alone was attributed to impairment of enzyme(s) concerned with transport of soluble sugars, photosynthate, and other regulatory substances needed for rooting (Bhattacharya et al. 1977). Auxin treatment is known to increase Amy activity (Venis 1964, Nanda and Anand 1970) and rooting of stem cuttings.

The appearance of new isoenzymes of α and β Amy in etiolated segments of *Populus nigra* cultured in glucose plus IAA (which promoted rooting) and their absence in the presence of inhibitors of nucleic acid and protein metabolism (which inhibited rooting) suggested involvement of Amy isoenzymes in rooting (Bhattacharya et al. 1976b).

94   Nikhil C. Bhattacharya

Invertase (Sucrase)

The requirements for simple sugars during rooting might also be met by hydrolysis of sucrose, in addition to or in lieu of starch hydrolysis. Moreover, sucrose may be particularly important because it is the predominant form of translocatable sugar, which makes it redistributable in cuttings during rooting. Changes in INV activity and its relation with anatomical differentiation of root primordia was investigated in cuttings of *Phaseolus mungo* treated with purine and pyrimidine bases singly as well as in combination with sucrose plus IAA (Bhattacharya and Nanda 1978). The exogenous application of purine and pyrimidine bases in combination with auxin and sucrose hastened and enhanced rooting in the cuttings (Table 1). The previous authors concluded that pyrimidines are associated with the regulation of enzyme activity responsible for mobilization of food reserves to the bases of cuttings. As an example, INV activity (Fig. 2) showed two distinct peaks. The upsurge in INV activity prior to rooting appeared to "balance" the level of soluble sugars with auxin level and thereby promote rooting. It therefore appeared that an increase in INV activity resulted in hydrolysis of sucrose, producing glucose and fructose to be utilized for rooting. INV is a substrate induced enzyme.

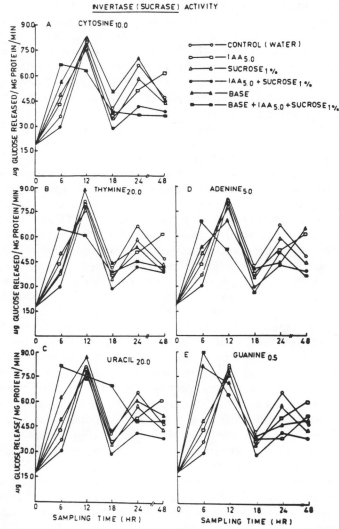

Figure 2. Periodic changes in the activity of invertase in rooting hypocotyl cuttings of *Phaseolus mungo* treated in water (control), sucrose, indole-3-acetic acid (IAA), and sucrose+IAA; and in either adenine, guanine, cytosine, uracil, or thymine alone or in combination with sucrose+IAA. Cuttings were maintained in the respective solution for 7 d in the light for rooting. [From Bhattacharya and Nanda (1978)]

Table 1. Effect of indole-3-acetic acid (IAA), sucrose (Suc), and purines and pyrimidines on rooting by hypocotyl cuttings of *Phaseolus mungo* after 7 d propagation in the solutions described below. Hypocotyl cuttings were prepared from light grown seedlings and were propagated in the light. Data for number of roots per cuttings are means ± S.E. (n=10). Treatments which yielded the greatest number of roots per cuttings also reduced the time for root emergence from 72 h (control) to 40–48 h.

Treatment (% or mg·l⁻¹)	No. Rooted	No. Roots per Cutting
Water (Control)	10	5.8 ± 0.1
IAA, 5.0	9	7.0 ± 0.3
Suc, 1.0	8	10.5 ± 0.4
IAA, 5.0 + Suc, 1.0%	9	15.8 ± 0.3
Adenine, 5.0	10	6.2 ± 0.3
IAA, 5.0 + Suc, 1.0% + adenine, 5.0	8	72.0 ± 2.1
Guanine, 0.5	10	6.5 ± 0.2
IAA, 5.0 + Suc, 1.0% + guanine, 0.5	10	39.5 ± 0.9
Cytosine, 10.0	9	6.8 ± 0.3
IAA, 5.0 + Suc, 1.0% + cytosine, 10.0	10	63.0 ± 2.0
Thymine, 20.0	10	6.0 ± 0.2
IAA, 5.0 + Suc. 1.0% + thymine, 20.0	9	81.5 ± 2.1
Uracil, 20.0	10	6.0 ± 0.3
IAA, 5.0 + Suc, 1.0% + uracil, 20.0	9	52.0 ± 1.2

Phosphatases

Origins of adventitious roots and locales of root initiation in explants have been reviewed (Haissig 1974a). The morpho-physiological events leading to emergence of roots through various tissues have been associated with crushing or hydrolysis of cells and tissues. For example, Bell and McCulley (1970) have shown that cortical cells ahead of an advancing root tip separate (after enzymic digestion of the middle lamella) or collapse, and become devoid of proteins and nucleic acids. Such hydrolytic activities would require enzymes (pectinase, protease, nuclease, etc.) and associated bioenergetics suitable for overcoming the constraints of physical barriers to emerging roots. In a test of the foregoing hypothesis, Bhattacharya and Nanda (1979) measured activities of AP, ALK, and ATPase during rooting in *Phaseolus mungo* cuttings. Their findings indicated that phosphatases were involved in the processes of root emergence from cuttings. For example, the activities of phosphatases remained low during early stages of rooting and increased thereafter, which suggested increased energy requirements as rooting progressed. The high energy requirement at the later stage was associated with the emergence of roots, and associated cell displacement and crushing. Nanda et al. (1978) reported that 2,4-dinitrophenol treatment enhanced rooting in hypocotyl cuttings made from dark-grown seedlings of *Phaseolus mungo*. Rooting was increased by treatment with IAA plus sucrose. In the presence of IAA plus sucrose, rooting was also enhanced by sodium azide and ammonium sulphate treatments, two non-phenolic uncouplers of oxidative phosphorylation. It appeared that the enhancement of rooting by those treatments was due to increased respiration triggered by uncoupling of phosphorylation. In stimulating respiration, the treatments would probably have reduced endogenous ATP levels. This postulate was also supported by the observation that rooting in this system was depressed by cobalt and sodium azide treatments, which counteract the uncoupling effect of 2,4-dinitrophenol, and also by exogenous ATP treatment (Nanda and Dhawan 1976, Nanda et al. 1978). The physiological significance of AP, ALK, and ATPase in rooting has been previously discussed (Bhattacharya et al. 1976a).

Nucleases

Quantitative and qualitative changes in macromolecules prior to rooting have been determined in many plant species (see Haissig 1974b, 1986). The necessity for a specific protein pool size (Nanda et al. 1973c) and *de novo* synthesis of mRNA (Böttger and Lüdemann 1964) for rooting have also been reported (Nanda and Bhattacharya 1973). There are primarily two different cellular pathways for synthesis of nucleotides. One is a *de novo* pathway in which ribose phosphate, certain amino acids, CO_2, and NH_3 are combined in successive reactions to form the nucleotides. Neither the free nitrogenous bases (adenine, guanine, cytosine, uracil, and thymine) nor the corresponding nucleosides (or deoxyribonucleosides) are intermediates in the *de novo* pathway. In essence, nucleotides may be synthesized without passing through a stage or pool containing free bases or nucleosides.

In marked contrast to the *de novo* pathway, there are mechanisms for converting free bases and nucleosides, produced by hydrolysis of nucleic acids, back into nucleotides. Kornberg (1974) described this second pathway and termed it the "Salvage Pathway."

A cell that lacks essential enzyme activity in the *de novo* pathway requires a salvage pathway for survival. Lack of an intact *de novo* pathway may be due to the inherent genetic makeup of the cell or to a deficiency resulting from disease, drugs, or poisons, or physiological stresses (Kornberg 1974). In any case, salvage of purine and pyrimidine bases and nucleosides produced from hydrolysis of nucleic acids becomes vital for the plant when there is inadequate *de novo* synthesis of nucleotides.

Operation of a salvage pathway in nucleic acid metabolism has not been explored during rooting because of limited data on RNase and DNase. However, utilization of precursor pools for nucleic acid anabolism during rooting is indicated by some studies. For example, exogenously applied nucleic acid bases (Bhattacharya et al. 1978b), pentose sugars (Bhattacharya et al. 1976d), and even DNA and RNA (Bhattacharya et al. 1976c) may stimulate rooting of *Phaseolus mungo* and *Impatiens balsamina* cuttings. In the latter tests, applied DNA and RNA may have been hydrolyzed to their respective purine and pyrimidine bases, and pentose sugars, which contributed to precursor pools for subsequent nucleic acid anabolism. The foregoing results led to investigations of DNase and RNase activities in cuttings of the preceding species. These studies used purine and pyrimidine treatments, and measured DNA, RNA, and protein contents during rooting. It was learned that DNase and RNase activities increased during early stages (18–24 h). These results supported the hypothesis that free DNA or RNA could be used, after partial catabolism by nucleases, for the synthesis of new DNA and RNA needed for rooting. The role of RNase in rooting has been discussed elsewhere (Bhattacharya et al. 1976a, Upadhyaya et al. 1986).

GLUTAMATE DEHYDROGENASE

Nanda et al. (1973c) postulated that the magnitude of rooting by stem segments of *Populus nigra* depended on the size of the protein pool available at the time rooting began. In this species, quantitative and qualitative changes in NAD-dependent GDH activity have been associated with rooting. For example, in etiolated stem segments of *Populus nigra,* the low activity of GDH in all rooting media tested corresponded with primordium initiation, which occurred at about 24 h according to microscopic analysis (Bhattacharya et al. 1975a). On the contrary, GDH activity increased significantly after AC treatment, which completely inhibited rooting. Thus, low GDH activity appeared to be associated with an enhanced rate of cellular growth leading to root differentiation. Inhibition of rooting by AC treatment may have been due to a concomitant decrease in the protein pool caused by increased GDH mediated deaminations. Pahlich and Joy (1971) postulated that GDH catalyzes deamination and is particularly involved in interconversions of amino acids derived from storage proteins. Furthermore, CY treatment inhibited rooting and synthesis of new GDH isoenzymes in etiolated stem segments of *Populus nigra,* compared to glucose plus IAA treatments, which promoted rooting. These results suggested the positive association of GDH isoenzymes with rooting (Bhattacharya et al. 1975a).

CONCLUSION

Rooting seems to be mediated through a chain of biochemical reactions localized in the nucleus and cytoplasm. It also appears that the genetic programming which controls rooting depends on the effect of interacting external and internal factors. Thus, variation in rooting of cuttings has been associated with changes in levels of endogenous growth regulators and metabolites in cuttings, and sometimes the stock plants. Such regulatory processes are controlled through qualitative and quantitative changes in enzymes. Although some research has been conducted on enzymes during rooting only limited information is available about macromolecules associated with different pathways prior to and during rooting. Some of the apparently key enzymes like Amy, PO, IAA-O might be used to predict the success of rooting and to determine optimum conditions of stock plant growth and propagation of cuttings. PO is being studied extensively during rooting of cuttings of numerous plant species. PO activity has sometimes been found suitable as a biochemical marker for use in screening plants for rooting ability. Along with enzymes, use of nucleic acid, pentose sugar, and purine and pyrimidine base treatments has demonstrated the possibilities of these biomolecules as predictive markers during rooting. However, more research is needed on respiratory enzymes, nucleases, phosphatases, and biophysical phenomenon involved in rooting of different genotypes. Only such research will help us to understand the sequential changes in the genes that are responsible for adverse changes in macromolecular synthesis in difficult-to-root species during rooting. For example, genetic engineering might be used to influence transcription or translation thereby causing deletion or substitution of critical metabolites, which can be precisely studied.

ACKNOWLEDGMENTS

I thank Dr. Sheila Bhattacharya, Carver Research Foundation, Tuskegee University, and Dr. A. W. Naylor, Duke University, for helpful suggestions during the preparation of this chapter; and Ms. Ritika Bhattacharya, Ms. Sandra Bowling, and Ms. Mary L. M. Brown for typing the manuscript. I acknowledge financial support from the Department of Agricultural Sciences.

REFERENCES

Avers, C. J. 1958. Histochemical localization of enzyme activity in the root epidermis of *Phleum pratense*. L. *Amer. J. Bot.* 45:609–613.

Bansal, M. P. and K. K. Nanda. 1981. IAA oxidase activity in relation to adventitious root formation on stem cuttings of some tree species. *Experientia* 37:1273–1274.

Bassuk, N. L., L. D. Hunter and B. H. Howard. 1981. The apparent involvement of polyphenol and phloridzin in the production of apple rooting cofactors. *J. Hortic. Sci.* 56:313–322.

Bell, J. K. and M. E. McCully. 1970. A histological study of lateral root initiation and development on *Z. mays. Protoplasma.* 70:179–205.

Bhattacharya, N. C. 1978. Physiology of adventitious root formation in relation to photosynthates. In *Winter School in Recent Advances in Physiology of Crop Production* (C. P. Malik, A. K. Srivastava and N. C. Bhattacharya, eds). Dept. Bot., Punjab Agric. Univ., India, Mar 20–22. pp. 15–17.

_____ A. Kumar. 1980. Physiological and biochemical studies associated with adventitious root formation in *Phaseolus mungo* L. in relation to auxin-phenol synergism. *Biochem. Physiol. Pflanzen* 175:421–436.

_____ S. Bhattacharya and K. K. Nanda. 1975a. Changes in glutamate dehydrogenase isoenzyme patterns in relation to rooting of etiolated stem segments of *Populus nigra. Biochem. Physiol. Pflanzen* 167:113–117.

_____ N. P. Kaur and K. K. Nanda. 1975b. Transients in isoperoxidases in rooting of etiolated stem segments of *Populus nigra. Biochem. Physiol. Pflanzen* 167:159–164.

_____ S. Bhattacharya and C. P. Malik. 1976a. Transients in the activities of some hydrolases and their relationship with adventitious root formation in hypocotyl cuttings of *Phaseolus mungo. Plant Sci.* 8:38–44.

98 Nikhil C. Bhattacharya

_____ S. S. Parmar and K. K. Nanda. 1976b. Isoenzyme polymorphism of amylase and catalase in relation to rooting etiolated stem segments of *Populus nigra*. *Biochem. Physiol. Pflanzen* 170:133–142.

_____ U. Parmar, C. P. Malik and S. Bhattacharya. 1977. Paradoxical effects of actinomycin-D in rooting hypocotyl cuttings of *Phaseolus mungo*. *Z. Pflanzen Physiol.* 85:377–382.

_____ S. Bhattacharya and K. K. Nanda. 1978a. Isoenzyme polymorphism of peroxidases. IAA-oxidase, catalase and amylase in rooting etiolated stem segments of *Salix tetrasperma*. *Biochem. Physiol. Pflanzen* 172:439–452.

Bhattacharya S. and K. K. Nanda. 1978. Stimulatory effect of purine and pyrimidine bases and their role in the mediation of auxin action through the regulation of carbohydrate metabolism during adventitious root formation in hypocotyl cuttings of *Phaseolus mungo*. *Z. Pflanzen Physiol.* 88:283–293.

_____ 1979. Promotive effect of purine and pyrimidine bases and their role in the mediation of auxin action through regulation of oxidases and phosphatases in rooting cuttings of *Phaseolus mungo*. *Indian J. Exp. Biol.* 17:40–45.

_____ N. C. Bhattacharya and K. K. Nanda. 1976c. Effect of exogenous application of nucleic acids and auxin on the rooting of hypocotyl cuttings of *Impatiens balsamina*. An evidence for the uptake of information molecules. *Experientia* 32:1301–1302.

_____ _____ 1976d. Synergistic effect of sugar moieties of nucleic acids of ribose and 2-deoxyribose with nutrition and auxin in rooting hypocotyl cuttings of *Phaseolus mungo*. *Plant Cell Physiol.* 17:399–402.

_____ _____ _____ 1978b. The promotive effect of purine and pyrimidine bases in rooting hypocotyl cuttings of *Phaseolus mungo* in relation to auxin and nutrition. *Physiol. Plant.* 42:391–394.

Borthwick, H. A., K. C. Hamner and M. W. Parker. 1937. Histological and microchemical studies of the reaction of tomato plants to indole-3-acetic acid. *Bot. Gaz.* 98:491–519.

Böttger, I. and I. Lüdemann. 1964. Über die Bildung einer stoffwechsel-aktiven Ribonucleinsäurefraktion in isolierten Blättern von *Euphorbia pulcherima* zu Beginn der Wurzelregeneration. *Flora* 155:331–340.

Brunner, H. 1978. Einfluss verschiedener Wuchsstoffe und Stoffwechselgifte auf wurzel-regenerierendes Gewebe von *Phaseolus vulgaris* L. Veränderungen des Wuchsstoffgehaltes sowie der Peroxydase- und der IAA-Oxydase-Aktivität. *Z. Pflanzenphysiol.* 88:13–23.

Chandra, G. R., L. E. Gregory and J. F. Worley. 1971. Studies on the initiation of adventitious roots on mung beans hypocotyl. *Plant Cell Physiol.* 12:317–324.

Chibbar, R. N., K. Gurumurti and K. K. Nanda. 1979. Changes in IAA oxidase activity in rooting hypocotyl cuttings of *Phaseolus mungo* L. *Experientia* 35:202–203.

_____ _____ _____ 1980. Effect of maleic hydrazide on peroxidase isoenzymes in relation to rooting hypocotyl cuttings of *Phaseolus mungo*. *Biol. Plant.* 29:1–6.

DeJong, D. W. 1967. An investigation of the role of plant peroxidase in cell wall development by the histochemical method. *J. Histochem. Cytochem.* 15:335–346.

Dhaliwal, G., N. C. Bhattacharya and K. K. Nanda. 1974. Promotion of rooting by cycloheximide on hypocotyl cuttings of *Impatiens balsamina* and associated changes in the pattern of isoperoxidases. *Indian J. Plant Physiol.* 17:73–81.

Dhawan, R. S. and K. K. Nanda. 1982. Stimulation of root formation on *Impatiens balsamina* L. cuttings by coumarin and the associated biochemical changes. *Biol. Plant.* 24:177–182.

Druart, P., C. Kevers, P. Boxus and T. Gaspar. 1982. *In vitro* promotion of root formation on apple shoots through darkness effect on endogenous phenols and peroxidases. *Z. Pflanzenphysiol.* 108:429–436.

Fabijan, D., E. Yeung, I. Mukherjee and D. M. Reid. 1981. Adventitious rooting in hypocotyls of sunflower (*Helianthus annuus*) seedlings. I. Correlative influences and developmental sequence. *Physiol. Plant.* 53:578–588.

Fadl, M. S. and H. T. Hartmann. 1967. Effect of reciprocal bud graft transfers between Old Home and Bartlett pears and centrifugation on translocation of endogenous growth substances in hard wood cuttings. *Physiol. Plant.* 20:802–813.

Frenkel, C. and C. E. Hess. 1974. Isoenzymic changes in relation to root initiation in mung bean. *Can. J. Bot.* 52:295–297.

Galston, A. W. and P. J. Davies. 1969. Hormonal regulation in higher plants. *Science* 163:1288–1297.

Gaspar, T. 1980. Rooting and flowering, two antagonistic phenomena from a hormonal point of view. In *Aspects and Prospects of Plant Growth Regulators* (B. Jeffcoat, ed.). British Plant Growth Regulator Group, Wantage, England. Monograph 6. pp. 34–49.

_____ C. Penel, J. C. Federico and H. Greppin. 1985. A two step control of basic and acidic peroxidases and its significance for growth and development. *Physiol. Plant.* 64:418–423.

Gordon, A. R. and N. A. Alldridge 1971. Cytohistochemical localization of peroxidases in developing stem tissues of extreme dwarf tomato. *Can. J. Bot.* 49:1487–1496.

Gurumurti, K. and K. K. Nanda. 1974. Changes in peroxidase isoenzymes of *Phaseolus mungo* hypocotyl cuttings during rooting. *Phytochemistry* 13:1089–1093.

_____ R. N. Chibbar and K. K. Nanda. 1973. Evidence for the mediation of indole-acetic acid effects through its oxidation products. *Experientia* 30:997–998.

Haissig, B. E. 1974a. Origins of adventitious roots. *N. Z. J. For Sci.* 4:299–310.

_____ 1974b. Influences of auxins and auxin synergists on adventitious root primordium initiation and development. *N. Z. J. For Sci.* 4:311–323.

_____ 1974c. Metabolism during adventitious root primordium initiation and development. *N. Z. J. For Sci.* 4:324–327.

_____ 1986. Metabolic processes in adventitious rooting of cuttings. In *New Root Formation in Plants and Cuttings* (M. B. Jackson, ed). Martinus Nijhoff Pub., Dordrecht/ Boston/Lancaster. pp. 141–189. ISBN-90-247-3260-3.

Hoyle, M. C. 1974. The hypothetical case for methylene-oxindole as a plant growth arrestor. In *Mechanisms of Regulation of Plant Growth* (R. L. Bielski, A. R. Ferguson and M. M. Cresswell, eds). *Royal Sci. of N. Z. Bull.* 12:659–664.

Jankiewicz, L. S., T. Bojarczuk and M. G. Piatkowski. 1973. The effect of rutin and pyrogallol upon rooting of soft wood cuttings of magnolia and of *Syringa meyeri* Schnied. *Acta Agrobot.* 26:277–283.

Kaminski, C. 1959. Recherches sur les phénoxy-dases dans les hypocotyles de *Impatiens balsamina* L. *Bull. Acad. R. Belg. Cl. Sci.* 65:154–168, 299–315.

Kornberg, A. 1974. *DNA Synthesis.* W. H. Freeman and Co., San Francisco, CA, USA. ISBN 0716705869.

Krul, W. R. 1968. Increased root initiation in pinto bean hypocotyls with 2,4-dinitrophenol. *Plant Physiol.* 43:439–441.

Legrand, B., T. Gaspar, C. Penel and H. Greppin. 1976. Light and hormonal control of phenolic inhibitors of peroxidase in *Cichorium intybus* L. *Plant Biochem. J.* 3:119–127.

Meudt, W. J. 1967. Studies of the oxidation of indole-3-acetic acid by peroxidase enzymes. *Ann. N. Y. Acad. Sci.* 144:118–128.

_____ 1971. Oxidation of indole-3-acetic acid by peroxidase enzyme. II. Sulfite and manganous ion interaction with intermediate oxidation products. In *Plant Growth Substances* (D. J. Carr, ed). Springer Verlag, Berlin, W. Germany. pp. 110–116.

Molnar, J. M. amd L. J. LaCroix. 1972a. Studies of the rooting of cuttings of *Hydrangea macrophylla:* Enzyme changes. *Can. J. Bot.* 50:315–322.

_____ _____ 1972b. Studies on the rooting of cuttings of *Hydrangea macrophylla:* DNA and protein changes. *Can. J. Bot.* 50:387–392.

Moncousin, C. H. and T. Gaspar. 1983. Peroxidase as a marker for rooting improvement in *Cynara scolymus* L. cultured *in vitro. Biochem. Physiol. Pflanzen* 178:263–271.

Nanda, K. K. 1975. Physiology of adventitious root formation. *Indian J. Plant Physiol.* 18:80–89.

_____ 1979. Adventitious root formation in stem cuttings in relation to hormones and nutrition. In *Recent Researches in Plant Sciences* (S. S. Bir, ed). Kalyani Publishers, New Delhi, India. pp. 461–492.

_____ V. K. Anand. 1970. Seasonal changes in auxin effects on rooting of stem cuttings of *Populus* nigra and its relationship with mobilization of starch. *Physiol. Plant.* 23:99–107.

_____ N. C. Bhattacharya. 1973. Electrophoretic separation of ribonucleic acids on polyacrylamide gels in relation to rooting of etiolated stem segments of *Populus nigra*. *Biochem. Physiol. Pflanzen* 164:632–35.

_____ A. K. Dhawan. 1976. A paradoxical effect of 2,4-dinitrophenol in rooting hypocotyl cuttings of *Phaseolus mungo*. *Experientia* 32:1167–1168.

_____ M. K. Jain. 1972. Utilization of sugars and starch as carbon sources in the rooting of etiolated stem segments of *Populus nigra*. *New Phytol.* 71:825–828.

_____ A. N. Purohit and A. Bala. 1967. Effect of photoperiod, auxins and gibberrellic acid on rooting of stem cuttings of *Bryophyllum tubiflorum* Harv. *Physiol. Plant.* 20:1096–1102.

_____ _____ V. K. Kochhar. 1969. Effect of auxins and light on rooting stem cuttings of *Populus nigra* and its relationship with mobilization of starch. *Physiol. Plant.* 23:99–107.

_____ _____ P. Kumar. 1970. Some investigations of auxin effect on rooting of stem cuttings of forest plants. *Indian Forester* 96:171–87.

_____ N. C. Bhattacharya, K. G. Murti and N. P. Kaur. 1972. Studies on isoperoxidases and their relationship with rooting stem and hypocotyl cuttings of *Populus nigra* and *Phaseolus mungo*. In *Symp. on Curr. Trends in Plant Sci.* (H. Y. Mohan Ram, ed). Delhi Univ., New Delhi, India, Oct 3–10. pp. 28–30.

_____ _____ N. P. Kaur. 1973a. Mechanism of morphactin action in rooting hypocotyl cuttings of *Impatiens balsamina*. *Plant Cell Physiol.* 14:207–11.

_____ _____ _____ 1973b. Disc electrophoretic studies of IAA oxidase and their relationship with rooting of etiolated stem segments of *Populus nigra*. *Physiol. Plant.* 29:442–444.

_____ M. K. Jain and N. C. Bhattacharya. 1973c. Rooting response of etiolated stem segments of *Populus nigra* to antimetabolites in relation to auxin and nutrition. *Biol. Plant.* 15:412–418.

_____ N. C. Bhattacharya and V. K. Kochhar. 1973d. Some studies on rooting of stem cuttings. In Prof. M. R. Saxena Commemoration Volume. *J. Andhra Pradesh Acad. Sci.* 11:75–99.

_____ _____ _____ 1974. Biochemical basis of adventitious root formation on stem cuttings. *N. Z. J. For Sci.* 4:347–58.

_____ K. Gurumurti and R. N. Chibbar. 1975. Evidence for the allosteric nature of IAA-oxidase system in *Phaseolus mungo* hypocotyl. *Experientia* 31:635–636.

_____ G. L. Bansal, V. K. Kochhar and N. C. Bhattacharya. 1978. Effect of some metabolic inhibitors of oxidative phosphorylation on rooting of hypocotyl cuttings of *Phaseolus mungo*. *Ann. Bot.* 42:659–663.

Ockerse, R. J., J. Waber and M. F. Mescher. 1970. The promotion of IAA-oxidase by GA_3 in terminal pea buds. *Plant Physiol.* (Suppl.) 46:47.

Pahlich, E. and K. W. Joy. 1971. Glutamate dehydrogenase from pea roots, purification and properties. *Can. J. Biochem.* 49:127–138.

Pingel, U. 1976. Der Einfluss phenolischer Aktivatoren und inhibitoren der IES-Oxydase-Aktivität auf die Adventivbewurzelung bei *Tradescantia albiflora* Z. *Pflanzenphysiol.* 79:109–120.

Poapst, P. A. and A. B. Durkee. 1967. Root-differentiating properties of some simple aromatic substances of the apple and pear fruit. *J. Hortic. Sci.* 42:429–438.

_____ _____ F. B. Johnston. 1970. Root differentiating properties of some glycosides and polycylic phenolic compounds found in apple and pear fruits. *J. Hortic. Sci.* 45:69–74.

Saito, Y. and R. Ogasawara. 1960. Studies on rooting of cuttings and changes of growth substances in *Salix gracilistyla* Mqi. *J. Jap. For. Soc.* 42:331–334.

Siegel, B. L. and A. W. Galston. 1967. Indole-acetic-acid oxidase activity of apoperoxidase. *Science* 157:1557–1559.

Stenlid, G. 1963. The effect of flavonoid compounds on oxidative phosphorylation and on the enzymatic destruction on indole-acetic acid. *Physiol. Plant.* 16:110–120.

Thorpe, T. A., M. T. Than Van and T. Gaspar. 1978. Isoperoxidase in epidermal layers of tobacco and changes during organ formation *in vitro*. *Physiol. Plant.* 44:388–394.

Tomaszewski, M. 1959. Chlorgenic acid-phenolase as a system inactivating auxin isolated from leaves of some of *Prunus* L. species. *Bull. Acad. Pol. Sci. Series Biol.* 7:127–130.

_____ K. V. Thimann. 1966. Interactions of phenolic acids, metallic ions and chelating agents on auxin-induced growth. *Plant Physiol.* 41:1443–1454.

Tuli, V. and H. S. Moyed. 1969. The role of 3-methylenoxindole in auxin action. *J. Biol. Chem.* 2:4916–4920.

Turetskaya, R. K. and E. M. Kof. 1965. Dynamics in the changes in auxins and inhibitors in green and etiolated bean cuttings in the process of root formation. *Dokl. Bot. Sci.* 165:118–121.

Upadhyaya, A., T. D. Davis and N. Sankhla. 1986. Some biochemical changes associated with paclobutrazol-induced adventitious root formation on bean hypocotyl cuttings. *Ann. Bot.* 57:309–315.

Venis, M. A. 1964. Induction of enzymatic activity by indole-3-acetic acid and its dependence on synthesis of ribonucleic acid. *Nature* 202:900–901.

Wirth, K. 1960. Experimentelle Beeinflussung der Organbildung an *in vitro* kultivierten Blattstücken von *Begonia rex. Planta* 54:265–293.

CHAPTER 8

Water Relations and Adventitious Rooting

K. Loach

Institute of Horticultural Research, Worthing Road,
Littlehampton, West Sussex, BN17 6LP, England

INTRODUCTION 102
CHANGES IN WATER STATUS DURING ROOTING 103
    Psychrometric Studies 103
    Gravimetric Studies 106
    Stomatal Changes 108
WATER UPTAKE BY CUTTINGS 109
    Through Leaves 109
    Through the Stem Base 110
FACTORS AFFECTING WATER RELATIONS OF CUTTINGS 111
    Irradiance 111
    Carbon Dioxide 112
    Humidity 112
    Temperature 112
    Wetting of Foliage 113
STOCK PLANT WATER RELATIONS 113
ROLE OF WATER IN ROOTING 113
CONCLUSION 114
ACKNOWLEDGMENTS 115
REFERENCES 115

Additional key words: propagation, mist, fog, water potential, stomata, evaporimeter.
Abbreviations: ABA, abscisic acid; ARF, adventitious root formation; EM, polyethylene-enclosed mist; EP, non-misted polyethylene enclosure; OM, open-bench mist (in glasshouse); P, turgor (pressure) potential; PAR, photosynthetically active radiation (400–700 μm) (unless PAR is specified, radiation figures quoted are total short-wave radiation); PEG, polyethylene glycol; r_{leaf}, leaf resistance to water vapor transfer; s.d., saturation deficit; V, water vapor pressure (V_{air}, in air; V_{leaf}, in leaf); ψ, water potential; π, osmotic (solute) potential.

INTRODUCTION

A central feature of the behavior of leafy cuttings is that lacking roots, they readily develop water deficits. It has long been recognized that turgor must be maintained in the cuttings to achieve good rooting. Evans (1952) wrote that, "even a slight water deficit which may be insufficient to cause any visual symptoms of distress results in considerable delay or reduction in the rooting response." Surprisingly, there was little quantitative evidence to support this statement at the time. The purpose of this chapter is to review current knowledge relating to water status and its physiological role in rooting.

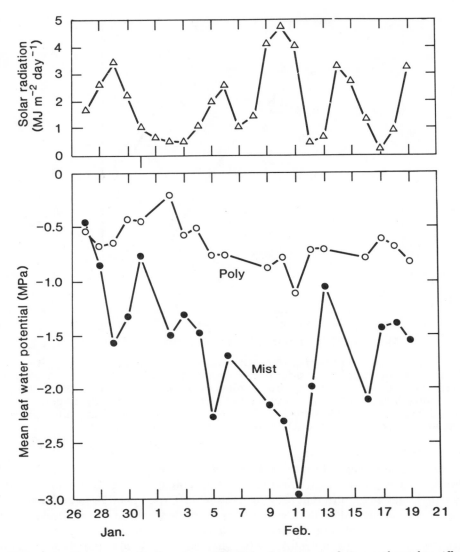

Figure 1. Changes in leaf water potential of cuttings of *Ceanothus thyrsiflorus* Esch. during propagation under mist or clear polyethylene (poly) covers, in relation to solar radiation. ψ values are means of readings at 0900 and 1400 h.

CHANGES IN WATER STATUS DURING ROOTING

Psychrometric Studies

Loach (1977) measured ψ in the leaves of cuttings under mist and shaded polyethylene covers. Water potential was consistently lower (more negative) in the cuttings under mist and daily variations in ψ related to radiation levels (Fig. 1). Characteristically, ψ decreased from a relatively high value in the morning to a low value around noon and recovery began in the late afternoon and continued overnight. Following a bright day, recovery in the water status of cuttings under mist was often incomplete and ψ started from a relatively low value on the succeeding day. In cuttings of *Ceanothus thyrsiflorus* Esch., ψ fell as low as −3.9 MPa under mist and −1.2 MPa under polyethylene in the early afternoon, and at 0900 h averaged −1.4 and −0.54 MPa, respectively in the two systems. Fifty-six percent of the cuttings rooted under mist compared to 96% under shaded polyethylene.

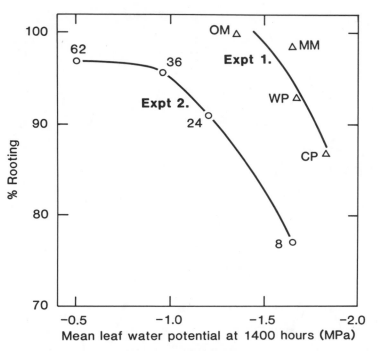

Figure 2. Percent rooting in cuttings of *Hebe* 'Caledonia' in relation to measured leaf water potentials, averaged over the period prior to root emergence. Experiment 1 compared four propagation systems: open mist (OM), mist applied over a cloth mesh above the cuttings (MM), clear polyethylene cover (CP), and white polyethylene cover (WP). Experiment 2 used a polyethylene cover with additional shading to give four light treatments. Figures indicate the percent shading factor. [Data from Loach and Whalley (1978)]

Further evidence that rooting is influenced by the degree of water stress developed by the cuttings was obtained in a comparison of water potential and rooting under four propagation systems (Fig. 2, Experiment 1), and under different shading treatments (Experiment 2). In both experiments, rooting percentages for cuttings of *Hebe* 'Caledonia' were closely related to the measured leaf ψ, averaged over the period before rooting occurred (Fig. 2). Note that in Experiment 1, bright weather (April–June) gave lower ψ under clear polyethylene than in mist—a reversal of the situation in Fig. 1—but best rooting occurred in cuttings with the highest ψ (least water stress) in each case, irrespective of propagation system.

While leaf ψ provides a useful guide to the water status of cuttings and frequently relates to rooting, it is not necessarily the most pertinent measure in this situation. For growth processes in tissue regularly subjected to water deficits, P is more significant physiologically than ψ (Hsiao et al. 1976). Estimation of P requires measurement of π, and P is then derived by: $P = \psi - \pi$ [neglecting matric potential which is likely to be insignificant (Tyree and Jarvis 1982)].

The relevance of P is illustrated in Table 1 which compares water status of cuttings under OM vs. EM, both in the glasshouse. Leaf water potential was appreciably lower in cuttings under OM but

Table 1. Mean water (ψ), osmotic (π), and pressure (P) potentials of leaf tissue in open and polyethylene-enclosed mist systems, for 30 species ($\pm$ S.E. in parenthesis). [Data from Grange and Loach (1984)]

	Potentials (MPa)		
	ψ	π	P
Open mist	−0.77 (0.10)	−1.62 (0.08)	0.85 (0.07)
Enclosed mist	−0.34 (0.05)	−1.15 (0.07)	0.81 (0.06)

π_{leaf} was also lower in this system, possibly because elevated temperatures in the enclosure increased respiration and reduced net photosynthesis thereby reducing the amount of osmotically active solutes. In consequence, P was similar in both systems as were the average rooting percentages. These data are averages for 30 species, but for those propagations using soft, young cuttings with thin leaves, ψ was substantially lower in open mist (−1.20 vs. −0.46 MPa), so that turgor and rooting were greater in the enclosure.

A further problem in using ψ_{leaf} measurements as an indicator of the water status of cuttings, is that roots are initiated in the stem base of the cutting whose water status may differ from that of the leaves. Kemp (1952) described an extreme example in cuttings of *Pilea grandis* L. allowed to dry slowly in a propagation case. The tissues at the base remained fresh and even produced roots long after the internodal tissues had severely shrunk.

Grange and Loach (1985) measured tissue potentials in both the leaves and bases of cuttings of *Forsythia* × *intermedia* 'Lynwood' in two shade treatments. Strips of bark were removed from the lower 30 mm of stem to provide values for the bases. The results in Fig. 3 show that under the highest irradiance, leaf turgor fell to as low as 0.2 MPa after two weeks. The basal tissue however, showed

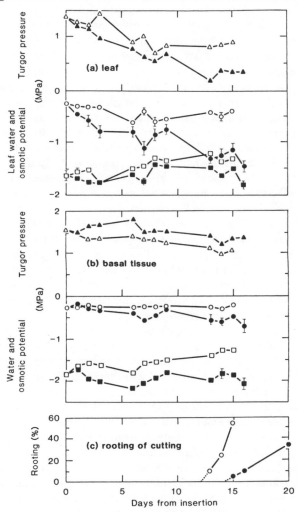

Figure 3. Changes in ψ ($\bigcirc$), π ($\square$), and P ($\triangle$) of (a) leaf and (b) basal tissue in cuttings of *Forsythia* × *intermedia* 'Lynwood'. Open symbols, cuttings in low irradiance (0.08 MJ · m⁻² · d⁻¹ total short-wave); closed symbols, cuttings in high irradiance (4.1 MJ·m⁻²·d⁻¹). Bars show ± S.E. if >0.05 MPa. Part (c) gives rooting percentages for the two irradiance treatments at several intervals after insertion.

little change in P from insertion, averaging around 1.6 MPa in the higher irradiance and 1.3 MPA in the heavily shaded treatment, where π increased slightly, i.e. there was solute (probably sugar) loss. Interestingly, rooting occurred fastest in the latter treatment with greatest leaf but lowest basal turgor; it seems that in this instance, basal turgor was adequate in both light treatments and differences in π may have been more important (i.e. least solute gave best rooting).

Because tissue water potentials are so dynamic (see Fig. 1), characterization of the water status of cuttings during propagation requires many replicated measurements, not easily achieved unless a multi-chamber psychrometer unit is available. Furthermore, measurement of ψ in basal tissues is not straightforward because of the varied geometry of the sample and its influence on water vapor exchange between sample and air within a psychrometer chamber. For these reasons there is value in the simpler but more integrative, gravimetric methods for measuring water status.

Gravimetric Studies

Kemp (1952) weighed cuttings during propagation in a heavily shaded propagation case and demonstrated a pronounced diurnal fluctuation in their water content, with an evident deficit occurring each day. Hess and Snyder (1955) compared weight changes of cuttings in mist and under double-glass. Misted cuttings gained substantially more water over 35 d than did those under double-glass, and rooted correspondingly better (99% vs. 22%). It is not wholly clear from these data whether the improved water status of misted cuttings was the cause or the result of better rooting, since differences in weight were much less pronounced before rooting had occurred.

Changes in fresh and dry weights of cuttings between two harvests gives useful information on the condition of the cuttings during rooting. As in standard growth analysis procedures, precision is improved if the fresh and dry weights of a particular sample of cuttings at any harvest is known at the start of the preceding interval. Fresh weights can be measured directly but dry weight must be estimated from regressions based on a destructively harvested sample of similar cuttings. Weight changes in the foliar and basal portions of the cutting can be measured separately if reliable regressions for estimating the initial weights of the bases can be obtained. Harvesting must be carried out systematically in time across all treatments, to make reliable comparisons. Spomer (1985) discusses further precautions and methods of expressing the results.

Illustrative results are shown in Fig. 4 which presents mean data for six species propagated in three different systems during summer and a further four species in winter. Cuttings under OM in the glasshouse gained dry weight substantially faster than those in EM or EP, where temperatures were higher (which is consistent with data in Table 1). In both seasons, EM gave greatest gain in water content, while cuttings in EP lost water, especially in summer. Rooting was scored in all three systems when any one had given a reasonable "take." In summer, the percentage rooting was greatest in EM and poorest in EP, but in winter this result was reversed.

The foregoing experiment illustrates the important principle that in summer, maintenance or gain in water by the cuttings may be a prime determinant of successful rooting. Under lower irradiance during winter, the relationship between water gain and rooting was reversed: maintenance of turgor was no longer a difficulty so other factors affected rooting. There was no evident relationship between rooting and dry matter gain in either season.

A further experiment gave an indication of the nature of the factors influencing rooting in winter. Cuttings of Chamaecyparis lawsoniana 'Ellwoodii' were positioned along a fogged polyethylene tent within the glasshouse, at stations ranging from 0–8.5 m from a single, sonic fogging nozzle. Pots of cuttings were either protected by an additional polyethylene "umbrella," supported 100 mm above them to prevent direct fallout of fog droplets onto the cuttings, or they were left uncovered. This dual set of experimental arrangements achieved a range of wetness in the cuttings and media.

The foliar and basal (15 mm) portions of the cuttings were weighed individually. After 6 weeks, rooting showed no evident relationship to the change in water content of the top portion of the cuttings ($r = -0.398$). However, a close inverse relationship was observed between the eventual rooting score and the water gain in the bases measured before rooting had occurred (Fig. 5). This and further experiments using rooting media with different water contents (Loach 1985) suggest that when transpiration rates are low the bases of the cuttings can become waterlogged. Oxygen diffusion to

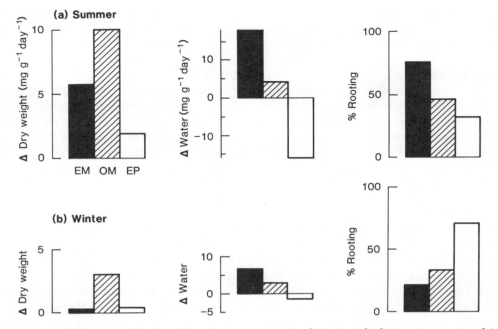

Figure 4. Changes in dry weight and water content of cuttings before roots emerged (2–5 weeks) and the eventual rooting percentages, for leafy cuttings in three different propagation systems in a glasshouse equipped with automatic shading. EM = polyethylene-enclosed mist, OM = open-bench mist, EP = non-misted polyethylene enclosure. (a) Summer: means for six species, *Rhododendron* 'Lapwing', *Corylus avellana* L. 'Aurea', *Pittosporum tobira* Ait., *Parrotia persica* (DC.) C. A. Mey., *Cotoneaster* 'Cornubia', and *Garrya elliptica* Lindl. (b) Winter: means for four species, *Cupressus glabra* Gord., *Cryptomeria japonica* 'Elegans', *Rhododendron* 'Fastuosum Flore Pleno', and *Rhododendron* 'Joseph Whitworth'.

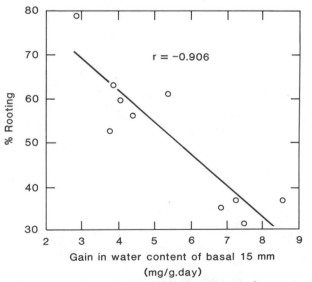

Figure 5. Percent rooting of cuttings of *Chamaecyparis lawsoniana* Parl. 'Ellwoodii' in relation to changes in water content of the basal portion of the cutting (mg·g^{-1} initial water content·d^{-1}) prior to root emergence. Cuttings were propagated in a dense fog system, in the winter. They were placed in a polyethylene enclosure at varying distances from a single fogging nozzle, and were either protected by an additional clear polyethylene umbrella to prevent direct fall-out of fog droplets onto the cuttings or were left uncovered.

developing root initials is probably obstructed and toxic accumulations of metabolic products (e.g. ethylene or carbon dioxide) may occur. Browning of the stem base often occurs and in severe cases, the cortical tissues break down completely.

Stomatal Changes

We should expect that changes in the degree of stomatal opening in the foliage of cuttings will reflect their water status, but surprisingly few relevant studies have been reported. Porometer measurements suggested that the stomata of cuttings close within a few hours of preparation and insertion and remain virtually closed until new roots have emerged (Gay and Loach 1977). However, these measurements were made on cuttings taken from the propagation bench and transferred to a controlled environment cabinet for the duration of the measurements. This procedure simplified calibration of the diffusion potometer but the rapid flow of air needed to achieve a stable temperature under artificial lighting (up to 100 $W \cdot m^{-2}$) may have inadvertently encouraged stomatal closure in unrooted cuttings by increasing the rate of water loss.

Subsequently, Butcher and Loach (unpublished) compared the rate of water loss from cuttings inserted in potometers, to that from sensitive evaporimeters. The evaporimeters consisted of a filter paper surface (1000 mm^2), attached to a horizontally mounted, water-filled, 2 ml pipette. The upper surface of the filter paper was covered with black, waterproof tape supported by a thin, wire former (see Loach 1983 for details). Water loss from leaves is restricted by several resistances (collectively r_{leaf}), of which the major variable component is stomatal resistance. Loss from evaporimeters suffers no such restriction, except for the boundary layer resistance which is of minor importance. The ratio between simultaneous evaporimeter and potometer measurements gives an indication of r_{leaf}. Results from these studies showed that the stomata of cuttings were not fully closed but adopted a degree of opening related to ambient evaporative conditions. Thus, when cuttings of *Hebe elliptica* (Forst. f.) Pennell were transferred at 2 h intervals between humid and dry polyethylene tents, r_{leaf} increased, and then decreased again with the reverse transfer. In a comparative study of three glasshouse propagation systems, r_{leaf} and evaporation were measured over 6 h intervals on four separate days. Fig. 6 shows that there was a linear relationship between the two.

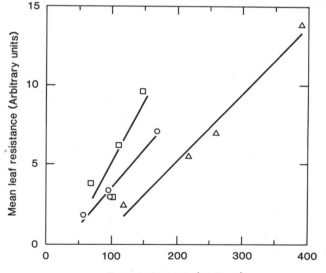

Figure 6. Leaf resistances in cuttings of *Hebe elliptica* (Forst. f.) Pennell measured over 6 h periods on four individual days, in relation to evaporation conditions in the propagation bench on those days. Simultaneous measurements were made in three systems: open mist operated by a Pye controller (O), open mist operated by an EDAC controller (□), and open bench without mist (△). Correlation coefficients were 0.96, 0.81, and 0.99, respectively.

Studies of stomatal mechanisms (Seybold 1961/62, reviewed in Cowan 1977) support the hypothesis that stomata respond rapidly to ambient humidity through direct cuticular water loss from guard and subsidiary cells—peristomatal transpiration. Localized changes in the water balance of the peristomatal region effect a rapid change in stomatal opening, without necessarily involving changes in the bulk water status of the leaf. Interestingly, the system with the lowest evaporation rates (Fig. 6, □), gave the steepest increase in r_{leaf} per unit increment in evaporation. A likely explanation is that the general level of turgor in the epidermal cells in this system generated greatest pressure on the stomatal cells and forced closure more effectively.

Leaf resistance was also estimated more directly by measurement of the rate of water loss into an atmosphere of known humidity in the laboratory. Cuttings were taken from OM, EM, and fog systems, and weight changes were followed for up to 10 min after cuttings were transferred to the laboratory (probably long enough for some closure to have occurred but nevertheless useful for comparative purposes). Calculated values for r_{leaf} are given in Table 2. These data indicate that the stomata were open in the fog and EM systems but partially closed in OM. The results accord with the notion of stomatal adjustment to ambient evaporative conditions, for measured evaporation rates were low and similar in the fog and EM but greater in OM.

Table 2. Leaf resistances to water vapor loss (r_{leaf}) in cuttings from three propagation systems. (Single measurements only for *Ulmus*, S.E. for four cuttings given in parentheses for *Cupressocyparis*).

	r_{leaf} (s·mm^{-1})		
	Open Mist	**Enclosed Mist**	**Fog**
× *Cupressocyparis leylandii*	1.50 (0.07)	0.38 (0.05)	0.47 (0.06)
Ulmus pumila 'Hansen'	1.68	0.31	0.40

It follows that measurements of the degree of stomatal opening in the leaves can provide a useful indication of the ability of a propagation system to sustain adequate turgor in the leaves of cuttings. However, it is simpler and probably more useful to measure evaporation rates directly, using evaporimeters.

WATER UPTAKE BY CUTTINGS

The water status of cuttings represents a balance between rates of water uptake and loss. The problem of uptake in cuttings, which lack roots and must therefore rely on absorption through a cut stem base or through intact leaf and stem surfaces, will now be considered.

Through Leaves

Kemp (1952) used solutions of dye to study water uptake in the semi-transparent stems of species such as *Impatiens holstii* Engler & Warb. In freshly inserted cuttings, movement of water across the cut stem surface was rapid but soon slowed as the film of water covering the wound surface contracted. Water columns within the xylem were in a state of tension. Kemp believed that condensate on the leaf surfaces diffused through the cuticle and contributed to the relief of water stress but the relative importance of foliar absorption was not assessed.

Wetzel (1924), in a study of over 100 different plants, concluded that wilted leaves were able to absorb water but that the amount conducted away from the leaves to the stem was very small. Again, the relative significance of foliar versus stem absorption was not determined. Others (see Brierley 1934) have made similar observations.

Grange and Loach (1983b) measured water absorption rates in detached, wilted leaves of nine woody ornamental species. For completely wetted leaves, uptake averaged 0.6 mg·m^{-2}·s^{-1} through adaxial and 0.9 mg·m^{-2}·s^{-1} through abaxial surfaces but there were marked differences between species. These values would be lower for more turgid or less fully wetted foliage. In comparison, uptake through the stem bases was around 8 mg·m^{-2}·s^{-1} for fresh cuttings in water and with 0.5 kPa ambient saturation deficit, but only 2 mg·m^{-2}·s^{-1} for cuttings which had been in the propagation bench for 7 d (see below). (Note: to facilitate comparisons, figures in this section are all expressed on a single surface, leaf area basis). Clearly, foliar absorption of water is of sufficient magnitude to make some contribution to the water balance in cuttings, though probably not a major one for most species.

Cameron and Rook (1974) considered four alternative pathways for entry of water into cuttings of *Pinus radiata* D. Don set into the nursery bed with all their foliage intact and with or without a mist spray: 1) the cut base; 2) through the foliage immersed in moist soil; 3) through foliage exposed to air and wetted by rain and dew; and 4) through the bark of the stem below soil level. By weighing the cuttings and by placing tritiated water on foliage, they concluded that (1) and (2) were the most important entry routes. Cremer and Svensson (1979) reached the same conclusion for this species. They applied a continuous, intense mist to detached shoots whose cut end was sealed. Uptake of water occurred through the shoot surfaces but was much slower than normal transpirational loss of water. However, in drought conditions, where transpiration was very restricted, they suggested that foliar absorption from natural fog could perhaps aid survival.

Through the Stem Base

A significant, though variable resistance to water entry into cuttings is represented by the degree of contact between the cut stem and water films around particles of the rooting medium. By measuring uptake rates under standard conditions, from a series of media prepared with sequentially increased water contents, Grange and Loach (1983b) showed that uptake increased in direct proportion to the volumetric water content of the medium, i.e. the wetter the medium the better the uptake. However, excessive wetness restricts aeration and causes other associated problems. Wounding cuttings increases the area of contact between the cutting base and the medium, so improving water uptake whether cuttings were fresh or not (Table 3).

Table 3. Water uptake by wounded and non-wounded cuttings of *Ilex* × *altaclarensis* 'Golden King' from a peat/perlite (1/1, v/v) rooting medium. (Control figures are uptake from water by unwounded cuttings; mean rates for 10 replicate cuttings ± S.E. in parentheses.) [Data from Grange and Loach (1983b)]

	Water Uptake (mg·m^{-2}·s^{-1})			
	Control	Unwounded	1 Wound	2 Wounds
Fresh cuttings	6.6 (0.6)	1.4 (0.2)	1.9 (0.2)	2.2 (0.3)
"7 d" cuttings	2.2 (0.1)	0.8 (0.1)	1.8 (0.2)	1.8 (0.2)

Grange and Loach also made simultaneous measurements of transpiration and water uptake through the stem base, for cuttings with their bases in water but subjected to a range of ambient s.d. within a plexiglass chamber. For 7-d-old cuttings, uptake lagged behind transpiration when the s.d. exceeded 0.1 kPa, and tissue water deficits developed (Fig. 7). If however the basal 5 mm of the stem was removed, then uptake increased substantially and remained in balance with transpiration as the s.d. was increased to 0.6 kPa. It was concluded that the week-old cutting had developed a significant resistance to water uptake in the basal 5 mm portion, perhaps analogous to that observed in cut flower stems.

Ikeda and Suzaki (1986) measured reductions in the hydraulic conductance of cuttings following insertion which were accompanied by a decline in xylem pressure potential. Examination by scanning electron microscopy showed that the reduced hydraulic conductance was caused by blockage of xylem vessels with tyloses in *Populus carolinensis* Moench and by the occlusion of bordered pits in *Cryptomeria japonica* D. Don. In the latter species conductance through the base of the cutting was zero after 30 d and it was suggested that water then entered through the bark and moved through a network of intercellular spaces in the xylem.

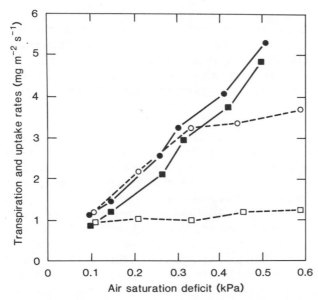

Figure 7. Changes in transpiration ($\bigcirc$) and water uptake rates ($\square$) at increasing ambient saturation deficits for cuttings of *Garrya elliptica* Lindl., 7 d after insertion. Treatments were none (open symbols) and removal of basal 5 mm of stem (closed symbols).

Loach and Grange (unpublished) treated cuttings with a wide range of biocidal agents and compounds commonly used for preservation of cut flowers, without substantially reducing the occurrence of stem blockage. This suggests that the blockages have a physical rather than a biological origin, e.g. cavitation of tracheids.

FACTORS AFFECTING WATER RELATIONS OF CUTTINGS

In the preceding sections we have shown that a variety of factors influence the water status of cuttings. It is useful here to summarize these findings in a more formal manner and include others not yet mentioned, which bear on the problem.

The rate of water loss from foliage is determined by the V gradient from the leaves to the surrounding air and by the resistances (r_{leaf} comprising cuticular, stomatal, and boundary layer components) to its transfer (T):

$$T = (V_{leaf} - V_{air})/r_{leaf}$$

A range of environmental factors operate on these parameters.

Irradiance

Irradiance exerts its influence primarily through raising the temperature of the leaf and hence the vapor pressure within its tissues (V_{leaf}). If the foliage is wetted, then part of the energy gain by illuminated leaves is used to evaporate surface water and the increase in temperature is less than when leaves are dry. An indication of the magnitude of this effect is provided by Grange and Loach (1983a) who measured a leaf temperature rise of around 1 °C per 100 W·m^{-2} (total short-wave irradiance) in a clear polyethylene tent, a corresponding figure of 0.3 when the polyethylene was laid directly on top of the cuttings, and only 0.1 for misted foliage in the open glasshouse.

Irradiance also influences the stomatal component of r_{leaf}. Its effect is complex, operating via its influence on carbon dioxide levels within the leaf on the one hand, and through its effects on cellular turgor (particularly in the peri-stomatal region), on the other. No detailed studies regarding these interactions have been carried out under propagation conditions.

Carbon Dioxide

Carbon dioxide enrichment has been shown to reduce transpiration in cuttings by inducing stomatal closure (Forschner and Reuther 1984). The improved rooting that sometimes occurs through enrichment may result from the improved water status of the tissues, from stimulated photosynthesis, or through inhibition of ethylene formation at the elevated carbon dioxide concentrations. Discriminative experiments have not been reported but would be worthwhile.

Humidity

The humidity of the air surrounding cuttings has a major effect on the water status of the cuttings so most propagation systems aim to maintain a high degree of saturation in the atmosphere. High vapor pressures are maintained either by enclosing the cuttings in water-retentive enclosures (usually polyethylene) or providing additional finely divided water droplets (humidification or fogging), or both (e.g. Whitcomb et al. 1982). Contrary to common belief, a polyethylene cover does not ensure a permanently saturated atmosphere around cuttings. With increasing irradiance, the saturation deficit of the air increases due to the associated rise in temperature and the inability of evaporation from wet surfaces within to keep pace with evaporative demand. For example, Grange and Loach (1983a) measured an increase in saturation deficit of about 0.5 kPa per 100 $W \cdot m^{-2}$ increase in irradiance under polyethylene. Shading is thus essential to operate polyethylene-enclosed systems effectively.

Using intermittent mist, the decrease in humidity (towards glasshouse ambient) between bursts of mist subjects those parts of the foliage incompletely covered by a film of water to an appreciable saturation deficit. Further information regarding this phenomenon is given in Chapter 18 by Loach.

Temperature

The effects of temperature on the water relations of cuttings are complex and often indirect, due to the interacting influences of irradiance and humidity already discussed. A rise in temperature occasioned by increased irradiance raises V_{leaf}, decreases V_{air}, and increases the rate of water loss from the cuttings. A rise in leaf temperature of 1 °C above ambient is roughly equivalent to a 6% drop in relative humidity, in terms of the leaf-to-air vapor pressure gradient.

A given leaf temperature elevation (above ambient) has a much greater effect on water loss in warm than in cool glasshouse conditions. For example calculation suggests that a 2 °C rise in leaf temperature causes 3.5 times more water loss into a saturated atmosphere at 40 °C than at 15 °C (assuming no change in r_{leaf}). There is considerable merit therefore in avoiding high temperatures during propagation both to reduce water stress and the intrinsically damaging, biological effects of high temperature.

Ambient temperatures can be moderated in dry climates by ventilating the propagation enclosure with air cooled through humidification (Milbocker and Wilson 1979). Evaporative cooling of the foliage is of course reduced at high humidities but convective loss of heat to the cooled air helps to reduce leaf temperatures.

Other effects of temperature are probably less significant. The rate of water uptake from the medium will be increased slightly at increased temperature due to a reduction of the viscosity of water at higher temperatures (27% reduction between 15 and 35 °C). Resistances to water flux in the stem and leaf are not noticeably temperature dependent (Jensen and Taylor 1961). Stomatal resistance is often minimal between 20–30 °C but increases at higher or lower temperatures (Hofstra and Hesketh 1969). However, these effects are likely to be insignificant in comparison with the effect of change in the saturation deficit.

At night, radiative heat loss from the foliage of cuttings decreases leaf temperature below that of the surrounding air, by 0.5–2 °C according to measurements of Grange and Loach (1983a). In propagation enclosures, the saturation deficit is near zero at night and water condenses on the foliage, which aids recovery from any daytime water deficit.

Wetting of Foliage

The benefits of operating with wet foliage have already been outlined, i.e. foliar absorption of water may occur in some species and evaporative cooling reduces leaf temperature and V_{leaf}. Most importantly, wet foliage provides "insurance." If the evaporative demand of the environment is satisfied by externally applied water (e.g. as mist) rather than water originating from within the leaves themselves, then severe tissue water deficits are less likely to occur. Spatial variations in evaporative demand and in leaf temperatures cannot entirely be avoided in a propagation house but wet foliage reduces their influence. Over-wetting however, brings other problems—foliar leaching, a poorly aerated rooting medium, and spread of disease, as already discussed.

STOCK PLANT WATER RELATIONS

It is often recommended that cuttings be taken from fully turgid stock plants, i.e. in the early morning or in cloudy weather (Hartmann and Kester 1983). Circumstantial evidence indicates that cuttings taken from stock plants in a wet season root better than those taken in a dry season (Cameron and Rook 1974). The inference, that cuttings fully turgid at the outset root best, is likely to be true but has not been rigorously tested.

Evans (1952) obtained poor results with cuttings taken during the dry season, when water in the vascular tissue was in a state of tension. Mere cutting of the shoot produced gas bubbles in the vessels which interfered with the upward movement of water, and resulted in wilting and leaf death. Kemp (1952) observed that air penetration during preparation of cuttings occurred only so far as the first perforation plate. Subsequently, some elimination of air embolisms from the vessels may occur but water flow through the cutting must be impaired, at least initially.

Rajagopal and Andersen (1980a) placed stock plants of Pisum sativum L. 'Alaska' in solutions of polyethylene glycol (PEG 6000) for up to 6 h before preparing cuttings. Subsequent rooting responses to these stress treatments depended on the irradiance during propagation under mist. Under low irradiance (16 W·m⁻²), stressing the stock plant reduced the mean number of roots per cutting but at high irradiance (38 W·m⁻²), 2 h of stock plant stress increased root number. They also examined the stress-induced changes in ABA content and ethylene production (Rajagopal and Andersen 1980b). An increase in ABA in response to stress seemed primarily responsible for the observed improvement in rooting at 38 W·m⁻² with 2 h of stress, though overall, the correlation between rooting and ABA was not close. These experiments illustrate the complexity of the factors relating to root formation and stock plant condition.

ROLE OF WATER IN ROOTING

The most immediately evident effect of water stress on the cutting is closure of the stomata. The threshold water potential value for onset of closure in many species is around −1 MPa (Hsiao 1973) and leaf water potentials of this magnitude frequently occur in cuttings. Stomatal closure affects carbohydrate gain through photosynthesis directly by reducing diffusion of carbon dioxide to the chloroplasts, and indirectly by causing a rise in leaf temperature. The direct effect is probably the more important, but in most propagation conditions, the bases of cuttings show an appreciable gain in dry weight preceding root initiation. It seems unlikely therefore, that in woody cuttings, carbohydrate supply commonly limits initiation of primordia; e.g. in Fig. 3b and c, best rooting occurred with least basal accumulation of solute. However, once roots are initiated, their further growth is heavily dependent upon a good supply of photosynthate. For more information on the role of carbohydrate in ARF, see Chapter 5 by Veierskov.

Alongside and possibly directly linked with stomatal closure, a rise in endogenous ABA occurs in water-stressed cuttings. As indicated above (and reviewed by Rasmussen and Andersen 1980), ABA has been shown to influence rooting in a wide range of species, but promotional effects, inhibitions, and lack of response are reported in the literature (see Chapter 13 by Davis and Sankhla for more information). Such diversity of response is common in hormone-mediated processes, probably because of the complex interactions of different classes of endogenous hormones and a dependence on their relative balance. It is possible that an increase in endogenous ABA, brought about by water stress in cuttings, is involved in ARF, but its precise role will be difficult to determine. Similarly, it has been thought likely that ethylene is involved in root induction, particularly in wet conditions which may entrap the gas (Kawase 1981) and especially due to its close relationship to auxin metabolism (Nordström and Eliasson 1984). Again, as with ABA, rooting responses to exogenously applied ethylene-producing agents differ markedly (see Chapter 11 by Mudge for more information).

Adventitious root formation clearly involves cell growth and synthesis of new components, both affected by water stress in a number of ways. Turgor pressure is essential in providing the necessary force for cell expansion; its measurement and an appreciation of its importance were discussed at the beginning of this chapter. Cell expansion is obviously important during growth of the initiated roots through the cortical tissues but even the earliest stages of initiation involve enlargement of "competent" cells in the peri-vascular region, prior to their division.

In another context Green (1968) showed that cell growth involves the extensibility of the cell as well as the generation of a critical turgor pressure (i.e. the threshold turgor below which extension will not occur). There is evidence that hormones affect extensibility, e.g. auxin (Boyer and Wu 1978), whose endogenous levels are affected by tissue water status (Darbyshire 1971). However, changes in turgor caused by reduced osmotic potentials are probably more important. The translocation of sugars from the leaves and their accumulation in the cells at its base may contribute in a major way to this reduced osmotic potential.

Other component processes of ARF known to be affected by water stress are cell wall synthesis and cell division (reviewed by Hsiao 1973). Cause and effect relationships are difficult to assess in this context. Whether these processes are directly reduced by stress or simply follow from the decreased cell expansion caused by insufficient turgor is not yet clear. The effects of water stress on protein synthesis on the other hand, are rapid, reversible, and severe, which suggests a more direct mechanism. Molnar and Lacroix (1972) showed that protein synthesis preceded cell division in the further development of preformed root initials of *Hydrangea macrophylla* (Thunb.) Ser.

In intact plants, the translocation of photosynthate from the leaves remains active under moderate or even severe stress (Boyer 1976). It seems unlikely that in cuttings the rate of translocation of photosynthate to developing root primordia seriously limits ARF, though this proposition has not been critically tested. However, except under very low irradiances, a steady gain in dry weight of the basal portion of the cutting usually occurs, even before roots form (Loach unpublished).

It is conceivable that water stress in the foliage may diminish the supply of leaf cofactors, known to synergize the action of auxin in promoting ARF at the base of the cutting (Hess 1969). No relevant studies have been reported, probably because the nature and action of the cofactors themselves are so poorly understood.

It is fully evident that this discussion of the role of water stress in ARF relies heavily on relevant studies of material other than leafy cuttings. Because sizeable variations in ψ and P are a daily feature in the life of unrooted cuttings, such material provides a useful system for studying the effects of water stress. We hope that in the future it attracts more attention from research workers.

CONCLUSION

Evidence has been presented that success in rooting cuttings depends to a considerable extent on maintaining a satisfactory water balance in the tissues. Many different factors, both internal and environmental, influence water status. Of these, the environmental influences are the most amenable to manipulation, especially those regulating transpirational loss of water, since uptake through a cut stem base is inherently limited. Control of irradiance and temperature through shading, of ambient humidity through misting or fogging, and of foliar wetting, are vital practical considerations.

Simple means of comparing the effectiveness of these practices, through measuring evaporation rates, are available, though development of automatic monitoring equipment is to be greatly desired. To date, information gained in this way has principally been used to compare systems. It should prove possible to specify the optimal range of conditions for different times of year, weather, and types of cutting material. This has not been very practical in the past because the propagation systems used have not provided a closely controlled environment. However, with improvements in automatic shading and the development of high capacity humidification equipment, closer regulation of propagation conditions is technically feasible and could substantially reduce costly propagation failures, which are presently commonplace.

We might question the cost effectiveness of such a "high-tech" approach to an operation where intrinsic biological variation in the cutting material is inevitably present and important. But, in seeking to define optimal conditions, our knowledge of the critical features of the environment will be improved, and this should contribute substantially to more effective management, even of simple propagation systems.

ACKNOWLEDGMENTS

Dr. R. I. Grange and Dr. A. R. Rees kindly reviewed this chapter, and I am grateful to them for their helpful comments and suggestions. I also thank the New Zealand Research Advisory Council which supported portions of the work reported during my tenure of a Senior Research Fellowship at the Levin Horticultural Research Centre. Mr. Stephen Butcher shared the work on evapometric measurements in propagation systems and on stomatal behavior in cuttings. Numerous colleagues have assisted in these studies over several years and I greatly value their help and advice.

REFERENCES

Boyer, J. S. 1976. Photosynthesis at low water potentials. *Phil. Trans. Royal Soc. London. B.* 273:501–512.

_____ G. Wu. 1978. Auxin increases the hydraulic conductivity of auxin-sensitive hypocotyl tissue. *Planta* 139:227–237.

Brierley, W. G. 1934. Absorption of water by the foliage of some common fruit species. *Proc. Amer. Soc. Hortic. Sci.* 32:277–83.

Cameron, R. J. and D. A. Rook. 1974. Rooting stem cuttings of *radiata* pine: environmental and physiological aspects. *N. Z. J. For Sci.* 4:291–8.

Cowan, I. R. 1977. Stomatal behaviour and environment. *Adv. Bot. Res.* 4:117–228.

Cremer, K. W. and J. G. P. Svensson. 1979. Changes in length of *Pinus radiata* shoots reflecting loss and uptake of water through foliage and bark surfaces. *Aust. For. Res.* 9:163–72.

Darbyshire, B. 1971. Changes in indoleacetic acid oxidase activity associated with plant water potential. *Physiol. Plant.* 25:80–84.

Evans, H. 1952. Physiological aspects of the propagation of cacao from cuttings. In *Proc. 13th Int. Hortic. Cong.* 2:1179–90.

Forschner, W. and G. Reuther. 1984. Photosynthese und Wasserhaushalt von Pelargonium-Stecklingen während der Bewurzelung unter dem Einfluss verschiedener Licht- und CO_2-Bedingungen. *Gartenbauwissenschaft* 49:182–190.

Gay, A. P. and K. Loach. 1977. Leaf conductance changes on leafy cuttings of *Cornus* and *Rhododendron* during propagation. *J. Hortic. Sci.* 52:509–516.

Grange, R. I. and K. Loach. 1983a. Environmental factors affecting water loss from leafy cuttings in different propagation systems. *J. Hortic. Sci.* 58:1–7.

_____ _____ 1983b. The water economy of unrooted leafy cuttings. *J. Hortic. Sci.* 58:9–17.

_____ _____ 1984. Comparative rooting of eighty-one species of leafy cuttings in open and polythene-enclosed mist systems. *J. Hortic. Sci.* 59:15–22.

_____ _____ 1985. The effect of light on the rooting of leafy cuttings. *Scientia Hortic.* 27:105–111.

Green, P. B. 1968. Growth physics in *Nitella*: a method for continuous *in vivo* analysis of extensibility based on a micro-manometer technique for turgor pressure. *Plant Physiol.* 43:1169–1184.

Hartmann, H. T. and D. E. Kester. 1983. *Plant Propagation: Principles and Practices.* Prentice Hall, NJ, USA. 4th ed. p. 258. ISBN 0-13-681007-1.

Hess, C. E. 1969. Internal and external factors regulating root initiation. Root Growth. In *Univ. Nottingham 15th Easter School in Agric. Sci. 1968* (W. J. Whittington, ed). pp. 42–53.

_____ W. E. Snyder. 1955. A physiological comparison of the use of mist with other propagation procedures used in rooting cuttings. In *Proc. 14th Hortic. Congress,* The Hague, The Netherlands. 2:1133–1139.

Hofstra, G. and J. D. Hesketh. 1969. The effect of temperature on stomatal aperture in different species. *Can. J. Bot.* 47:1307–1310.

Hsiao, T. C. 1973. Plant responses to water stress. *Ann. Rev. Plant Physiol.* 24:519–70.

_____ E. Avecedo, E. Fereres and D. W. Henderson. 1976. Water stress, growth, and osmotic adjustment. *Phil. Trans. Royal Soc. London. B.* 273:479–500.

Ikeda, T. and T. Suzaki. 1985. Influence of hydraulic conductance of xylem on water status in cuttings. *Can. J. For. Res.* 16:98–102.

Jensen, R. D. and S. A. Taylor. 1961. Effect of temperature on water transport through plants. *Plant Physiol.* 36:639–642.

Kawase, M. 1981. Anatomical and morphological adaptation of plants to waterlogging. *HortScience* 16:30–34.

Kemp, E. E. 1952. The water economy of unrooted cuttings. In *Proc. 13th Int. Hortic. Cong., London, England.* 1:490–497.

Loach, K. 1977. Leaf water potential and the rooting of cuttings under mist and polythene. *Physiol. Plant.* 40:191–197.

_____ 1983. Propagation systems in New Zealand and a means of comparing their effectiveness. *Proc. Int. Plant Prop. Soc.* 33:291–294.

_____ 1985. Rooting of cuttings in relation to the propagation medium. *Proc. Int. Plant Prop. Soc.* 35:472–485.

_____ D. N. Whalley. 1978. Water and carbohydrate relationships during the rooting of cuttings. *Acta Hortic.* 79:161–168.

Milbocker, D. C. and R. Wilson. 1979. Temperature control during high humidity propagation. *J. Amer. Soc. Hortic. Sci.* 104:123–126.

Molnar, J. M. and L. J. La Croix. 1972. Studies of rooting of cuttings of *Hydrangea macrophylla*: enzyme changes. *Can. J. Bot.* 50:315–322.

Nordström, A. C. and L. Eliasson. 1984. Regulation of root formation by auxin-ethylene interaction in pea stem cuttings. *Physiol. Plant.* 64:298–302.

Rajagopal, V. and A. S. Andersen. 1980a. Water stress and root formation in pea cuttings. I. Influence of the degree and duration of water stress on stock plants grown under two levels of irradiance. *Physiol. Plant.* 48:144–149.

_____ A. S. Andersen. 1980b. Water stress and root formation in pea cuttings. III. Changes in the endogenous level of abscisic acid and ethylene production in the stock plants under two levels of irradiance. *Physiol. Plant.* 48:155–160.

Rasmussen, S. and A. S. Andersen. 1980. Water stress and root formation in pea cuttings. II. Effect of abscisic acid treatment of cuttings from stock plants grown under two levels of irradiance. *Physiol. Plant.* 48:150–154.

Spomer, L. A. 1985. Techniques for measuring plant water. *HortScience.* 20:1021–1038.

Tyree, M. T. and P. G. Jarvis. 1982. Water in tissues and cells. In *Encyc. of Plant Physiol.* New Series 12B. Springer Verlag, Berlin, W. Germany. pp. 35–77.

Wetzel, K. 1924. Die Wasseraufnahme der höheren Pflanzen gëmassigter Klimate durch oberirdische Organe. *Flora* 117:221–269.

Whitcomb, C. E., C. Gray and B. Cavanaugh. 1982. Propagating under a wet tent. *Aust. Hortic.* May: 97–98.

CHAPTER 9

Auxin Metabolism During Adventitious Rooting

Thomas Gaspar and Michel Hofinger

Institut de Botanique B 22
Université de Liège—Sart Tilman
B-4000 Liège, Belgium

INTRODUCTION...117
NATURAL AUXINS AND THEIR NATURAL CONJUGATES 118
    Biosynthesis of Auxins..118
        Indolyl-3-acetic acid *118*
        4-Chloroindolyl-3-acetic acid *119*
        Indolyl-3-acrylic acid *120*
    Conjugation of Auxins 120
        Amide conjugates...*120*
        Ester conjugates *120*
    Catabolism of Auxins 120
ENDOGENOUS AUXIN LEVELS AND ROOTING.........................122
    Changes in Free IAA-Like Substances...........................122
    Changes in Auxin Conjugates..................................123
ENZYMIC IAA OXIDATION AND ROOTING...........................123
CONCLUSION 125
ACKNOWLEDGMENTS 126
REFERENCES..126

Additional key words: auxin biosynthesis, auxin catabolism, auxin conjugates, basic peroxidase, developmental physiology.
Abbreviations: 2,4-D, 2,4-dichlorophenoxyacetic acid; IAA, indolyl-3-acetic acid; 4-Cl-IAA, 4-chloroindolyl-3-acetic acid; IAc, indolyl-3-acetaldehyde; IAcrA, indolyl-3-acrylic acid; IAN, indolyl-3-acetonitrile; IAox, indolyl-3-acetaldoxime; IBA, indolyl-3-butyric acid; IPyA, indolyl-3-pyruvic acid; NAA, naphthylacetic acid; TNH_2, tryptamine; Tol, tryptophol (indolyl-3-ethanol); Try, L-tryptophan.

INTRODUCTION

The relatively specific root promoting properties of applied natural and synthetic auxins lead us to believe that these substances play some crucial roles in the processes of rooting by cuttings. A high level of endogenous auxin has been causally related to the initiation of adventitious root primordia. However, none of the recent book chapters dealing with auxin metabolism in relation to rooting (Gurumurti et al. 1984, Haissig 1986, Jarvis 1986, Gaspar and Coumans 1987) have considered auxin as the sole determinant. As will be seen from the literature below, there are few available data relating changes in endogenous auxin level to a cutting's ability to initiate roots. For example, no data have indicated an immediate increase in the level of endogenous natural auxin after an application of a root inducing natural or synthetic auxin. Nor is there direct evidence that a synthetic auxin might substitute for a natural one in cells. The physiology of rooting thus must be disembodied from prejudgements; the role of auxin should be reconsidered in view of actually demonstrated developmental physiology and hormonology.

First, in certain species of cuttings, rooting has been described to occur in at least six successive interdependent phases: induction (the period preceding cell division); transverse first division of pericycle cell(s); longitudinal first divisions of daughter cells; continued cell divisions without an increase in gross volume of the meristematic cluster(s); volume increase of cell cluster(s) by cell expansion; and root protrusion (Mitsuhashi-Kato et al. 1978ab, Gaspar 1981). Each of these anatomical phases probably has a specific associated physiology. The need for auxin in each phase is not known but may be different because at least two different phases of sensitivity to IAA are known (Imaseki 1985). Second, it is now becoming clear (Trewavas and Cleland 1983) that sensitivity to growth substances, not concentrations of growth substances, may be the limiting factor in development. If we are to understand the relations between plant hormones and the control of plant development, we must be willing to consider all options, whether they be changes in hormone concentration, changes in sensitivity to hormones, or a combination of these. Third, characterization of the phases of rooting has most often come from analyses of the whole cutting including the leaves. Such analyses produced some types of valid information. For example, aspects of the physiological state characterizing induced or non-induced plants can be simultaneously obtained in all plant parts (Thorpe et al. 1978) as a result of rapid inter-organ communication (Gaspar et al. 1985, Penel et al. 1985). It must be kept in mind, however, that further refinement of techniques is still to be desired for measurements of biochemical changes in the precise tissues and cells that form roots.

The present chapter reviews the literature concerning variation of the levels of endogenous auxins and auxin conjugates in the course of adventitious rooting, including background information relative to anabolism, catabolism, and conjugation of auxins.

NATURAL AUXINS AND THEIR NATURAL CONJUGATES

Biosynthesis of Auxins

The following naturally occurring auxins are known in higher plants: IAA, 4-Cl-IAA, and IAcrA. Their biosynthetic pathways are briefly as follows.

Indolyl-3-acetic acid

IAA is generally regarded as the major auxin, perhaps universally found in higher plants. Other indolic compounds originating from metabolism of IAA have been identified in some plants. Numerous publications about the biosynthesis of IAA indicated that Try is the universal primary precursor of IAA (Sembner et al. 1980). Three main biosynthetic pathways lead from Try to IAA as shown from many biosynthetic studies in various taxa of higher plants (Fig. 1).

Pathway 1. In the IPyA biosynthetic pathway for IAA, Try is first transaminated to IPyA by a Try aminotransferase (EC 2.6.1.27, Forest and Wightman 1972), which is widely distributed in higher plants (Truelsen 1973). IPyA is a labile compound susceptible to oxidation and acids; therefore, its presence as a native compound in plants is difficult to establish. In the second step, IPyA is decarboxylated to IAc (Moore and Shaner 1968, Rajagopal 1968, Gibson et al. 1972b, Purves and Brown 1978). The oxidation of IAc to IAA can be catalyzed by two different enzymes, depending on the species: 1) an NAD-dependent indolylacetaldehyde dehydrogenase in mung bean [Vigna radiata (L.) R. Wilczek] seedlings (Wightman and Cohen 1968); or 2) an NAD-independent (using O_2 as acceptor) IAc oxidase (EC 1.2.3.7) in Avena coleoptile, Pisum spp., and Nicotiana spp. (Wightman and Cohen 1968, Rajagopal 1971, Liu et al. 1978, Miyata et al. 1981).

Pathway 2. A Try decarboxylase (EC 4.1.1.28) first decarboxylates Try to TNH_2 (Sherwin 1970, Gibson et al. 1972a) that an amine oxidase (EC 1.4.3.) then converts to IAc (Lantican and Muir 1967, Muir and Lantican 1968, Sherwin and Purves 1969). TNH_2 is apparently present in many families of plants (Schneider et al. 1972, Smith 1977). In some plants, e.g. mung bean, IAc can be reversibly reduced to Tol by an NAD-dependent alcohol dehydrogenase (EC 1.1.1.1, Wightman and Cohen

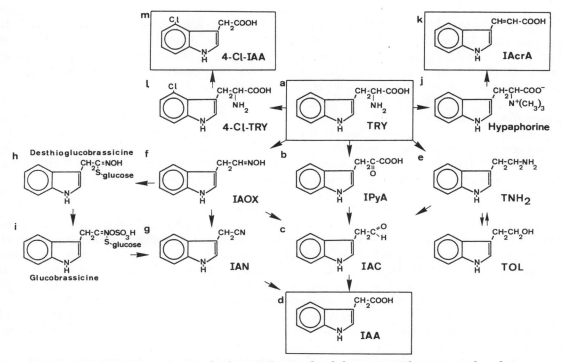

Figure 1. Schematic pathways for biosynthesis of indole auxins. The routes a-b-c-d, a-e-c-d, a-f-c-d, and a-f-g-d lead to indolyl-3-acetic acid in most higher plants. The route a-f-h-l-g-d is typical of the Brassicaceae family. The pathway a-j-k has been established for *Vicia lens (Lens culinaris)*. The pathway a-l-m is restricted to seeds of the Viciae tribe.

1968). An alcohol oxidase (EC 1.1.3.13), which catalyses the irreversible oxidation of Tol to IAc and H_2O_2, was demonstrated in cucumber (*Cucumis sativus* L.) shoots (Vickery and Purves 1972). Tol is assumed to be an important storage product involved in the regulation of IAA biosynthesis equally in plants using pathways 1 or 2 (Percival et al. 1973).

Pathway 3. In this pathway, IAox is formed from Try. Two possible sub-pathways lead from IAox: 1) hydrolysis of IAox may lead to IAc (Rajagopal and Larsen 1972) which is then oxidized to IAA or reduced to Tol; or 2) IAox may be converted by an indoleacetaldoxime dehydratase (EC 4.2.1.29) to IAN, which a nitrilase (EC 3.5.5.1) hydrolyzes to IAA. This second subpathway seems more general (Libbert et al. 1970, Schneider and Wightman 1974). Nitrilase seems to be limited to plant families such as Brassicaceae, Poaceae, and Musaceae (Mahadevan and Thimann 1964, Thimann and Mahadevan 1964). In Brassicaceae an ancillary pathway may convert IAox to the glucosinolates desthioglucobrassicin and glucobrassicin, the latter of which is converted to IAN by myrosinase (EC 3.2.3.1). The physiological significance of this pathway is uncertain (Schraudolf and Weber 1969).

4-Chloroindolyl-3-acetic acid

4-Cl-IAA was identified in immature seeds of some Fabaceae (Marumo et al. 1968, Hofinger and Böttger 1979, Engvild et al. 1981). This compound exhibits strong auxin activity (Marumo et al. 1974), but its physiological significance remains unclear. 4-Cl-IAA is probably specific to the Viciae tribe, although its distribution has not been systematically investigated. It could be involved in seed physiology. 4-Cl-IAA and its methyl ester (Hofinger and Böttger 1979, Engvild et al. 1981), and D-4-chlorotryptophan derivatives (Marumo and Hattori 1970) were isolated from immature seeds of some Fabaceae of the Viciae tribe. 4-Cl-IAA may be formed from D-4-chlorotryptophan but no biosynthetic research has been published.

Indolyl-3-acrylic acid

IAcrA was found in lentil (*Lens culinaris* Med.) seedlings and shown to exhibit auxin activity (Hofinger 1969, Darimont et al. 1971) but its distribution in the plants was not examined nor is its physiological role established. IAcrA was shown to be formed *in vitro* and *in vivo* from Try through hypaphorin [N-trimethyl L-tryptophan (Kutacek et al. 1971, Hofinger et al. 1975)]. Hypaphorin occurs naturally in some plants of the Fabaceae family (ref. in Hofinger et al. 1975).

Conjugation of Auxins

Auxins are often covalently bonded (i.e. conjugated) through their carboxyl group to low molecular weight compounds. The resulting conjugates are divided into two categories, depending upon whether they are formed through an amide link with an amino acid (e.g. indolyl-3-acetylaspartate) or an ester link with a sugar or inositol (e.g. indolyl-3-acetylglucose).

Amide conjugates

Indolyl-3-acetylaspartate was the first described major conjugate of IAA (Andreae and Good 1955). Its enzymatic formation can be induced by auxin treatment in all plants studied (Mollan et al. 1972, Epstein and Lavee 1977). It is also widely distributed as a natural compound in untreated plants (Klämbt 1960, Row et al. 1961). The formation of indolyl-3-acetylaspartate needs a lag period during which the synthesis of the responsible enzyme must be induced. This induction may only be realized by auxin treatment (Venis 1972). Light stimulates formation of this conjugate (Muir 1972). From a physiological standpoint, indolyl-3-acetylaspartate is regarded as a detoxification product of exogenous auxins (Zenk 1964, Andreae 1967). It might also be involved in the regulation of the level of the native auxins, because of a reversible conjugation (Feung et al. 1977).

4-Chloroindolyl-3-acetylaspartate methyl ester has been isolated from immature seeds of *Pisum* spp. (Hattori and Marumo 1972). Other amino acid (glycine, alanine, valine, and glutamic acid) conjugates of IAA were found after IAA treatment of *Parthenocissus* spp. callus (Feung et al. 1976).

IAA can also be linked by an amide bond to higher molecular weight compounds, e.g. proteins (Schneider and Wightman 1974, Percival and Bandurski 1976).

Ester conjugates

Ester conjugates of IAA with various sugars [arabinose in maize (*Zea mays* L.) (Steward and Shantz 1959)] and rhamnose in *Peltophorium* spp. (Ganguly et al. 1974)] have been isolated, but glucose is the most common sugar conjugated to IAA (Klämbt 1961, Zenk 1961, 1964, Keglevic and Pokorny 1969, Davies 1972). The formation of IAA-glucose does not need a lag phase and is thus considered to be a rapid detoxification mechanism (Zenk 1964).

Maize kernels are particularly rich in ester conjugates of IAA. Three isomeric glucosyl esters (2′, 4′, and 6′ isomers) have been isolated (Ehmann 1974). In this material a special group of IAA conjugates was isolated which contained the polyol inositol. One inositol molecule may bind one, two, or three IAA molecules (Ehmann and Bandurski 1974). An additional sugar (arabinose, galactose) can be bound by the inositol moiety of IAA-inositol (Ueda and Bandurski 1974). Physiologically, the inositol derivatives of IAA might be storage forms of IAA in desiccated kernels (Ueda and Bandurski 1974) and a transport form of IAA (its transport rate is 1000 fold that of IAA). They could also be involved in the light control of native IAA levels by conjugating or releasing IAA (Bandurski et al. 1977).

Catabolism of Auxins

IAA is a labile compound that may be degraded either chemically by factors such as acidic pH, heat, light, oxidants, peroxides, or heavy metals; or enzymatically by certain metalloproteins. The so-called IAA-oxidase system, functioning with or without H_2O_2, manganese, and/or monophenols as cofactors, is now believed to consist of one or a group of isoperoxidases (EC 1.11.1.7). Any type of peroxidase is able to destroy IAA *in vitro* but *in vivo* only basic isoperoxidases (high isoelectric point, migrating to a cathode) are thought to be IAA-oxidases (Gaspar et al. 1982).

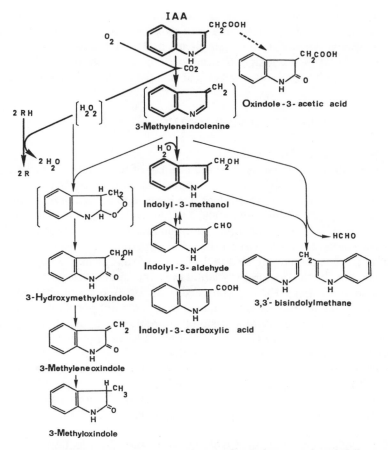

Figure 2. Schematic pathways for (per)oxidative degradation of indolyl-3-acetic acid (IAA). Reactions occurring in the presence of phenolic compounds (RH) have been accentuated with heavy lines. Substances within brackets have not been directly identified. [Adapted from Grambow and Langenbeck-Schwich (1983)]

Depending upon physical and biochemical conditions, IAA can be degraded by peroxidase through different pathways leading to 3-methyleneoxindole or indolyl-3-aldehyde as the main final oxidation products (BeMiller and Colilla 1982, Grambow and Langenbeck-Schwich 1983). Oxindole-3-acetic acid was also identified as a catabolic product of IAA. A methyleneindolenine intermediate probably is the common precursor for the formation of indolyl-3-methanol or 3-hydroxymethyl-oxindole (Fig. 2). H_2O_2 is produced during IAA decarboxylation as a result of oxidase action. Indolyl-3-methanol is virtually the only product of IAA metabolism if H_2O_2 is consumed by peroxidase action in the presence of phenolic compounds. Indolyl-3-methanol can be reversibly oxidized to indolyl-3-aldehyde, a natural compound in some plants, which itself can be irreversibly oxidized to indolyl-3-carboxylic acid. At a low concentration of phenols, however, excess H_2O_2 is consumed by reaction with the methyleneindolenine intermediate, leading to 3-hydroxymethyloxindole and 3-methylene-oxindole, which can be reduced to 3-methyloxindole by a 3-methyleneoxindole reductase. In biological systems this reaction may function to remove "excess" H_2O_2, thus giving detoxifying properties to the IAA-phenolic co-oxidizing system. As a result, endogenous and exogenous phenolic compounds might indirectly influence rooting through auxin metabolism. For further discussion of auxin metabolizing enzymes, see Chapter 7 by Bhattacharya.

The question of whether auxins are active by themselves or become active only through their metabolism remains unanswered. In particular, methyleneoxindole has been found to exhibit a strong auxinic activity [up to 10-fold compared to IAA (Tuli and Moyed 1969, Hofinger et al. 1980, Bhattacharya et al. 1986] but the reason for this extra activity is unclear (Evans and Ray 1972,

Marumo et al. 1974). IAA contamination could not explain an activity 10-fold higher than IAA; contrariwise, any absence of activity might be due to enzymatic destruction of 3-methyleneoxindole. In relation to the foregoing, we have found methyleneoxindole in the β-inhibitor fraction of lentil roots. Physical or chemical treatment which enhanced auxin destruction *in situ* caused root growth inhibition that corresponded with increasing activity of this β-inhibitor fraction (Gaspar 1973).

ENDOGENOUS AUXIN LEVELS AND ROOTING

Changes In Free IAA-Like Substances

Pioneering research has suggested that the number of roots initiated per cutting may be a function of the amount of auxin-like substances in the regeneration zone (Hemberg 1954, Bastin 1966). In a number of cases, endogenous auxin content has been reported to increase in the base of cuttings during the root inducing period (Blahova 1969, Michniewicz and Kriesel 1970). A relation has been shown between the endogenous acidic auxins and the rooting response to added root promoting substances (Odom and Carpenter 1965), which suggested that optimum levels of auxin were required. A reduced level of auxin has been implicated in the failure of rooting in a number of species of plant cuttings (Cooper 1935, Smith and Wareing 1972ab). However, in other cases auxin levels did not appear to be limiting (Biran and Halevy 1973, Greenwood et al. 1976). For example, Stoltz (1968) found that in *Chrysanthemum* spp. endogenous auxin concentration did not positively correlate with rooting. In contrast, the number of lateral primordia arising from roots of *Pisum sativum* increased significantly after decapitation; however, a change in the level of IAA was not detected (Böttger 1978). Further complexities were introduced by the observation that, in some easy-to-root cuttings, the endogenous levels of auxin decreased appreciably over the course of rooting (Bose et al. 1973). In another test, auxins in extract fractions that co-chromatographed with IAA were quantified by bioassays at 24 h intervals after pruning *Quercus rubra* seedlings (Carlson and Larson 1977). Auxin activity sharply increased in the first 24 h and then decreased to prepruning levels. New laterals appeared 4–5 d after pruning.

Despite the development of refined techniques for the identification and estimation of auxins, relatively little progress beyond the foregoing observations has been made. By means of gas chromatography, Brunner (1978) identified IBA and IAA in hypocotyl cuttings of *Phaseolus vulgaris* L. In the control cuttings and all the auxin treated cuttings, a distinct increase in the contents of IAA and IBA became evident in the region of root regeneration in the first 24 h. With progressing regeneration, levels of the auxins decreased in all cases. Nakano et al. (1980) also showed that auxin activity at the base of the cuttings diminished during propagation and was lower in well-rooted than in unrooted cuttings. They assumed that the auxin may have been metabolized before root differentiation.

Seasonal changes in hormone levels have been determined in stem cuttings of the easy-to-root *Rhododendron ponticum* L. and the difficult-to-root *Rhododendron* 'Britannia'. The amount of IAA (and cytokinin) in the stem tissue of both groups was similar, with little seasonal variation in either (Wu and Barnes 1981).

Tréfois and Brunner (1982) found a good positive correlation between endogenous auxin-like activity in *Prunus* spp. cuttings and percent of rooting of cuttings only in instances where auxin level was initially high at the time of auxin treatment. There was no effect of endogenous auxin treatment on rooting when the endogenous auxin level was low at the time cuttings were prepared.

Dunberg et al. (1981) found that IAA content at the bases of IBA treated cuttings of *Pinus sylvestris* L. was three times higher than in untreated cuttings. They did not find differences in IAA metabolism or transport between the treated and untreated cuttings, and concluded that applied IBA was converted to IAA by the cuttings. The foregoing results were confirmed by Epstein and Lavee (1984) in olive (*Olea europaea* L.) and grape (*Vitis vinifera* L.) cuttings through the use of radioactive IBA. Using a solid phase enzyme immunoassay, Weigel et al. (1984) have found a large increase of auxin in the apical parts of *Chrysanthemum* spp. shoots after removing them as terminal stem cuttings from stock plants. Auxin level remained high until about when roots were formed. Synthesis of auxin probably occurred because the total amount increased, not only free auxin derived from bound forms. However, basipetal transport of IAA would have been arrested at the cut end of the cutting, which may have lead to an endogenous, basal accumulation. In this system, prolonged high irradiance of stock plants (40 W·m⁻²) delayed an IAA increase in the cuttings and concomitantly decreased the number of roots per cutting compared to controls (4.5 W·m⁻²). However, root growth as determined by measuring root length or fresh weight were not affected in the latter test. A distinct relation was found between IAA content of stock plants at the time when cuttings were taken and the number of roots formed by the cuttings 20 d later.

Similarly, rooting percentage correlated positively with IAA concentrations in *Cotinus coggygria* Scop. cuttings (Blakesley et al. 1985a): IAA concentration in stem tissue of spring cuttings (which rooted well) was approximately 10 times the concentration in autumn cuttings (which rooted poorly). The concentration of IAA fell markedly after excision and before root appearance (Blakesley et al. 1985a). A transient increase in the concentration of IAA was also discovered in *Phaseolus aureus* hypocotyl cuttings, where root initiation occurs during the first 15 h (Blakesley et al. 1985b).

Changes in Auxin Conjugates

Good rooting of cuttings has sometimes been related to the endogenous level of auxin conjugates. For example, the concentration of IAA esters in poor-rooting autumn *Cotinus* spp. cuttings was 10–20 times higher than that of good-rooting spring cuttings (Blakesley et al. 1985c). Thus, the ratio of free to total IAA was much higher in spring than in autumn cuttings.

In stem cuttings of the easy-to-root *Rhododendron ponticum* and the difficult-to-root *Rhododendron* 'Britannia', Wu and Barnes (1981) measured the highest levels of free and conjugated IAA in summer. However, the content of conjugated IAA decreased between summer and autumn much more in *R.* 'Britannia' than in *R. ponticum*; free IAA on the contrary decreased much more in *R. ponticum*. In another study, terminal stem cuttings of *Chrysanthemum morifolium* Ramat. accumulated both IAA and ester IAA immediately after their removal, compared to levels in stock plants, with a subsequent decline. The maximum total IAA level was reached when the first roots were seen (Weigel et al. 1984). Collet and Le (1987) recently showed that best rooting (rate of rooting, number of roots, absence of callus) of microcuttings of *Rosa* spp. and *Malus* spp. shoots that were produced *in vitro* was achieved by a brief pretreatment of the cut ends with a high concentration of IAA. In the foregoing study, prior to rooting, there was first an accumulation of IAA in the rooting zone, followed by the rapid disappearance of IAA and the appearance of IAA conjugates (IAA-glucosyl, proved by hydrolysis by β-glucosidase). This kind of physiological metabolism-mobilization of "active" compounds was compared by Collet and Le (1987) with the previously demonstrated transformation of radioactively labeled IBA into IAA in the cuttings of *Vitis* spp. and *Olea* spp. (Epstein and Lavee 1984).

In summary, literature dealing with the variation of free and bound auxins in relation to adventitious rooting is very confusing. This confusion has resulted because it has proven difficult to identify the cells which ultimately generate primordia in cuttings and tissue cultures. Frequently, anatomical observations have not been reported in support of experimental results and, understandably so, because primordium initiation is often asynchronous. Another reason is that the physiological state of the donor plants from which cuttings originated has not often been considered. Several strong themes, however, can be gleaned from the literature: 1) there is a good positive correlation between the endogenous free auxin content and the percent of rooting when auxin level is high at the time cuttings are made; 2) free auxin increases in the rooting zone prior to rooting; and 3) the rapid decrease in free IAA level immediately preceding root initiation coincides with the formation of IAA conjugates.

ENZYMIC IAA OXIDATION AND ROOTING

In this section we have excluded papers dealing with changes in IAA-oxidase activity in relation to rooting because this subject is reviewed elsewhere in the present book (see Chapter 7 by Bhattacharya). This section considers only changes that we have observed in peroxidases themselves in the course of rooting.

Physicochemical studies of peroxidases, and our own physiological investigations relating types and activities of peroxidases to growth and morphogenetic processes, have implicated the basic (cationic) isoperoxidases in *in vivo* auxin catabolism (Gaspar et al. 1982) and regulation of endogenous auxin level. Therefore, we have investigated changes in total peroxidase activity and in the isoperoxidase spectrum during the course of rooting (and flowering) by different plants and explants cultured *in vitro* (Quoirin et al. 1974, Gaspar and Van Hoof 1976, Van Hoof and Gaspar 1976, Gaspar et al. 1977b, Thorpe et al. 1978, Druart et al. 1982, Moncousin and Gaspar 1983).

In summary, our observations indicated that total peroxidase activity during rooting and flowering undergo inverse variations (Fig. 3B and D): rooting occurs after the cutting has reached and passed a peak of maximum peroxidase activity, whereas flowering begins after the plant has passed a point of minimum peroxidase activity (Gaspar 1981). The activity or number of acidic isoperoxidases, or both, increases continuously during the course of both processes, indicating that the total peroxidase activity variations are due solely to the basic isoenzymes. The data are characteristic of cuttings from all species analyzed and also correspond relatively well to results from similar investigations on rooting and flowering by others, even though some discrepancies might appear on first examination of the data (Gaspar 1981). Only a few authors (Haissig 1971, Gaspar et al. 1977a, Mitsuhashi-Kato et al. 1978ab) have distinguished an inductive or preparatory phase (no discernable morphological nor histological changes) from an initiative phase (first cell divisions organizing the root primordia), i.e. before phases of root elongation and growth. The physiological necessity of an inductive phase for flowering is, however, generally recognized. In addition, explants (cuttings) may have undergone their inductive phase of rooting or flowering, which quantitatively and qualitatively modified peroxidases, while still attached to the donor plants. In these instances (Fig. 3A and C), peroxidase changes characterizing the initiative phase may take place directly after placing the

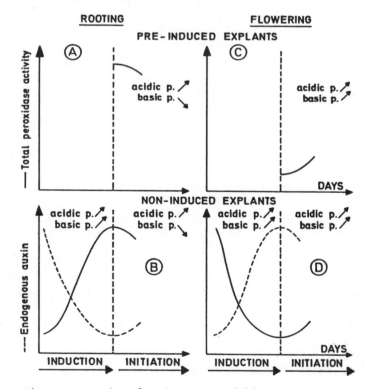

Figure 3. Schematic representation of variations in soluble peroxidase activity (——) and auxin concentration (----) during the induction and initiation phases of rooting (A, B) and flowering (C, D) by preinduced (A, C) and non-preinduced (B, D) explants or plants. The arrows accompanying acidic and basic per(oxidases) indicate an increase or decrease during induction and initiation phases. [Adapted from Gaspar et al. (1982)]

explants in favorable conditions. With regard to endogenous auxin levels, inverse variations in basic isoperoxidases during the inductive and initiative phases of rooting and flowering would also result in inverse variations of auxin content (Fig. 3). Evidence for this hypothesis in flowering has been discussed elsewhere (Gaspar 1981, Gaspar et al. 1985) and seems to apply equally to rooting as described above.

The apparent antagonism that we have described between rooting and flowering have been demonstrated in other tests. For example, antagonistic effects of rooting on the ability of plants to flower were convincingly demonstrated in several species (Miginiac 1971, Sotta and Miginiac 1975, Bismuth et al. 1979), which may explain the promotive effect of root removal on flowering (Krekule and Seidlova 1973) and the reverse (Deletang 1973). The existence of such correlations together with our observations that the biochemical events associated with rooting and flowering characterize the whole plant, cutting, or explant, suggest pronounced hormonal control of these physiological processes.

CONCLUSION

Our analysis of the literature has identified new opportunities and approaches for future research. Past rooting research has somewhat superficially or simplistically studied influences of auxin anabolism and catabolism, in the absence of a needed anatomical or developmental focus. However, it now seems that there are no simple relations between auxin levels in cuttings and rooting. Thus, future research should wherever possible stress auxin anabolism and catabolism in those tissues and cells that initiate root primordia. In addition, data should be interpreted in the context of differential cellular sensitivity to auxin, about which much must be learned.

In terms of the whole cutting, a brief period of metabolic flux precedes rooting. During this period cuttings have a high level of activity of basic (i.e. IAA-oxidase) peroxidases and a low level of auxin, compared to the period of primordium initiation (Gaspar 1981, 1988). However, much additional research is needed in order to determine whether these peroxidase and auxin relations apply equally to preprimordial (target) cells and, if so, whether they directly relate to primordium initiation. There are also other needs for cellular level research concerned with auxin-rooting relations. For example, recent studies indicate that the target cells can be identified in some organs (Chriqui 1985). Exogenous IAA induces rooting from such target cells, starting with initiation of DNA synthesis in G_1 cells. Simultaneously, the G_2 cells enter mitosis. However, the exact relations between endogenous auxin and nucleic acid metabolism in the target cells is poorly understood. In comparison, more is known about teratological growth resulting from treatment with high dosages of 2,4-D and NAA (Nougaréde et al. 1985).

Studies of auxin anabolism and catabolism will be the most meaningful if the degree of cellular sensitivity to auxin is also known. Studies of membrane-based auxin binding protein may help to unravel the cause of differential cellular sensitivity to auxin. For example, a recent study indicated that auxin and cytokinin controlled the presence or absence of auxin binding protein, which was positively correlated with rooting ability (Maan et al. 1985).

Genetic transformation is a sophisticated cellular approach to studying auxin-rooting relations. Auxin biosynthesizing genes from the T-DNA of the bacterium *Agrobacterium* (Liu and Kado 1979) can be inserted into the genome of higher plant cells (Chilton et al. 1982), which then become autonomous for auxin (Black et al. 1986). Such transformation of higher plant tissues by *Agrobacterium rhizogenes* may result in proliferation of adventitious roots ("hairy root" disease, see Chapter 20 by Strobel and Nachmias). One study has shown that the "hairy root" characteristic was stable during plantlet regeneration and associated with significant modifications of isoperoxidases (Benvenuto et al. 1983). Further studies that use *Agrobacterium* transformation may help reveal how variations in peroxidases and auxin metabolism modify cellular differentiation and result in rooting.

ACKNOWLEDGMENTS

We are greatly indebted to Dr. Bruce Haissig for major suggestions during the preparation of this review.

REFERENCES

Andreae, W. A. 1967. Uptake and metabolism of indoleacetic acid, naphthaleneacetic acid and 2,4-dichlorophenoxyacetic acid by pea root segments in relation to growth inhibition during and after auxin application. *Can. J. Bot.* 45:737–753.

_____ N. E. Good. 1955. The formation of indole-acetyl aspartic acid in pea seedlings. *Plant Physiol.* 30:380–382.

Bandurski, R. S. 1978. Chemistry and physiology of myoinositol esters of indole-3-acetic acid. In *Cyclitols and the Phosphoinositides* (W. W. Wells and F. Elsenberg, eds). Pergamon Press, NY, USA, London, England. pp. 35–54.

_____ A. Schulze and J. D. Cohen. 1977. Photo-regulation of the ratio of ester to free indole-3-acetic acid. *Biochem. Biophys. Res. Commun.* 79:1219–1223.

Bastin, M. 1966. Root initiation, auxin level and biosynthesis of phenolic compounds. *Photochem. Photobiol.* 5:423–429.

BeMiller, J. N. and W. Colilla. 1972. Mechanism of corn indole-3-acetic acid oxidase *in vitro*. *Phytochemistry* 11:3393–3402.

Benvenuto, E., G. Ancora, L. Spano and P. Costantino. 1983. Morphogenesis and isoperoxidase characterization in tobacco "hairy root" regenerants. *Z. Pflanzenphysiol.* 110:239–245.

Bhattacharya, R. N., K. K. Chattopadhyay and P. S. Basu. 1986. Auxin activity of 3-hydroxymethyl oxindole and 3-methylene oxindole in oat. *Biol. Plant.* 28:219–226.

Biran, I. and A. H. Halevy. 1973. Endogenous levels of growth regulators and their relationship to the rooting of dahlia cuttings. *Physiol. Plant.* 28:244–247.

Bismuth, F., J. Brulfert and E. Miginiac. 1979. Mise à fleurs de l'*Anagallis arvensis* L. en cours de rhizogénèse. *Physiol. Vég.* 17:477–482.

Black, R. C., G. A. Kuleck and A. N. Binns. 1986. The initiation of auxin autonomy in tissue from tobacco plants carrying the auxin biosynthesizing genes from the T-DNA of *Agrobacterium tumefaciens*. *Plant Physiol.* 80:145–151.

Blahova, M. 1969. Changes in the level of endogenous gibberellins and auxin preceding the formation of adventitious roots on isolated epicotyls of pea plants. *Flora* 160:493–499.

Blakesley, D., P. J. Alderson and G. D. Weston. 1985a. Rooting studies on *Cotinus*. In *Abst. 12th Int. Conf. Plant Growth Sub.*, Heidelberg, W. Germany. p. 86.

_____ J. F. Hall, G. D. Weston and M. C. Elliott. 1985b. Endogenous plant growth regulators and the rooting of *Phaseolus aureus* cuttings. In *Abst. 12th Int. Conf. Plant Growth Sub.*, Heidelberg, W. Germany. p. 87.

_____ G. Weston, P. Alderson, J. Hall and M. C. Elliott. 1985c. The effect of time of year and the induction of etiolation on rooting and the concentration of IAA and cytokinin in cuttings of *Cotinus coggygria* royal purple. *News Bull. British Plant Growth Regul. Group* 7:51–52.

Brunner, H. 1978. Einfluss verschiedener Wuchsstoffe und Stoffwechselgifte auf wurzel-regenerierendes Gewebe von *Phaseolus vulgaris* L. Veränderungen des Wuchsstoffgehaltes sowie der Peroxydase- und der IAA-oxydase-Aktivität. *Z. Pflanzenphysiol.* 88:13–23.

Bose, T. K., S. Basu and R. N. Basu. 1973. Changes in rooting factors during the regeneration of roots on cuttings of easy- and difficult-to-root cultivars of *Bougainvillea* and *Hibiscus*. *Indian J. Plant Physiol.* 16:127–139.

Böttger, M. 1978. Levels of endogenous indole-3-acetic acid and abscisic acid during the course of the formation of lateral roots. *Z. Pflanzenphysiol.* 86:283–286.

Carlson, W. C. and M. M. Larson. 1977. Changes in auxin and cytokinin activity in roots of red oak, *Quercus rubra*, seedlings during lateral root formation. *Physiol. Plant.* 41:162–166.

Chilton, M. D., D. A. Tepfer, A. Petit, C. David, F. Casse-Delbart and J. Tempe. 1982. *Agrobacterium rhizogenes* inserts T-DNA into the genomes of the host plant root cells. *Nature* 295:432–434.

Chriqui, D. 1985. Induction de prolifération des cellules prérhizogènes: auxine et polyamines. *Bull. Soc. bot. Fr., Actual. bot.* 132:127–141.

Cohen, J. D. and R. S. Bandurski. 1982. Chemistry and physiology of the bound auxins. *Ann. Rev. Plant Physiol.* 33:403–430.

Collet, G. F. and C. L. Le. 1987. Role of auxin during in vitro rhizogenesis of rose and apple-trees. *Acta Hortic.* 212:273–280.

Cooper, W. C. 1935. Hormones in relation to root formation on stem cuttings. *Plant Physiol.* 10:789–794.

Darimont, E., Th. Gaspar and M. Hofinger. 1971. Auxin-kinetin interaction on the lentil root growth in relation to indoleacrylic acid metabolism. *Z. Pflanzenphysiol.* 64:232–240.

Davies, P. J. 1972. The fate of exogenously applied indoleacetic acid in light grown stems. *Physiol. Plant.* 27:262–270.

Deletang, J. 1973. Influence exercée par les bourgeons floraux sur la rhizogénèse de *Nicotiana tabacum. C. R. Acad. Sci. Paris* 277:189–192.

Druart, Ph., C. Kevers, Ph. Boxus and Th. Gaspar. 1982. *In vitro* promotion of root formation by apple shoots through darkness effect on endogenous phenols and peroxidases. *Z. Pflanzenphysiol.* 108:429–436.

Dunberg, A., S. Hsihan and G. Sandberg. 1981. Auxin dynamics and the rooting of cuttings of *Pinus sylvestris. Plant Physiol.* (Suppl.) 67:5.

Ehmann, A. 1974. Isolation of 2-0-(indole-3-acetyl)-D-glucopyranose, 4-0-(indole-3-acetyl)-D-glucopyranose, and 6-0-(indole-3-acetyl)-D-glucopyranose from kernels of *Zea mays* by gas liquid chromatography-mass spectrometry. *Carbohydr. Res.* 34:99–114.

_____ R. S. Bandurski 1974. The isolation of di-0-(indole-3-acetyl)-myo-inositol and tri-0-(indole-3-acetyl)-myo-inositol from mature kernels of *Zea mays. Carbohydr. Res.* 36:1–12.

Engvild, K. C., H. Egsgaard and E. Larsen. 1981. Determination of 4-chloroindoleacetic acid methyl ester in *Vicia* species by gas chromatography-mass spectrometry. *Physiol. Plant.* 53:79–81.

Epstein, E. and S. Lavee. 1977. Uptake, translocation, and metabolism of IAA in olive *(Olea europaea)* II. The fate of exogenously applied (2-^{14}C) IAA to detached ''Manzanilla'' olive leaves. *J. Exp. Bot.* 28:629–635.

_____ _____ 1984. Conversion of IBA to IAA by cuttings of grapevine *(Vitis vinifera)* and olive *(Olea europea). Plant Cell Physiol.* 25:697–703.

Evans, M. L. and P. M. Ray. 1972. Inactivity of 3-methyleneoxindole as mediator of auxin action on cell elongation. *Plant. Physiol.* 52:186–189.

Feung, C. S., R. H. Hamilton and R. O. Mumma. 1976. Metabolism of indole-3-acetic acid. III. Identification of metabolites isolated from crown gall. *Plant Physiol.* 58:666–669.

_____ _____ _____ 1977. Metabolism of indole-3-acetic acid. IV. Biological properties of amino acid conjugates. *Plant Physiol.* 59:91–93.

Forest, J. C. and F. Wightman. 1972. Amino acid metabolism in plants. III. Purification and some properties of a multispecific aminotransferase isolated from bushbean seedlings *(Phaseolus vulgaris* L.). *Can. J. Biochem.* 50:813–829.

Ganguly, T., S. N. Ganguly, P. K. Sircar and S. M. Sirkar. 1974. Rhamnose bound indole-3-acetic acid in the floral parts of *Peltophorium ferrugineum. Physiol. Plant.* 31:330–332.

Gaspar, Th. 1973. Inhibition of root growth as a result of methyleneoxindole formation. *Plant Sci. Lett.* 1:115–118.

_____ 1981. Rooting and flowering, two antagonistic phenomena from a hormonal point of view. In *Aspects and Prospects of Plant Growth Regul.* (B. Jeffcoat, ed). British Plant Growth Regul. Group, Wantage, England. Monograph No. 6. pp. 39–49. ISBN 0-906673-04-6.

_____ 1988. Aspects physiologiques de l'organogénèse in vitro. In *Cultures in vitro chez les végétaux. Méthodes et applications* (J. P. Zryd, ed). (In Press)

_____ M. Coumans. 1987. Root formation. In *Cell and Tissue Culture in Forestry* (J. M. Bonga and

D. J. Durzan, eds). Martinus Nijhoff Pub., Dordrecht, The Netherlands. Vol. 2. pp. 202–217. ISBN 90-247-3431-2.

_____ P. Van Hoof. 1976. Application d'un test peroxydasique dans le choix des plantes d'Asperge a propager in vitro. Revue Agric. 3:583–592.

_____ D. Smith and T. A. Thorpe 1977a. Arguments supplémentaires en faveur d'une variation inverse du niveau auxinique endogène au cours des deux premières phases de la rhizogénèse. C. R. Acad. Sci. Paris 185:327–330.

_____ T. A. Thorpe and M. Tran Than Van. 1977b. Changes in isoperoxidases during differentiation of cultured tobacco epidermal layers. Acta Hortic. 78:61–73.

_____ Cl. Penel, T. Thorpe and H. Greppin (eds). 1982. Peroxidases 1970–1980. A Survey of Their Biochemical and Physiological Roles in Higher Plants. Univ. Genève-Centre Bot. Pub., Switzerland.324 pp.

_____ _____ Cl. Roduit, C. Moncousin and H. Greppin. 1985. The role of auxin level and sensitivity in floral induction. Biol. Plant. 27:325–329.

Gibson, R. A., G. Barrett and F. Wightman. 1972a. Biosynthesis and metabolism of indol-3yl-acetic acid. J. Exp. Bot. 23:775–786.

_____ E. A. Schneider and F. Wightman. 1972b. Biosynthesis and metabolism of indol-3yl-acetic acid. J. Exp. Bot. 23:381–399.

Grambow, H. J. and B. Langenbeck-Schwich. 1983. The relationship between oxidase activity, peroxidase activity, hydrogen peroxide, and phenolic compounds in the degradation of indole-3-acetic acid in vitro. Planta 157:131–137.

Greenwood, M. S., O. R. Atkinson and H. W. Yawney. 1976. Studies of hard- and easy-to-root ortets of sugar maples: difference not due to endogenous auxin content. Plant Propagator 22:3–6.

Gurumurti, K., B. B. Gupta and A. Kumar. 1984. Hormonal regulation of root formation. In Hormonal Regulation of Plant Growth and Development (S. S. Purohit, ed). Agro Bot. Pub., Bikaner, India. Vol. I. pp. 387–400.

Haissig, B. E. 1971. Influence of indole-3-acetic acid on incorporation of ^{14}C-uridine by adventitious root primordia in brittle willow. Bot. Gaz. 132:263–267.

_____ 1986. Metabolic process in adventitious rooting of cuttings. In New Root Formation in Plants and Cuttings (M. B. Jackson, ed). Martinus Nijhoff Pub., Dordrecht/Boston/Lancaster. pp. 141–189. ISBN 90-247-3260-3.

Hattori, H. and S. Marumo. 1972. Monomethyl-4-chloroindolyl-3-acetyl-L-aspartate and absence of indolyl-3-acetic acid in immature seeds of Pisum sativum. Planta 102:85–90.

Hemberg, T. 1954. The relation between the occurrence of auxin and the rooting of hypocotyls in Phaseolus vulgaris L. Physiol. Plant. 7:323–331.

Hofinger, M. 1969. L'acide 3-(3-indolyl)-acrylique: sa présence dans les plantules de quelques Fabaceae, son activité auxinique. Arch. Int. Physiol. Biochim. 77:225–230.

_____ M. Böttger. 1979. Identification by GC-MS of 4-chloroindolylacetic acid and its methyl ester in immature Vicia faba seeds. Phytochemistry 18:653–654.

_____ X. Monseur, M. Pais and F. X. Jarreau. 1975. Further confirmation of the presence of indolylacrylic acid in lentil seedlings and identification of hypaphorine as its precursor. Phytochemistry 14:475–477.

_____ Th. Gaspar and D. Ménard. 1980. Effects de l'acide indolylacrylique, de la kinétine, de l'acide abscissique et du méthylèneoxindole sur la croissance et la production d' éthylène par des racines de Lentille. C. R. Acad. Sci. Paris 290:139–142.

Imaseki, H. 1985. Hormonal control of wound-induced responses. In Hormonal Regulation of Development, III (R. P. Pharis and D. M. Reid, eds). Encyc. Plant Physiol. Springer Verlag, Berlin, W. Germany. New Ser. II. pp. 485–512. ISBN 3-540-10197-7.

Jarvis, B. C. 1986. Endogenous control of adventitious rooting in non-woody cuttings. In New Root Formation in Plants and Cuttings (M. B. Jackson, ed) Martinus Nijhoff Pub., Dordrecht/Boston/Lancaster. pp 191–222. ISBN 90-247-3260-3.

Keglevic, D. and M. Pokorny. 1969. The chemical synthesis of 1-0-(indol-3'-ylacetyl)-β-D-glucopyranose. Biochem. J. 114:827–832.

Klämbt, H. D. 1960. Indol-3-acetylasparginsäure, ein natürlich vorkommendes Indolderivat. *Naturwissenschaften* 47:397.

_____ 1961. Wachstumsinduktion and Wuchsstoffmetabolismus im Weizenkoleoptilzlinder. II. Stoffwechselprodukte der Indol-3-essigsäure und der Benzoesäure. *Planta* 56:618–631.

Krekule, J. and F. Seidlova. 1973. Treatments that enhance flowering in the post-inductive period of a short-day plant, *Chenopodium rubrum* L. *Ann. Bot.* 37:615–623. ·

Kutacek, M., M. Hofinger and M. Krălová, 1971. A contribution to the study of indole compounds contained in *Lens culinaris* Med. *Arch. Int. Physiol. Biochim.* 79:669–709.

Lantican, B. P. and R. M. Muir. 1967. Isolation and properties of the enzyme system forming indole-acetic acid. *Plant Physiol.* 42:1158–1160.

Libbert, E., N. Erdmann and U. Schiewer. 1970. Auxinbiosynthese. *Biol. Rundsch.* 8:369–390.

Liu, S. T. and C. I. Kado. 1979. Indoleacetic acid production: a plasmid function of *Agrobacterium tumefaciens* C58. *Biochem. Biophys. Res. Comm.* 90:171–178.

_____ C. D. Katz and C. A. Knight. 1978. Indole-3-acetic acid synthesis in tumorous and nontumorous species of *Nicotania. Plant Physiol.* 61:743–747.

Maan, A. C., P. C. G. Van Derlinde, P. A. A. Harkes and K. R. Libbenga. 1985. Correlation between the presence of membrane-bound auxin binding and root regeneration in cultured tobacco cells. *Planta* 164:376–378.

Mahadevan, S. and K. V. Thimann. 1964. Nitrilase. II. Substrate specificity and possible mode of action. *Arch. Biochem. Biophys.* 107:62–68.

Marumo, S. and H. Hattori. 1970. Isolation of D-4-chlorotryptophane derivatives as auxin-related metabolites from immature seeds of *Pisum sativum. Planta* 90:208–211.

_____ _____ H. Abe and K. Munakata. 1968. Isolation of 4-chloroindolyl 3-acetic acid from immature seeds of *Pisum sativum. Nature* 219:959–960.

_____ _____ A. Yamamoto. 1974. Biological activity of 4-chloroindolyl-3-acetic acid. In *Plant Growth Substances 1973.* Hirokawa Pub. Co., Tokyo, Japan. pp. 419–428.

Michniewicz, M. and K. Kriesel. 1970. Dynamics of auxins, gibberellin-like substances and growth inhibitors in the rooting process of black poplar cuttings (*Populus nigra* L.). *Acta Soc. Bot. Pol.* 39:383–390.

Miginiac, E. 1971. Influence des racines sur le développement végétatif ou floral des bourgeons cotylédonaires chez le *Scrofularia arguta:* rôle possible des cytokinines. *Physiol. Plant.* 25:234–239.

Mitsuhashi-Kato, M., H. Shibaoka and M. Shimokoriyama. 1978a. Anatomical and physiological aspects of development processes of adventitious root formation in *Azukia* cuttings. *Plant Cell Physiol.* 19:393–400.

_____ _____ _____ 1978b. The nature of the dual effect of auxin on root formation in *Azukia* cutting. *Plant Cell Physiol.* 19:1535–1542.

Miyata, S., Y. Suzuki, S. Kamisaka and Y. Masuda. 1981. Indole 3-acetaldehyde oxidase of pea seedlings. *Physiol. Plant.* 51:402–406.

Moore, T. C. and C. A. Shaner. 1968. Synthesis of indoleacetic acid from tryptophan via indolepyruvic acid in cell-free extracts of pea seedlings. *Arch. Biochem. Biophys.* 127:613–621.

Mollen, R. C., D. M. X. Donnelly and M. A. Harmey. 1972. Synthesis of indole-3-acetylaspartic acid. *Phytochemistry* 11:1485–1488.

Moncousin, Ch. and Th. Gaspar. 1983. Peroxidase as a marker for rooting improvement of *Cynara scolymus* L. cultured *in vitro. Biochem. Physiol. Pflanzen.* 178:263–271.

Muir, R. M. 1972. The control of growth by the synthesis of IAA and its conjugation. In *Plant Growth Substances 1970* (D. J. Carr, ed). Springer Verlag, Berlin, Heidelberg, W. Germany, NY, USA. pp. 96–101. ISBN 3-540-05850-8.

_____ B. P. Lantican. 1968. Purification and properties of the enzyme system forming indoleacetic acid. In *Biochemistry and Physiology of Plant Growth Substances* (F. Wightman and G. Setterfield, eds). Runge Press, Ottawa, Canada. pp. 259–272.

Nakano, M., E. Yuda and S. Nakagawa. 1980. Studies on rooting of hardwood cuttings of grapevine, cv. Delaware. *Jap. Soc. Hortic. Sci.* 48:385–394.

Nougaréde, A., D. Chriqui and J. Bercetche. 1985. Natural and experimental expression of rhizogenesis behaviour of target and non-target cells. In *Abstr. (II. Posters) Symp. In Vitro Problems Related to Mass Propagation of Hortic. Plants,* Gembloux, France, Sept. 1985. p. 49.

Odom, R. E. and W. J. Carpenter. 1965. The relationship between endogenous indole auxins and the rooting of herbaceous cuttings. *Proc. Amer. Soc. Hortic. Sci.* 87:494–501.

Penel, C., Th. Gaspar and H. Greppin. 1985. Rapid interorgan communications in higher plants with special reference to flowering. *Biol. Plant.* 27:334–338.

Percival, F. W. and R. S. Bandurski. 1976. Esters of indole-3-acetic acid from *Avena* seeds. *Plant Physiol.* 58:60–67.

_____ W. K. Purves and L. E. Vickery. 1973. Indole-3-ethanol oxidase: kinetics, inhibition, and regulation by auxins. *Plant Physiol.* 51:739–743.

Purves, W. K. and H. M. Brown. 1978. Indoleacetaldehyde in cucumber seedlings. *Plant Physiol.* 61:104–106.

Quoirin, M., Ph. Boxus and Th. Gaspar. 1974. Root initiation and isoperoxidases of stem tip cuttings from mature *Prunus* plants. *Physiol. Vég.* 12:165–174.

Rajagopal, R. 1968. Occurrence and metabolism of indoleacetaldehyde in certain higher plant tissues under aseptic conditions. *Physiol. Plant.* 21:378–385.

_____ 1971. Metabolism of indole-3-acetaldehyde. III. Some characteristics of the aldehyde oxidase of *Avena* coleoptiles. *Physiol. Plant.* 24:272–281.

_____ P. Larsen. 1972. Metabolism of indole-3-acetaldoxime in plants. *Planta* 103:45–54.

Row, V. V., W. W. Sanford and A. E. Hitchcock. 1961. Indole-3-acetyl-D,L-aspartic acid as a naturally-occurring indole compound in tomato seedlings. *Contr. Boyce Thompson Inst.* 21:1

Schneider, E. A. and F. Wightman. 1974. Metabolism of auxin in higher plants. *Ann. Rev. Plant Physiol.* 25:487–513.

_____ R. A. Gibson and F. Wightman. 1972. The native indoles of barley and tomato shoots. *J. Exp. Bot.* 23:152–170.

Schraudolf, H. and H. Weber. 1969. IAN-Bildung aus Glucobrassicin: pH-Abhängigkeit und waschstumphysiologische Bedeutung. *Planta* 88:136–143.

Sembner, G., D. Gross, H.-W. Liebisch and G. Schneider. 1980. Biosynthesis and Metabolism of Plant Hormones. *Encyc. of Plant Physiol,* New Series. 9:281–444. ISBN 3-540-10161-6.

Sherwin, J. E. 1970. A tryptophan decarboxylase from cucumber seedlings. *Plant Cell Physiol.* 11:865–872.

_____ W. K. Purves. 1969. Tryptophan as an auxin precursor in cucumber seedlings. *Plant Physiol.* 44:1303–1309.

Smith, D. R. and P. F. Wareing. 1972a. Rooting of hardwood-cuttings in relation to bud dormancy and the auxin content of the excised stems. *New Phytol.* 71:63–80.

_____ _____ 1972b. The rooting of actively growing and dormant leafy cuttings in relation to the endogenous hormone levels and photoperiod. *New Phytol.* 71:483–500.

Smith, T. A. 1977. Tryptamine related compounds in plants. *Phytochemistry.* 16:171–175.

Sotta, B. and E. Miginiac. 1975. Influence des racines et d'une cytokinine sur le développement floral d'une plante de jours courts, le *Chenopodium polyspermum* L. *C.R. Acad. Sci. Paris* 281:37–40.

Steward, F. C. and E. M. Shantz. 1959. The chemical regulation of growth (Some substances and extracts which induce growth and morphogenesis). *Ann. Rev. Plant Physiol.* 10:379–404.

Stoltz, L. P. 1968. Factors influencing root initiation in an easy- and a difficult-to-root Chrysanthemum. *Proc. Amer. Soc. Hortic. Sci.* 92: 622–626.

Thimann, K. V. and S. Mahadevan. 1964. Nitrilase. I. Occurrence, preparation and general properties of the enzyme. *Arch. Biochem. Biophys.* 105:133–141.

Thorpe, T. A., M. Tran Than Van and Th. Gaspar. 1978. Isoperoxidases in epidermal layers of tobacco and changes during organ formation *in vitro. Physiol. Plant.* 44:388–394.

Tréfois, R. and T. Brunner. 1982. Influence du contenu auxinique endogène sur la réponse au bouturage et sur l'effet nanifiant de quelques *Prunus. Bot. Kozlem.* 69:197–204.

Trewavas, A. J. and R. E. Cleland. 1983. Is plant development regulated by changes in the concentration of growth substances or by changes in the sensitivity to growth substances? *Trends*

Biochem. Sci. 8:354–357.

Truelsen, T. A. 1973. Indole-3-pyruvic acid as an intermediate in the conversion of tryptophan to indole-3-acetic acid. II. Distribution of tryptophan transaminase activity in plants. *Physiol. Plant.* 28:67–70.

Tuli, V. and H. S. Moyed. 1969. The role of 3-methyleneoxindole in auxin action. *J. Biol. Chem.* 244:4916–4920.

Ueda, M. and R. S. Bandurski. 1974. Structure of indole-3-acetic acid myoinositol esters and pentamethyl myoinositols. *Phytochemistry.* 13:243–253.

Van Hoof, P. and Th. Gaspar. 1976. Peroxidase and isoperoxidase changes in relation to root initiation of *Asparagus* cultured *in vitro. Scientia Hortic.* 4:27–31.

Venis, M. A. 1972. Auxin-induced conjugation systems in peas. *Plant Physiol.* 49:24–27.

Vickery, L. E. and W. K. Purves. 1972. Isolation of indole-3-ethanol oxidase from cucumber seedlings. *Plant Physiol.* 49:716–721.

Weigel, U., W. Horn and B. Hock. 1984. Endogenous auxin levels in stem cuttings of *Chrysanthemum morifolium* during adventitious rooting. *Physiol. Plant.* 61:422–428.

Wightman, F. and D. Cohen. 1968. Intermediary steps in the enzymatic conversion of tryptophan to IAA in cell-free systems from mung bean seedlings. In *Biochemistry and Physiology of Plant Growth Sub.* (F. Wightman and G. Setterfield, eds). Runge Press, Ottawa, Canada. pp. 273–288.

Wu, F. T. and M. F. Barnes. 1981. The hormone levels in stem cuttings of difficult-to-root and easy-to-root rhododendrons. *Biochem. Physiol. Pflanzen.* 176:13–22.

Zenk, M. H. 1961. 1-(indole-3-acetyl)-β-D-glucose, a new compound in the metabolism of indole-3-acetic acid in plants. *Nature* 191:493–494.

_____ 1964. Isolation, biosynthesis and function of indoleacetic acid conjugates. In Rég. Nat. Croiss. Vég. (J. P. Nitsch, ed). C.N.R.S., Paris, France. pp. 241–249.

CHAPTER 10

Chemicals and Formulations Used to Promote Adventitious Rooting

Frank A. Blazich

Department of Horticultural Science
North Carolina State University
Raleigh, NC 27695-7609

INTRODUCTION. 132
CHEMICALS USED TO PROMOTE ROOTING. 134
RELATIVE EFFICACY OF AUXINS 134
TREATING CUTTINGS WITH ROOT PROMOTING
    COMPOUNDS 136
    Benefits of Treatment 136
    Treatment Techniques 137
        Auxin-talcum powder mixtures 137
        Dilute-solution soaking. 137
        Concentrated-solution dip 137
        Other methods 138
FACTORS INFLUENCING RESPONSE TO APPLIED AUXIN. 139
    Direct Studies 139
    Indirect Studies. 140
STOCK PLANT TREATMENTS TO ENHANCE RESPONSE TO
    AUXIN. 140
CONCLUSION 141
REFERENCES. 147

Additional key words: auxin(s), propagation, cuttings, chemical rooting formulations.
Abbreviations and chemical names used: CO, carbon monoxide; C_2H_2, acetylene; C_2H_4, ethylene; C_3H_6, propylene; 2,4-D, 2,4-dichlorophenoxyacetic acid; DMF, dimethyl formamide; ethazol, 5-ethoxy-3-(trichloromethyl)-1,2,4-thiadiazole; IAA, indole-3-acetic acid; IBA, indole-3-butyric acid; IPrA, indole-3-propionic acid; K-salt, potassium salt; NAA, α-naphthaleneacetic acid; β-NAA, β-naphthaleneacetic acid; NAM, α-naphthaleneacetamide; NaOH, sodium hydroxide; NaOCl, sodium hypochlorite; NP-IBA, N-phenyl indolyl-3-butyramide; P-IBA, phenyl indole-3-butyrate; P-ITB, phenyl indole-3-thiolobutyrate; thiram, tetramethyl thiuramdisulfide; 2,4,5-T, 2,4,5-trichlorophenoxyacetic acid.

INTRODUCTION

Treatment of cuttings with various substances to promote adventitious rooting is an old concept, undoubtedly dating to the earliest attempts to vegetatively propagate plants. Although early successes were often reported, repeatability was usually impossible and positive results were limited to easy-to-root species, but even then only when proper environmental conditions were used. Luck was also an important factor in successful propagation.

In early attempts to promote rooting, some unusual practices were used, such as embedding grain seeds in the split ends of cuttings. This unusual technique used by early European and Middle Eastern propagators was later found to have a sound scientific basis because germinating grain seeds produce IAA (Hartmann and Kester 1983) (Fig. 1). Such techniques were sometimes effective but often they were unreliable and generally made rooting of cuttings more art than science. However, the unpredictable nature of rooting was removed or at least reduced in 1934–35 as a result of the chemical identification and elucidation of IAA as a rooting promoter (Thimann and Went 1934, Thimann and Koepfli 1935). These landmark papers in the history of plant propagation led to auxin[1] treatment of cuttings to promote rooting, and made it possible to consistently root large quantities of cuttings from plants that previously had been very difficult if not impossible to propagate vegetatively.

This review is a discussion of chemical compounds used to promote rooting, primarily auxins. Various topics related to auxin treatment of cuttings to stimulate rooting are considered from both practical and theoretical standpoints to encourage research directed towards improved rooting of cuttings.

Figure 1. Examples of common auxins.

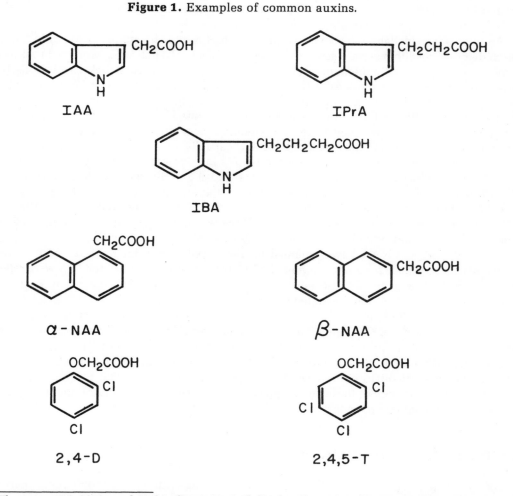

---

[1]The term auxin(s) as used in this discussion is defined as "organic substances which at low concentrations (<0.001 mol·l^{-1}) promote growth (cell enlargement) along the longitudinal axis, when applied to shoots of plants freed as far as practical from their own inherent growth promoting substance and inhibit the elongation of roots" (Thimann 1969). Despite being involved in many physiological processes which the definition does not address, that role of auxin of greatest interest here is the initiation of adventitious roots.

CHEMICALS USED TO PROMOTE ROOTING

Following the discovery that IAA promoted adventitious rooting, the search began for other naturally occurring auxins. Also, compounds with structures similar and dissimilar to IAA were examined for root promoting properties. Searches for naturally occurring auxins were unsuccessful. Currently, it is generally agreed that IAA is the major (Moore 1979) if not the only (Thimann 1969) naturally occurring auxin found in plants, existing in both free and bound (conjugated) forms (Cohen and Bandurski 1982). Tests of IAA analogues, however, were more successful and in 1935 the first report appeared which indicated that the synthetic auxins IBA and NAA had strong root promoting properties (Zimmerman and Wilcoxon 1935) (Fig. 1). Reports of additional synthetic compounds also classified as auxins and having root promoting activity appeared in 1942 (Hitchcock and Zimmerman 1942, Zimmerman and Hitchcock 1942). These substances included the phenoxy compounds (e.g. 2,4-D; Fig. 1) which in later years would be extensively utilized as herbicides.

Since 1942, additional naturally occurring and synthetic, auxin and nonauxin compounds have been reported to stimulate rooting in cuttings (Hartmann and Kester 1983). Commercial use, however, has been limited principally to IBA and NAA, and less to IAA and phenoxy compounds such as 2,4-D and 2,4,5-T (Fig. 1). Preference for IBA and NAA is illustrated by the large number of commercial rooting formulations containing one or both of them (Table 1).

An interesting sidelight to discovery of promotion of rooting by auxins occurred in 1933, when research demonstrated that certain gases such as C_2H_2, CO, C_2H_4, and C_3H_6 promoted rooting in cuttings of various species (Zimmerman et al. 1933, Zimmerman and Hitchcock 1933). In herbaceous cuttings, the promotion of rooting by these gases was through greater root initiation and/or root development; for woody cuttings, only development of existing root primordia was greater (Zimmerman et al. 1933). Little of this knowledge has ever been used commercially. The role of C_2H_4 in adventitious rooting has continued to receive attention since an early study suggested that auxin induced production of C_2H_4 may account for the capacity of auxin to stimulate root initiation (Zimmerman and Wilcoxon 1935). However, attempts to correlate adventitious rooting with auxin mediated C_2H_4 production have been inconclusive (Mudge and Swanson 1978, Geneve and Heuser 1982). Variable results have also been reported as to the effects of applied C_2H_4-generating compounds on rooting (Krishnamoorthy 1970, Samananda et al. 1972, Mudge and Swanson 1978, Geneve and Heuser 1983, Robbins et al. 1983). Relations between auxin, C_2H_4, and adventitious rooting are complex and poorly understood (see Chapter 11 by Mudge).

Although IBA and NAA have remained for many years the principal auxins for rooting cuttings, the search continues for other compounds. In 1979, aryl esters of IAA and IBA were reported to be superior to the free acids of these same compounds in promoting root initiation (Haissig 1979). Subsequent research in 1983 demonstrated that two aryl esters (P-IBA and P-ITB, Fig. 2a) and an aryl amide (NP-IBA, Fig. 2a) were more efficient in inducing rooting of cuttings of bean (*Phaseolus vulgaris* L. cv. Top Crop) and jack pine (*Pinus banksiana* Lamb.) than IBA (Haissig 1983). Studies with other plant species have confirmed previous findings that the aryl esters of IBA may have merit in rooting cuttings (Dirr 1986, Struve and Arnold 1986). The principal benefit, however, may not be greater effectiveness than IBA but reduced toxicity (Haissig 1983, Dirr 1986). Overall, more research is needed to test the efficacy of these newly developed aryl esters and amides of indole auxins and to seek even more potent compounds.

RELATIVE EFFICACY OF AUXINS

As reports appeared indicating that treatment of cuttings with various compounds, particularly auxins, promoted rooting so did information regarding the relative efficacy of these compounds. IPrA was reported to be less effective than IAA, α-NAA more effective than β-NAA, and IBA and α-NAA more effective than IAA (Zimmerman and Wilcoxon 1935) (Fig. 1). Differences between IBA and NAA were established by rooting studies on a wide range of plant species of varying rooting ability. Although IBA and NAA had been judged to be superior to IAA, IBA was found to be more effective

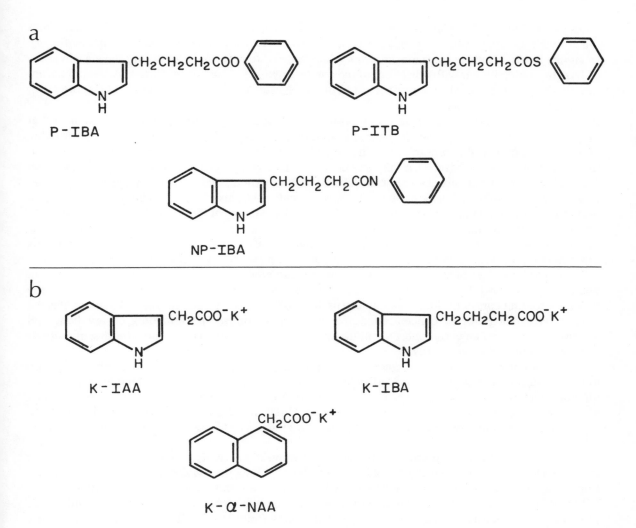

Figure 2. Aryl esters (P-IBA and P-ITB) and an aryl amide (NP-IBA) of IBA (a) and K-salts of IAA, IBA and NAA (b).

than NAA for most species (Hitchcock and Zimmerman 1939). The same report also showed that the K-salts of IAA, IBA, and NAA were consistently more effective than the corresponding acids (Fig. 2b). This knowledge has been ignored and through the years the acids have been used much more than their salts.

Although IBA was found to be more effective than NAA (Hitchcock and Zimmerman 1939), comparison of the two auxins is not simple because many factors influence auxin efficacy. For example, other factors being equal, that aspect of formulation consisting of the actual preparation in which the auxin resides (e.g. auxin-talcum powder mixture vs. an auxin solution) can influence efficacy (Hitchcock and Zimmerman 1936, 1939, Heung and McGuire 1973, Bonaminio and Blazich 1983). This aspect of formulation is not to be confused with another dealing with the chemical form of the auxin, for example, acid vs. salt. Also, a particular species may react differently when treated with equivalent concentrations (mol·l^{-1}) of different auxins (Proebsting 1984).

Due to the considerable variation in relative auxin efficacy, studies have been conducted to determine the physiological bases for these differences. These data plus problems associated with using particular auxins explain why some auxins are used with greater frequency than others. For example, reduced root promoting activity of applied IAA, compared to IBA and NAA, has been attributed to the fact that plant tissues possess several metabolic mechanisms that remove IAA from the growth regulating system (Leopold and Kriedemann 1975, see Chapter 9 by Gaspar and Hofinger). In simple terms, plants possess mechanisms which operate to reduce and/or nullify the effectiveness of IAA by conjugating it with other compounds or destroying it. Also, nonsterilized IAA solutions are rapidly destroyed by microorganisms (Hartmann and Kester 1983) and strong sunlight (Moore 1979, Hartmann and Kester 1983).

Phenoxy compounds, such as 2,4-D, promote rooting in cuttings of many species when used at very low concentrations. However, phenoxy compounds have not been extensively used for rooting cuttings even though they are relatively light-stable and are more resistant to microbial decomposition. There are several reasons for this, one being that following rooting these materials inhibit shoot formation, possibly because they are translocated to buds (Anon. 1983). Also, applied phenoxy compounds often stimulate callus growth on the bases of cuttings. This growth is usually associated with numerous short roots that are bent, thickened, and stubby. Masses of these short roots often appear to be fused. This aberrant form of root system markedly contrasts with the strong fibrous root systems produced after IBA treatment of cuttings. Roots produced after treatment with phenoxy compounds often develop slowly. Slow root system development reduces the overall growth rate of cuttings.

In addition, the concentration of a phenoxy compound that promotes rooting often results in necrosis of the basal portion of the cutting. If the concentration used is somewhat greater than the optimum rooting concentration, severe injury or death of the cuttings may result, which illustrates the narrow, effective range of concentrations for these compounds. The foregoing observations and experiences confirm an early report about use of phenoxy compounds to promote rooting of cuttings (Hitchcock and Zimmerman 1942).

Despite problems with using phenoxy compounds, they may be effective in combination with one or more non-phenoxy auxins. Studies have shown that if a phenoxy compound is used in combination with an auxin such as IBA, NAA, or both IBA and NAA, the rooting response is enhanced in comparison to use of a phenoxy compound alone (Hitchcock and Zimmerman 1942, Ellyard 1981). A similar improvement was obtained by using mixtues of IBA and NAA (Hitchcock and Zimmerman 1940).

The excellent root promoting properties of IBA and NAA, which were established soon after they were found to promote rooting, continue to be borne out by their widespread use for rooting cuttings. Another reason for their extensive use is that they are not easily degraded during storage. Whether initially purchased as reagent grade chemical from which particular rooting formulations are prepared, or purchased in the form of commercial rooting formulations, IBA and NAA are presently the two most widely used auxins (Table 1).

TREATING CUTTINGS WITH ROOT PROMOTING COMPOUNDS

Benefits of Treatment

Countless references can be found on the stimulatory influence of auxin treatment in rooting cuttings of difficult-to-root species. Although many such species positively respond to auxin treatment, some do not, illustrating that positive physiological response to auxin treatment is not universal. However, the benefit of auxin treatment is obvious in those species where rooting would not otherwise occur. In addition, four specific advantages associated with auxin treatment have long been recognized: 1) increasing the percentage of cuttings that form roots; 2) hastening root initiation; 3) increasing the number and quality of roots produced per cutting; and 4) increasing the uniformity of rooting (Hitchcock and Zimmerman 1936). The first, second, and fourth advantages need no explanation. However, the third regarding the number and quality of roots per cutting is intriguing because countless studies have noted it as a benefit of auxin treatment, yet published reports on the influence of root numbers on subsequent cutting growth are lacking.

Treatment Techniques

Based on reports in the 1930s that cuttings could be treated with auxins to promote rooting, various methods were tried for treatment. From this early research the main techniques for treating cuttings with root promoting compounds were developed: 1) application of auxin-talcum powder mixtures; 2) dilute-solution soaking; and 3) concentrated-solution dip (Hitchcock and Zimmerman 1939, Hartmann and Kester 1983).

Auxin-talcum powder mixtures

This method consists of dipping the basal portion (e.g. 1–2 cm) of a cutting into an auxin-talcum powder mixture. With this type of formulation, the active ingredient(s) are carried in the inert talcum powder. Auxin-talcum powder mixtures can be purchased commercially, or prepared by the user if reagent grade auxin is available. When preparing auxin-talcum powder mixtures, thorough blending of the auxin(s) and talcum powder is very important (Stoutemyer 1938, see Table 2 for procedures). Auxin-talcum powder mixtures often contain more than one auxin and may also contain a fungicide (Table 1). Treatment of cuttings with a fungicide, whether incorporated into the auxin-talcum powder mixture or used in other ways has been shown to protect newly formed roots from fungal attack, increase survival, and increase overall quality of the rooted cuttings (Doran 1952, Wells 1963, Hansen and Hartmann 1968, Fiorino et al. 1969). In addition to fungicides, other substances such as sucrose have also been added to these preparations (Hare 1977).

For the most effective application of auxin-talcum powder preparations, it is desirable to have some moisture at the base of the cutting so the powder will adhere (Hitchcock and Zimmerman 1939). After treatment, the cutting is lightly tapped to remove any excess powder and immediately inserted into the rooting medium. Before insertion, a trowel or similar device should be used to make a trench in the rooting medium into which the bases of the cuttings are placed. The trench should be sufficiently wide so that the powder is not rubbed off when cuttings are inserted (Stoutemyer 1938). After insertion, the rooting medium should be gently firmed around each cutting. A dibble (e.g. a wooden plant label) to make holes can also be used when inserting cuttings into the rooting medium.

When treating cuttings, it is advisable to remove a small quantity of stock material containing the auxin(s) and place it in another container that will be used for treatment. Any excess left after treatment should be discarded. Cuttings should never be dipped into the entire stock of powder because this may result in microbial contamination of the stock and deterioration of the auxin(s).

The major advantages of using this procedure include: 1) ready availability of commercial formulations; and 2) simplicity and ease of application. On the other hand, disadvantages are: 1) difficulty of obtaining uniform results due to potentially varying amounts of powder applied to individual cuttings; and 2) high cost of commercial powder preparations when treating extremely large quantities of cuttings. The first disadvantage also deters treating more than one cutting at a time. Thus, it is best to treat single cuttings when using auxin-talcum powder mixtures.

Dilute-solution soaking

This technique involves soaking the basal portion (e.g. 1–3 cm) of stem cuttings in a dilute solution (e.g. 20–500 mg·l^{-1}) of one or more auxins for varying lengths of time (e.g. 2–3 up to 24 h). Following soaking, the cuttings are inserted into the rooting medium (see Table 3 for preparation of solutions).

The dilute-solution soaking method is not popular because it only has disadvantages: 1) it is slow; 2) special equipment is needed for soaking the cuttings; and 3) results can be variable because the amount of solution absorbed depends upon surrounding environmental conditions during treatment. In order to minimize (3), cuttings should be consistently treated at about 20 °C under indirect illumination (Hartmann and Kester 1983).

Concentrated-solution dip

This popular procedure is often referred to as the concentrated-dip or, simply, quick-dip. It consists of briefly (e.g. 1–5 s) dipping the basal portion (e.g. 0.5–2 cm) of a stem cutting into a concentrated (e.g. 500–30,000 mg·l^{-1}) solution of one or more auxins, followed by insertion of the cutting into the rooting medium. Solutions can be purchased commercially or prepared (see procedures in Table 4).

Advantages associated with this technique include: 1) greater economy if reagent grade auxin is purchased and used to prepare solutions, compared to purchasing the same amount of a commercial rooting formulation; 2) speed and ease of use; 3) uniform treatment of bundled cuttings; and 4) generally uniform results. Disadvantages are: 1) difficulty in solution preparation, particularly if non-water soluble auxin is used (i.e. acid form); 2) requires practical experience; and 3) lack of commercial preparations, compared to auxin-talcum powder formulations.

The first disadvantage concerning solution preparation warrants further explanation. The reagent grade of most auxins is synthesized as an acid. In this form, the chemical is generally insoluble in water and must be dissolved in an organic solvent such as alcohol or a strong base, e.g. aqueous NaOH. Strong bases have the disadvantage of greatly raising the pH of the solution, unless it is buffered. High pH might injure cuttings, but preparation of buffers may be inconvenient and costly. Thus, use of buffered auxin solutions has only been used in research (Haissig 1979). Ethyl, methyl, and isopropyl alcohol are all satisfactory solvents and are commonly used. It is best to avoid using full or nearly full strength (95–100%) alcohol because it may dehydrate and injure basal stem tissue. If possible, alcohol should be diluted by 50% with distilled or deionized water. However, 50% alcohol may not dissolve the very large amounts of auxin required to produce concentrated stock or treatment solutions. Also, it is often necessary to use a mechanical stirring device to completely dissolve auxins, which are only slowly soluble in alcohol. Salts of some auxins, particularly K-salts, are available commercially and should be dissolved in distilled or deionized water as opposed to tap water. Salts are usually much more expensive than acids.

Lack of practical experience with this technique can result in killing or severely injuring cuttings during treatment. Solution concentrations must be carefully adjusted to the species and degree of lignification of the cuttings; duration of treatment must also not be too long.

When using the concentrated-solution dip some precautions should be taken. For example, evaporation of alcohol can markedly elevate the concentration of auxin in solution. Excessive concentration can cause injury or death of cuttings. Dilution causing a reduction in effectiveness can also occur if the cuttings have external moisture (Blazich 1978). Concentrated (95–100%) alcohol solutions may also absorb water from humid air, resulting in dilution. To avoid changes in concentration, it is advisable to dip cuttings in only a suitably small portion of the stock solution. Any solution remaining after use should be discarded, not returned to the stock. On a large production scale, the best way to avoid an increase or decrease in concentration is to periodically use a fresh quantity of solution. Stock solutions should be kept in tightly capped bottles and stored in the dark under refrigeration.

Other methods

Although cuttings are generally treated with auxin applied either as an auxin-talcum powder mixture, or as a dilute or concentrated solution, there are other treatment methods which may have merit—e.g. methods that have been used for inducing rooting of seedlings of difficult-to-transplant species. One method has consisted of inserting toothpicks impregnated with auxin into the severed tap root and main lateral roots of recently dug plants. The toothpicks were impregnated with auxin by soaking in an auxin solution for 24 h. This procedure has proven successful in stimulating root initiation in pecan [*Carya illinoensis* Koch (Romberg and Smith 1938)], pear [*Pyrus communis* L. cv. Bartlett (Looney and McIntosh 1968)], and scarlet oak [*Quercus coccinea* Muenchh. (Struve and Moser 1984)]. Reasons for the promotive influence of this treatment may be: 1) the auxin is placed in close proximity to those cells involved in adventitious rooting; 2) a slow release effect of the auxin over time; and 3) the combination of (1) and (2) (Struve and Moser 1984).

Success with the toothpick method with seedlings suggests that it may have potential in rooting cuttings, particularly for difficult-to-root species. Unfortunately, except for one report utilizing cuttings of eastern white pine [*Pinus strobus* L. (Struve and Blazich 1982)], this technique has not been investigated. The eastern white pine study involved a comparison of application of auxin applied as a talcum powder formulation, a concentrated solution, or impregnated in toothpicks. The toothpick method increased percent rooting and induced more roots over a longer time period than basal treatment with IBA in talc or as a concentrated solution. Inducement by the toothpick method of more adventitious roots over a longer period of time suggested that application of auxin in a slow release form may have merit in promoting adventitious rooting. Thus, gradual release of auxin to cuttings warrants further investigation.

For example, instead of using toothpicks, it may be possible to apply auxin mixed with a starch-polyacrylate gel (Starbuck and Preczewski 1986). Mixing auxin with a viscous material followed by application to cuttings is not a new concept; in one of the first techniques used, an auxin-lanolin mixture was applied to cuttings which promoted rooting (Zimmerman and Wilcoxon 1935). However, a later study utilizing cuttings of various woody species found auxin solutions to be more effective (Hitchcock and Zimmerman 1936).

Another approach for auxin application might involve use of vacuum treatment. This could possibly have merit for species and types of cuttings in which uptake of auxin solution is not easily accomplished. This was attempted but results were variable (Butterfield and McClintock 1939).

FACTORS INFLUENCING RESPONSE TO APPLIED AUXIN

Given the stimulatory response achieved by treating cuttings of various species with auxins, it is surprising how few studies have been conducted to identify factors influencing the response of cuttings to auxin application. When such investigations have been undertaken, they have been designed primarily to define how an aspect of a particular treatment technique (e.g. duration or depth of treatment associated with the concentrated-solution dip) influenced rooting response. For the purposes of this discussion, such investigations will be termed "direct studies." That term will also be used for experiments designed to examine factors also related to treatment of cuttings such as auxin efficacy and formulation. On the other hand, examination of data from other rooting experiments allows identification of additional factors influencing auxin response and these will be termed "indirect studies."

Direct Studies

One of the first factors to be studied was efficacy. As mentioned previously, an early report showed that treatments with IBA or NAA were more effective than IAA (Zimmerman and Wilcoxon 1935); IBA treatment was later found to be more effective than NAA for most species (Hitchcock and Zimmerman 1939).

Although efficacy is unquestionably important, formulation can also have a pronounced influence on rooting. There are, however, two ways to view formulation. One involves the actual chemical form of the auxin, for example, acid vs. K-salt. The second is the preparation in which the auxin resides that will be used to treat cuttings, e.g. auxin-talcum powder compared to concentrated solution.

It is difficult to discuss formulation only from a chemical standpoint. If one were comparing the acid form of IBA to the K-salt, both prepared as solutions, comparison would be confounded because the acid form is generally only soluble in an organic solvent whereas the K-salt is water soluble. Even if striking treatment differences were noted one could argue that such differences were due to the carrier, and not the chemical formulation. In any event, an early study reported that the K-salts of IAA, IBA, and NAA were more effective than the acids (Hitchcock and Zimmerman 1939).

A test to eliminate such confusion would compare the acid form of the auxin applied as an auxin-talcum powder mixture and as a concentrated solution in an organic solvent such as alcohol. Results of this type of test have generally demonstrated the superiority of the concentrated solutions over talc formulations (Heung and McGuire 1973, Bonaminio 1983, Bonaminio and Blazich 1983). A common finding of these studies has been that much less auxin is needed to obtain good rooting when concentrated solutions are used, compared to talc. Greater effectiveness of concentrated solutions has been attributed to increased uptake (Heung and McGuire 1973). Another factor which may have a bearing on the rooting response, particularly when the carrier is an organic solvent, is that solvents such as ethanol, methanol, and acetone can stimulate rooting (Bhattacharya et al. 1985); high concentrations, however, may inhibit rooting (Middleton et al. 1978, Bhattacharya et al. 1985). Such findings also show the importance of proper solvent control treatments when conducting rooting experiments (Middleton et al. 1978).

On the other hand, talc preparations may have merit for species where slow auxin uptake at low concentrations may be beneficial (Shibaoka 1971, Heung and McGuire 1973). More experiments are needed, however, to compare effects of talcum powder and alcohol formulations on rooting.

Other experiments have defined how various aspects of the preparation and utilization of auxin formulations influence auxin response. For example, when preparing talc formulations, the manner of mixing auxin with talcum powder influences efficacy. Talc formulations of auxin prepared by initially dissolving the auxin in alcohol are superior in promoting rooting, compared to formulations prepared by initially grinding IAA crystals with talc (Heung and McGuire 1973). Other factors that influence efficacy of auxin-talc formulations include the amount of moisture at the base of a cutting, which influences adherence of the powder (Hitchcock and Zimmerman 1939); loss of powder when inserting cuttings into the rooting medium (Stoutemyer 1938); and length of the basal stem that is treated (Stoutemyer 1938).

Studies have also examined factors influencing the response of cuttings to auxin applied in dilute or concentrated solution, e.g. concentration (Nahlawi 1970); duration of treatment (Nahlawi 1970, Howard 1974); time elapsed between cutting collection and treatment (Howard 1974); length of basal stem that is treated (Nahlawi 1970, Howard 1974, Blazich 1979); position of auxin applied to the basal stem (Nahlawi 1970); wounding (Nahlawi 1970); external moisture on cuttings prior to treatment (Blazich 1978); treatment with the surface sterilant NaOCl prior to auxin treatment (Blazich 1978); and drying of solution applied to the basal stem prior to insertion of cuttings into rooting medium (Blazich 1978).

Indirect Studies

Of numerous factors influencing rooting, wounding and bottom heat have consistently been shown to modify auxin induced rooting in cuttings of many species. Wounding cuttings of particular species is known to stimulate rooting but with the exception of one previously mentioned study (Nahlawi 1970), which viewed wounding as a factor influencing auxin response, most investigators have failed to recognize wounding as a factor enhancing auxin treatment. This may have resulted because: 1) most species do not require any deliberate wounding for satisfactory rooting; or 2) stimulation by a combination of wounding plus auxin treatment for those species where wounding is not required is only slightly better than auxin treatment alone (Blazich and Bonaminio 1983). Examination of data from various rooting studies concerning the influence of wounding shows that the greatest response from wounding was generally achieved when wounding was followed by auxin treatment (Hinesley and Blazich 1981). The interaction of wounding and auxin in promoting rooting may be due to increased exposure of those cells or tissues that initiate roots as a result of auxin stimulation.

Bottom heat has also been overlooked as a factor which influences auxin response. Bottom heat involves maintaining the rooting medium at 18–24 °C. Bottom heat is generally acknowledged to stimulate rooting of cuttings, but available evidence suggests that combined bottom heat and auxin treatment better promote rooting than either treatment alone (Hinesley and Blazich 1981). Elevated temperature at the base of the cutting may enhance the rate of metabolic processes associated with rooting that are triggered by auxin. For more information on basal heating, see Chapter 18 by Loach.

STOCK PLANT TREATMENTS TO ENHANCE RESPONSE TO AUXIN

Use of particular stock plant treatments to enhance the rooting response of cuttings to applied auxin may have significance but is mostly unstudied. However, data have shown that such treatments may have an effect. For example, severe stock plant pruning may enhance auxin response

(Howard and Harrison-Murray 1985). Two other treatments that produce a positive response are girdling, and etiolation and/or blanching.

Girdling a stem results in destruction of the phloem at the point of girdling, which blocks the basipetal translocation of carbohydrates, plant hormones (e.g. auxin), and other substances that may be critical for rooting (e.g. rooting cofactors). Thus, these substances may accumulate in the stem just above the girdle and promote rooting. Girdling has promoted rooting in many species but the bases for the response are unknown. Girdling is a standard treatment when propagating various species by air layering (Hartmann and Kester 1983). Auxin treatment further improves rooting of girdled stems (Sparks and Chapman 1970), which suggests that the accumulation of root inducing factors at the point of girdling interacts additively or synergistically with auxin to promote rooting. Evidence for accumulation of root promoting substances was demonstrated in another investigation which reported that the level of a rooting cofactor increased substantially above the girdle in an easy-to-root clone of hibiscus [*Hibiscus rosa-sinensis* L. (Stoltz and Hess 1966)].

Etiolation, the total development of stock plants or their parts in the absence of light (partial development is termed blanching) has long been known to stimulate rooting (see Chapter 2 by Maynard and Bassuk). Although enhancement of rooting by these treatments is without question, factors responsible for the promotive influence are uncertain. One of the first studies aimed at elucidating the underlying physiological basis for this phenomenon concluded that increased rooting was due to a complex of interacting factors (Herman and Hess 1963). These factors included increased levels of endogenous auxin and rooting cofactors, and various anatomical changes conducive to rooting. A subsequent investigation also reported anatomical changes induced by etiolation, corresponding to increased rooting (Doud and Carlson 1977). However, another study noted that etiolation did not significantly change the auxin concentration and results suggested that etiolation stimulated rooting by increasing the sensitivity of the tissue to auxin (Kawase and Matsui 1980). It is apparent from the few studies reported that promotion of rooting by etiolation and blanching are poorly understood and need further investigation.

CONCLUSION

Auxins were discovered years ago and are still the only applied compounds that consistently enhance adventitious rooting in cuttings that have at least some natural capacity to root. In the interim there has been little improvement in the efficacy of auxins beyond that achieved with IBA, NAA, and the phenoxy compounds such as 2,4-D, which also were first tested years ago. Why, then, have new, more effective auxins not been discovered? Possibly, the apogee of auxin promoted rooting has been attained with the existing compounds, but that seems unlikely (cf. Haissig 1979, 1983). It is more likely that lack of development of root promoting auxins has resulted from general disinterest in the study of adventitious rooting, lack of a clear understanding of its economic importance, and difficulties and high cost of synthetic auxin research and development. Such research and development requires the combined skills of horticulturists, foresters, plant physiologists, organic chemists, and numerous cooperators who will conduct broad scale, practical tests and freely share their results. Based on past evidence such an approach may take years with only a small probability of success. However, the economic importance of adventitious rooting in horticulture and forestry worldwide is sufficient justification for undertaking such research, which hopefully will be done.

Additional research is also needed to determine whether there are "rooting cofactors" (auxin synergists) that, when applied alone or with auxins, markedly stimulate rooting. Evidence collected over 30 yr suggests that such compounds exist, but they have remained elusive. In part, the research cited has not been structured and implemented precisely enough to provide unequivocal evidence and, most importantly, pure, chemically characterized compounds for practical testing. Thus, future research should stress this critical area of fundamental and practical importance.

Until either more effective auxins are discovered, or the rooting cofactors are chemically characterized for practical testing, the principal improvement in efficient use of auxins will result from improvements in proprietary rooting formulations. Here too, research is needed to facilitate auxin uptake, minimize auxin destruction, and generally encourage the health and vigor of cuttings during the critical period that precedes rooting. In such research many previously studied influences on rooting will need to be retested under modern conditions of propagation, such as fog tunnels, heated benches, media, etc. Physiological condition of stock plants must also be considered in rating proprietary rooting treatments. How stock plants are grown before cuttings are taken and how the cuttings are subsequently treated and maintained profoundly influence auxin mediated rooting responses.

Table 1. Trade names of some commercial rooting formulations, formulators, nature of formulation (powder, liquid, liquid-gel, or tablet), and active and inert ingredients (carriers, if known).

Trade Name	Formulator	Formulation	Active and Inert Ingredients[1]
CHRYZO-Series®	ACF Chemiefarma N.V. P.O. Box 5 Straatweg 2 3600 AA Maarssen The Netherlands	Powder	CHRYZOPON (0.1% IBA + 99.9% inert ingredients) CHRYZOTEK (0.4% IBA + 99.6% inert ingredients) CHRYZOSAN (0.6% IBA + 99.4% inert ingredients) CHRYZOPLUS (0.8% IBA + 99.2% inert ingredients)
C-mone®	Coor Farm Supply Service, Inc. P.O. Box 525 Smithfield, NC 27577 USA	Liquid	C-mone (1.0% IBA + 50% isopropyl alcohol) C-mone (2.0% IBA + 50% isopropyl alcohol)
DIP'N GROW®	ALPKEM Corp. P.O. Box 1260 Clackamas, OR 97015 USA	Liquid	DIP'N GROW (1.0% IBA + 0.5% NAA + 98.5% inert ingredients)
Hormex®	Brooker Chemical Corp. P.O. Box 9335 North Hollywood, CA 91605 USA	Powder	Hormex No. 1 (0.1% IBA + 99.9% inert ingredients) Hormex No. 3 (0.3% IBA + 99.7% inert ingredients) Hormex No. 8 (0.8% IBA + 99.2% inert ingredients) Hormex No. 16 (1.6% IBA + 98.4% inert ingredients) Hormex No. 30 (3.0% IBA + 97.0% inert ingredients) Hormex No. 45 (4.5% IBA + 95.5% inert ingredients)
Hormodin®	MSD-AGVET Division of MERCK & CO., INC. Rahway, NJ 07065 USA	Powder	Hormodin 1 (0.1% IBA + 99.9% inert ingredients) Hormodin 2 (0.3% IBA + 99.7% inert ingredients) Hormodin 3 (0.8% IBA + 99.2% inert ingredients)

Trade Name	Formulator	Formulation	Active and Inert Ingredients[1]
Hormo-Root®	Hortus Products Co. P.O. Box 275 Newfoundland, NJ 07435 USA	Powder	Hormo-Root A (0.1% IBA + 15.0% thiram[2] + 84.9% inert ingredients) Hormo-Root B (0.4% IBA + 15.0% thiram + 84.6% inert ingredients) Hormo-Root C (0.8% IBA + 15.0% thiram + 84.2% inert ingredients)
RHIZOPON®	ACF Chemiefarma N.V. P.O. Box 5 Straatweg 2 3600 AA Maarssen The Netherlands	Powder	RHIZOPON A (0.5% IAA + 99.5% inert ingredients) RHIZOPON A (0.7% IAA + 99.3% inert ingredients) RHIZOPON A (1.0% IAA + 99.0% inert ingredients)
		Water-soluble tablet	RHIZOPON A (50 mg IAA)
		Powder	RHIZOPON B (0.1% NAA + 99.9% inert ingredients) RHIZOPON B (0.2% NAA + 99.8% inert ingredients)
		Water-soluble tablet	RHIZOPON B (25 mg NAA)
		Powder	RHIZOPON AA (0.5% IBA + 99.5% inert ingredients) RHIZOPON AA (1.0% IBA + 99.0% inert ingredients) RHIZOPON AA (2.0% IBA + 98.0% inert ingredients) RHIZOPON AA (4.0% IBA + 96.0% inert ingredients) RHIZOPON AA (8.0% IBA + 92.0% inert ingredients)
		Water-soluble tablet	RHIZOPON AA (50 mg IBA)
Rootone®	RHONE-POULENC AG CO. P.O. Box 12014 2 T.W. Alexander Drive Research Triangle Park, NC 27709 USA	Powder	Rootone (0.1% IBA + 0.2% NAM + 4.04% thiram[2] + 95.66% inert ingredients)
Roots®	Wilson Laboratories, Inc. Dundas, Ontario Canada L9H 3H3	Liquid-gel	Roots (0.4% IBA + 0.01% ethazol[2] + 99.59% inert ingredients)
WOOD'S ROOTING COMPOUND®	EARTH SCIENCE PRODUCTS CORP. P.O. Box 327 Wilsonville, OR 97070 USA	Liquid	WOOD'S ROOTING COMPOUND (1.03% IBA + 0.51% NAA + 78.46% ethanol SD 3A + 20% DMF)

[1]To convert percent to parts per million (ppm), multiply by 10,000.
[2]Fungicides.

Table 2. Procedures for preparing auxin-talcum powder mixtures: amounts of auxin and talc needed to prepare a rooting powder of a particular total amount, containing a specific concentration of auxin. See instructions at end of table.

Concn. (ppm)[1]	Total Weight (mg) of Auxin + Talc									
	1000	2000	3000	4000	5000	6000	7000	8000	9000	10,000
	mg auxin									
1000	1	2	3	4	5	6	7	8	9	10
1500	1.5	3	4.5	6	7.5	9	10.5	12	13.5	15
2000	2	4	6	8	10	12	14	16	18	20
2500	2.5	5	7.5	10	12.5	15	17.5	20	22.5	25
3000	3	6	9	12	15	18	21	24	27	30
3500	3.5	7	10.5	14	17.5	21	24.5	28	31.5	35
4000	4	8	12	16	20	24	28	32	36	40
4500	4.5	9	13.5	18	22.5	27	31.5	36	40.5	45
5000	5	10	15	20	25	30	35	40	45	50
5500	5.5	11	16.5	22	27.5	33	38.5	44	49.5	55
6000	6	12	18	24	30	36	42	48	54	60
6500	6.5	13	19.5	26	32.5	39	45.5	52	58.5	65
7000	7	14	21	28	35	42	49	56	63	70
7500	7.5	15	22.5	30	37.5	45	52.5	60	67.5	75
8000	8	16	24	32	40	48	56	64	72	80
8500	8.5	17	25.5	34	42.5	51	59.5	68	76.5	85
9000	9	18	27	36	45	54	63	72	81	90
9500	9.5	19	28.5	38	47.5	57	66.5	76	85.5	95
10000	10	20	30	40	50	60	70	80	90	100
12500	12.5	25	37.5	50	62.5	75	87.5	100	112.5	125
15000	15	30	45	60	75	90	105	120	135	150
17500	17.5	35	52.5	70	87.5	105	122.5	140	157.5	175
20000	20	40	60	80	100	120	140	160	180	200

Instructions for using Table 2.
1. Select from the left column the concentration (ppm) of auxin needed.
2. Select from the top of the table the total weight (mg) of the rooting preparation needed.
3. The value given where the concentration and total weight intersect is the quantity of auxin (in mg) needed.
4. The indicated quantity of auxin is adjusted to the selected total weight with talcum powder (Total weight in mg − mg auxin = mg talcum powder needed).
5. Thorough blending of the auxin and talcum powder is necessary to avoid inconsistent results and potential toxicity. The most efficient way of doing this is to dissolve the particular weight of auxin in a small quantity of 95% ethanol and then pour the solution over the prescribed weight of talcum powder. After mixing to a slurry, the mixture is dried. An excellent way to dry the mixture is use of a forced air oven at 25 °C. When preparing auxin-talcum powder mixtures use the acid formulation of the auxin.

---

[1]To convert parts per million (ppm) to percent, divide by 10,000.

Table 3. Procedures for preparing dilute auxin solutions: amounts of auxin and solvent needed to prepare a solution of a particular total amount, containing a specific concentration of auxin. See instructions at end of table.

Concn. (ppm)[1]	Total Volume (ml)									
	100	200	300	400	500	600	700	800	900	1000
					mg auxin					
20	2	4	6	8	10	12	14	16	18	20
40	4	8	12	16	20	24	28	32	36	40
60	6	12	18	24	30	36	42	48	54	60
80	8	16	24	32	40	48	56	64	72	80
100	10	20	30	40	50	60	70	80	90	100
120	12	24	36	48	60	72	84	96	108	120
140	14	28	42	56	70	84	98	112	126	140
160	16	32	48	64	80	96	112	128	144	160
180	18	36	54	72	90	108	126	144	162	180
200	20	40	60	80	100	120	140	160	180	200
220	22	44	66	88	110	132	154	176	198	220
240	24	48	72	96	120	144	168	192	216	240
260	26	52	78	104	130	156	182	208	234	260
280	28	56	84	112	140	168	196	224	252	280
300	30	60	90	120	150	180	210	240	270	300
320	32	64	96	128	160	192	224	256	288	320
340	34	68	102	136	170	204	238	272	306	340
360	36	72	108	144	180	216	252	288	324	360
380	38	76	114	152	190	228	266	304	342	380
400	40	80	120	160	200	240	280	320	360	400
420	42	84	126	168	210	252	294	336	378	420
440	44	88	132	176	220	264	308	352	396	440
460	46	92	138	184	230	276	322	368	414	460
480	48	96	144	192	240	288	336	384	432	480
500	50	100	150	200	250	300	350	400	450	500

Instructions for using Table 3.
1. Select from the left column the concentration (ppm) of auxin needed.
2. Select from the top of the table the volume (ml) of auxin solution needed.
3. The value given where the concentration and volume intersect is the quantity of auxin (in mg) needed.
4. The indicated quantity of auxin is dissolved in approximately 10 ml of solvent (alcohol or distilled water) and adjusted to the necessary volume with distilled water.
5. Keep solutions in a tightly capped bottle, stored in the dark, under refrigeration.

---

[1]To convert parts per million (ppm) to percent, divide by 10,000. Also, mg/l = ppm.

Table 4. Procedures for preparing concentrated auxin solutions: amounts of auxin and solvent needed to prepare a solution of a particular total amount, containing a specific concentration of auxin. See instructions at end of table.

Concn. (ppm)[1]	Total Volume (ml)									
	100	**200**	**300**	**400**	**500**	**600**	**700**	**800**	**900**	**1000**
	mg auxin									
500	50	100	150	200	250	300	350	400	450	500
1000	100	200	300	400	500	600	700	800	900	1000
1500	150	300	450	600	750	900	1050	1200	1350	1500
2000	200	400	600	800	1000	1200	1400	1600	1800	2000
2500	250	500	750	1000	1250	1500	1750	2000	2250	2500
3000	300	600	900	1200	1500	1800	2100	2400	2700	3000
3500	350	700	1050	1400	1750	2100	2450	2800	3150	3500
4000	400	800	1200	1600	2000	2400	2800	3200	3600	4000
4500	450	900	1350	1800	2250	2700	3150	3600	4050	4500
5000	500	1000	1500	2000	2500	3000	3500	4000	4500	5000
5500	550	1100	1650	2200	2750	3300	3850	4400	4950	5500
6000	600	1200	1800	2400	3000	3600	4200	4800	5400	6000
6500	650	1300	1950	2600	3250	3900	4550	5200	5850	6500
7000	700	1400	2100	2800	3500	4200	4900	5600	6300	7000
7500	750	1500	2250	3000	3750	4500	5250	6000	6750	7500
8000	800	1600	2400	3200	4000	4800	5600	6400	7200	8000
8500	850	1700	2550	3400	4250	5100	5950	6800	7650	8500
9000	900	1800	2700	3600	4500	5400	6300	7200	8100	9000
9500	950	1900	2850	3800	4750	5700	6650	7600	8550	9500
10000	1000	2000	3000	4000	5000	6000	7000	8000	9000	10000
12500	1250	2500	3750	5000	6250	7500	8750	10000	11250	12500
15000	1500	3000	4500	6000	7500	9000	10500	12000	13500	15000
17500	1750	3500	5250	7000	8750	10500	12250	14000	15750	17500
20000	2000	4000	6000	8000	10000	12000	14000	16000	18000	20000
22500	2250	4500	6750	9000	11250	13500	15750	18000	20250	22500
25000	2500	5000	7500	10000	12500	15000	17500	20000	22500	25000
27500	2750	5500	8250	11000	13750	16500	19250	22000	24750	27500
30000	3000	6000	9000	12000	15000	18000	21000	24000	27000	30000

Instructions for using Table 4.

1. Select from the left column the concentration (ppm) of auxin needed.
2. Select from the top of the table the volume (ml) of solution needed.
3. The value given where the concentration and volume intersect is the quantity of auxin (in mg) needed.
4. The indicated quantity of auxin is dissolved in a small quantity of solvent and adjusted to the necessary volume with additional solvent.
5. Keep solutions in a tightly capped bottle, stored in the dark, under refrigeration.

---

[1]To convert parts per million (ppm) to percent, divide by 10,000. Also, mg/l = ppm.

REFERENCES

Anon. 1983. *Herbicide Handbook.* Weed Sci. Soc. of Amer., Champaign, IL, USA. 5th ed. pp. 128–134. ISBN 0-911733-01-9.

Bhattacharya, S., N. C. Bhattacharya and V. B. Bhatnagar. 1985. Effect of ethanol, methanol and acetone on rooting etoliated cuttings of *Vigna radiata* in presence of sucrose and auxin. *Ann. Bot.* 55:143–145.

Blazich, F. A. 1978. Effects of Clorox used as a surface sterilant on stem cuttings prior to auxin treatment. In *Proc. Southern Nurserymen's Assoc. Res. Conf. 23rd Ann. Rept.* pp. 159–160.

_____ 1979. Effect of auxin dipping depth on the rooting of 'Helleri' holly cuttings. In *Proc. Southern Nurserymen's Assoc. Res. Conf. 24th Ann. Rept.* pp. 225–227.

_____ V. P. Bonaminio. 1983. Effects of wounding and auxin treatment on rooting stem cuttings of Fraser's photinia. *J. Environ. Hortic.* 1:104–106.

Bonaminio, V. P. 1983. Comparison of IBA quick-dips with talc for rooting cuttings. *Proc. Int. Plant Prop. Soc.* 33:565–568.

_____ F. A. Blazich. 1983. Response of Fraser's photinia stem cuttings to selected rooting compounds. *J. Environ. Hortic.* 1:9–11.

Butterfield, N. W. and J. A. McClintock. 1939. New method of treating cuttings. *Proc. Amer. Soc. Hortic. Sci.* 37:1077–1079.

Cohen, J. D. and R. S. Bandurski. 1982. Chemistry and physiology of the bound auxins. *Ann. Rev. Plant Physiol.* 33:403–430.

Dirr, M. A. 1986. New root promoting compounds and formulations. In *Proc. Southern Nurserymen's Assoc. Res. Conf. 31st Ann. Rept.* pp. 204–206.

Doran, W. L. 1952. Effects of treating cuttings of woody plants with both a root-inducing substance and a fungicide. *Proc. Amer. Soc. Hortic. Sci.* 60:487–491.

Doud, S. L. and R. F. Carlson. 1977. Effects of etiolation, stem anatomy, and starch reserves on root initiation of layered *Malus* clones. *J. Amer. Soc. Hortic. Sci.* 102:487–491.

Ellyard, R. K. 1981. Effect of auxin combinations on the rooting of *Persoonia chamaepitys* and *P. pinifolia* cuttings. *Proc. Int. Plant Prop. Soc.* 31:251–255.

Fiorino, P., J. N. Cummins and J. Gilpatrick. 1969. Increased production of rooted *Prunus besseyi* Bailey softwood cuttings with preplanting soak in benomyl. *Proc. Int. Prop. Soc.* 19:320–329.

Geneve, R. L. and C. W. Heuser. 1982. The effect of IAA, IBA, NAA, and 2,4-D on root promotion and ethylene evolution in *Vigna radiata* cuttings. *J. Amer. Soc. Hortic. Sci.* 107:202–205.

_____ _____ 1983. The relationship between ethephon and auxin on adventitious root initiation in cuttings of *Vigna radiata* (L.) R. Wilcz. *J. Amer. Soc. Hortic. Sci.* 108:330–333.

Haissig, B. E. 1979. Influence of aryl esters of indole-3-acetic and indole-3-butyric acids on adventitious root primordium initiation and development. *Physiol. Plant.* 47:29–33.

_____ 1983. N-phenyl indolyl-3-butyramide and phenyl indole-3-thiolobutyrate enhance adventitious root primordium development. *Physiol. Plant.* 57:435–440.

Hansen, C. J. and H. T. Hartmann. 1968. The use of indole butyric acid and captan in the propagation of clonal peach and peach-almond hybrid rootstocks by hardwood cuttings. *Proc. Amer. Soc. Hortic. Sci.* 92:135–140.

Hare, R. C. 1977. Rooting of cuttings from mature water oak. *Southern J. Appl. For.* 1(2):24–25.

Hartmann, H. T. and D. E. Kester. 1983. *Plant Propagation: Principles and Practices.* Prentice-Hall, Inc., Englewood Cliffs, NJ, USA. 4th ed. pp. 235–342. ISBN 0-13-681007-1.

Herman, D. E. and C. E. Hess. 1963. The effect of etiolation upon the rooting of cuttings. *Proc. Int. Plant Prop. Soc.* 13:42–62.

Heung, S.-L. R. and J. J. McGuire. 1973. Effect of formulation on uptake of 3-indoleacetic acid in cuttings. *Proc. Int. Plant Prop. Soc.* 23:296–304.

Hinesley, L. E. and F. A. Blazich. 1981. Influence of postseverance treatments on the rooting capacity of Fraser fir stem cuttings. *Can. J. For. Res.* 11:316–323.

Hitchcock, A. E. and P. W. Zimmerman. 1936. Effect of growth substances on the rooting response of cuttings. *Contr. Boyce Thompson Inst.* 8:63–79.

148 Frank A. Blazich

_____ _____ 1939. Comparative activity of root-inducing substances and methods for treating cuttings. *Contr. Boyce Thompson Inst.* 10:461–480.

_____ _____ 1940. Effects obtained with mixtures of root-inducing and other substances. *Contr. Boyce Thompson Inst.* 11:143–160.

_____ _____ 1942. Root-inducing activity of the phenoxy compounds in relation to their structure. *Contr. Boyce Thompson Inst.* 12:497–507.

Howard, B. H. 1974. Factors which affect the response of cuttings to hormone treatments. *Proc. Int. Plant Prop. Soc.* 24:142–143.

_____ R. S. Harrison-Murray. 1985. Optimising the rooting response of stem cuttings to applied auxin. In *Growth Regulators In Horticulture.* British Plant Growth Regulator Group, Wantage, England. Monograph 13. pp. 101–111.

Kawase, M. and H. Matsui. 1980. Role of auxin in root primordium formation in etiolated 'Red Kidney' bean stems. *J. Amer. Soc. Hortic. Sci.* 105:898–902.

Krishnamoorthy, H. N. 1970. Promotion of rooting in mung bean hypocotyl cuttings with ethrel, an ethylene releasing compound. *Plant Cell Physiol.* 11:979–982.

Leopold, A. C. and P. E. Kriedemann. 1975. *Plant Growth and Development.* McGraw-Hill Book Co., NY, USA. 2nd ed. pp. 109–135. ISBN 0-07-037200-4.

Looney, N. E. and D. L. McIntosh. 1968. Stimulation of pear rooting by pre-plant treatment of nursery stock with indole-3-butyric acid. *Proc. Amer. Soc. Hortic. Sci.* 92:150–154.

Middleton, W., B. C. Jarvis and A. Booth. 1978. The effects of ethanol on rooting and carbohydrate metabolism in stem cuttings of *Phaseolus aureus* Roxb. *New Phytol.* 81:279–285.

Moore, T. C. 1979. *Biochemistry and Physiology of Plant Hormones.* Springer Verlag, NY, USA. pp. 32–89. ISBN 0-387-90401-8.

Mudge, K. W. and B. T. Swanson. 1978. Effect of ethephon, indole butyric acid and treatment solution pH on rooting and on ethylene levels within mung bean cuttings. *Plant Physiol.* 61:271–273.

Nahlawi, N. 1970. The effect of dipping depth and duration of auxin treatment on the rooting of cuttings. *Proc. Int. Plant Prop. Soc.* 20:292–300.

Proebsting, W. M. 1984. Rooting of Douglas-fir stem cuttings: relative activity of IBA and NAA. *HortScience* 19:854–856.

Robbins, J. A., S. J. Kays and M. A. Dirr. 1983. Enhanced rooting of wounded mung bean cuttings by wounding and ethephon. *J. Amer. Soc. Hortic. Sci.* 108:325–329.

Romberg, D. L. and C. L. Smith. 1938. Effects of indole-3-butyric acid in the rooting of transplanted pecan trees. *Proc. Amer. Soc. Hortic. Sci.* 36:161–170.

Samananda, N., D. P. Ormrod and N. O. Adepipe. 1972. Rooting of chrysanthemum stem cuttings as affected by (2-chlorethyl) phosphonic acid and indolebutyric acid. *Ann. Bot.* 36:961–965.

Shibaoka, H. 1971. Effects of indoleacetic, p-chlorophenoxyisobutyric and 2,4,6-trichlorophenoxyacetic acids on three phases of rooting in *Azukia* cuttings. *Plant Cell Physiol.* 12:193–200.

Sparks, D. and J. W. Chapman. 1970. The effect of indole-3-butyric acid on rooting and survival of air-layered branches of the pecan, *Carya illinoensis* Koch, cv. 'Stuart'. *HortScience* 5:445–446.

Starbuck, C. J. and J. L. Preczewski. 1986. Effect of root-applied IBA on root and shoot growth of dwarf peach trees. *J. Environ. Hortic.* 4:80–82.

Stoltz, L. P. and C. E. Hess. 1966. The effect of girdling upon root initiation: auxin and rooting cofactors. *Proc. Amer. Soc. Hortic. Sci.* 89:744–751.

Stoutemyer, V. T. 1938. Talc as a carrier of substances inducing root formation in softwood cuttings. *Proc. Amer. Soc. Hortic. Sci.* 36:817–822.

Struve, D. K. and M. A. Arnold. 1986. Aryl esters of IBA increase rooted cutting quality of red maple 'Red Sunset' softwood cuttings. *HortScience* 21:1392–1393.

_____ F. A. Blazich. 1982. Comparison of three methods of auxin application on rooting of eastern white pine stem cuttings. *For. Sci.* 28:337–344.

_____ B. C. Moser. 1984. Auxin effects on root regeneration of scarlet oak seedlings. *J. Amer. Soc. Hortic. Sci.* 109:91–95.

Thimann, K. V. 1969. The auxins. In *The Physiology of Plant Growth and Development* (M. B. Wilkins, ed). McGraw-Hill Pub. Co., London, England. pp. 1–45.

_____ J. B. Koepfli. 1935. Identity of the growth-promoting and root-forming substances of plants. *Nature* 135:101–102.

_____ F. W. Went. 1934. On the chemical nature of the root forming hormone. *Proc. K. Ned. Akad. Wet. Amst.* 37:456–459.

Wells, J. S. 1963. The use of captan in the rooting of rhododendrons. *Proc. Int. Plant Prop. Soc.* 13:132–135.

Zimmerman, P. W., W. Crocker and A. E. Hitchcock. 1933. Initiation and stimulation of roots from exposure of plants to carbon monoxide gas. *Contr. Boyce Thompson Inst.* 5:1–17.

_____ A. E. Hitchcock. 1933. Initiation and stimulation of adventitious roots caused by unsaturated hydrocarbon gases. *Contr. Boyce Thompson Inst.* 5:351–369.

_____ _____ 1942. Substituted phenoxy and benzoic acid growth substances and the relation of structure to physiological activity. *Contr. Boyce Thompson Inst.* 12:321–343.

_____ F. Wilcoxon. 1935. Several chemical growth substances which cause initiation of roots and other responses in plants. *Contr. Boyce Thompson Inst.* 7:209–229.

CHAPTER 11

Effect of Ethylene on Rooting

Kenneth W. Mudge

Department of Floriculture & Ornamental Horticulture
Cornell University
Ithaca, NY 14853

INTRODUCTION..150
EFFECTS OF APPLIED ETHYLENE...152
FACTORS AFFECTING RESPONSES TO ETHYLENE.........................153
ROLE OF ENDOGENOUS ETHYLENE IN ROOTING...........................154
STUDIES WITH ETHYLENE INHIBITORS AND PRECURSORS 155
ETHYLENE-AUXIN INTERACTIONS 156
MECHANISTIC CONSIDERATIONS 157
CONCLUSION 158
ACKNOWLEDGMENTS 158
REFERENCES..159

Additional key words: ethephon, cutting, ethylene inhibitors, flooding.
Abbreviations: ACC, 1-aminocyclopropane-1-carboxylic acid; AVG, aminoethoxyvinyl-glycine; BITC, benzylisothiocyanate; 2,4-D, 2,4-dichlorophenoxyacetic acid; IAA, indole-3-acetic acid; IBA, indole-3-butyric acid; STS, silver thiosulfate; 2,4,5-TP, 2-(2,4,5-trichloro-phenoxy) propionic acid.

INTRODUCTION

The plant hormone, ethylene, primarily known for its senescence related effects, including abscission and fruit ripening, also displays formative effects on plant development. Included in this category are the promotive effects of ethylene on the initiation and subsequent development of root systems. Specifically, ethylene stimulates the initiation of adventitious roots on shoots (see refs. in Table 1), the initiation of lateral roots (Zimmerman and Hitchcock 1933, Zobel 1973, Graham and Linderman 1980) and dichotomous branching of pine (*Pinus* spp.) roots (Rupp and Mudge 1985, Wilson and Field 1984). The discovery that ethylene and ethylene analogues stimulate adventitious root formation was made in the 1930s by scientists at the Boyce Thompson Institute for Plant Research (Zimmerman et al. 1933, Zimmerman and Hitchcock 1933). This finding was made prior to the discovery that auxin had similar root promoting activity (Thimann and Went 1934). Although the original reports of ethylene stimulated rooting have been frequently corroborated, there are numerous reports to the contrary (Table 1). Unlike the auxins, neither ethylene nor ethylene-releasing chemicals have achieved widespread use in commercial plant propagation. The finding that auxins can induce ethylene biosynthesis in most plant tissues has led to the suggestion that auxin stimulated rooting may be mediated by ethylene. This review will examine the effect of exogenously applied ethylene on the formation and development of adventitious roots on plant shoots, as well as the role of endogenously produced ethylene in the process of root formation when it is induced by other factors such as auxin application or flooding.

Table 1. Summary of the effects of ethylene on adventitious rooting of cuttings or intact plants. No distinction is made between experiments involving ethylene alone vs. simultaneous application of auxin. + = stimulation of rooting; − = inhibition of rooting; 0 = no response (equivalent to control).

Plant Species	Response	Reference
Amorpha fragrans Sweet	+	Swanson (1974)
Begonia semperflorens Link & Otto	+	Zimmerman and Hitchcock (1933)
Bryophyllum pinnatum Kurz.	+	Zimmerman and Hitchcock (1933)
Coleus blumei Benth.	+	Zimmerman and Hitchcock (1933)
Cosmos bipinnatus Cav.	+	Zimmerman and Hitchcock (1933)
Cosmos sulphureus Cav.	+	Zimmerman and Hitchcock (1933)
Cotoneaster racemiflorus (Desf.) J. R. Booth ex Bosse.	+	Swanson (1974)
Chrysanthemum morifolium Ramat.	0	Shanks (1969)
	0	Samananda et al (1972)
Dianthus caryophyllus L.	0	Shanks (1969)
Euphorbia pulcherrima Willd. ex Kotzsch	0	Shanks (1969)
Forestiera neomexicana A. Gray.	+	Swanson (1974)
	0	Mudge (1976)[1]
Fuchsia hybrida Hort. ex Vilm.	+	Zimmerman and Hitchcock 1933
Galinsoga parviflora Cav.	+	Zimmerman and Hitchcock 1933
Heliotropium peruvianum L.	+	Zimmerman and Hitchcock 1933
Helianthus annuus L. (sunflower)	0	Wample and Reid (1979)
	+	Fabijan et al. (1981)
Hibiscus rosa-sinensis L.	+	Jackson and Hamilton (1977)
Hydrangea macrophylla (Thunb.) Ser.	+	Zimmerman and Hitchcock (1933)
Impatiens balsamina L.	+	Zimmerman and Hitchcock (1933)
Juglans nigra L.	−	Carpenter (1975)
Juniperus scopulorum Sarg.	0	Swanson (1974)
Lycopersicon esculentum L. (tomato)	−	Coleman et al. (1980)
	+	Hitchcock (1940)
	+	Orion and Minz (1969)
	+	Phatak et al. (1981)
	−	Roy et al. (1972)
	+	Zimmerman and Hitchcock (1933)
Malus spp. Mill. (apple)	+	Cummins and Forino (1969)
Mangifera indica L.	+	Dhua et al. (1983)
	+	Sadhu and Bose (1980)
Nicotiana tabacum L.	+	Zimmerman and Hitchcock (1933)
Pelargonium × *hortorium* L. H. Bailey	+	Tsujita and Harney (1978)
Phaseolus vulgaris L.	+	Curtis and Fellenberg (1972)
	+	Linkins et al. (1973)
	+	Mandal and Basu (1982)
Pisum sativum L. (pea)	+	Michner (1935)
	−	Nordström and Eliasson (1984)
Populus nigra L.	+	Schmid and Schurter (1973)
Populus tremuloides Michx.	0	Schier (1975)
	0	Schmid and Schurter (1973)
Potentilla fruiticosa L.	+	Mudge (1976)[1]
Protea neriifolia R. Br.	+	Criley and Parvin (1979)
Prunus tomentosa Thunb.	+	Swanson (1974)
Psidium guajava L.	+	Sadhu and Bose (1980)
Rhamnus cathartica L.	±	Swanson (1974)
Rhododendron spp. L.	0	Nell and Sanderson (1972)
Ribes alpinum L.	0	Mudge (1976)[1]

[1]K. W. Mudge. 1976, M.S. Thesis, Colorado State Univ., Fort Collins, CO, USA

Plant Species	Response	Reference
Ribes nigrum L.	+	Lipecki and Selwa (1973)
Rosa hybrida	0	Mudge (1976)[1]
Rosa laxa Retz.	0	Mudge (1976)[1]
Rosa spp. L.	−	Shanks (1969)
Salix alba L. (white willow)	0	Mudge (1976)[1]
Salix babylonica L.	+	Zimmerman and Hitchcock (1933)
Salix caprea L.	+	Mudge (1976)[1]
Salix fragilis L.	+	Kawase (1971)
Sedum rubrotinctum R. T. Clausen	+	Boe et al. (1972)
Sinapis spp. (Brassica spp. L.)	−	Moore (1975)
Solanum tuberosum L.	−	Mingo-Castel (1976)
Syzygium javanaca L.	0	Sadhu and Bose (1980)
Tagetes erecta L. (marigold)	+	Zimmerman and Hitchcock (1933)
	+	Mullins (1972)
	+	Roy et al. (1972)
Tagetes patula L.	+	Zimmerman and Hitchcock (1933)
Vigna radiata (L.) R. Wilcz. (mung bean)	0	Batten and Mullins (1978)
	−	Geneve and Heuser (1983)
	+	Krishnamoorthy (1970)
	+	Krishnamoorthy (1972a)
	+	Krishnamoorthy (1972b)
	0	Mudge and Swanson (1978)
	−	Mullins (1972)
	+	Robbins et al. (1983)
	+	Robbins et al. (1985)
	0	Roy et al. (1972)
Zea mays L.	+	Zimmerman and Hitchcock (1933)

EFFECTS OF APPLIED ETHYLENE

The initial discovery that ethylene stimulates adventitious root formation occurred as part of a comprehensive characterization of the physiological effects of ethylene on plant growth and development. Zimmerman and Hitchcock (1933) exposed 27 different species or varieties of plants to relatively high concentrations of ethylene, propylene or acetylene gases ($10–2000 \, \mu l \cdot l^{-1}$). The latter two are now known to be ethylene analogues. Fifteen species responded by formation of adventitious roots on above-ground shoots and also, in several species, on leaf petioles and the underside of leaves. Unfortunately, the authors did not identify the 12 species which did not respond to ethylene treatment. Fourteen of the 15 species which did respond were intact herbaceous plants. Salix babylonica was the only woody plant in the group responding to ethylene, and it was treated as a cutting rather than as an intact plant. Subsequently, Zimmerman and Wilcoxon (1935) demonstrated that application of the auxin, IAA, to tomato shoots resulted in "gaseous emanations" which caused epinasty (an ethylene response) in adjacent marigolds. They speculated, correctly, that the volatile emanation from the auxin treated plant was ethylene gas, although at that time no method was available to detect this gas at biologically relevant concentrations. This was, in effect, the first (indirect) evidence of auxin stimulated ethylene production.

Verification of this widespread phenomenon did not occur until almost three decades later (Morgan and Hall 1964), along with a general upsurge of research in ethylene physiology. This was made possible by the development of gas chromatography with flame ionization detection, as a practical means of qualitatively and quantitatively analyzing biologically relevant concentrations of ethylene (Burg and Stolwijk 1959, Burg and Thimann 1959). Another technical problem which delayed any significant follow-up on the work of Zimmerman and coworkers was the inconvenience of applying gaseous ethylene to plants. This was overcome in 1963 with the discovery that ethephon (2-chloroethylphosphonic acid) decomposes in aqueous solution to release ethylene (Maynard and Swan 1963). Since then, most research on the effects of ethylene on rooting has involved the use of this compound rather than gaseous ethylene. In most propagation studies, ethephon has been applied in aqueous solution as a basal dip or soak at concentrations ranging from 10–5000 mg·l^{-1} (e.g. Nell and Sanderson 1972, Schmid and Schurter 1973, Swanson 1974, Mudge and Swanson 1978, Criley and Parvin 1979, Sadhu and Bose 1980). Less frequently, it has been applied as a foliar spray to intact plants (Johnson and Hamilton 1977, Tsujita and Harney 1978, Phatak et al. 1981). It should be noted that solution pH influences the rate at which ethephon decomposes to ethylene, with little or no decomposition occurring below pH 4 (Anon. 1967). Because the pH of a 10 mg·l^{-1} solution of ethephon is 3.7, and the pH of most plant tissues is close to neutrality, unbuffered ethephon solutions release ethylene primarily after uptake into a cutting rather than in the external treatment solution (deWilde 1971). Mudge and Swanson (1978) demonstrated that mung bean cuttings treated with ethephon solutions buffered to pH 7.4 had substantially higher tissue ethylene levels after 12 h than cuttings treated with unbuffered solutions of equivalent ethephon concentration. Confounding of experiments due to variation in the pH of treatment solutions or plant tissues can be avoided to some extent by the use of buffered solutions, but this precaution is rarely taken.

Since the resurgence in interest in the horticultural applications and physiology of ethylene in the 1960s brought about by the development of ethephon and of gas chromatography, other researchers have tested the effect of ethylene or ethephon on the rooting of cuttings of several of the same or related plants used by Zimmerman and Hitchcock in 1933. Of the 14 herbaceous species which rooted in response to ethylene in the original study, only marigold and tomato have been reinvestigated. Positive results have been consistently reported with marigold either as cuttings (Roy et al. 1972) or as intact plants (Mullins 1972). With tomato, however, there have been reports of both stimulation (Hitchcock and Zimmerman 1940, Orion and Minz 1969, Phatak et al. 1981) and inhibition (Roy et al. 1972, Coleman et al. 1980) of rooting. Similar to the S. babylonica cuttings tested by Zimmerman and Hitchcock (1933), the rooting of S. fragilis (Kawase 1971) and S. caprea (K. W. Mudge. 1976, M.S. Thesis, Colorado State Univ., Fort Collins, CO, USA) cuttings was promoted by ethylene, while the rooting of S. alba was unaffected.

FACTORS AFFECTING RESPONSES TO ETHYLENE

Table 1 summarizes the results of rooting experiments using ethylene (or ethephon) with a total of 52 plant species, across a wide range of experimental conditions and taxonomic groups. As indicated above, most experiments have been performed using ethephon at concentrations ranging from 10–5000 mg·l^{-1}. In the relatively few cases where ethylene gas has been applied directly, promotion of rooting has been reported at concentrations ranging from 1000 μl·l^{-1} for marigold (Mullins 1972); 13–25 μl·l^{-1} for mung bean (Robbins et al. 1985); and 10–2000 μl·l^{-1} for a range of species (Zimmerman and Hitchcock 1933). Unfortunately, the latter study, which was the most comprehensive with respect to the number of species investigated, did not reveal which concentration(s) were effective with which species.

Sixty-eight percent of the experiments summarized in Table 1 reported that ethylene or ethephon stimulated rooting. The percentage of positive responses increases to 73% if only herbaceous plants are considered, and drops to 56% if the compilation is restricted to woody plants. This comparison suggests that the degree of tissue lignification might be a factor in the response to ethylene. A test of the effect of ethylene on the rooting of cuttings taken from a single species varying in degree of lignification would be helpful in resolving this question. However, in such an experiment it would be difficult to avoid the confounding effect of simultaneous variation in starch content and/or other factors which tend to vary with degree of lignification.

Another consideration which might influence the rooting response to ethylene, but which has rarely been addressed experimentally, is the stage of root development at the time of exposure to exogenous ethylene. It is well known that plants such as tomato, coleus, willow, apple, and others initiate adventitious roots early in ontogeny of the shoot (Carlson 1950, Haissig 1974). These so-called preformed or latent root initials lie dormant until some perturbation of the shoot, such as severing it from the rest of the plant by taking a cutting, stimulates the root initials to elongate and emerge from the stem. Eighty-five percent of experiments involving plants with preformed root initials reported stimulation of rooting by ethylene whereas only 62% of experiments involving plants without preformed roots reported stimulation. The effect of ethylene on the elongation of preformed roots can be demonstrated by applying ethephon (1000 mg·l^{-1} in lanolin paste) to the preformed root initials which can be found at the nodes of *Zebrina pendula* Schnizl. (Mudge unpublished).

Another contrast which is revealed in the reports summarized in Table 1 is the substantial difference between the effect of ethylene on rooting of intact plants vs. cuttings. Ninety-five percent of the experiments involving intact plants reported a stimulation of rooting by ethylene, whereas promotion of rooting was reported in only 56% of experiments involving cuttings. The phenomenon of wound ethylene production in response to the excision of cuttings might provide an explanation for this difference between intact plants and cuttings. If endogenous (wound) ethylene were high enough to saturate an ethylene response in cuttings, no further stimulation of rooting from exogenous ethylene would occur. On the other hand, the relatively low endogenous levels of ethylene in intact shoots might be below the threshold for stimulation of rooting, resulting in a positive response to exogenous ethylene application. If this hypothesis is correct, treatment of cuttings or pretreatment of stock plants with an ethylene synthesis inhibitor such as AVG should inhibit rooting and make cuttings more responsive to exogenously applied ethylene. In fact, Jusaitis (1986b) has recently reported that treatment of intact mung bean seedling hypocotyls with AVG had no effect on rooting when applied alone, but when applied with IBA, rooting was inhibited. Similarly, the ethylene precursor ACC stimulated rooting from intact hypocotyls when applied with IBA, but not when applied alone. On the other hand, in another study by the same author involving mung bean hypocotyl cuttings rather than intact seedlings, both ACC and AVG inhibited rooting. These contrasting results with intact hypocotyls vs. cuttings of the same species are consistent with the hypothesis stated above, i.e. a low threshold level of endogenous ethylene is necessary for rooting, at least auxin stimulated rooting, but that excision of cuttings may induce sufficient wound ethylene to saturate that response.

While interspecific differences in the response to ethylene (Table 1) are not particularly surprising, the inconsistent results reported by different investigators working on the same species are more difficult to reconcile. Investigations involving the mung bean rooting bioassay are a case in point (see Chapter 19 by Heuser). More research has been done on the effects of ethylene on rooting using this system than any other. Ethylene (or ethephon) has been reported to promote rooting of mung bean cuttings by Krishnamoorthy (1970, 1972ab) and Robbins et al. (1983, 1985). Inhibition of rooting has been reported by Mullins (1972), and Geneve and Heuser (1982), whereas Roy et al. (1972), Mudge and Swanson (1978), and Batten and Mullins (1978) have reported that ethylene has no effect. Although these inconsistent results may be due to variation in the bioassay conditions employed by different researchers, there is no obvious relationship between the observed response to ethylene and differences in ethylene concentration, stock plant lighting (including etiolation), timing or duration of treatment, cutting age, treatment solution pH, wounding, source of ethylene (gas or ethephon), or presence or absence of exogenous auxin. Unfortunately, no single investigation has systematically evaluated the contribution of more than one or two of these variables to the ethylene response.

ROLE OF ENDOGENOUS ETHYLENE IN ROOTING

In addition to the effects of exogenous ethylene on rooting, the possibility must be considered that endogenous ethylene may be affecting the response of cuttings or intact plants to other root inducing stimuli. Several factors can contribute to elevation of endogenous ethylene levels in cuttings or intact plants which are destined to form adventitious roots. The wounding which naturally occurs when a cutting is excised from a stock plant results in so-called wound ethylene production (Abeles 1973, Fabijan et al. 1981). Environmental stress such as drought can trigger ethylene production in intact stock plants (McMichael et al. 1972, Abeles 1973, Coker et al. 1985) and also, presumably, in drought stressed cuttings. Flooding is another environmental stress which typically results in ethylene accumulation in the shoots of intact plants (Kawase 1971, 1972, 1974, Jackson and Campbell 1975ab, Wample and Reid 1979). It has been suggested that this might be the cause of flooding induced adventitious rooting, radial thickening of shoots, and other responses to flooding in sensitive species such as sunflower, tomato, and willow. Attempts to mimic flooding-induced rooting of intact sunflower seedlings by the application of ethephon or ethylene to nonflooded shoots has not been successful (Wample and Reid 1979, Jackson 1985), but ethephon does stimulate rooting of sunflower hypocotyl cuttings (Fabijan et al. 1981). Similarly, ambiguous responses have been reported for other species with respect to the ability of ethylene to mimic flooding induced rooting, including tomato (Roy et al. 1972, Phatak et al. 1981) and willow (Kawase 1972; K. W. Mudge. 1976, M.S. Thesis, Colorado State Univ., Fort Collins, CO, USA) (summarized in Table 1). Evidence for an accumulation of endogenous auxin in the hypocotyls of flooded sunflower seedlings, and inhibition of basipetal auxin transport by ethylene in other systems (Beyer and Morgan 1969, Abeles 1973), has led Wample and Reid (1979) to speculate that ethylene, produced in response to flooding, may stimulate rooting by causing the accumulation of auxin in the shoot (i.e. ethylene inhibition of auxin transport). This is consistent with the observation of Geneve and Heuser (1982) that ethephon caused adventitious root initiation further up the hypocotyl of mung bean cuttings, although no net increase in rooting occurred.

STUDIES WITH ETHYLENE INHIBITORS AND PRECURSORS

Further evidence of a role for endogenous ethylene in adventitious rooting comes from studies involving the use of inhibitors of ethylene action or synthesis. Silver ion, a potent inhibitor of ethylene action (Beyer 1979), applied to sunflower hypocotyl cuttings as silver nitrate during the first 12 h after excision promoted, rather than inhibited, rooting (Fabijan et al. 1981). This suggests that endogenous ethylene was inhibitory to rooting. This interpretation was supported by the result of another experiment by the same authors using an inhibitor of ethylene synthesis, BITC (Patil and Tang 1974), which, like silver nitrate, also promoted rooting when applied during the same time period. The application of either ethylene synthesis inhibitor more than 12 h after excision slightly inhibited rooting or had no effect. STS is a more effective ethylene action inhibitor than silver nitrate in some situations, due to its increased mobility and reduced phytotoxicity (Veen 1983). This compound had no effect on the rooting of either control or ethylene treated mung bean cuttings in a study conducted by Robbins et al. (1985), but in a similar experiment conducted in my laboratory (Table 2), rooting was inhibited by STS. Like STS, norbornadiene is an inhibitor of ethylene action (Sisler and Goren 1981). In contrast to their results with STS, Robbins et al. (1985) reported that norbornadiene did inhibit rooting of either ethylene treated or control cuttings. The authors' suggestion that STS had no effect on rooting because it may have failed to penetrate to the site of adventitious root formation is unconvincing. The site of rooting (hypocotyl) was completely immersed in STS solution for 15 min, and uptake was sufficient to reverse two other ethylene responses (decrease in root and bud dry weight). An earlier study by Robbins and coworkers (1983) demonstrated that the ethylene synthesis inhibitor AVG inhibited rooting of mung bean cuttings, regardless of whether or not they were also treated with ethephon or auxin. Although they interpreted this as evidence for a role of endogenous ethylene in rooting, their

Table 2. Effect of silver thiosulfate (STS) on rooting of 8-d-old, light grown mung bean cuttings.

Concentration (mmol·l^{-1})	Roots/Cutting	SE·$t_{\alpha=.05}$
0 (water control)	8.1	0.6
0.1	7.8	0.7
0.5	5.5	1.0
1.1	4.1	1.2

conclusion must be considered tentative since it is based on the effect of a single concentration (100 μ mol·l^{-1}) of only one ethylene inhibitor. The authors assumed, but did not demonstrate, that this AVG treatment reduced endogenous ethylene levels in their system. This is suggested by the reversal of AVG inhibited rooting by ethylene. In a similar study by Jusaitis (1986a), AVG also inhibited rooting of mung bean cuttings, but again, the effect of this compound on ethylene production was not measured. The immediate biosynthetic precursor of ethylene, ACC, did not reverse AVG inhibited rooting, except in the presence of a high concentration of auxin (10^{-4} mol·l^{-1} IBA). ACC alone (in the absence of AVG and/or auxin) inhibited rooting. In the presence of auxin (minus AVG), ACC inhibited rooting of mung bean cuttings in this study, but markedly stimulated rooting in another study by the same author (Jusaitis 1986b), which involved rooting of intact mung bean hypocotyls rather than hypocotyl cuttings. In a study involving pea cuttings, Nordström and Eliasson (1984) reported that ACC inhibited rooting. Clearly the interpretation of inhibitor and precursor studies is clouded by failure to report the effects of these compounds on endogenous levels of ethylene. Geneve and Heuser (1982) did monitor the effect of AVG on both ethylene production and rooting of mung bean cuttings, and reported that auxin induced ethylene production was reduced by AVG, but in contrast to the results of Robbins et al. (1983) and of Jusaitis (1986ab), this inhibitor promoted rooting by 50% in the absence of auxin or had no effect on rooting in the presence of auxin. Thus, in general, it is difficult to define a clear role for endogenous ethylene in rooting based upon the results of experiments with inhibitors, because no clear trends are evident in the literature to date. Furthermore, inhibitor studies should be interpreted cautiously because of the possibility that a given inhibitor may have secondary effects on metabolism in addition to the expected one. For example, AVG is not only an inhibitor of ethylene biosynthesis, but also a general inhibitor of pyridoxal phosphate-requiring enzymes (Giovanelli et al. 1971).

ETHYLENE-AUXIN INTERACTIONS

Table 1 includes not only experiments involving application of ethylene alone, but also ethylene in combination with various auxins. In many instances, the combination of ethylene and auxin has been reported to stimulate rooting above that achieved with auxin alone, either additively (Schmid and Schurter 1973, Criley and Parvin 1979) or synergistically (Michener 1935, Krishnamoorthy 1972ab, Schmid and Schurter 1973; K. W. Mudge. 1976, M.S. Thesis, Colorado State Univ., Fort Collins, CO, USA; Mandal and Basu 1982). There are several reports, however, where auxin stimulated rooting has been unaffected by ethylene (Linkins et al. 1973, Swanson 1974, Carpenter 1975, Mudge and Swanson 1978, Johnson and Hamilton 1977, Batten and Mullins 1978, Robbins et al. 1983) or even antagonized by ethylene (Roy et al. 1972; K. W. Mudge. 1976, M.S. Thesis, Colorado State Univ., Fort Collins, CO, USA; Coleman et al. 1980, Geneve and Heuser 1983).

Relevant to this discussion of auxin ethylene interaction is the suggestion made by several authors that auxin stimulated rooting, even in the absence of exogenous ethylene, might be mediated by auxin stimulated ethylene production (Zimmerman and Wilcoxon 1935, Abeles 1973, Linkins et al. 1973, Robbins et al. 1983). This hypothesis is based on the evidence for auxin stimulated ethylene production by the vegetative tissues of most higher plants (Zimmerman and Wilcoxon 1935, Morgan and Hall 1964, Abeles 1973), including auxin treated cuttings of a number of species used in rooting experiments (Mullins 1972, Linkins et al. 1973, Batten and Mullins 1978, Coleman et al. 1980, Geneve and Heuser 1982). Also, it has been demonstrated that ethylene is an intermediary or "second messenger" in a number of other developmental auxin responses to auxin, including flowering in Bromeliaceae, flower fading in Orchidaceae (Abeles 1973), gravitropism and lateral root formation in tomato (Zobel 1973), and dichotomous branching of Pinus spp. roots (Rupp and Mudge 1985). A similar mechanism might be involved in the auxin stimulated adventitious rooting of shoots. Evidence in support of this hypothesis has come from experiments by Linkins et al. (1973), who reported that auxin was ineffective in stimulating rooting of Phaseolus vulgaris cuttings when auxin stimulated ethylene was kept at a low concentration (<0.02 nl·l^{-1}) through absorption by a mercuric perchlorate trap. Inhibitor studies mentioned above involving mung beans suggest that endogenous ethylene is involved in the rooting of auxin treated hypocotyls, based on inhibition of auxin stimulated rooting by AVG (Robbins et al. 1983, Jusaitis 1986ab). Contrasting results from a number of other investigations suggest that ethylene is not involved in auxin stimulated rooting, and even that it may actually be antagonistic to the process. For example, Coleman et al. (1980) reported that auxin induced rooting of tomato leaf discs *in vitro* was stimulated by reduction of ethylene levels with a mercuric perchlorate trap, which is in sharp contrast to the inhibition of rooting of French bean shoot cuttings in response to the same treatment (Linkins et al. 1983). Nordström and Eliasson (1984) reported that rooting of pea cuttings is inhibited by a low (10 μmol$\cdot$l^{-1}) concentration of auxin, and that this inhibition is correlated with and possibly a consequence of auxin stimulated ethylene production.

If ethylene is involved in auxin stimulated rooting, a reasonable hypothesis is that the magnitude of rooting stimulation by a range of auxin analogues should be correlated with the relative amount of ethylene synthesis induced. Several attempts to test this hypothesis have consistently shown that there is no such correlation between the two. Using etiolated mung bean cuttings, Mullins (1972) and Batten and Mullins (1978) found that while IBA was more effective in promoting rooting than IAA, NAA, or 2,4,5-TP, it was the least effective auxin in terms of stimulating ethylene production. Geneve and Heuser (1982) obtained a similar poor correlation with the mung bean rooting bioassay, except that they used light-grown rather than etiolated mung bean cuttings, and 2,4-D replaced 2,4,5-TP. Coleman et al. (1980) tested seven different auxins for their effect on rooting and on ethylene production by tomato discs and found no correlation between ethylene production and rooting. These consistently negative results could be misleading, however, if a very low saturating concentration for ethylene stimulated rooting exists. Auxin stimulated ethylene production above this level might then be without additional effect or inhibitory, especially under conditions which restrict diffusion and dissipation of ethylene gas. Thus, a poor correlation between ethylene production and rooting might occur despite the existence of a causal relationship between the two. Furthermore, the possibility exists that auxin might stimulate rooting by two separate mechanisms—directly, and, in addition, by its ability to stimulate (low level) ethylene production.

MECHANISTIC CONSIDERATIONS

Clearly, under some conditions, ethylene plays a role in the developmental sequence leading to adventitious root formation in at least some species of higher plants. The great variation in this response evident in the literature, as compared to nearly ubiquitous auxin stimulated rooting, suggests that ethylene is less directly involved in the process than auxin, and/or that it is less often a limiting factor.

One finding which may explain some of the variation in the literature is the existence of differential responses to ethylene, depending upon the time and duration of treatment. Robbins et al. (1985) reported that ethylene stimulated rooting of mung bean cuttings did not occur unless cuttings were exposed to the gas for at least 48 h after excision, and rooting increased with increasing duration of exposure, up to 4 d. In a second experiment, they demonstrated that a single 24 h pulse of ethylene gas was only effective if it was applied from 48–72 h or from 72–96 h post-excision, with the first interval being the more effective. In experiments conducted by Fabijan et al. (1981) with sunflower hypocotyl cuttings, the ethylene action inhibitors silver nitrate and BITC promoted rooting if applied in 3 h pulses during the first 12 h after excision, but slightly inhibited or had no effect on rooting if applied 24 h or more after excision. The authors proposed a biphasic response to ethylene. During an early phase (approximately the first 15 h of the rooting period) rooting was inhibited by ethylene, and during a later phase, rooting was promoted by ethylene. This interpretation is consistent with the peak in ethylene effectiveness at day 3 observed by Robbins et al. (1985). Furthermore, it might explain some of the variation in the results of the numerous experiments performed with ethephon and ethylene gas. If conditions were favorable for a rapid release of ethylene gas from ethephon (relatively high tissue pH, etc.) or rapid uptake of exogenously applied ethylene gas, the exposure to the gas during an early inhibitory phase might depress rooting compared to controls. On the other hand, where conditions favored minimal early release of ethylene from ethephon (low solution and/or tissue pH), or slow uptake of exogenously applied ethylene, the rooting might be promoted by the relatively late exposure to ethylene gas during a more responsive phase.

In contrast to this interpretation (increasing sensitivity to ethylene with time after cutting excision) are the results of Linkins et al. (1973). Based on removal of auxin stimulated ethylene by means of a mercuric perchlorate trap, they found that rooting of *Phaseolus vulgaris* cuttings was most sensitive to auxin stimulated ethylene during the first 44 h after excision. Removal of ethylene prior to this time completely blocked auxin stimulated rooting, whereas rooting was unaffected if ethylene was removed later. Differing reports of the period of maximum sensitivity to ethylene may merely reflect differences between mung bean (Robbins et al. 1985) and sunflower (Fabijan et al. 1981), where rooting is promoted by relatively late application, as compared to *P. vulgaris,* where rooting is promoted by earlier application (Linkins et al. 1973). At any rate, changes in ethylene responsiveness over time have been noted with another system, viz. formation of an abscission zone in leaf petioles (Osborne and Sargent 1976). A careful consideration of this variable in future experimentation might result in some further clarification of the role of ethylene in the rooting process.

Although the information which is currently available suggests that timing may be critical, there is very little to suggest the mechanism by which ethylene might exert its effect. The study mentioned above by Linkins et al. (1973) is an exception. Their findings suggest a mechanism somewhat analogous to the mechanism thought to be involved in ethylene stimulated leaf abscission, i.e. activation of cellulase resulting in cell wall hydrolysis (Osborne and Sargent 1976). Linkins et al. (1973) found that auxin and/or ethylene treatment of *P. vulgaris* petioles resulted in an increase in cellulase activity. Cellulase isozymes were separated by isoelectric focusing, and it was shown that auxin treatment preferentially increased the activity of a cellulase with an acidic isoelectric point (pI). Addition of exogenous ethylene (or stimulation of endogenous auxin stimulated ethylene) resulted in a stimulation of a basic pI cellulase. Based on time-course studies, it was determined that the auxin stimulated acidic cellulase was associated with cell division and xylem differentiation, whereas the ethylene stimulated basic cellulase was associated with degradation of cortical and pith tissues. The authors suggest that the auxin stimulated acidic cellulase is involved in root primordium initiation (cell division, xylem differentiation) and that the ethylene stimulated basic cellulase is involved in cortical degradation, which facilitates root initial development and emergence. Such a mechanism might also account for reports in which ethylene stimulated the emergence of preformed adventitious roots, in species such as willow, marigold, and tomato (Table 1).

CONCLUSION

The effect of ethylene on adventitious root formation is highly variable depending upon plant species, and the prevailing environmental and physiological conditions. Promotion of rooting by ethylene has been reported more frequently in intact plants than cuttings; herbaceous rather than woody plants; and plants with preformed root initials rather than those without. Conflicting results of various studies involving cuttings may be due to changes in the responsiveness of plant tissues to ethylene over time. Although a consensus does not exist, a preponderance of evidence suggests that endogenous ethylene is not directly involved in auxin induced rooting of cuttings or flood induced rooting of intact plants.

ACKNOWLEDGMENTS

The author would like to thank Dr. Arthur C. Cameron, Dr. Nina L. Bassuk, and Brian K. Maynard for reviewing this manuscript.

REFERENCES

Anon. 1967. *Amchem 66–329, A New Plant Growth Regulator.* Amchem Products, Inc. Information Sheet No. 37.

Anon. 1976. *Hortus Third.* Macmillan Pub. Co., NY, USA.

Abeles, F. B. 1973. *Ethylene in Plant Biology.* Academic Press, NY, USA. ISBN 0-12-041450-3.

Batten, D. J. and M. G. Mullins. 1978. Ethylene and adventitious root formation in hypocotyl segments of etiolated mung-bean (*Vigna radiata* (L.) Wilczek) seedlings. *Planta* 138:193–197.

Beyer, E. M., Jr. 1979. Effect of silver ion, carbon dioxide, and oxygen on ethylene action and metabolism. *Plant Physiol.* 63:169–173.

_____ P. W. Morgan. 1969. Ethylene modification of an auxin pulse in cotton stem sections. *Plant Physiol.* 44:1690–1694.

Boe, A. A., R. B. Stewart and T. J. Banko. 1972. Effects of growth regulators on root and shoot development on sedum leaf cuttings. *HortScience* 7:404–405.

Burg, S. P. and J. A. Stolwijk. 1959. A highly sensitive katharometer and its application to the measurement of ethylene and other gases of biological importance. *J. Biochem. Microbiol. Technol. Eng.* 1:245–259.

_____ K. B. Thimann. 1959. The physiology of ethylene formation in apples. *Proc. Nat. Acad. Sci. USA.* 45:335–344.

Carlson, M. C. 1950. Nodal adventitious roots in willow stems of different ages. *Amer. J. Bot.* 37:555–561.

Carpenter, S. B. 1975. Rooting black walnut cuttings with ethephon. *Tree Planter's Notes* 26(3):3.

Coker, T., S. Mazek and J. E. Thompson. 1985. Effect of water stress on ethylene production and on membrane microviscosity in carnation flowers. *Scientia Hortic.* 27:317–324.

Coleman, W. K., T. J. Huxter, D. M. Reid and T. A. Thorpe. 1980. Ethylene as an endogenous inhibitor of root regeneration in tomato leaf discs cultures *in vitro. Physiol. Plant.* 48:519–525.

Criley, R. A. and P. E. Parvin. 1979. Promotive effects of auxin, ethephon, and daminozide on the rooting of *Protea neriifolia* cuttings. *J. Amer. Soc. Hortic. Sci.* 104:592–596.

Cummins, J. N. and P. Forino. 1969. Preharvest defoliation of apple nursery stock using Ethrel. *HortScience* 4:339–341.

Curtis, R. W. and G. Fellenberg. 1972. Effect of malformin on adventitious root formation and metabolism of indoleacetic acid-2-^{14}C by *Phaseolus vulgaris. Plant Cell Physiol.* 13:715–726.

deWilde, R. C. 1971. Practical applications of (2-chloroethyl) phosphonic acid in agricultural production. *HortScience* 6:12–18.

Dhua, R. S., S. K. Mitra, S. K. Sen and T. K. Bose. 1983. Changes in endogenous growth substances, cofactors and metabolites in the rooting of mango cuttings. *Acta Hortic.* 134:147–161.

Fabijan, D., E. Yeung, I. Mukherjee and D. M. Reid. 1981. Adventitious rooting in hypocotyls of sunflower (*Helianthus annuus*) seedlings. I. Correlative influences and developmental sequence. *Physiol. Plant.* 53:578–588.

Geneve, R. L. and C. W. Heuser. 1982. The effect of IAA, IBA, NAA, and 2,4-D on root promotion and ethylene evolution in *Vigna radiata* cuttings. *J. Amer. Soc. Hortic. Sci.* 107:202–205.

_____ _____ 1983. The relationship between ethephon and auxin on adventitious root initiation in cuttings of *Vigna radiata* (L.) R. Wilcz. *J. Amer. Soc. Hortic. Sci.* 108:330–333.

Giovanelli, J., L. D. Owens and S. H. Mudd. 1971. Mechanism of inhibition of β-cystathionase by rhizobitoxine. *Biochem. Biophys. Acta* 227:671–684.

Graham, J. H. and R. G. Linderman. 1981. Effect of ethylene on root growth, ectomycorrhiza formation and *Fusarium* infection of Douglas fir. *Can. J. Bot.* 59:149–155.

Haissig, B. E. 1974. Origins of adventitious roots. *N.Z. J. For. Sci.* 4:299–310.

Hitchcock, A. E. and P. W. Zimmerman. 1940. Effects obtained with mixtures of root-inducing and other substances. *Contr. Boyce Thompson Inst.* 11:155–159.

Jackson, M. B. 1985. Ethylene and response of plants to soil waterlogging and submergence. *Ann. Rev. Plant Physiol.* 36:145–286.

_____ D. J. Campbell. 1975a. Movement of ethylene from roots to shoots, a factor in the response of tomato plants to waterlogged soil conditions. *New Phytol.* 74:397–406.

———— ———— 1975b. Ethylene and waterlogging effects in tomato. *Appl. Biol.* 81:102–105.

Johnson, C. R. and D. F. Hamilton. 1977. Rooting of *Hibiscus rosa-sinensis* L. cuttings as influenced by light intensity and ethephon. *HortScience* 12:39–40.

Jusaitis, M. 1986a. Rooting response of mung bean cuttings to 1-aminocyclopropane-1-carboxylic acid and inhibitors of ethylene biosynthesis. *Scientia Hortic.* 29:77–85.

———— 1986b. Rooting of intact mung bean hypocotyls stimulated by auxin, ACC, and low temperature. *HortScience* 21:1024–1025.

Kawase, M. 1971. Causes of centrifugal root promotion. *Physiol. Plant.* 25:64–70.

———— 1972. Effect of flooding on ethylene concentration in horticultural plants. *J. Amer. Soc. Hortic. Sci.* 97:584–588.

———— 1974. Role of ethylene in induction of flooding damage in sunflower. *Physiol. Plant.* 31:29–38.

Krishnamoorthy, H. N. 1970. Promotion of rooting in mung bean hypocotyl cuttings with ethrel, an ethylene releasing compound. *Plant Cell Physiol.* 11:979–982.

———— 1972a. Effect of ethrel, auxins and maleic hydrazide on the rooting of mung bean hypocotyl cuttings. *Z. Pflanzenphysiol.* 66:273–276.

———— 1972b. Effect of ethrel on the rooting of mung bean hypocotyl cuttings. *Biochem. Physiol. Pflanzen.* 163:505–508.

Linkins, A. E., L. N. Lewis and R. L. Palmer. 1973. Hormonally induced changes in the stem and petiole anatomy and cellulase enzyme patterns in *Phaseolus vulgaris* L. *Plant Physiol.* 52:554–560.

Lipecki, J. and J. Selwa. 1973. The effect of ethrel on the rooting of black currant hardwood cuttings. *Acta Agrobot* 26:229–235.

Mandal, K. and R. N. Basu. 1982. Involvement of ethylene in synergism between indoleacetic acid and indole in adventitious root formation. *Indian J. Exp. Biol.* 20:147–151.

Maynard, J. A. and J. M. Swan. 1963. Organophosphorus compounds. I. 2-chloroalkylphosphonic acids as phosphorylating agents. *Aust. J. Chem.* 16:596–608.

McMichael, B. L., W. R. Jordan and R. D. Powell. 1972. An effect of water stress on ethylene production by intact cotton petioles. *Plant Physiol.* 49:658–660.

Michener, H. D. 1935. Effects of ethylene on plant growth hormone. *Science* 82:551–552.

Mingo-Castel, A. M., O. E. Smith and J. Kumamoto. 1976. Studies on the CO_2 promotion and ethylene inhibition of tuberization in potato explants cultured *in vitro*. *Plant Physiol.* 57:480–485.

Moore, K. G. 1975. Role of ethylene in the effects of sucrose and other sugars on detached *Sinapis* cotyledons. *Proc. Assoc. Appl. Biol.* 81:102.

Morgan, P. W. and W. C. Hall. 1964. Accelerated release of ethylene by cotton following application of indolyl-3-acetic acid. *Nature* 201:99.

Mudge, K. W. and B. T. Swanson. 1978. Effect of ethephon, indole butyric acid, and treatment solution pH on rooting and on ethylene levels within mung bean cuttings. *Plant Physiol.* 61:271–273.

Mullins, M. G. 1972. Auxin and ethylene in adventitious root formation in *Phaseolus aureus* (Roxb.). In *Plant Growth Substances 1970* (D. J. Carr, ed). Springer Verlag, Berlin, W. Germany. pp. 526–533.

Nell, T. A. and K. C. Sanderson. 1972. Effect of several growth regulators on the rooting of three azalea cultivars. *Florist's Rev.* 150:21–22.

Nordström, A. and L. Eliasson. 1984. Regulation of root formation by auxin-ethylene interaction in pea stem cuttings. *Physiol. Plant.* 61:298–302.

Orion, D. and G. Minz. 1969. The effect of ethrel (2-chloroethane phosphonic acid) on the pathogenicity of the root knot nematode *Meloidogyne javanica*. *Nematologica* 15:608–614.

Osborne, D. J. and J. A. Sargent. 1976. The positional differentiation of abscission zones during the development of leaves of *Sambucus nigra* and the response of the cells to auxin and ethylene. *Planta* 132:197–204.

Patil, S. S. and C. S. Tang. 1974. Inhibition of ethylene evolution in papaya pulp tissue by benzyl isothiocyanate. *Plant Physiol.* 53:585–588.

Phatak, S. C., C. A. Jaworski and A. Liptay. 1981. Flowering and adventitious root growth of tomato cultivars as influenced by ethephon. *HortScience* 16:181–182.

Robbins, J. A., S. J. Kays and M. A. Dirr. 1983. Enhanced rooting of wounded mung bean cuttings by wounding and ethephon. *J. Amer. Soc. Hortic. Sci.* 108:325–329.

_____ M. S. Reid, J. L. Paul and T. L. Rost. 1985. The effect of ethylene on adventitious root formation in mung bean *(Vigna radiata)* cuttings. *J. Plant Growth Regul.* 4:147–157.

Roy, B. N., R. N. Basu and T. K. Bose. 1972. Interaction of auxins with growth-retarding, -inhibiting and ethylene-producing chemicals in rooting of cuttings. *Plant Cell Physiol.* 13:1123–1127.

Rupp, L. A. and K. W. Mudge. 1985. Ethephon and auxin induce mycorrhiza-like changes in the morphology of root organ cultures of Mugo pine. *Physiol. Plant.* 64:316–322.

Sadhu, M. K. and S. Bose. 1980. Effect of ethylene on rooting of cuttings and air layers of mango, guava, and waterapple. *Indian J. Hortic.* 37:335–337.

Samananda, N., D. P. Ormrod and N. O. Adepipe. 1972. Rooting of chrysanthemum stem cuttings as affected by (2-chloroethyl) phosphonic acid and indolebutyric acid. *Ann. Bot.* 36:961–965.

Schier, G. A. 1975. Effect of ethephon on adventitious root and shoot development in aspen. *Plant Physiol.* (Suppl.) 56:71.

Schmid, A. and R. Schurter. 1973. Influence de l'éthylène sur la rhizogénèse de boutures lignifiées de *Populus tremula* L. et de *Populus nigra* L. *C.R. Acad. Sci. Paris* 276:1293–1296.

Shanks, J. B. 1969. Some effects and potential uses of ethrel on ornamental crops. *HortScience* 4:56–58.

Sisler, E. C. and R. Goren. 1981. Ethylene binding—the basis for hormone action in plants? *What's New in Plant Physiol.* 12:37–40.

Swanson, B. T. 1974. Ethrel as an aid in rooting. *Proc. Int. Plant Prop. Soc.* 24:351–361.

Thimann, K. V. and F. W. Went. 1934. On the chemical nature of the root forming hormone. *Proc. K. Ned. Akad. Wet. Amst.* 37:456–459.

Tsujita, M. J. and P. M. Harney. 1978. The effects of Florel and supplemental lighting on the production and rooting of geranium cuttings. *J. Hortic. Sci.* 53:349–350.

Veen, H. 1983. Silver thiosulphate: an experimental tool in plant science. *Scientia Hortic.* 20:211–224.

Wample, R. L. and D. M. Reid. 1979. The role of endogenous auxins and ethylene in the formation of adventitious roots and hypocotyl hypertrophy in flooded sunflower plants *(Helianthus annuus).* *Physiol. Plant.* 45:219–226.

Wilson, E. R. and R. J. Field. 1984. Dichotomous branching in lateral roots of pine: the effect of 2-chloroethylphosphonic acid on seedlings of *Pinus radiata* D. Don. *New Phytol.* 98:465–473.

Zimmerman, P. W. and A. E. Hitchcock. 1933. Initiation and stimulation of adventitious roots caused by unsaturated hydrocarbon gases. *Contr. Boyce Thompson Inst.* 5:351–369.

_____ F. Wilcoxon. 1935. Several chemical growth substances which cause initiation of roots and other responses in plants. *Contr. Boyce Thompson Inst.* 7:209–229.

_____ W. Crocker and A. E. Hitchcock. 1933. Initiation and stimulation of roots from exposure of plants to carbon monoxide gas. *Contr. Boyce Thompson Inst.* 5:1–17.

Zobel, R. W. 1973. Some physiological characteristics of the ethylene-requiring mutant, diageotropica. *Plant Physiol.* 52:385–389.

CHAPTER 12

Influence of Gibberellins on Adventitious Root Formation

Jürgen Hansen

Institute of Glasshouse Crops
Kirstinebjergvej 10
DK-5792 Aarslev
Denmark

INTRODUCTION..162
ENDOGENOUS GIBBERELLIN-LIKE SUBSTANCES AND ROOT
   FORMATION...163
RESPONSES TO EXOGENOUS GIBBERELLIN 163
SENSITIVITY TO GIBBERELLIN...163
INFLUENCE OF LIGHT 166
     Light Quality 166
     Irradiance...166
     Photoperiod 166
ROLE OF BUDS AND SHOOT ELONGATION 167
INTERACTION WITH OTHER PLANT GROWTH REGULATORS.................168
     Auxins..168
     Abscisic Acid..168
     Gibberellin Antagonists 169
MODE OF GIBBERELLIN ACTION 169
CONCLUSION 170
ACKNOWLEDGMENTS 170
REFERENCES. .\...170

Additional key words: abscisic acid, antagonist, auxin, irradiance, light quality, mode of action, photoperiod, sensitivity.
Abbreviations: ABA, abscisic acid; GA_1, gibberellin A_1; GA_3, gibberellic acid; GA_{4+7}, gibberellin A_{4+7}; IAA, indole-3-acetic acid; IBA, indole-3-butyric acid; NAA, naphthalene-acetic acid; RNA, ribonucleic acid; DNA, deoxyribonucleic acid.

INTRODUCTION

Gibberellins are naturally occurring plant growth substances which affect plant growth and development (Brian 1959ab). Shortly after the isolation of the first gibberellin, GA_3 (Curtis and Cross 1954), experiments were made to compare the effects of GA_3 with those of auxins (Brian et al. 1955, Kato 1958). In these studies it became evident that GA_3, in many physiological processes, had effects opposite to those obtained with auxins. Adventitious root formation, which was stimulated by auxin, was inhibited by GA_3 (Brian et al. 1955, Kato 1958).

Many experiments involving a number of different plant species have later confirmed that rooting is generally inhibited by GA_3. Since the early 1970s, several investigations have, however, shown that GA_3 may in some cases stimulate adventitious root formation.

Some of the most commonly investigated aspects of gibberellin action on adventitious root formation will be discussed in this review. Of the more than 70 different naturally occurring gibberellins (Sponsel 1985), GA_3 is the only gibberellin that has been extensively studied in this respect. Very few rooting experiments have dealt with other gibberellins. In experiments with *Phaseolus vulgaris* L., GA_1 and GA_3 both inhibited rooting to a similar extent (Brian et al. 1960). Similar inhibitory effects on root formation were also reported for GA_{4+7} and GA_3 in *Phaseolus aureus* Roxb. (Krishnamoorthy 1972a). The present review thus mainly deals with the effects of GA_3 though the broader term, gibberellin, is frequently used in the text.

ENDOGENOUS GIBBERELLIN-LIKE SUBSTANCES AND ROOT FORMATION

Few investigations concerning changes in content of endogenous gibberellins during rooting have been made. Bláhová (1969) reported that endogenous gibberellin activity in the basal part of cuttings from etiolated pea plants was reduced 24 h after excision followed by a gradual increase over the next 48 h. Thus, the early events in root formation in *Pisum sativum,* which take place over the initial 4 d (Eriksen 1973), are associated with a temporary decrease in endogenous gibberellin activity.

Measurements of endogenous gibberellin activity in relation to root formation have also been made in tissue cultures of *Malus pumila* Mill. (Takeno et al. 1982/83). In that experiment an inverse relationship between the endogenous gibberellin level and rooting ability was found. With increasing number of subcultures the endogenous level of gibberellin-like substances decreased while the percentage of rooted cuttings increased. This increase in rooting percentage could be due to changes in factors other than gibberellin content but the correlation is worth noting. Additional investigations are needed to verify the possibility of a relationship between endogenous gibberellin content and ability to form roots.

RESPONSES TO EXOGENOUS GIBBERELLIN

Exogenous gibberellin inhibits adventitious root formation in many species (see Table 1 and review by Batten and Goodwin 1978). Generally the inhibition is increased with increasing concentration higher than 10^{-6} mol·l^{-1}. In several species and under certain environmental conditions, however, gibberellin enhances root formation (Table 1). These discrepancies may be due to the use of different plant species, but since inhibition as well as stimulation can be observed in the same species and cultivars, it is likely that other factors are involved in the regulation of the rooting response to gibberellin.

Changes in response to applied gibberellin during the process of root formation indicate that different developmental stages of root formation have different degrees of sensitivity to gibberellin. This accentuates the importance of the presence or absence of gibberellin at a particular time or time interval. Furthermore, differences in application technique and amount of applied active substance may affect the response to gibberellin. This implies that many factors have to be considered when comparisons between different experiments are made.

SENSITIVITY TO GIBBERELLIN

Inhibition of root formation by gibberellin is dependent on the time of application. Several studies concerning GA_3 application at different times or over specific time intervals during root formation have demonstrated that GA_3 must be present during the early stages of root formation in order to be inhibitory (Brian et al. 1960, Reinert and Besemer 1967, Kato and Hongo 1974, Smith and

Table 1. Effect of exogenous gibberellin on adventitious root formation of different plant species.

Species	Inhibition	Stimulation	Reference
Abelmoschus esculentus Moench		+	Bhattacharya et al. (1978)
Acer rubrum L.	+		Bachelard and Stowe (1963)
Azukia angularis	+		Mitsuhashi-Kato et al. (1978)
Begonia rex Putz.	+		Schraudolf and Reinert (1959)
	+		Reinert and Besemer (1967)
Begonia × *cheimantha* Everett	+		Heide (1969)
Brassica juncea L. (Czern. & Coss.)	+		Saxena and Singh (1982)
Bryophyllum tubiflorum Harv.	+	+	Nanda et al. (1967)
Calystegia sepium R. Br.	+		Krelle and Libbert (1969)
Citrus sinensis Osbeck		+	Kochba et al. (1974)
Cucumis melo L.	+		Krishnamoorthy (1972b)
Helianthus annuus L.	+		Libbert and Krelle (1966)
	+		Fabijan et al. (1981)
Helianthus tuberosus L.	+	+	Gautheret (1969)
Heloniopsis orientalis	+		Kato and Hongo (1974)
Ipomoea fistulosa		+	Anand et al. (1972)
		+	Nanda et al. (1972)
Lycopersicon esculentum Mill.	+		Jansen (1967, 1971)
	+	+	Coleman and Greyson (1976, 1977ab)
Nicotiana tabacum L.	+		Murashige (1964)
Panicum maximum Jacq.		+	Felippe (1980)
Phaseolus vulgaris L.	+		Brian et al. (1960)
	+		Libbert and Krelle (1966)
		+	Varga and Humphries (1974)
		+	Haddon and Northcote (1976)
Phaseolus aureus Roxb.	+		Fernqvist (1966)
	+		Chin et al. (1969)
	+		Krishnamoorthy (1972a)
	+		Kefford (1973)
Pinus radiata D. Don	+	+	Smith and Thorpe (1975b)
Pinus sylvestris L.	+		Ernstsen and Hansen (1986)
Pisum sativum L.	+		Kato (1958)
	+		Brian et al. (1960)
	+	+	Leroux (1967)
	+		Fellenberg (1969)
	+	+	Eriksen (1971, 1972)
	+	+	Hansen (1975, 1976)
		+	Adhikari and Bajracharya (1978)
Populus nigra L.	+		Nanda et al. (1968)
Rhododendron spp.	+		Pierik and Steegmans (1975)
Salix fragilis L.	+		Haissig (1972)
Salix purpurea L.	+		Gundersen (1958)
Salix viminalis L.	+		Kriesel (1975)
Sinapis alba L.	+		Pfaff and Schopfer (1980)
Thevetia peruviana (Pers.) Nerill.	+		Kumar and Sharma (1978)

Thorpe 1975b, Coleman and Greyson 1977a). In *Phaseolus vulgaris*, GA$_3$ must be applied either within 48 h prior to cutting excision or within the first 24 h after cutting excision to be inhibitory (Brian et al. 1960). In *Begonia rex* Putz., GA$_3$ inhibited rooting only when applied within the first 10 d of rooting (Reinert and Besemer 1967) whereas inhibition was observed in *Pinus radiata* D. Don (Fig.

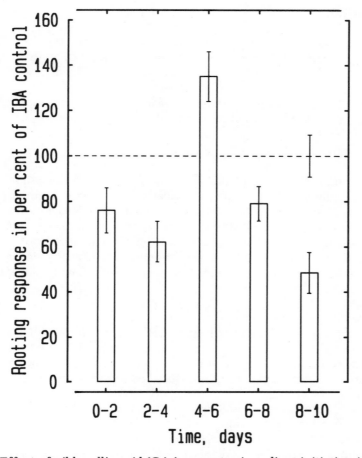

Figure 1. Effect of gibberellic acid (GA_3) on root primordium initiation in hypocotyls of *Pinus radiata* D. Don. GA_3 at the concentration of 10^{-5} mol·l^{-1} was applied only over the days indicated. IBA at a concentration of 1.48×10^{-5} mol·l^{-1}, pH 4.5, was present throughout the 10 d assay period. Number of root primordia per cutting is given relative to the control which only received IBA. Vertical bars denote S.E. [After Smith and Thorpe (1975b)]

1) when GA_3 at a concentration of 10^{-5} mol·l^{-1} was supplied for the first 4 d (Smith and Thorpe 1975b).

The detailed investigation by Smith and Thorpe (1975b) further demonstrated a period during which the same GA_3 concentration of 10^{-5} mol·l^{-1} significantly stimulated the formation of root primordia. This occurred when GA_3 was supplied from days 4 to 6 after the planting of cuttings (Fig. 1). When GA_3 was supplied at later time intervals, rooting was inhibited (Smith and Thorpe 1975b). For other species such as *Pisum sativum* (Fellenberg 1969) and *Lycopersicon esculentum* Mill. (Jansen 1967, Coleman and Greyson 1977a) GA_3 was most inhibitory when applied within the first 3 d of root formation.

The existence of a GA_3 sensitive phase during which GA_3 inhibits root formation thus seems well documented. It is also clear that this phase coincides with the early events in root formation because GA_3 generally is most inhibitory during the initial stages of the rooting period. Furthermore, the work by Smith and Thorpe (1975b) indicates that another GA_3 sensitive phase exists during which rooting is stimulated. Thus far, this is the only investigation which has demonstrated an application-time-dependent inhibition and stimulation of root formation.

INFLUENCE OF LIGHT

Light has a strong impact on the reaction of plants to exogenously supplied gibberellins (Bown et al. 1975, Hansen 1975). The metabolism of gibberellins is affected differently in light and darkness and may further depend on the type of gibberellin that is studied (Bown et al. 1975, Lance et al. 1976).

In order to present detailed information on the reaction of plants to supplied GA_3 under different light treatments, the influence of light is dealt with as the separate influences of light quality, irradiance, and photoperiod.

Light Quality

Light quality has not been widely studied in connection with root formation. It appears, however, that those wavelengths which stimulate the formation of the active form of phytochrome (P_{fr}) also stimulate the formation of adventitious roots in isolated cotyledons and hypocotyl cuttings of *Sinapis alba* L. (Pfaff and Schopfer 1974). The involvement of phytochrome in gibberellin regulation *in vitro* is also fairly well established (Evans and Smith 1976). Nevertheless, Pfaff and Schopfer (1980) concluded from their experiment with exogenously supplied GA_3 that for *Sinapis alba,* there is no causal link between phytochrome mediated root formation and changes in gibberellin levels.

Irradiance

Experiments where GA_3 was supplied to light or dark grown cultures of *Helianthus tuberosus* L. tissue revealed that rooting was promoted in the dark, but inhibited in the light (Gautheret 1969). Similar observations were made by Leroux (1967) with *Pisum sativum* stem segments grown *in vitro*. From these investigations, however, it cannot be deduced whether the results reflect the influence of photoperiod or irradiance.

Hansen (1975, 1976) demonstrated with *Pisum sativum* cuttings that the irradiance at which the stock plants were grown influenced the rooting response to gibberellin. Stimulation of root formation by GA_3 concentrations of 10^{-9} to 10^{-6} mol·l^{-1} was only observed with cuttings from low-irradiance (e.g. 16 W·m^{-2}, 400–700 nm) treated stock plants (Fig. 2). Furthermore, Hansen (1976) found that the stock plant irradiance, at which the low GA_3 concentrations could stimulate root formation, differed with year. The results presented in Fig. 2 were based on cuttings from seeds harvested during 1972. In order to demonstrate the stimulation of root formation by GA_3 in cuttings from seeds harvested during 1973, these stock plants had to be grown at an irradiance of 3 or 6 W·m^{-2} (Hansen 1976). The same low GA_3 concentrations had no effect on cuttings from stock plants grown at 38 W·m^{-2}. GA_3 concentrations higher than 10^{-6} mol·l^{-1} always inhibited root formation and the inhibition was relatively stronger in cuttings from stock plants grown under low irradiance (Hansen 1975, 1976).

Photoperiod

Photoperiodic influences on GA_3 mediated root formation have been studied by Nanda et al. (1967) in *Bryophyllum tubiflorum* Harv. a long-short-day species. Cuttings from short-day (8 h) treated stock plants had a higher rooting percentage and produced more roots than cuttings from natural-day (11–14 h) or long-day (24 h) treated plants. The photoperiod during propagation influenced rooting in a similar manner. GA_3 at a concentration of 3×10^{-5} or 3×10^{-4} mol·l^{-1} increased the number of roots and rooting percentage in short-day treated cuttings from short-day treated stock plants, whereas it failed to induce rooting in long-day treated cuttings from long-day treated stock plants (Nanda et al. 1967).

In some investigations, which have demonstrated inhibitory gibberellin effects, it was observed that the inhibition was less pronounced when the stock plants had been grown under short-day rather than long-day conditions (Bachelard and Stowe 1963) or when root formation occurred in continuous darkness compared with continuous light (Coleman and Greyson 1977a). Jansen (1971) reported that the dry weight of adventitious roots on tomato cuttings was reduced by GA_3 over the concentration range from 3×10^{-9} to 3×10^{-5} mol·l^{-1}. The reduction, however, was more pronounced when rooting

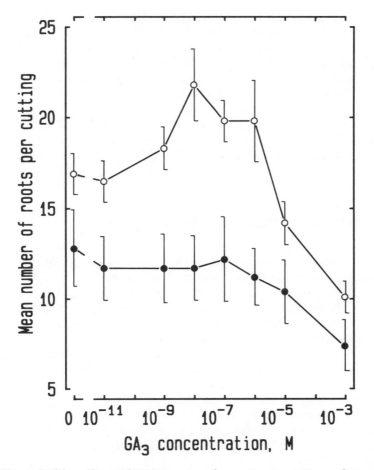

Figure 2. Effect of gibberellic acid (GA$_3$) on root formation in cuttings of *Pisum sativum* L. Stock plants were grown at an irradiance of 16 W·m^{-2} (O — O) or 38 W·m^{-2} (● — ●), the cuttings at 16 W·m^{-2}. Each cutting was treated with 10 μl of the GA$_3$ test solution. The number of roots was examined after 13 d. Vertical bars denote 95% confidence limits. [After Hansen (1975)]

occurred at a photoperiod of 20 h compared with a photoperiod of 8 h.

The light effects described above indicate that irradiance and photoperiod are two important factors in influencing the rooting response to GA$_3$. The results demonstrate that low irradiances to the stock plants, and short-day conditions to stock plants and cuttings, may reduce the inhibition caused by GA$_3$ and provide a basis for a stimulation by GA$_3$.

ROLE OF BUDS AND SHOOT ELONGATION

Exogenously supplied gibberellin affects the growth of intact plants. Some species show high sensitivity and specificity, and have been used in biological assays for gibberellins (Bailiss and Hill 1971).

Cuttings of certain species treated with gibberellin show an increase in stem elongation during root formation, e.g. *Phaseolus vulgaris* (Brian et al. 1960) and *Pisum sativum* (Hansen 1976). This effect is dose dependent and appears in *Pisum sativum* when the concentration of the 10 μl GA$_3$ exceeds 10^{-5} mol·l^{-1} (34 ng per cutting) (Hansen 1976).

The concomitant inhibition of root formation at such high GA$_3$ doses led Brian (1957) to suggest that the inhibition of root formation was due to a competitive diversion of respirable substrates for shoot extension and root formation. By excision of the apical bud and any lateral buds developing during the rooting period or by applying small doses of GA_3 to different parts of the cuttings, Brian et al. (1960) were experimentally able to dissociate the dual effect of GA_3, and it was concluded that the inhibition of root formation by GA_3 occurs independently of cutting extension. Later, however, Gamburg (1962) and Nanda et al. (1968) reported results in support of Brian's nutrient diversion theory.

The role of apical and lateral buds in the GA_3 mediated inhibition of root formation was studied further by Eriksen (1971) and Felippe (1980). Although Eriksen (1971) found a significant reduction in relative root number in decapitated cuttings of the pea cultivar 'Kelwo', the overall observations are in agreement with the findings of Brian et al. (1960) that the presence or absence of buds is not essential for GA_3 to exert its inhibitory effect.

INTERACTION WITH OTHER PLANT GROWTH REGULATORS

Auxins

Simultaneous treatment of cuttings with IAA or IBA may partially relieve the inhibition of rooting by GA_3 in *Pisum sativum* (Kato 1958) and in leaf cuttings of *Lycopersicon esculentum* (Coleman and Greyson 1976). Such an antagonism between auxin and GA_3 is noncompetitive as shown by Brian et al. (1960) with *Pisum sativum* and *Phaseolus vulgaris*. The GA_3 inhibition in *Azukia angularis* [*Vigna radiata* (L.) R. Wilcz.] cuttings can also be overcome by a subsequent IAA treatment 24 h after cutting excision (Mitsuhashi-Kato et al. 1978). With *Begonia* × *cheimantha* Everett leaf cuttings, however, NAA at a concentration of 2.5–40 × 10^{-6} mol·l^{-1} was unable to overcome the inhibitory effect of GA_3 on the number of roots formed, but a slight increase in rooting percentage was obtained (Heide 1969).

An interaction between IBA and GA_3 was also reported for *in vitro* grown cultures of *Lycopersicon esculentum* (Coleman and Greyson 1977a). Their data showed that the inhibition of root formation by GA_3 at a concentration of 10^{-8} to 10^{-4} mol·l^{-1} was more pronounced at the high IBA concentration of 10^{-5} mol·l^{-1} than at 10^{-6} mol·l^{-1}. In a detailed study of IAA and GA_3 interaction in root formation of tomato cuttings, Jansen (1967) reported that the inhibition by GA_3, measured as a decrease in relative dry weight of the roots, was increased with increasing IAA concentrations above 10^{-7} mol·l^{-1}.

Experiments with *Abelmoschus esculentus* Moench (Bhattacharya et al. 1978) and *Pisum sativum* (Adhikari and Bajracharya 1978) revealed that the stimulation of root formation by GA_3 was increased by simultaneous application of IAA. Bhattacharya et al. (1978) reported a synergistic effect of GA_3 and IAA whereas Adhikari and Bajracharya (1978) found an additive effect with respect to final rooting percentage. Further, the combination of both growth regulators accelerated root formation.

Although only a limited number of investigations on auxin-gibberellin interactions in root formation have been performed, and with partly conflicting observations, it may be concluded that auxin generally counteracts the inhibition of root formation by gibberellin. In addition, the stimulation of root formation by GA_3 which has so far only been observed in a few species may further be increased by simultaneous application of auxin.

Abscisic Acid

Chin et al. (1969) showed that the inhibitory effect of a 24 h GA_3 treatment during the initial rooting period of *Phaseolus aureus* could be overcome in part with a subsequent ABA treatment. This was, however, not possible with *Phaseolus vulgaris* (Krelle and Libbert 1969) and *Begonia* × *cheimantha* (Heide 1969) where the inhibition caused by GA_3 was further enhanced by ABA. This contrasts with other phenomena where GA_3 and ABA normally exhibit opposite effects. ABA applied alone had no effect on rooting of *Phaseolus vulgaris* (Krelle and Libbert 1969). Root formation in callus cultures of *Phaseolus vulgaris* was stimulated by GA_3 but inhibited by the simultaneous presence of ABA (Haddon and Northcote 1976).

Gibberellin Antagonists

Kefford (1973) showed that the gibberellin antagonist, ancymidol, counteracted the gibberellin inhibition of root formation in *Phaseolus aureus*. Ancymidol also stimulated root formation in the absence of exogenous gibberellin. The gibberellin antagonist, chlormequat chloride, did not have any effect on the GA_3 inhibition in *Helianthus* spp. (Libbert and Krelle 1966, Fabijan et al. 1981), *Phaseolus* spp. (Libbert and Krelle 1966), *Lycopersicon* spp. (Jansen 1967) and *Begonia* spp. (Heide 1969) but it inhibited rooting further in hypocotyl cuttings of *Cucumis melo* L. (Krishnamoorthy 1972b). Also the antagonist, chlorphonium chloride, had no effect on the GA_3 induced inhibition of rooting in *Begonia* spp. (Heide 1969). For more information on the effects of gibberellin antagonists, see Chapter 13 by Davis and Sankhla.

MODE OF GIBBERELLIN ACTION

Differences in gibberellin sensitivity during the process of root formation are fairly well documented. Brian et al. (1960) and Reinert and Besemer (1967) suggested that GA_3 inhibited root formation through an inhibition of those cell divisions in the stem tissue that precede the formation of organized root primordia. The detailed investigation of the GA_3 effect in relation to developmental sequences during rooting of *Pinus radiata* revealed that the inhibition was most pronounced during the pre-initiative phase, the phase preceding the formation of the meristematic locus (Smith and Thorpe 1975a). During the phase of cytoplasmic migration in peripheral cells with subsequent asymmetric cell divisions, GA_3 enhanced the formation and/or development of meristematic loci (Smith and Thorpe 1975b). During the subsequent phase of growth of meristemoids, GA_3 exerted a strong inhibition of rooting. This is in agreement with the work of Haissig (1972) who concluded that GA_3 reduced the number of cell divisions in already established primordia of *Salix fragilis* L.

The literature thus strongly indicates that the GA_3 sensitivity of the tissue varies during the process of root formation. The presence or absence of specific receptor molecules (Trewavas 1982) may explain the sequential differences in gibberellin sensitivity thus giving an explanation of the phase dependent inhibitory and stimulatory effects of GA_3 found by Smith and Thorpe (1975b). This possibility has yet to be investigated, however.

The interaction between gibberellin and auxin appears to be rather confusing. The stimulation of rooting by GA_3 in some cases may be related to changes in endogenous auxin content as suggested by Eriksen (1972), Varga and Humphries (1974), and Coleman and Greyson (1977b) since Valdovinos et al. (1967), Varga and Bitó (1968), and Andersen and Muir (1969) have demonstrated that GA_3 was able to increase the auxin level *in vivo*. A GA_3 stimulation of auxin synthesis is, however, not likely to be responsible for the commonly observed inhibition of root formation by GA_3.

The well known effect of gibberellin on the synthesis of certain hydrolytic enzymes in the aleurone layer of cereal seeds (Briggs 1973) and the effect on enzymes that regulate carbohydrate metabolism in, e.g. dwarf pea internodes (Broughton and McComb 1971) and bean (Morris and Arthur 1985) provides the basis for the different hypotheses of gibberellin induced, carbohydrate mediated effects on root formation (Nanda et al. 1968, Hansen 1976.). Carbohydrate effects on root formation have been reported for many years but the role of carbohydrate is still obscure (Veierskov et al. 1982, Ernstsen and Hansen 1986, also see Chapter 5 by Veierskov). Recently Ernstsen and Hansen (1986) reported that the inhibition of rooting by GA_3 in *Pinus sylvestris* L. coincided with mobilization of starch in the cuttings during rooting. However, no significant changes in content of the measured low-molecular weight carbohydrates were found.

For the moment the mechanism of gibberellin action at the molecular level is unclear. Probably all of the observed gibberellin effects on rooting are regulated through a common mechanism at the RNA (Fellenberg 1969) or DNA (Iwasaki et al. 1986) level.

CONCLUSION

Inhibition of root formation by gibberellin is dependent upon the time of application. The strongest inhibition is usually obtained when gibberellin is applied during the early stages of the rooting process, typically within 3–4 d after the excision of the cutting. Light conditions during stock plant growth and root formation can modify the gibberellin effect. A relatively low irradiance or short photoperiod during stock plant growth or a short photoperiod during rooting may, in at least some species, provide the basis for a stimulation of root formation by gibberellin. Auxin generally counteracts the inhibition by gibberellin and further stimulates rooting in those species where gibberellin alone stimulates rooting. Investigations with other growth regulators are too scarce to allow any general conclusion.

The mode of gibberellin action in root formation is still obscure. Although the inhibition of cell division, the gibberellin induced auxin synthesis, and the effect of gibberellin on carbohydrate metabolism demonstrate that gibberellin may exert its action at different biochemical levels, it is possible that the mechanism at the molecular level is the same. It is likely that gibberellin acts via its effect on transcription.

The inhibitory effect of GA_3 on root formation has been known for many years, whereas observations of GA_3 stimulation of root formation have been made only over the last 20 yr. The resulting effects of GA_3 treatments have been observed and described and, in some investigations, valuable anatomical studies have been made. Nevertheless, we know very little about the mode of action of GA_3 in adventitious root formation, particularly at the molecular level. It is of utmost importance to acquire basic information on the sequential steps during the process of root formation in order to understand the shifts in sensitivity to GA_3 during root formation. It is also necessary to investigate the fundamental influence of the different light parameters, in particular the light conditions prevailing during stock plant growth. Otherwise we will not be able to explain why gibberellin may inhibit as well as stimulate rooting and so explain the apparently conflicting observations reported.

We also need to know more about the interaction between GA_3 and other plant growth substances in the rooting process. A more detailed knowledge of the interaction with auxins with respect to the time of application of the growth substances in relation to the different phases during rooting may give a valuable contribution to the understanding of the rooting process. Finally, we need more information about the role of other gibberellins than GA_3 and to get reliable data on the endogenous content of specific gibberellins during root formation. Measurements of endogenous gibberellin content in relation to anatomical events during the process of root formation would be very valuable.

ACKNOWLEDGMENTS

The valuable discussions with Drs. A. Skytt Andersen and E. Nymann Eriksen, Institute of Horticulture, Royal Veterinary and Agricultural University, Copenhagen, are gratefully acknowledged. I also thank Ms. Margit R. Simonsen for typing the manuscript.

REFERENCES

Adhikari, U. K. and D. Bajracharya. 1978. Interaction of gibberellic and indole-3-acetic acid on root formation in pea (*Pisum sativum* L.) epicotyl cuttings. *Planta* 143:331–332.

Anand, V. K., R. N. Chibbar and K. K. Nanda. 1972. Effects of GA_3 and IBA on rooting and on the sprouting of buds on stem cuttings of *Ipomoea fistulosa*. *Plant Cell Physiol.* 13:917–921.

Andersen, A. S. and R. M. Muir. 1969. Gibberellin induced changes in diffusible auxins from savoy cabbage. *Physiol. Plant.* 22:354–363.

Bachelard, E. P. and B. B. Stowe. 1963. Rooting of cuttings of *Acer rubrum* L. and *Eucalyptus camaldulensis* Dehn. *Aust. J. Biol. Sci.* 16:751–767.

Bailiss, K. W. and T. A. Hill. 1971. Biological assays for gibberellins. *Bot. Rev.* 37:437–479.

Batten, D. J. and P. B. Goodwin. 1978. Phytohormones and the induction of adventitious roots. In *Phytohormones and Related Compounds—A Comprehensive Treatise*. (D. S. Letham, P. B. Goodwin and T. J. V. Higgins, eds). Elsevier/North-Holland Biomedical Press, Amsterdam, The Netherlands. Vol. II. pp. 137–173. ISBN 0-444-80054-9.

Bhattacharya, S., N. C. Bhattacharya and C. P. Malik. 1978. Synergistic effect of gibberellic acid and indole-3-acetic acid on rooting in stem cuttings of *Abelmoschus esculentus* Moench. *Planta* 138:111–112.

Bláhová, M. 1969. Changes in the level of endogenous gibberellins and auxins preceding the formation of adventitious roots on isolated epicotyls of pea plants. *Flora,* Abt. A 160:493–499.

Bown, A. W., D. R. Reeve and A. Crozier. 1975. The effect of light on the gibberellin metabolism and growth of *Phaseolus coccineus* seedlings. *Planta* 126:83–91.

Brian, P. W. 1957. The effects of some microbial metabolic products on plant growth. *Symp. Soc. Exp. Biol.* 11:166–182.

_____ 1959a. Effects of gibberellins on plant growth and development. *Biol. Rev. Cambridge Phil. Soc.* 34:37–84.

_____ 1959b. Morphogenetic effects of the gibberellins. *J. Linn. Soc. London. Ser. Bot.* 56:237–248.

_____ H. G. Hemming and D. Lowe. 1960. Inhibition of rooting of cuttings by gibberellic acid. *Ann. Bot.* 24:407–419.

_____ _____ M. Radley. 1955. A physiological comparison of gibberellic acid with some auxins. *Physiol. Plant.* 8:899–912.

Briggs, D. E. 1973. Hormones and carbohydrate metabolism in germinating cereal grains. In *Biosynthesis and its Control in Plants. Ann. Proc. Phytochem. Soc.* (B. V. Milborrow, ed). Academic Press, London, England and NY, USA. No. 9. pp. 219–277. ISBN 0-12-496150-9.

Broughton, W. J. and A. J. McComb. 1971. Changes in the pattern of enzyme development in gibberellin-treated pea internodes. *Ann. Bot.* 35:213–228.

Chin, T.-Y., M. M. Meyer, Jr. and L. Beevers. 1969. Abscisic-acid-stimulated rooting of stem cuttings. *Planta* 88:192–196.

Coleman, W. K. and R. I. Greyson. 1976. Root regeneration from leaf cuttings of *Lycopersicon esculentum* Mill.: Application of the leaf plastochron index and responses to exogenous gibberellic acid. *J. Exp. Bot.* 27:1339–1350.

_____ _____ 1977a. Analysis of root formation in leaf discs of *Lycopersicon esculentum* Mill. cultured *in vitro. Ann. Bot.* 41:307–320.

_____ _____ 1977b. Promotion of root initiation by gibberellic acid in leaf discs of tomato (*Lycopersicon esculentum*) cultured *in vitro. New Phytol.* 78:47–54.

Curtis, P. J. and B. E. Cross. 1954. Gibberellic acid. A new metabolite from the culture filtrates of *Gibberella fujikuroi. Chem. Ind.* 1066.

Eriksen, E. N. 1971. Promotion of root initiation by gibberellin. *Royal Vet. Agric. Univ. Denmark, Yearbook* 1971:50–59.

_____ 1972. Root initiation in pea cuttings: Interactions between tryptophane and gibberellin. *Royal Vet. Agric. Univ. Denmark, Yearbook* 1972:115–126.

_____ 1973. Root formation in pea cuttings. I. Effects of decapitation and disbudding at different developmental stages. *Physiol. Plant.* 28:503–506.

Ernstsen, A. and J. Hansen. 1986. Influence of gibberellic acid and stock plant irradiance on carbohydrate content and rooting in cuttings of Scots pine seedlings (*Pinus sylvestris* L.). *Tree Physiol.* 1:115–125.

Evans, A. and H. Smith. 1976. Localization of phytochrome in etioplasts and its regulation *in vitro* of gibberellin levels. *Proc. Nat. Acad. Sci.* USA 73:138–142.

Fabijan, D., J. S. Taylor and D. M. Reid. 1981. Adventitious rooting in hypocotyls of sunflower (*Helianthus annuus*) seedlings. II. Action of gibberellins, cytokinins, auxins and ethylene. *Physiol. Plant.* 53:589–597.

Felippe, G. M. 1980. Bud breaking and adventitious root formation in *Panicum maximum* Jacq. *Biol. Plant.* 22:392–395.

172   Jürgen Hansen

Fellenberg, G. 1969. Der Einfluss von Gibberellinsäure und Kinetin auf die auxininduzierte Wurzelneubildung und auf das Nucleoproteid von Erbsenepikotylen. Z. Pflanzenphysiol. 60:457–466.

Fernqvist, I. 1966. Studies on factors in adventitious root formation. Ann. Agric. Coll. Sweden 32:109–244.

Gamburg, K. Z. 1962. Influence of gibberellin and auxin on the growth of the first leaf and on the root formation in maize segments. Dokl. Akad. Nauk SSSR 145:941–944.

Gautheret, R. J. 1969. Investigations on the root formation in the tissues of Helianthus tuberosus cultured in vitro. Amer. J. Bot. 56:702–717.

Gundersen, K. 1958. Some experiments with gibberellic acid. Acta Horti Gotoburgensis 22:87–110.

Haddon, L. and D. H. Northcote. 1976. The influence of gibberellic acid and abscisic acid on cell and tissue differentiation of bean callus. J. Cell Sci. 20:47–55.

Hansen, J. 1975. Light dependent promotion and inhibition of adventitious root formation by gibberellic acid. Planta 123:203–205.

_____ 1976. Adventitious root formation induced by gibberellic acid and regulated by the irradiance to the stock plants. Physiol. Plant. 36:77–81.

Haissig, B. E. 1972. Meristematic activity during adventitious root primordium development. Influences of endogenous auxin and applied gibberellic acid. Plant Physiol. 49:886–892.

Heide, O. M. 1969. Non-reversibility of gibberellin-induced inhibition of regeneration in Begonia leaves. Physiol. Plant. 22:671–679.

Iwasaki, T., H. Fukuda and H. Shibaoka. 1986. Inhibition of cell division and DNA synthesis by gibberellin in isolated Zinnia mesophyll cells. Plant Cell Physiol. 27:717–724.

Jansen, H. 1967. Die Wirkung von Gibberellinsäure und Indolylessigsäure auf die Wurzelbildung von Tomatenstecklingen. Planta 74:371–378.

_____ 1971. Weitere Untersuchungen über die Gibberellin- und Auxinwirkung auf die Adventivwurzelbildung bei Tomaten. Gartenbauwissenschaft 36:149–153.

Kato, J. 1958. Studies on the physiological effect of gibberellin. II. On the interaction of gibberellin with auxins and growth inhibitors. Physiol. Plant. 11:10–15.

Kato, Y. and M. Hongo. 1974. Regulation of bud and root formation in excised leaf segments of Heloniopsis orientalis (Liliaceae). Phytomorphology 24:273–279.

Kefford, N. P. 1973. Effect of a hormone antagonist on the rooting of shoot cuttings. Plant Physiol. 51:214–216.

Kochba, J., J. Button, P. Spiegel-Roy, C. H. Bornman and M. Kochba. 1974. Stimulation of rooting of Citrus embryoids by gibberellic acid and adenine sulphate. Ann. Bot. 38:795–802.

Krelle, E. and E. Libbert. 1969. Interactions between abscisic acid and gibberellic acid regarding root formation. Flora, Abt. A 160:299–300.

Kriesel, K. 1975. Investigations on the role of gibberellic acid (GA$_3$) and chlorocholine chloride (CCC) in the process of rooting of willow (Salix viminalis L.) cuttings. Bull. Acad. Polon. Sci. Sér. Sci. Biol. 23:203–208.

Krishnamoorthy, H. N. 1972a. Effect of ethrel, auxins and maleic hydrazide on the rooting of mung bean hypocotyl cuttings. Z. Pflanzenphysiol. 66:273–276.

_____ 1972b. Effect of growth retardants and abscisic acid on the rooting of hypocotyl cuttings of musk melon (Cucumis melo cv. Kutana). Biochem. Physiol. Pflanzen 163:513–517.

Kumar, P. and R. A. Sharma. 1978. Effects of some growth regulators on rooting and regeneration of stem cuttings of Thevetia peruviana. Acta Bot. Indica 6:167–170.

Lance, B., R. C. Durley, D. M. Reid, T. A. Thorpe and R. P. Pharis. 1976. Metabolism of [³H] gibberellin A$_{20}$ in light- and dark-grown tobacco callus cultures. Plant Physiol. 58:387–392.

Leroux, R. 1967. Interaction de l'auxine, de la gibbérelline et de la lumière sur la rhizogénèse de fragments de tiges de Pois (Pisum sativum L.) cultivés in vitro. C. R. Séances Soc. Biol. 161:2402–2408.

Libbert, E. and E. Krelle. 1966. Die Wirkung des "Gibberellinantagonisten" 2-Chloräthyltrimethylammoniumchlorid (CCC) auf die Stecklingsbewurzelung windender und nicht windender Pflanzen. Planta 70:95–98.

Mitsuhashi-Kato, M., H. Shibaoka and M. Shimokoriyama. 1978. The nature of the dual effect of auxin on root formation in *Azukia* cuttings. *Plant Cell Physiol.* 19:1535–1542.

Morris, D. A. and E. D. Arthur. 1985. Invertase activity, carbohydrate metabolism and cell expansion in the stem of *Phaseolus vulgaris* L. *J. Exp. Bot.* 36:623–633.

Murashige, T. 1964. Analysis of the inhibition of organ formation in tobacco tissue culture by gibberellin. *Physiol. Plant.* 17:636–643.

Nanda, K. K., A. N. Purohit and A. Bala. 1967. Effect of photoperiod, auxins and gibberellic acid on rooting of stem cuttings of *Bryophyllum tubiflorum. Physiol. Plant.* 20:1096–1102.

_____ _____ K. Mehrotra. 1968. Effect of sucrose, auxins and gibberellic acid on rooting of stem segments of *Populus nigra* under varying light conditions. *Plant Cell Physiol.* 9:735–743.

_____ V. K. Anand and R. N. Chibbar. 1972. The promotive effect of gibberellic acid on the production of adventitious roots on stem cuttings of *Ipomoea fistulosa. Planta* 105:360–363.

Pfaff, W. and P. Schopfer. 1974. Phytochrom-induzierte Regeneration von Adventivwurzeln beim Senfkeimling (*Sinapis alba* L.). *Planta* 117:269–278.

_____ _____ 1980. Hormones are no causal links in phytochrome-mediated adventitious root formation in mustard seedlings (*Sinapis alba* L.). *Planta* 150:321–329.

Pierik, R. L. M. and H. H. M. Steegmans. 1975. Analysis of adventitious root formation in isolated stem explants of *Rhododendron. Scientia Hortic.* 3:1–20.

Reinert, J. and J. Besemer. 1967. Gibberellinsäure, ein Inhibitor morphogenetischer Prozesse. *Wiss. Zeitschr. Univ. Rostock, Mat. Nat.* 16:599–604.

Saxena, H. K. and L. Singh. 1982. Effect of IAA and GA on growth and rooting of apical fragments of *Brassica juncea* L. (Czern et Coss). *Indian J. Plant Physiol.* 25:330–335.

Schraudolf, H. and J. Reinert. 1959. Interaction of plant growth regulators in regeneration processes. *Nature* 184:465–466.

Smith, D. R. and T. A. Thorpe. 1975a. Root initiation in cuttings of *Pinus radiata* seedlings. I. Developmental sequence. *J. Exp. Bot.* 26:184–192.

_____ _____ 1975b. Root initiation in cuttings of *Pinus radiata* seedlings. II. Growth regulator interactions. *J. Exp. Bot.* 26:193–202.

Sponsel, V. M. 1985. Gibberellins in *Pisum sativum*—their nature, distribution and involvement in growth and development of the plant. *Physiol. Plant.* 65:533–538.

Takeno, K., J. S. Taylor, S. Sriskandarajah, R. P. Pharis and M. G. Mullins. 1982/83. Endogenous gibberellin- and cytokinin-like substances in cultured shoot tissues of apple, *Malus pumila* cv. Jonathan, in relation to adventitious root formation. *Plant Growth Regul.* 1:261–268.

Trewavas, A. J. 1982. Growth substance sensitivity: The limiting factor in plant development. *Physiol. Plant.* 55:60–72.

Valdovinos, J. G., L. C. Ernest and J. E. Perley. 1967. Gibberellin effect on tryptophan metabolism, auxin destruction, and abscission in *Coleus. Physiol. Plant.* 20:600–607.

Varga, M. and M. Bitó 1968. On the mechanism of gibberellin-auxin interaction. I. Effect of gibberellin on the quantity of free IAA and IAA conjugates in bean hypocotyl tissues. *Acta Biol. Acad. Sci. Hung.* 19:445–453.

_____ E. C. Humphries. 1974. Root formation on petioles of detached primary leaves of dwarf bean (*Phaseolus vulgaris*) pretreated with gibberellic acid, triiodobenzoic acid, and cytokinins. *Ann. Bot.* 38:803–807.

Veierskov, B., A. S. Andersen and E. N. Eriksen. 1982. Dynamics of extractable carbohydrates in *Pisum sativum*. I. Carbohydrate and nitrogen content in pea plants and cuttings grown at two different irradiances. *Physiol. Plant.* 55:167–173.

CHAPTER 13

Effect of Shoot Growth Retardants and Inhibitors on Adventitious Rooting

Tim D. Davis

Department of Agronomy and Horticulture
Brigham Young University
Provo, UT 84602

and

Narendra Sankhla

Department of Botany
University of Jodhpur
Jodhpur, India 342001

INTRODUCTION...174
ABSCISIC ACID..175
SYNTHETIC ANTIGIBBERELLINS...176
    Chlormequat Chloride...176
    Paclobutrazol..177
    Morphactins..179
    Daminozide 180
    Other Antigibberellins...180
TRIIODOBENZOIC ACID..181
CONCLUSION 181
ACKNOWLEDGMENTS 182
REFERENCES..182

Additional key words: abscisic acid, ancymidol, antigibberellins, chlormequat chloride, chlorphonium chloride, daminozide, flurprimidol, gonadotropin, hormones, morphactins, paclobutrazol, plant growth regulators, triiodobenzoic acid, XE-1019.
Abbreviations: ABA, abscisic acid; CCC, chlormequat chloride (cycocel); CF, chlorflurenol; GA_3, gibberellic acid; PP 333, paclobutrazol; SADH, daminozide; TIBA, triiodobenzoic acid.

INTRODUCTION

Plant scientists have been interested in the chemical control of adventitious rooting for many years. The first major discovery regarding the chemical control of rooting was that auxins could dramatically promote rooting. Although this finding was of considerable theoretical and practical importance, it soon became apparent that auxins did not promote rooting on all types of cuttings and under all circumstances. Hence, it appeared that other factors, in addition to auxin, were important for rooting. This finding, coupled with an increased understanding of plant growth substances in the past 20–25 yr seems to have led to the screening of many types of compounds for their ability to promote rooting.

A number of shoot growth retardants and inhibitors have been tested for their ability to influence rooting of cuttings. Although the rationale behind using these compounds has not always been clear, the justification for testing them has usually been that they either: 1) antagonize the activity or synthesis of gibberellins which normally inhibit rooting (Brian and Hemming 1960, Bachelard and Stowe 1963, Krelle and Libbert 1969, Krishnamoorthy 1972, Kefford 1973, Hartmann and Kester 1983, see also Chapter 12 by Hansen); or 2) reduce shoot growth which may compete with the base of the cutting for assimilates to the detriment of rooting (Eliasson 1971). In some cases the root promoting properties were discovered fortuitously during basic studies on responses of intact plants to growth retardants (Kefford 1973, Sankhla et al. 1985).

This chapter summarizes the influence of several types of shoot growth inhibitors on the rooting of cuttings. Unfortunately, few studies have determined if the supposed primary effect of these compounds (reduced gibberellin synthesis or shoot growth) is actually the important effect with regard to rooting. Hence, relatively little is known about the fundamental modes by which these compounds influence rooting.

ABSCISIC ACID

Most published studies on the effects of ABA on adventitious rooting have utilized exogenous applications rather than measuring endogenous ABA content in relation to rooting. Treatment of plant tissue with ABA may induce a number of responses pertaining to rooting of cuttings. In most experimental systems ABA opposes the action of gibberellins which, as mentioned previously, inhibit adventitious rooting. ABA may also influence stress tolerance of plants which may be significant in cuttings, because cuttings undergo considerable stress upon excision from the stock plant. Stomatal closure is also induced by ABA which might influence photosynthesis and transpiration by the cutting. Although unconfirmed, Chin et al. (1969) suggested that ABA might be the "rooting cofactor" (see Chapter 1 by Hackett) described as an oxygenated terpenoid by Hess (1964). In this regard, ABA isolated from mango (*Mangifera indica* L.) shoots was reported to be a mung bean [*Vigna radiata* (L.) R. Wilcz.] rooting cofactor (Basu et al. 1968). Because of these properties, a number of investigators have tested the influence of applied ABA on the rooting of cuttings.

In a number of studies, applied ABA (2.5–30 mg·l⁻¹) has been found to promote rooting (Chin et al. 1969, Basu et al. 1970, Rajagopal et al. 1971, Blazich et al. 1977, Hartung et al. 1980, Rasmussen and Andersen 1980). In other studies, however, applied ABA has been reported to inhibit (Heide 1968, Krishnamoorthy 1972, Venverloo 1976) or have no effect (Krelle and Libbert 1969, Read and Hoysler 1971, Biran and Halevy 1973) on rooting. Rasmussen and Andersen (1980) suggested that, depending upon the experimental protocol, almost any rooting response to ABA can be obtained. They reported that factors such as ABA concentration, rooting period length, and stock plant growth conditions all influenced the response of pea (*Pisum sativum* L.) cuttings to ABA. It was also observed that ABA treatment prolonged the rooting period such that if rooting of cuttings were evaluated too early, one might have concluded that ABA treatment reduced rooting when in reality only root emergence was delayed. Hartung et al. (1980) reported that the rooting response to ABA treatment was species dependent; rooting increased in most species but decreased in others as a result of ABA treatment. The above observations may, at least in part, explain the apparent discrepancies in the literature regarding the influences of exogenous ABA on rooting.

Exogenous ABA has generally been found to partially overcome GA_3 induced inhibition of adventitious rooting (Chin et al. 1969, Coleman and Greyson 1976, Hartung et al. 1980). These observations are consistent with the opposing nature of ABA and GA_3 in other developmental phenomena such as bud and seed dormancy. In contrast, ABA had no effect on kinetin induced inhibition of rooting of mung bean cuttings (Chin et al. 1969). When applied in conjunction with auxins, ABA treatment promotes rooting. For example, a combination of ABA and auxin has generally had additive effects on rooting (Basu et al. 1970) whereas synergistic interactions have been observed with IBA in some cases (Basu et al. 1970, Biran and Halevy 1973).

In addition to opposing gibberellin action and possibly modifying stress response, applied ABA may also affect rooting in other ways, e.g. Rasmussen and Andersen (1980) found that ABA reduced apical shoot growth on pea cuttings, which might improve rooting by reducing competition for assimilates normally transported to and metabolized at the cutting base. In this respect, it has been observed that ABA treatment caused an accumulation of assimilates, especially sucrose, in the base of runner bean (*Phaseolus vulgaris* L.) cuttings (Hartung et al. 1980). This accumulation of assimilates was accompanied by increased rooting but it is difficult to ascertain whether this accumulation was a cause or an effect of the increased rooting.

The promotive effects of applied ABA on rooting have been too small and inconsistent to be of any commercial value. Recently, however, it has been suggested that ABA may have practical value in reducing injury and deterioration during the shipment and storage of leafy *Pelargonium* spp. cuttings (Arteca et al. 1985). More work is needed to substantiate this possibility.

Only a few investigators have measured endogenous concentrations of ABA in relation to rooting. This is somewhat surprising because ABA content of plant tissue is relatively easily determined, especially compared to other plant growth substances. Rajagopal and Anderson (1980) found that a stress induced and light mediated increase in rooting of pea cuttings due to stock plant treatments was positively correlated with ABA content of the leaves. The physiological significance of this correlation is not clear. Wu and Barnes (1981) reported that *Rhododendron* spp. cuttings rooted best at times of the year when endogenous ABA concentrations in the stems were highest and that contents of no other endogenous hormones were correlated with rooting capacity, which suggested that ABA may play an important role in regulating rooting of rhododendron. Similar findings and conclusions were made more recently regarding ABA and the rooting of hybrid poplar and aspen cuttings (*Populus* spp.) (Blake and Atkinson 1986). However, a mere correlation between ABA content and rooting does not necessarily establish a cause-effect relationship, especially in the case of woody species where seasonal effects could have simultaneously influenced a host of physiological parameters. It would be interesting to know if exogenous ABA stimulates rooting of *Rhododendron* spp., hybrid poplar, and aspen at times of the year when endogenous ABA and rooting capacity are low.

SYNTHETIC ANTIGIBBERELLINS

A number of synthetic antigibberellins have been tested for their ability to influence adventitious rooting (Fig. 1). These compounds inhibit the biosynthesis of gibberellins and hence may lower the endogenous gibberellin content of treated tissue. It has been suggested that the reduction of endogenous gibberellins in cuttings should promote rooting (Hartmann and Kester 1983). Because of their effect on gibberellin content, antigibberellins reduce shoot elongation and many have been developed commercially for use as growth retardants. Antigibberellins have thus far not been found to be of commercial value in inducing rooting, but have been used to reduce endogenous gibberellin content and shoot growth of cuttings in basic rooting studies. It should be noted that, in addition to reducing gibberellin synthesis, some antigibberellins may influence other factors such as auxin content (Kuraishi and Muir 1963) or sterol metabolism (Henry 1985). Hence the mode of action of antigibberellins in rooting may not be solely related to reduced gibberellin content of the cuttings.

Chlormequat Chloride

CCC is probably the most widely used plant growth regulator in agriculture. It is used extensively in Europe to control lodging of wheat and throughout the world to control shoot elongation on many ornamental crops. The effect of CCC treatment on rooting has been somewhat variable. Most investigators have reported that CCC treatment promoted rooting (Libbert and Krelle 1966, Krelle and Libbert 1969, Roy et al. 1972, Krishnamoorthy 1972) but others have reported no effect (Beck and Sink 1974) or inhibition (Read and Hoysler 1969, 1971). These discrepancies might be explained by differences in the methods employed to supply CCC to the cuttings. The investigators who have reported that CCC promoted rooting placed cuttings in solutions of 15 mg·l⁻¹ or less for 24 h whereas those who reported that CCC inhibited or had no effect on rooting, dipped cuttings in solutions of 100–2500 mg·l⁻¹ for 15 s. It is possible that the concentrated dips supplied excessive CCC to the base of the cuttings.

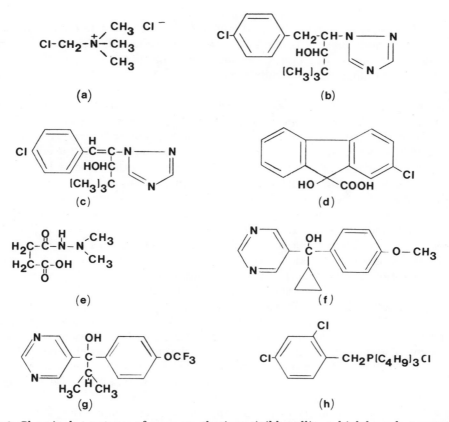

Figure 1. Chemical structures of some synthetic antigibberellins which have been reported to influence adventitious rooting: (a) chlormequat chloride; 2-chloroethyltrimethyl ammonium chloride; M.W.=158.1. (b) paclobutrazol; (2RS,3RS)-1-(4-chlorophenyl)-4-4-dimethyl-2-1,4-triazol-yl-pentan-3-ol; M.W.=293.8. (c) XE-1019; (E)-1-(4-chlorophenyl)-4-4-dimethyl-2-1,4-triazol-yl-penten-3-ol; M.W.=291.8. (d) chlorflurenol; 2-chloro-9-hydro-xyfluorene-9-carboxylic acid; M.W.=262.7. (e) daminozide; butanedioic acidmono(2,2-dimethylhydrazide) M.W.=160. (f) ancymidol; alpha-cyclopropyl-alpha-(4-methoxyphenyl)-5-pyrimidine-methanol; M.W.=256.3. (g) flurprimidol; alpha-(1-methylethyl)-alpha-[4-(trifluoromethoxy)phenyl]-5-pyrimidine-methanol; M.W.=312.3. (h) chlorphonium chloride; 2,4-dichlorobenzyltributylphosphonium chloride; M.W.=397.8.

CCC does not reverse GA_3 induced inhibition of rooting (Krelle and Libbert 1969, Krishnamoorthy 1972). When CCC and auxins have been applied together to cuttings, they have generally had additive (Krishnamoorthy 1972) or synergistic (Roy et al. 1972) effects on rooting. Read and Hoysler (1969), however, reported that 15 s dips in relatively high concentrations of CCC reversed IBA induced promotion of rooting of several species. Kinetin and TIBA have been reported to reduce the effectiveness of CCC in promoting rooting of *Cucumis melo* L. (Krishnamoorthy 1972).

Paclobutrazol

The triazole compound, PP 333, exhibits strong growth controlling properties in a range of plant species and induces a number of physiological responses including reduced gibberellin and sterol biosynthesis, increased chlorophyll concentration, altered carbohydrate status, increased stress tolerance, and delayed senescence (Davis et al. 1986b). We initially observed that soybean (*Glycine max* (L.) Merr.) seedlings treated with PP 333 formed numerous adventitious roots at the soil level (Sankhla et al. 1985), which led to tests of PP 333 effects on the rooting of cuttings.

The effect of PP 333 on rooting appears to be species dependent. Placement of cuttings in solutions containing PP 333 for 24 h increased the number of roots formed on *Coleus blumei* Benth.,

Plectranthus australis R. Br., *Prunus laurocerasus* L., *Salix discolor* Muhl., *Vitis labrusca* L., (Davis et al. 1986a) and *Phaseolus vulgaris* (Davis et al. 1985) but had no effect on *Ficus benjamina* L., *Ficus pumila* L., *Hibiscus rosa-sinensis* L., and *Zebrina pendula* Schnizl. (Davis et al. 1986a) or *Pinus banksiana* Lamb. cuttings (B. E. Haissig, personal communication). The mean length of the longest root formed on the cuttings was generally reduced by PP 333, but the degree of reduction varied among species.

PP 333 treatment almost completely stopped shoot elongation on cuttings which otherwise continued to grow during the rooting period (Davis et al. 1985, 1986a). This growth inhibition did not appear to be related to the rooting response. For example, rooting of *Prunus laurocerasus* L. was promoted by PP 333 treatment even though shoot growth did not occur on the control or treated cuttings. Likewise rooting of some leafless hardwood cuttings has been promoted by PP 333 treatment even though actively growing shoots were not present on the cuttings. Furthermore, rooting of *Zebrina pendula* was unaffected by PP 333 treatment although shoot growth during rooting was strongly inhibited by PP 333. These observations do not support a hypothesis that PP 333 promotes rooting by reducing shoot growth and thereby increasing the partitioning of assimilates to the base of the cuttings. Post-propagation shoot growth of PP 333 treated cuttings generally returns to normal within 10 weeks or less (Davis et al. 1986a).

Enhancement of rooting on bean hypocotyl cuttings by PP 333 treatment is accompanied by changes in the activities of a number of enzymes which have been postulated to be involved in rooting (Upadhyaya et al. 1986). Basal portions of treated cuttings exhibited increased activities of catalase (EC 1.11.1.6), peroxidase (EC 1.11.1.7), polyphenol oxidase (EC 1.10.3.1), malate dehydrogenase (EC 1.1.1.37), protease (EC 3.4.24.4), and RNase (EC 3.1.4.22) compared to untreated controls. In contrast, the activity of amylase (EC 3.2.1.1) in the base of the cuttings was reduced by PP 333 treatment. The PP 333 induced changes in the activities of the above enzymes were generally greatest during the latter portion of the rooting period (3–5 d after excision), suggesting that these enzymes might have been involved in the development rather than the initiation of roots.

GA$_3$ and PP 333 treatments exhibited opposing effects on rooting of *Phaseolus vulgaris* hypocotyl cuttings (Table 1). The promotion of rooting by PP 333 was reversed by incubating cuttings in GA$_3$ for 24 h subsequent to PP 333 treatment. When applied to *P. vulgaris* cuttings at 12 mg·l^{-1}, the effect of GA$_3$ was somewhat stronger than that of PP 333 as evidenced by the fact that cuttings treated with both compounds formed fewer roots than untreated controls. These results suggested that modification of the gibberellin status of *P. vulgaris* hypocotyl cuttings may have a profound effect on rooting. PP 333 and GA$_3$ treatments both reduced the length of the longest root present on the cuttings, although GA$_3$ treatment reduced this parameter to a much greater extent than did PP 333

Table 1. Effect of paclobutrazol (PP 333) and gibberellic acid (GA$_3$) treatment on rooting and shoot elongation of bean hypocotyl cuttings. After 24 h treatment, and sometimes retreated for 24 h, cuttings were rooted in peat-perlite in a growth chamber for 6 d. Data are means ± S.E. (n ≥ 25).

| Parameter | Treatment | | | | |
	Control (water only)	PP 333	GA$_3$[1]	PP 333 Then GA$_3$	GA$_3$ Then PP 333
Roots per cutting	29.6 ± 2.3	41.5 ± 2.3	1.8 ± 0.3	8.9 ± 1.0	5.5 ± 0.7
Length of longest root (cm)	7.4 ± 0.3	6.5 ± 0.2	2.6 ± 0.5	4.6 ± 0.4	4.2 ± 0.4
Shoot length (cm)	13.1 ± 0.3	10.2 ± 0.3	17.1 ± 0.7	13.2 ± 0.5	15.3 ± 0.6

[1]PP 333 and GA$_3$ concentrations were 12 mg·l^{-1}.

treatment. PP 333 and GA_3 treatment had opposing effects on shoot elongation on the cuttings, similar to that observed with intact plants of several species (e.g. Wample and Culver 1983, Steffens et al. 1985, Braun and Garth 1986).

We have recently observed that XE-1019, an experimental triazole growth retardant structurally related to PP 333, is also active in promoting rooting of bean hypocotyl cuttings (Table 2). Like PP 333, XE-1019 inhibited shoot growth during rooting. At concentrations up to 5 mg·l^{-1}, XE-1019 had little effect on root length. When compared against PP 333, XE-1019 appeared to be slightly more active in promoting rooting of bean hypocotyls (Davis 1986).

Table 2. Effect of treatment with the experimental anti-gibberellin, XE-1019, on rooting and shoot elongation of bean hypocotyl cuttings. Cuttings were placed in aqueous XE-1019 solutions for 24 h prior to insertion into peat-perlite in a growth chamber for 7 d. Data are means ± S.E. (n=35).

Parameter	XE-1019 Concentration (mg·l^{-1})		
	0	0.05	5.0
Number of roots per cutting	24.5 ± 2.0	29.9 ± 2.0	41.8 ± 3.3
Length of longest root (cm)	3.9 ± 0.3	3.9 ± 0.3	4.1 ± 0.4
Shoot length (cm)	13.4 ± 0.2	11.6 ± 0.3	11.4 ± 0.2

Morphactins

Morphactins are synthetic derivatives of fluorene-9-carboxylic acid. Treatment of plants with these compounds induces a number of physiological and morphogenic responses including inhibition of mitosis in apical meristems, reduction of apical dominance, promotion of branching, abrogation of polarity, inhibition of lateral root growth, interference with gravitropism and phototropism, and disturbance of correlative phenomena (Schneider 1970). One morphactin, CF, is marketed commercially as a growth retardant for woody plants.

When applied to cuttings, CF enhanced cell division in the bases and increased the number of root primordia (Schneider 1970 and references therein). The primordia, however, were disorganized and abnormal, and subsequent root growth was retarded. Hence the number of visible roots per cutting has often been reduced by CF application (Kochhar et al. 1972, Nanda et al. 1973, Jarvis and Shaheed 1986). Morphactin treatment of cuttings has not overcome GA_3 induced inhibition of rooting and has antagonized auxin induced rooting (Schneider 1970 and references therein, Roy et al. 1972).

When applied to the apex of pea stock plants at relatively low concentrations (13 mg·l^{-1} or less), CF promoted subsequent rooting (Khan et al. 1977). At higher concentrations, CF application to stock plants reduced subsequent rooting (Punjabi et al. 1974, Khan et al. 1977). Khan et al. (1977) suggested that the effect of morphactins on rooting might be related to auxin and carbohydrate availability at the rooting zone. More recently, Jarvis and Shaheed (1986) reported that inhibition of root formation in mung bean by CF treatment was accompanied by decreased movement of ^{14}C-IAA from the leaves to the hypocotyl which supported the hypothesis of Khan et al. (1977).

The rooting response of *Impatiens* spp. cuttings to morphactins has depended upon seasonal influences on stock plants (Sankhla et al. 1975 and refs. therein). During April, morphactins promoted rooting, but in all other months rooting was inhibited. The inhibition caused by morphactin was overcome by IBA treatment, although with differing effectiveness throughout the year (Nanda et al. 1972). Nanda et al. (1973) reported that morphactin induced inhibition of rooting of *Impatiens* spp. was accompanied by an alteration in the spectrum of peroxidase isozymes. Several isozymes which were present in control cuttings were absent in the morphactin treated cuttings which failed to root. It is not known whether these morphactin induced changes in peroxidase isozymes were responsible for reduced rooting or merely were a coincidental response.

Daminozide

SADH is a hydrazine derivative which is used commercially to regulate fruiting and post-harvest quality of tree fruits, and as a growth retardant on a number of herbaceous ornamental species. SADH has been reported to be a strong promoter of adventitious rooting in several species. Read and Hoysler (1969, 1971) found that a 15 s dip in solutions containing 1000–5000 mg·l^{-1} SADH increased root number, weight, and length on cuttings from several herbaceous species. In the latter test, treated cuttings rooted faster than controls. Krishnamoorthy (1972) and Roy et al. (1972) likewise reported that incubating herbaceous cuttings in dilute solutions (ca. 1–15 mg·l^{-1}) of SADH for 24 h promoted rooting. In contrast to these findings, Beck and Sink (1974) reported that SADH had no effect on rooting of *Euphorbia* spp. cuttings when applied as a 15 s dip at a concentration of 2500 mg·l^{-1}. The reason for this discrepancy is not clear, but Beck and Sink (1974) noted that their studies were conducted primarily during times of the year when conditions were conducive to good rooting. It may be that SADH induced effects on rooting are dependent upon seasonal influences.

Other Antigibberellins

Treatment of *Begonia semperflorens* Link & Otto and *Vitis vinifera* L. cuttings with human chorionic gonadotropin hormone (a complex glycoprotein composed of about 30% carbohydrates) increased the number of roots per cutting by 50–100% compared to untreated controls (Leshem and Lunenfeld 1968). Endogenous gibberellin content in the basal sections of treated cuttings was considerably less than that of controls; endogenous auxin concentrations were unaffected by the treatments.

The gibberellin antagonist, ancymidol, has been reported to increase rooting by cuttings of a number of species (Nell and Sanderson 1972, Kefford 1973). When mung bean cuttings were rooted in ancymidol solutions, root formation was increased dramatically and at optimal concentrations (ca. 0.8 mg·l^{-1}) the physical limits for the accommodation of root primordia in the hypocotyl appeared to have been reached (Kefford 1973). In the same study the promotion of rooting by ancymidol treatment was reversed by application of gibberellin. In contrast to the above findings, Beck and Sink (1974) reported that dipping *Euphorbia* spp. cuttings in ancymidol (5–20 mg·l^{-1}) for 15 s did not effect root number or dry weight of roots. It is possible that such an application did not allow for sufficient uptake of the compound by the cuttings at the concentrations used.

We recently tested the effect of the experimental growth retardant, flurprimidol, on the rooting of bean hypocotyl cuttings (Table 3). At 12 mg·l^{-1}, this compound increased the number of roots formed by about 80% compared to controls. Roots on treated cuttings were shorter and thicker than those on controls. Flurprimidol almost completely inhibited shoot elongation during the rooting period.

In contrast to the general promotive nature of most antigibberellins on rooting, chlorphonium chloride at 0.1–10 mg·l^{-1} has been reported to reduce both root number and length on muskmelon cuttings when applied during a 24 h incubation immediately following excision (Krishnamoorthy 1972). The chlorphonium chloride induced inhibition of rooting was evident when cuttings were rooted either in light or dark.

Table 3. Effect of flurprimidol treatment on rooting and shoot elongation of bean hypocotyl cuttings. Cuttings were placed in their respective solutions for 24 h prior to insertion into peat-perlite in a growth chamber for 7 d. Data are means; asterisk indicates that treatment mean is significantly different (P < 0.05) from control mean (n=38).

Parameter	Flurprimidol Concentration (mg·l^{-1})	
	0	12
Number of roots per cutting	32.6	58.6*
Length of longest root (cm)	5.4	4.2*
Shoot length (cm)	12.6	11.5*

TRIIODOBENZOIC ACID

The plant growth regulator, TIBA, is generally considered to be an auxin antagonist (Niedergang-Kamien and Leopold 1967, Steward and Krikorian 1971). TIBA may have practical value in promoting flower initiation and annual bearing in apple trees (Abdel-Rahman et al. 1981) and retards shoot elongation of conifers (Ross et al. 1983). Because of its anti-auxin activity, a number of investigators have used TIBA as a tool in basic studies of adventitious rooting. TIBA treatment has generally been reported to inhibit rooting presumably by reducing basipetal auxin transport within the cutting (Katsumi et al. 1969, Haissig 1972, Krishnamoorthy 1972, Varga and Humphries 1974, Jarvis and Shaheed 1986). As pointed out by Batten and Goodwin (1981), however, the rooting response of cuttings to TIBA may be influenced by application methods. For example, contrasting results were obtained depending upon whether cuttings were floated on or stood in TIBA treatment solutions. It should also be noted that it is possible that TIBA influences rooting by modes of action other than by inhibition of auxin transport because it also alters ion absorption, respiration, and membrane permeability in treated tissues (Jacobson and Jacobson 1981).

We studied the influence of TIBA on rooting and carbohydrate transport in leafy pea cuttings (Table 4). TIBA (1%, w/w in lanolin) was dissolved in 80% ethanol and then added to boiling lanolin. The mixture was stirred for about 10 min to evaporate the ethanol and then cooled prior to use. Application of a TIBA/lanolin ring to the main stem just below the apex of the cuttings, reduced rooting to about 40% that of the controls. Pre-treatment of the cuttings with 20 mg·l⁻¹ IBA for 24 h before TIBA application increased rooting to a level above that of the controls. This finding is consistent with the hypothesis that TIBA inhibited rooting by interfering with polar transport, possibly auxin transport, to the base of the cuttings. The TIBA/lanolin ring did not influence dry weight, sucrose, or glucose contents in the base of the cuttings which were measured 7 d after excision just prior to root emergence. These results suggested that TIBA did not interfere with carbohydrate transport to the base of the cuttings.

Haissig (1972), using ^{14}C-IAA, found that polar auxin transport to the base of *Salix* spp. cuttings was reduced by TIBA/lanolin rings applied to the stems. The treatment also decreased cambial cell division and the mean number of cells per primordium. Established primordia (those formed before treatment) were less affected than primordia which were initiating at the time of treatment.

Table 4. Influence of triiodobenzoic acid (TIBA) and combined TIBA/indolebutyric acid (IBA) treatments on rooting, dry weight, and sugar content in the base of leafy pea cuttings. TIBA (1%, w/w) was applied as a lanolin ring just below the apex. The number of roots was counted 10 d after excision. Dry weights and sugar contents were determined 7 d after excision, just prior to root emergence. Data are means. Means within a row having the same lower case letter did not differ significantly (P < 0.05; n >20 for rooting data; n=8 for dry weight and sugar data).

| | Treatment | | |
Parameter	Control	TIBA Ring	Pre-Treated With 20 mg·l⁻¹ (24 h) IBA Then TIBA Ring Application
Number of roots per cutting	8.9b	3.4c	12.4a
Dry weight basal 3 cm (mg)	16.3a	16.5a	—
Sucrose content basal 3 cm (% dry weight)	8.4a	9.0a	—
Glucose content basal 3 cm (% dry weight)	7.6a	5.4a	—

CONCLUSION

Thus far, the major contribution of research on the effects of shoot growth inhibitors on rooting has been academic. It is of interest to note that compounds which antagonized gibberellins and/or reduced shoot elongation during rooting generally have promoted rooting. There are some exceptions to this generality, however, and reducing shoot growth on cuttings has not necessarily always promoted rooting. Although there does not seem to be any general commercial value of these compounds as propagation aids at present, future research might reveal specialized uses such as treatment of cuttings to improve quality during shipping and storage.

Unfortunately, there has not been much focus to the work that has been done relative to growth retardant effects on rooting. Because a wide array of growth retarding compounds have been found to promote rooting, it is difficult to postulate any structure-activity relationships which might be useful in designing compounds for use as commercial rooting agents. In order to improve our understanding along these lines, future work should address the mode of action by which these compounds promote rooting. At present, it is not known whether the supposed primary effect of these compounds (i.e. reduced gibberellin content or shoot growth) is responsible for root promoting activity. It is possible that these compounds induce secondary effects which lead to increased rooting. The first step to sorting out primary versus secondary effects will be measurement of endogenous growth substance content in relation to exogenous applications. For example it is critical whether or not applied antigibberellins actually reduce the content of endogenous gibberellins in the rooting zone. It is as important to know how these compounds influence the content of other endogenous substances which influence rooting (e.g. auxins, carbohydrates). Without this information, it will remain unclear as to whether the wide array of growth retardants that influence rooting do so by a common mode of action.

ACKNOWLEDGMENTS

We thank Drs. Bruce E. Haissig and Jürgen Hansen for critically reviewing our manuscript.

REFERENCES

Abdel-Rahman, M., B. A. Schneider and J. R. Frank. 1981. *Plant Growth Regulator Handbook*. Plant Growth Regulator Working Group. Longmont, CO, USA. pp. 122–123.

Arteca, R. N., D. Tsai, and C. Schlagnhaufer. 1985. Abscisic acid effects on photosynthesis and transpiration in geranium cuttings. *HortScience* 20:370–372.

Bachelard, E. P. and B. B. Stowe. 1963. Rooting of cuttings of *Acer rubrum* L. and *Eucalyptus camuldulensis* Dehn. *Aust. J. Biol. Sci.* 16:751–767.

Basu, R. N., B. Ghosh and P. K. Sen. 1968. Naturally occurring rooting factors in mango (*Mangifera indica* L.). *Indian Agric.* 12:194–196.

_____ B. N. Roy and T. K. Bose. 1970. Interaction of abscisic acid and auxins in rooting of cuttings. *Plant Cell Physiol.* 11:681–684.

Batten, D. J. and P. B. Goodwin. 1981. Auxin transport inhibitors and the rooting of hypocotyl cuttings from etiolated mung-bean *Vigna radiata* (L.) Wilczek seedlings. *Ann. Bot.* 47:497–503.

Beck, G. R. and K. C. Sink, Jr. 1974. Rooting stimulation of poinsettia stem cuttings by growth regulators. *HortScience* 9:144–146.

Biran, I. and A. H. Halevy. 1973. Endogenous levels of growth regulators and their relationship to the rooting of dahlia cuttings. *Physiol. Plant.* 28:436–442.

Blake, T. J. and S. M. Atkinson. 1986. The physiological role of abscisic acid in the rooting of poplar and aspen stump sprouts. *Physiol. Plant.* 67:638–643.

Blazich, F. A. and C. W. Heuser. 1977. The effects of ABA, kinetin, NAA and selected putative inhibitors of RNA (actinomycin D and 6-methylpurine) or protein (cyclohexamide and puromycin) synthesis on adventitious root initiation of mung bean cuttings. *HortScience* 12:392 (abst.).

Braun, J. W. and J. K. L. Garth. 1986. Strawberry vegetative and fruit growth response to paclobutrazol. *J. Amer. Soc. Hort. Sci.* 111:364–367.

Brian, P. W., H. G. Hemming and D. Lowe. 1960. Inhibition of rooting of cuttings by gibberellic acid. *Ann. Bot.* 24:407–419.

Chin, T., M. M. Meyer, Jr. and L. Beevers. 1969. Abscisic-acid-stimulated rooting of stem cuttings. *Planta* 88:192–196.

Coleman, W. K. and R. I. Greyson. 1976. Root regeneration from leaf cuttings of *Lycopersicon esculentum* Mill.: Application of the leaf plastochron index and responses to exogenous gibberellic acid. *J. Exp. Bot.* 27:1339–1351.

Davis, T. D. 1986. Influence of triazole growth retardants on adventitious root formation in bean hypocotyl cuttings. *Proc. Plant Growth Regul. Soc. Amer.* 13:217–223.

_____ N. Sankhla, R. H. Walser and A. Upadhyaya. 1985. Promotion of adventitious root formation on cuttings by paclobutrazol. *HortScience* 20:883–884.

_____ _____ A. Upadhyaya. 1986a. Paclobutrazol: a promising plant growth regulator. In *Hormonal Regulation of Plant Growth and Development* (S. S. Purohit, ed). Agro Bot. Pub., Bikaner, India. pp. 311–331.

_____ R. H. Walser, K. Soerenson and N. Sankhla. 1986b. Rooting and subsequent growth of cuttings treated with paclobutrazol. *Plant Prop.* 32(1):7–9.

Eliasson, L. 1971. Adverse effect of shoot growth on root growth in rooted cuttings of aspen. *Physiol. Plant.* 25:268–272.

Haissig, B. E. 1972. Meristematic activity during adventitious root primordium development. Influences of endogenous auxin and applied gibberellic acid. *Plant Physiol.* 49:886–892.

Hartmann, H. T. and D. E. Kester. 1983. *Plant Propagation: Principles and Practices*. Prentice-Hall, Englewood Cliffs, NJ, USA. pp. 248. ISBN 0-13-681007-1.

Hartung, W., B. Ohl and V. Kummer. 1980. Abscisic acid and the rooting of runner bean cuttings. Z. *Pflanzenphysiol.* 98:95–103.

Heide, O. M. 1968. Stimulation of adventitious bud formation in begonia leaves by abscisic acid. *Nature* 219:960–961.

Henry, M. J. 1985. Plant growth and regulating activity of sterol and gibberellin biosynthesis inhibitors. *Bull. Plant Growth Regul. Soc. Amer.* 13:9–11.

Hess, C. W. 1964. Naturally occurring substances which stimulate root initiation. In Reg. Nat. de la Croiss. Végét. (J. P. Nitsch, ed). C.N.R.S, Paris, France. pp. 517–527.

Jacobson, A. and L. Jacobson. 1981. Inhibitory effects of 2,3,5-triiodobenzoic acid on ion absorption, respiration, and carbon metabolism in excised barley roots. *Plant Physiol.* 67:282–286.

Katsumi, M., Y. Chiba and M. Furuyama. 1969. The roles of the cotyledons and auxin in the adventitious root formation of hypocotyl cuttings of light-grown cucumber seedlings. *Physiol. Plant.* 22:993–1000.

Jarvis, B. C. and A. I. Shaheed. 1986. Adventitious root formation in relation to the uptake and distribution of supplied auxin. *New Phytol.* 103:23–31.

Kefford, N. P. 1973. Effect of a hormone antagonist on the rooting of shoot cuttings. *Plant Physiol.* 51:214–216.

Khan, A. R., A. S. Andersen and J. Hansen. 1977. Morphactin and adventitious root formation in pea cuttings. *Physiol. Plant.* 39:97–100.

Kochhar, V. V., V. K. Anand and K. K. Nanda. 1972. Effects of morphactin on rooting and sprouting of buds on stem cuttings of *Salix tetrasperma*. *Bot. Gaz.* 133:361–368.

Krelle, E. and E. Libbert. 1969. Interactions between abscisic acid and gibberellic acid regarding root formation. *Flora* 160:299–300.

Krishnamoorthy, H. N. 1972. Effect of growth retardants and abscisic acid on the rooting of hypocotyl cuttings of musk melon (*Cucumis melo* cv. Kutana). *Biochem. Physiol. Pflanzen* 163:513–517.

Kuraishi, S. and R. M. Muir. 1963. Mode of action of growth retarding chemicals. *Plant Physiol.* 38:19–24.

Leshem, Y. and B. Lunenfeld. 1968. Gonadotropin promotion of adventitious root formation on cuttings of *Begonia semperflorens* and *Vitis vinifera*. *Plant Physiol.* 43:313–317.

Libbert, E. and E. Krelle. 1966. The effect of the "gibberellin antagonist" (2-chloroethyl)trimethylammonium chloride (CCC) on the rooting of cuttings from twining and non-twining plants. *Planta* 70:95–98.

Nanda, K. K., N. C. Bhattacharya and N. P. Kaur. 1973. Effect of morphactin on peroxidases and its relationship to rooting hypocotyl cuttings of *Impatiens balsamina*. *Plant Cell Physiol.* 14:207–211.

_____ V. K. Kochhar and S. Gupta. 1972. Effects of auxins, sucrose, and morphactin in the rooting of hypocotyl cuttings of *Impatiens balsamina* during different seasons. In *Proc. All India Symp. Biol. Land Plants*. Meerut Univ., Meerut, India. pp. 181–187.

Nell, T. A. and K. C. Sanderson. 1972. Effect of several growth regulators on the rooting of three azalea cultivars. *Florist's Rev.* 150(3891):21–22, 52–53.

Niedergang-Kamien, E. and A. C. Leopold. 1967. Inhibitors of polar auxin transport. *Physiol. Plant.* 10:29–38.

Punjabi, B., N. C. Paria and R. N. Basu. 1974. Opposing effects of morphactin on the rooting of cuttings in the presence or absence of auxin. *J. Hortic. Sci.* 49:253–256.

Rajagopal, V. and A. S. Andersen. 1980. Water stress and root formation in pea cuttings. III. Changes in the endogenous level of abscisic acid and ethylene production in the stock plants under two levels of irradiance. *Physiol. Plant.* 48:155–160.

_____ M. R. K. Rao and I. M. Rao. 1971. Influence of indoleacetic acid and abscisic acid on the rooting of tomato shoot cuttings. *Indian J. Plant Physiol.* 14:91–96.

Rasmussen, S. and A. S. Andersen. 1980. Water stress and root formation in pea cuttings. II. Effect of abscisic acid treatment of cuttings from stock plants grown under two levels of irradiance. *Physiol. Plant.* 48:150–154.

Read, P. E. and V. C. Hoysler. 1969. Stimulation and retardation of adventitious root formation by application of B-nine and cycocel. *J. Amer. Soc. Hortic. Sci.* 94:314–316.

_____ 1971. Improving rooting of carnation and poinsettia cuttings with succinic acid-2,2-dimethylhydrazide. *HortScience* 6:350–351.

Ross, S. D., R. P. Pharis and W. D. Binder. 1983. Growth regulators and conifers: their physiology and potential uses in forestry. *In Plant Growth Regulating Chemicals* (L. G. Nickell, ed). CRC Press, Boca Raton, FL, USA. Vol. II pp. 35–78. ISBN 0-8493-5003-4.

Roy, B. N., R. N. Basu and T. K. Bose. 1972. Interaction of auxins with growth-retarding, -inhibiting, and ethylene-producing chemicals in rooting of cuttings. *Plant Cell Physiol.* 13:1123–1127.

Sankhla, N., S. P. Bohra, S. P. Vyas and D. Sankhla. 1975. Morphactins and plant morphogenesis. In *Form, Structure, and Function in Plants* (H. Y. Mohan, J. J. Ram, J. J. Shah and C. K. Shah, eds). Sarita Prakashan, Meerut, India. Prof. Johri Comm. Vol. pp. 255–264.

_____ T. D. Davis, A. Upadhyaya, D. Sankhla, R. H. Walser and B. N. Smith. 1985. Growth and metabolism of soybean as affected by paclobutrazol. *Plant Cell Physiol.* 26:913–921.

Schneider, G. 1970. Morphactins: physiology and performance. *Ann. Rev. Plant. Physiol.* 21:499–536.

Steffens, G. L., J. K. Byun, and S. Y. Wang. 1985. Controlling plant growth via the gibberellin biosynthesis system-I. Growth parameter alterations in apple seedlings. *Physiol. Plant.* 63:163–168.

Steward, F. C. and A. D. Krikorian. 1971. *Plants, Chemicals, and Growth*. Academic Press, NY, USA. pp. 95–96.

Upadhyaya, A., T. D. Davis and N. Sankhla. 1986. Some biochemical changes associated with paclobutrazol-induced adventitious root formation on bean hypocotyl cuttings. *Ann. Bot.* 57:309–315.

Varga, M. and E. C. Humphries. 1974. Root formation on petioles of detached primary leaves of dwarf bean (*Phaseolus vulgaris*) pretreated with gibberellic acid, triiodobenzoic acid, and cytokinins. *Ann. Bot.* 38:803–807.

Venverloo, C. J. 1976. The formation of adventitious organs. III. A comparison of root and shoot formation on *Nautilocalyx* explants. *Z. Pflanzenphysiol.* 80:310–322.

Wample, R. L. and E. B. Culver. 1983. The influence of paclobutrazol, a new growth regulator, on sunflowers. *J. Amer. Soc. Hortic. Sci.* 108:122–125.

Wu, F. T and M. F. Barnes. 1981. The hormonal levels in stem cuttings of difficult-to-root and easy-to-root rhododendrons. *Biochem. Physiol. Pflanzen* 176:13–22.

CHAPTER 14

Cytokinins and Adventitious Root Formation

J. Van Staden and A. R. Harty

UN/CSIR Research Unit for Plant Growth and Development
Department of Botany
University of Natal
Pietermaritzburg 3200
Republic of South Africa

INTRODUCTION 185
EFFECTS OF APPLIED CYTOKININS ON ROOTING OF
    CUTTINGS 186
FACTORS AFFECTING RESPONSES TO CYTOKININS. 188
    Foliar vs. Basal Application 188
    Stage of Development. 189
CHANGES IN ENDOGENOUS CYTOKININS IN RELATION TO
    ROOTING. 189
    Leaf Cuttings. 190
    Shoot Cuttings 191
    Root Cuttings 192
INTERACTIONS WITH OTHER PLANT GROWTH
    REGULATORS 193
    Auxins 193
    Other Growth Substances 194
CYTOKININS AND ROOT GROWTH 194
    Roots as Sites of Cytokinin Biosynthesis. 194
    Influences on Root Apical Dominance. 195
    Applied Cytokinins and Root Growth 196
CONCLUSION 196
ACKNOWLEDGMENT 197
REFERENCES 197

Additional key words: N^6-substituted adenine; root initiation; natural cytokinins; synthetic cytokinins; kinetin; zeatin; benzyladenine; cytokinin biosynthesis, transport, and metabolism; cytokinin-auxin interactions.
Abbreviations: ARF, adventitious root formation; LRF, lateral root formation; CK, cytokinin; 2,4-D, 2,4-dichlorophenoxyacetic acid.

INTRODUCTION

Cytokinins have been defined as a class of compounds which, in the presence of optimal auxin, induces cell division in tobacco (*Nicotiana tabacum* L.) pith or similar tissue cultures (Letham 1978). The effects of CKs are, however, not limited to cell division. These hormones have been shown to play a role in many other phases of plant development, including cell enlargement, cell differentiation, and the flow of assimilates and nutrients through the plant. Their diverse effects on plant metabolism impinge on the activities of enzymes, the biosynthesis of growth factors, and the appearance and disappearance of organelles. As a consequence, they influence a large number of physiological processes related to the growth and development of plants or plant parts (Skoog and Armstrong 1970). Although a wide range of CK-like substances participate in plant development, most attention has been paid to the naturally occurring and synthetic N^6-substituted adenines. Cytokinins are ubiquitous substances present in the tissues and cells of both higher and lower plants where they can occur both as free and bound forms, the latter being associated with tRNA (Skoog et al. 1966). The concentration and type of free CKs present within plants, and their component parts, vary considerably. This variation is influenced greatly by both the stage of development and the environmental conditions to which the plant is exposed (Letham 1978, Van Staden and Davey 1979). Of further significance is that different classes or groups of N^6-substituted adenines, e.g. the zeatin or iso-pentenyladenine groups, may be present in plant material. It has been suggested that iso-pentenyladenine acts as a precursor for zeatin, which then together with its derivatives (metabolites) appear to make up the bulk of free CKs detected in both vegetative and reproductive plant material. Zeatin appears to be the most active of the known CKs (Letham and Palni 1983). Many derivatives of zeatin have been isolated and identified in plant tissues and organs. As a result of metabolic studies, excellently reviewed by Letham and Palni (1983), it is clear that zeatin can be ribosylated at the 9-position of the purine ring to form a cytokinin riboside, which in turn can be phosphorylated to produce a nucleotide. The unsaturated side-chain of both the free base and the riboside can be saturated by reduction, producing dihydro derivatives which are stable against oxidation (Whitty and Hall 1974, Letham and Palni 1983). Both the saturated and unsaturated zeatin and ribosylzeatin derivatives can be glucosylated. Cytokinin glucosides are presently regarded as storage or inactivation products (Hewett and Wareing 1973a, Van Staden 1976a, Palmer et al. 1981, Letham et al. 1983).

With respect to root formation, CKs have typically been observed as inhibitors of this process. Since the classic organogenesis experiments of Skoog and Miller (1957), it has almost become dogma that high CK to auxin ratios in growth media and *in vivo* will favor shoot growth to the detriment of root development. Although the same general picture emerges when reviewing attempts to induce root formation on stem cuttings by applying CKs, it is nevertheless noteworthy that there have been exceptions to the general trend. In view of these and the fact that small amounts of CKs are often required for good root formation *in vitro*, even in the presence of auxin (Letham 1978), it becomes imperative that the role of CKs be re-evaluated. The case for these compounds as inhibitors of root formation is probably not as clear-cut nor as strong as is generally conveyed by the current literature.

EFFECTS OF APPLIED CYTOKININS ON ROOTING OF CUTTINGS

The fact that none of the commercially available formulations for root induction in cuttings contains CKs (Dirr 1981, Berry 1984) is perhaps the most relevant indicator of the lack of success in promoting ARF in cuttings through exogenous CK application. Inclusion of CKs in rooting trials is generally rare, but where attempted, the usual result is either an insignificant promotion of ARF over the control or clear inhibition of ARF. Instances where the latter occurred include trials with *Phaseolus vulgaris* L. (bean; Humphries 1960), *Acer rubrum* L. and *Eucalyptus camaldulensis* Denhardt (Bachelard and Stowe 1963), *Phaseolus aureus* Roxb. (mung bean; Chandra et al. 1973), *Zebrina pendula* Schnizl. (Delegher-Langohr 1974), *Pisum sativum* L. (pea; Eriksen 1974, Bollmark and Eliasson 1986), and *Pinus radiata* D. Don (Smith and Thorpe 1975).

Examples where CK application promoted ARF are rare. Notable exceptions are those of Meredith et al. (1970) and Nemeth (1979). The first study was conducted on the effects of indole-3-acetic acid and kinetin on the rooting of *Feijoa sellowiana* Berg. cuttings. An 8 h basal soak in kinetin $(4.6 \times 10^{-7} \text{ mol} \cdot l^{-1})$ promoted rooting in one clone, but had no effect on two others tested. Although kinetin and indole-3-acetic acid combined were more effective on one clone, their effects were apparently additive at the level used, rather than synergistic. In the study of Nemeth (1979), root formation was induced by benzyladenine at $10^{-6} \text{ mol} \cdot l^{-1}$ in *Prunus cerasifera* Ehrh., *P. avium* L., *P.*

persica × *P. amygdalus* Batsch., and *Cydonia oblonga* Mill. shoot cultures maintained without exogenous auxin. Although auxins were also active in inducing ARF, other CKs such as kinetin, zeatin, iso-pentenyladenine, and iso-pentenyladenosine were not.

The reason for this apparent discrepancy between CKs is unclear, but warrants further investigation with respect to the rooting process which can arbitrarily be subdivided into: 1) an induction and root meristem formation phase; and 2) an elongation phase (Maldiney et al. 1986). That greater emphasis on these aspects is necessary is also indicated by the exhaustive experiments of Wightman et al. (1980). These authors reported that four natural (zeatin, ribosylzeatin, isopentenyladenine, and isopentenyladenosine) and two synthetic (kinetin and ribosylkinetin) CKs had markedly different effects on lateral root initiation and emergence. All the CKs tested showed strong inhibition of root initiation and emergence at concentrations above 10^{-6} mol·l^{-1}, except for isopentenyladenine and kinetin which exhibited inhibition at 10^{-5} mol·l^{-1} and above. It is of interest that most CKs give an optimum response with respect to cell division in the range of 10^{-5} to 10^{-6} mol·l^{-1} (Forsyth and Van Staden 1986). The synthetic CKs (kinetin and ribosylkinetin) were less inhibitory than the natural CKs, being promotive at 10-fold higher concentrations. This effect may be due to a faster rate of metabolism. Zeatin, which is the most active CK in cell division bioassays, was the most inhibitory of the natural CKs investigated. It was significant, however, that in general all the CK ribosides tested were less inhibitory with respect to rooting than their free bases (Wightman et al. 1980). Not only were the ribosides of kinetin, isopentenyladenine, and zeatin less inhibitory to root initiation and emergence, but also less so with respect to root elongation.

These results should be viewed in conjunction with a number of seemingly unrelated facts, all of which indicated that cytokinin ribosides may be of primary importance in the rooting process. First, there is the finding by Nemeth (1979) that benzyladenine stimulated rooting. This cytokinin, when applied to tissue, is rapidly metabolized to ribosylbenzyladenine (Van Staden et al. 1986). Second, ribosylzeatin is the most abundant CK exported from roots and is present in xylem sap (Van Staden and Davey 1979) and stem material (Van Staden and Brown 1977, Van Staden and Dimalla 1981) in relatively high levels. Third, centrifugation of *Salix* spp. cuttings prior to planting resulted in the accumulation of CKs, largely compounds which co-eluted with ribosylzeatin, in the basal portions of the cuttings which subsequently rooted better (Van Staden 1978). All these results argue for the involvement of CK ribosides in the rooting process. The differences observed with different CK bases (Nemeth 1979, Wightman et al. 1980) may well be related to the degree to which they are, at least initially, transported and metabolized.

The apparent inefficiency of CKs in the rooting process as reflected in the literature, could be due to a number of factors. First, it is rather alarming that in most trials only a single CK has been used, whereafter a blanket response for all CKs has been claimed. The results of Nemeth (1979) and Wightman et al. (1980) highlight the shortcomings of this approach. Second, it appears to be common for researchers to apply only one or a narrow range of concentrations of CKs in their trials. It is now becoming evident that, as is the case with other hormones, CKs have optimal levels which vary greatly from tissue to tissue and organ to organ. With respect to the rooting process, it would now appear that the lower end of the concentration range has very seldom been tested. Typically, Humphries (1960) applied kinetin to *Phaseolus vulgaris* petioles in the range 4.6×10^{-7} to 4.6×10^{-5} mol·l^{-1}; Bachelard and Stowe (1963) in their studies on *Acer rubrum* used kinetin levels of 4.6×10^{-7} to 1.4×10^{-5} mol·l^{-1}; Chandra et al. (1973), when investigating rooting of mung bean hypocotyls, used benzyladenine at 10^{-5} mol·l^{-1} and higher; Eriksen (1974) studied root formation in pea cuttings using benzyladenine in the range 10^{-5} to 10^{-1} mol·l^{-1}. Interestingly, Wightman et al. (1980) induced initiation and emergence of lateral roots when using concentrations of CKs lower than 10^{-6} mol·l^{-1}. Biddington and Dearman (1982/83) found that zeatin concentrations of 3×10^{-10} and 10^{-9} mol·l^{-1} increased the numbers of lateral roots and lateral root extension growth of lettuce seedlings. Higher concentrations of zeatin were inhibitory to both these aspects of root growth. These results were confirmed by those of Finnie and Van Staden (1985), where elongation of cultured *Lycopersicon esculentum* Mill. (tomato) roots was optimal at zeatin concentrations of 10^{-8} mol·l^{-1}. Maximum lateral root emergence also occurred at 10^{-8} mol·l^{-1}, but both processes were inhibited at 10^{-6} mol·l^{-1} or higher. It is therefore clear that rooting experiments using lower levels of CKs are called for.

A third aspect is that greater consideration should be given to the stage of the sequential rooting process at which CKs could play a role. Hartmann and Kester (1983) divided the process of ARF into three stages: 1) cellular dedifferentiation followed by meristematic cell initiation (root initial formation); 2) development of the root initials into recognizable root primordia; and 3) formation of vascular connections with the conducting tissues of the cutting, and rupturing of other stem tissues to allow emergence of the new roots. An account of similar sequential stages of root initiation in cuttings of *Pinus radiata* seedlings was given by Smith and Thorpe (1975). The absence, presence, or relative strength of a compound, whether endogenous or exogenously applied, could have promotive or inhibitory effects on one or more of these stages. This is exemplified by the recent experiments on the levels of endogenous hormones in tomato cuttings (Maldiney et al. 1986). Obviously therefore, to assess the ability of a substance to induce ARF simply by applying it to the base of a cutting is in essence a very crude experiment. Notwithstanding the fact that more elegant treatments are probably not a viable alternative on a commercial basis, it is within the scope of the plant physiologist to determine more precisely the possible sequential actions of plant growth regulators on ARF and LRF.

FACTORS AFFECTING RESPONSES TO CYTOKININS

Foliar vs. Basal Application

The role of leaves and buds in stimulating ARF in cuttings is well documented (Van Overbeek and Gregory 1945, Batten and Goodwin 1978). Although auxins have been implicated as the causal agents in many of these studies, it is generally acknowledged that other naturally occurring root forming substances are also involved. Although some of these so-called "cofactors" have been identified (Hess 1965, Heuser and Hess 1972), others remain to be characterized. The possibility that CKs may be involved as cofactors cannot be discounted, but there is little evidence, circumstantial or otherwise, to suggest this at present. Where CKs do appear to play a role is in retarding senescence of leaves on cuttings, thus allowing transport of auxins and/or synergists out of leaf tissue to the sites of ARF.

Studies by Van Overbeek and Gregory (1945) and Van Overbeek et al. (1946) on *Hibiscus* spp. showed that grafting of easily rooted 'Brilliant' onto difficult-to-root 'Purity' cuttings induced rooting in the latter clone. It was noted, however, that ungrafted 'Purity' cuttings rapidly lost their leaves, whereas 'Brilliant' cuttings retained their leaves. When repeating these experiments, Ryan et al. (1958) were able to obtain high rooting percentages with ungrafted 'Purity' cuttings, apparently because under their experimental conditions, the cuttings retained their leaves for far longer.

The requirement of leaves for rooting of *Acer rubrum* was clearly shown by Bachelard and Stowe (1963). In the same study, kinetin (4.6×10^{-5} mol·l^{-1}) sprays applied to the leaves were shown to increase ARF, although applications (4.7×10^{-7} to 1.4×10^{-5} mol·l^{-1}) to the bases of cuttings were strongly inhibitory. It is unfortunate that no mention was made as to whether the foliar sprays of kinetin affected leaf retention or senescence. Similar results were obtained when benzyladenine (10^{-3} mol·l^{-1}) was applied above the second leaf of pea cuttings in a lanolin paste, provided they were not disbudded (Eriksen 1974). Application of benzyladenine to the bases of the cuttings was inhibitory. Similar results were obtained with difficult-to-root leafy peach (*Prunus persica* L.) cuttings (Gur et al. 1986). There are several questions related to these results which need clarification: to what extent were the foliar applied CKs metabolized; to what extent was the applied molecule or its metabolized derivative exported; and to what extent did the applied CKs affect export or import of molecules to or from the leaves? It is again disconcerting that these positive results have not received further attention, particularly as the advent of intermittent mist systems has allowed for greater retention of leaves on cuttings. The application of CKs to leaves of leafy cuttings could provide a breakthrough in propagating many difficult-to-root species. It is a technique which appears to have been relatively untouched thus far, and is a field wide open for research. Besides practical experimentation on a commercial basis, fundamental studies are needed to determine the action of CKs on leafy cuttings, and why these should be beneficial to ARF in at least some situations.

An interaction between CKs and auxins (applied or endogenous) in the rooting process also needs urgent attention. Reuveni and Raviv (1981) showed a positive correlation between rooting percentages of ten avocado (*Persea americana* Mill.) clones and the number of leaves retained on the cuttings under intermittent mist. Subsequently they demonstrated a significant delay in leaf shedding and an increase in rooting of 'Fuchs 20', a difficult-to-root avocado cultivar which drops its leaves under mist while still green, by applying an optimum foliar spray concentration of 10^{-4} mol·l^{-1} naphthaleneacetic acid plus 5×10^{-4} mol·l^{-1} benzyladenine (Raviv and Reuveni 1984). Similar results were reported for peach cuttings and it was suggested that the naphthaleneacetic acid possibly increased benzyladenine translocation (Gur et al. 1986).

Stage of Development

In almost all rooting experiments involving applied CKs the effect of these hormones was studied immediately after cutting preparation (Humphries 1960, Kaminek 1968, Fellenberg 1969). Eriksen (1974) recognized this shortcoming, and investigated the effect of CKs applied at intervals after taking cuttings of *Pisum sativum*. The results indicated that the influence of CKs was different at different stages of root development. High concentrations of benzyladenine (10^{-3} mol·l^{-1}) inhibited the early stages of ARF, but this inhibition decreased during the later stages (an anatomical definition of these stages was not provided, but their sequence was determined from days required for root emergence to occur). Lower concentrations of benzyladenine (10^{-4} to 10^{-5} mol·l^{-1}) were, however, promotive during the earlier stages. It was thus suggested that root primordia may become capable of controlling levels of active CK, and thus are not responsive to exogenous CK applications during the later stages of ARF. In repeating these experiments, Mohammed (1980) found similar results, and concluded that low levels of CK activity were necessary only during early ARF, whereas a higher level appeared to be more conducive at later stages or, as indicated by Bollmark and Eliasson (1986), high sensitivity may be restricted to the early phases of root development. Chandra et al. (1973) investigated the effect of benzyladenine on the initiation of adventitious roots on mung bean hypocotyls, including histological and isoenzyme studies. Cuttings incubated with 10^{-5} mol·l^{-1} or higher concentrations of benzyladenine showed extensive cell division and tracheid differentiation in the cortex, and root primordia did not form. Further data suggested that new isoperoxidases are formed only when the specialized phloem parenchyma cells are induced to differentiate into a primordium. However, benzyladenine prevented isoenzyme formation only when applied during the early period of root induction. No new isoperoxidases were formed in hypocotyls when these were induced by benzyladenine to form tracheids. Chandra et al. (1973) thus concluded that benzyladenine suppressed the differentiation of phloem parenchyma cells and the formation of isoperoxidase, thereby inhibiting the early processes of ARF. What becomes quite clear from the above experiments is that root initiation and root elongation are affected in different ways by CKs. To what extent tissue formation is controlled by these hormones needs to be established.

CHANGES IN ENDOGENOUS CYTOKININS IN RELATION TO ROOTING

Ease of rooting of a cutting can vary markedly depending on several morphological or environmental factors which may or may not involve changing CK levels. These include the aspects of juvenility and phase change; the inherent ease or difficulty of rooting, i.e. species, cultivar, and clonal differences; and the time of year at which cuttings are taken (Hartmann and Kester 1983).

Results obtained with *Citrus* spp. have indicated significant differences between the CK levels of juvenile and adult plants (Hendry et al. 1982b). It was suggested that these differences may be important at the stage of bud activation, prior to the breaking of dormancy. At this stage, polar CKs (cochromatographing with cytokinin glucosides) reached a higher level in juvenile plants. This aspect may be significant with respect to the known differences in rooting capacity between juvenile and mature cuttings.

The CKs present in juvenile and adult cuttings will certainly be different both quantitatively and qualitatively which may have considerable bearing on rooting ability. In *Citrus* spp., where juvenility is an important factor in rooting, the young leaves contained lower levels of CKs than mature leaves

(Hendry et al. 1982a). In the stems of willow (*Salix* spp.), considerable seasonal fluctuations in CK levels were detected in the bark, buds, and leaves, all of which may form part of shoot cuttings (Van Staden and Davey 1981).

The ability to rejuvenate mature tissue is of great benefit horticulturally, because of the ease with which roots may then be induced, without concomitant delay in subsequent flowering of regenerated plants (Zimmerman 1981). David et al. (1978) working on the rejuvenation of *Pinus pinaster* Ait. meristems found that high benzyladenine and low sucrose concentrations favored the appearance of juvenile characteristics. This indicates yet another area where the use of cytokinins may prove beneficial to rooting.

In vitro experiments with apple (*Malus pumila* L.) buds indicated that, when grown on a medium supplemented with benzyladenine, the buds attained a greatly improved rooting ability after a series of subcultures (Sriskandarajah et al. 1982). Unfortunately, the degree of CK uptake and metabolism was not determined. Takeno et al. (1982/83) investigated the CK and gibberellin content of cultured shoot tissues of apple. Most of the CK activity cochromatographed with zeatin and ribosylzeatin and the authors suggested that endogenous production of CKs did occur. In experiments with apple buds, a reduction in shoot diameter and changes in leaf shape accompanied the prolonged culture of the shoots, indicating that a partial reversion to the juvenile state occurred under the experimental in vitro conditions of continuous illumination and constant temperature (Sriskandarajah et al. 1982). Investigation of this subculture method, and a detailed monitoring of changing CK activity, may be a rewarding exercise if conducted on other hard-to-root species and cultivars.

Leaf Cuttings

Leaf cuttings of *Phaseolus vulgaris* have been the prime experimental material used in studying CK metabolism in relation to root growth and formation. Mothes and Engelbrecht (1956) considered a leaf cutting to represent a plant system in which the root to shoot ratio continuously increases, thus allowing a marked accumulation of root synthates in the lamina. Root initiation can be controlled and the effects of buds eliminated, while the leaf lamina provides a photosynthetic organ, making the explant independent as far as assimilates are concerned. Leaf cuttings of *P. vulgaris* are particularly suitable for this purpose, because of the ability of a single petiole to initiate more than 100 roots, all of the same age (Engelbrecht 1972).

CK activity was monitored in leaf cuttings of *P. vulgaris* by Engelbrecht (1972) over a period of 40 d, the explants being divided for purposes of analysis into laminae and petiole-plus-roots portions. A marked increase in activity occurred in the root-petiole portion, peaking after 8 d, whereafter activity stabilized, and began increasing in the lamina portion. CK levels rose much higher in laminae of leaf cuttings than in the equivalent laminae on intact plants. These findings support the general view that CKs are synthesized in the roots and transported to the leaves (Letham and Palni 1983). Later it will be shown that this generality is not always easy to demonstrate.

Featonby-Smith and Van Staden (1981) analyzed the CK content of primary leaves of *P. vulgaris* in relation to rooting ability. Leaves of 30-d-old plants contained higher CK levels than 10-d-old plants, occurring predominantly as CK glucosides. These cuttings rooted better than leaves of younger (10-d-old) plants. It was concluded that these higher levels of CK glucosides may have been used either in the process of ARF and/or for the retardation of leaf senescence. There is some evidence that CK glucosides are storage and inactivation forms (Van Staden 1976ab, Letham and Palni 1983), and although a few exceptions have been recorded, these conjugates are generally markedly less active than their corresponding bases in a number of CK bioassays (Letham et al. 1983). As yet, the effects of glucosides on root formation have not been investigated, partly because of difficulties in procuring sufficient amounts of these compounds (Letham et al. 1983). However, in a metabolic process where small, controlled amounts of CK are required to be released over a relatively long time period, it is highly probable that CK glucosides, ribosides, and other storage forms would act as slow release sources. Thus low total CK activity could be attained in a tissue by conversion of free bases to storage forms, thereby allowing CK sensitive processes such as root initiation to occur. Once the stage(s) of initiation susceptible to high free CK levels was complete, reconversion to free bases could take place, allowing these compounds to act in the remainder of the ARF-LRF process.

Bridglall and Van Staden (1984), also in studies on bean leaf cuttings, found that auxin application to the petioles first resulted in an increase in CK content, and second, in a redistribution of CKs within the explants. Auxin treatment appeared to concentrate CKs in the lamina, at the expense of levels in the petiole. The net effect was apparently to decrease CK activity in the petiole to a level where it was either promotory or no longer inhibitory to ARF. It is relevant that hormone analysis in this case was carried out prior to root primordia initiation. The latter process has been shown to stimulate CK production (Engelbrecht 1972, Featonby-Smith and Van Staden 1981), with synthesis and transport apparently beginning before the new adventitious roots rupture the petiole epidermis (Forsyth and Van Staden 1981). This could explain the increased levels of CK found by Carlson and Larson (1977) in the xylem exudate of root-pruned *Quercus* spp. seedlings before adventitious roots became visible. It could further explain the findings of Chen and Petschow (1978) and Van Staden and Choveaux (1980) that increased CK levels could be detected in rootless shoot explants of *Nicotiana tabacum* and *Salix babylonica* L., respectively. As in many other species, shoots of *Salix* spp. contain preformed root primordia to which Kriesel (1976) attributed increased CK levels detected prior to visible emergence of adventitious roots.

Böttger (1974) detected a significant increase in the number of lateral root primordia developing within 12 h after root tip removal of pea seedlings, and these results were later confirmed by Forsyth and Van Staden (1981). Thus, if CKs are produced by root primordia, a concurrent increase in CK activity could be expected fairly soon after preparation of cuttings or root segments. In fact, the results of Featonby-Smith and Van Staden (1981) indicate that CK synthesis and transport begin before the new adventitious roots rupture the petiole epidermis, which in this case occurred 6 d after the cuttings were prepared. Forsyth and Van Staden (1981) detected a rise in CK activity concurrent with significant LRF in decapitated pea seedling roots. In this study, increased activity occurred 8 h after decapitation, as the lateral roots began rupturing the primary root epidermis, and continued to rise with lateral root emergence and elongation.

The anomalous situation therefore appears to occur where applied CKs are inhibitory to root formation, but where these same compounds are produced by root primordia, perhaps only hours later. The obvious difference, however, between an intact plant and a rootless cutting is the absence of the root system, and in particular, of the root tips. It would be interesting to compare the metabolism of CKs by rootless petioles, petioles with root initials, and the root systems of a particular species.

Shoot Cuttings

When considering that CKs are almost universally accepted to be inhibitors of ARF, it is perhaps not surprising that very few investigations of endogenous CK levels in difficult- and easy-to-root plants has been conducted. Okoro and Grace (1978) monitored levels of CK at various stages of root formation in two species of poplar. Cuttings of difficult-to-root *Populus tremula* L. were found to contain higher CK activity than cuttings of easy-to-root *P.* × *euramericana,* a condition which persisted throughout the rooting period. Initially, CK activity decreased in both the upper and lower halves of cuttings of both species, but later increased. Increases in the upper halves of the cuttings were apparently due to bud burst and leaf expansion, processes which have been associated with increased CK activity (Hewett and Wareing 1973ab, Kannangara and Booth 1974). Root formation in *P.* × *euramericana* caused a large CK increase in the lower halves, while this increase was far less noticeable in the lower halves of *P. tremula,* which produced basal callus but did not root.

In contrast to these results, Wu and Barnes (1981) found very little variation between endogenous CK activity of stem tissue of easy-to-root *Rhododendron ponticum* L. vs. the difficult-to-root clone of *Rhododendron,* cv. *Britannia.* A marked seasonal variation in CK activity was also not detected in this study. However, changes in CK levels can occur within plants during changing environmental conditions (Van Staden and Davey 1981, Van Staden et al. 1981a). Increases in CK activity have been detected in spring sap of a number of woody plants prior to bud burst (Reid and Burrows 1968), while Skene (1972) showed that CK levels increased in sap during cold storage of grape (*Vitis* spp.) cuttings, later falling to low concentrations after bud break. Nakano et al. (1980) monitored hormone activities in hardwood cuttings and intact vines of grape. CK levels in growing buds of both intact and cutting material peaked at bud break, although the chromatographic patterns of active substances in the buds of intact vines were different from those of the cuttings. Nakano et al. (1980) concluded that CKs may not directly promote rooting of the cuttings, but could play an indirect role by promoting bud break. Hewett and Wareing (1973a) found similar results when determining CK levels in sap and vegetative buds of *Populus* × *robusta* during chilling and bud burst. In this case, however, maximum CK activity in the sap was detected two weeks prior to natural bud burst, while levels in the buds peaked a week after bud burst. Further studies on leaves and xylem sap of *Populus* × *robusta* (Hewett and Wareing 1973b) showed that marked variation in CK activity occurred throughout the season, with the maximum being detected in summer, and the minimum occurring in winter, once shoot growth had ceased. The work of Van Staden and Davey (1981) showed that this was also true with respect to roots, buds, and bark from intact plants.

CKs, and indeed probably all phytohormones, appear to exhibit peaks in activity which are dependent on external environmental stimuli. The time of season at which cuttings are struck therefore has an enormous impact on the degree of rooting success. Although nurserymen, through practical experience, have determined the optimal time of year for vegetative propagation of many plants, it would be interesting to attempt a correlation between hormonal levels at these times, and the degree of rooting obtained. It may become possible to apply stimuli such as chilling and heating to promote hormonal changes and thus induce ARF in cuttings. Alternatively, these processes could possibly be short circuited simply by applying the correct levels of hormone. Thus, for example, CKs could be applied to cuttings to promote bud break, and indirectly, ARF. Although attempts to induce bud burst in excised poplar shoots with benzyladenine have thus far proved unsuccessful (Hewett and Wareing 1973a), this may be due to insufficient uptake, lack of transport within the shoot, and/or extremely rapid degradative metabolism of the applied compound. In rose shoots it has been shown that basal, more dormant buds contained higher levels of extractable CKs than their more apical and less dormant counterparts (Van Staden et al. 1981b). The presence of these CKs in buds on cuttings, or in buds cultured *in vitro,* may be detrimental for the rooting process. Their utilization for bud extension, and the subsequent decrease in CK levels, may be an important factor with respect to the stimulation of rooting.

If a better understanding could be obtained with respect to the role of endogenous CKs in cuttings, these CK applications could open further exciting avenues in vegetative reproduction. Horticulturists have for many years recognized the need for cuttings to contain high carbohydrate levels in order for successful rooting to occur and many cultural practices, such as girdling of shoots, reducing nitrogen supply, and selecting tissues where carbohydrate accumulation occurs, are employed to meet this requirement (Hartmann and Kester 1983). Because of the strong assimilate mobilizing characteristics of CKs (Letham 1978), successful applications of these compounds several weeks prior to taking of cuttings could result in ideal propagating material. It is also possible that seaweed concentrates, which appear to be a source of inexpensive and relatively stable natural CKs (Featonby-Smith and Van Staden 1984), could be used to good effect in this regard.

Root Cuttings

Root segments lack the shoot, the mature root, and the root tip as potential sources of phytohormones (Goodwin 1978). Root cuttings also differ from other excised plant portions in that regeneration often requires formation of both adventitious roots and shoots (Hartmann and Kester 1983). As will be demonstrated later, removal of the root tip also involves removal of tissues which are a major area of CK biosynthesis. It can therefore be expected that root cuttings would contain relatively low levels of endogenous CK, unless preformed initials or primordia were present and acting as CK sources. As in experimentation on stem cuttings, care should be taken that roots produced in response to a treatment are in fact new adventitious organs, and not derivatives of existing initials. The lack of tissues capable of CK biosynthesis implies that CK activity is lowered after a period of time, and may therefore result in CK concentrations suitable for allowing ARF. This could explain why root cuttings are successfully used in vegetative propagation of species such as *Malus* and *Pyrus,* stem cuttings of which are not easily rooted (Hartmann and Kester 1983).

Very little concrete evidence exists in the literature concerning ARF and CK relationships in root cuttings. However, it can be speculated that CKs play a role similar to that in LRF in decapitated roots. Thus high levels of endogenous or applied CK bases would be expected to inhibit ARF in root segments, whereas low levels of bases and somewhat higher levels of ribosides could promote ARF. Conversely, CKs have been shown to stimulate bud formation when applied to root pieces (Danckwardt-Lillieström 1957). Bonnet and Torrey (1965) induced buds on root segments of *Convolvulus* spp. grown under sterile conditions by treating with 4.6×10^{-7} mol·l^{-1} kinetin. As it is common for root cuttings of many species to produce adventitious shoots prior to formation of a new root system (Hartmann and Kester 1983), it is possible that applied CKs could play an indirect role in propagation from roots by promoting bud break in contrast with applied auxins which tend to inhibit shoot formation in root cuttings. For successful regeneration from a root segment to occur, it may therefore be necessary for a degree of polarity to come about within the tissue, as appears to be the case in specialized underground structures such as tubers and tuberous roots (Letham 1978). The presence of CK glycosides may be necessary here for creating a delicate gradient of CK activity, thus allowing shoot and root formation in different portions of a root cutting.

INTERACTIONS WITH OTHER PLANT GROWTH REGULATORS

Auxins

Several instances where CK-auxin interactions have occurred in rooting systems have already been mentioned such as in organogenesis (Skoog and Miller 1957), rooting of hardwood (Meredith et al. 1970) and leafy (Bridglall and Van Staden 1984) cuttings, and foliar spraying of leafy cuttings to stimulate basal rooting (Raviv and Reuveni 1984, Gur et al. 1986).

Although it is true for many *in vitro* systems that high auxin concentrations in the presence of low CK levels promote root formation, there are certainly many cases where this does not apply. For example, Walker et al. (1978) found that transfer of alfalfa (*Medicago sativa* L.) callus from induction medium containing high levels of 2,4-D and low levels of kinetin to the regeneration medium resulted in the formation of shoots. Conversely, the transfer of callus from induction medium containing low levels of 2,4-D and high levels of kinetin resulted in the formation of roots. When reporting studies of this nature, it is advisable to distinguish between roots arising from explant tissue (which may still be undedifferentiated) and those arising from stable callus. Rhizogenesis refers, in the strict sense, to the development of complete root organs from callus. Many freshly established callus cultures tend to produce roots spontaneously for several subcultures, but subsequently produce only callus. This is probably due to the existence of preformed root primordia in the explant, and thus does not constitute genuine rhizogenesis (Gresshoff 1978). In fact, Gresshoff (1978) has further narrowed the cases of true rhizogenesis to exclude reports where the callus was not stable, where an original explant was present, or where a shoot had developed.

Many controversial results concerning the roles of CKs and auxins still occur, even for true cases of rhizogenesis. The two-phase model of Halperin (1969) used to describe root formation may help to explain some of these discrepancies. As is the case with ARF (Hartman and Kester 1983), organogenesis appears to occur in stages. The first phase produces a generalized primordium, and the second determines the organ type (root or shoot). Manipulation of CK and auxin levels during either phase can inhibit or promote root formation. Gresshoff (1978) has delineated two generalized phytohormone treatments for root induction: in the first, roots develop when callus is maintained on media containing almost equimolar concentrations of auxin and CK; the second treatment consists of withdrawal of CK and/or auxin. The first treatment apparently induces primordium formation, while providing sufficient auxin levels to allow further development resulting in a complete root. Withdrawal of exogenous phytohormones appears to allow for development of primordia into complete roots. In the latter case, primordia develop on normal callus inducing media, but are then suppressed from undergoing further differentiation by the presence of auxin and/or CK.

For cuttings of some species, callus formation from where root formation can occur is obligatory (Hartmann and Kester 1983). Although scant information is available on cytodifferentiation within basal callus of cuttings, it is probable that a sequence of events occurs similar to that of rhizogenesis in callus cultures. Minocha (1984) concluded that during one or more critical cell divisions in the presence of optimal levels of benzyladenine, a number of cells are induced or committed for later differentiation into tracheids. However, the high concentrations of benzyladenine required during induction are apparently not required during the intervening cell divisions, nor for the actual differentiation of the tracheary elements.

Some evidence exists that CKs may in certain instances protect auxins from degradation and thereby increase auxin activity in tissues. Noor-Saleh and Hemberg (1980) detected increased amounts of indoleacetic acid in seeds of bean, pine, and maize (Zea mays L.), and in young bean plants, which had been treated with kinetin. Further studies on indoleacetic acid content of roots of bean, maize, and oats (Avena spp.) (Noor-Saleh 1981) showed that significant increases occurred when the seedlings were watered twice with a 3.3×10^{-5} mol·l^{-1} kinetin solution. The CK treatment decreased indoleacetic acid oxidase activity in the root tissues, although the correlation between reduction in oxidase levels and increase in auxin concentration was not strong. Einset (1977) found that increasing supplies of CK to tobacco callus cultures resulted in less auxin being required to obtain the same growth. Syono and Furaya (1972) also found that kinetin treatment of tobacco callus increased the endogenous indoleacetic acid level. However, Gaspar and Xhaufflaire (1967) showed that kinetin activated indoleacetic acid oxidase in roots of Lens culinaris Medik., thus causing decreased indoleacetic acid levels in kinetin treated roots. These discrepancies may possibly be due to differing hormone concentrations (Lee 1971, 1974). CK treatments appear also to influence the predominance of certain isoenzymes of indoleacetic acid oxidase (EC 1.11.1.8). Tobacco cultures, grown on a medium supplemented with 2×10^{-6} mol·l^{-1} kinetin and 10^{-5} mol·l^{-1} indoleacetic acid, contained little or none of the isoenzyme A5 and none of A6 (Lee 1971). However, lowering the concentration of kinetin to 2×10^{-7} mol·l^{-1} resulted in the development of isoenzymes A5 and A6, and in particular, a 14-fold increase in A5 activity.

Thus application of a CK can have marked effects on the auxin levels in a cutting. The findings of Bachelard and Stowe (1963), Eriksen (1974), Reuveni and Raviv (1984), and Gur et al. (1986) that foliar applications increase ARF could be explained in terms of auxin protection. Further investigations of auxin/CK interactions using the bean leaf cutting system could provide some very interesting information. Jain et al. (1969) showed that 10^{-2} mol·l^{-1} kinetin applied to bean seedling shoot apices increased indoleacetic acid levels in the primary leaves to 194% above the control. Conversely, the findings of Bridglall and Van Staden (1984) that auxin application to bean leaf petioles increased CK content suggest that auxins may likewise stimulate or protect CKs.

Other Growth Substances

The interrelationships of CKs and the other three classes of phytohormones in the rooting process can only be speculated upon. The effects of gibberellins, abscisic acid, and ethylene on ARF are dealt with in detail in other chapters. There is little evidence to suggest that either gibberellins or abscisic acid interact with CKs in the same manner as auxins. However, because gibberellins are typically inhibitory to ARF, and abscisic acid is either promotory or inhibitory (depending on species) (Batten and Goodwin 1978), these compounds obviously affect the response to applied or endogenous CKs if present in sufficient amounts. The close relationship of auxins and ethylene in plant tissues implies that CK and ethylene levels may also have some bearing on promotion or inhibition of ARF. It would appear, however, that the overriding dogma of auxin/CK ratio has to some extent overshadowed and even discouraged research on other phytohormone interactions in the rooting process, and is a situation which requires remedying.

CYTOKININS AND ROOT GROWTH

Roots as Sites of Cytokinin Biosynthesis

It is now generally accepted that CKs are produced in the roots from where they are transported via the xylem to the above-ground portions of the plant (Letham and Palni 1983). Much of the evidence for the roots as sites of CK synthesis has been circumstantial (Van Staden and Davey 1979).

Recent results have however, indicated that both isolated (Van Staden and Forsyth 1984) and attached (Dickinson et al. 1986) tomato roots have the ability to incorporate ^{14}C-adenine into endogenous CKs although incorporation is extremely low.

Synthesis within the root appears to occur mostly in the root tip. Weiss and Vaadia (1965) found extracts of sunflower (*Helianthus annus* L.) root apices to have high CK activity, whereas extracts of older root tissues contained no detectable CK. Analysis of serial segments of young pea seedling roots for CK activity by Short and Torrey (1972) revealed that the terminal 1 mm of the root tip contained as much as 44 times more free CK on a fresh weight or per cell basis than the next 1–5 mm root segment. More proximal segments behind the tip (5–20 mm and 20–40 mm) contained no measurable free CK. Analysis of the terminal 1 mm of maize roots showed that approximately 20% of the CK activity occurred in the root cap (Feldman 1975). Measurements of CK activity in pea roots by Böttger (1974) showed that a 30 mm segment proximal to the 3 mm root tip had three to four times less activity if the root tip was removed, with significant decreases occurring within 12 h of decapitation. Thus the highest levels of free CKs in roots appear to occur in the meristematic tip region.

Goodwin and Morris (1979) applied a variety of growth regulators in lanolin paste to decapitated pea seedling roots, including benzyladenine, isopentenyladenine, and kinetin. While all the kinetin concentrations tested had no effect on LRF, the benzyladenine treatments caused a linear inhibition of LRF with increasing concentration. The lowest level applied (0.8 μg per root) was significantly inhibitory. However, isopentenyladenine at 0.8 μg per root caused a significant increase in LRF, although all higher concentrations were inhibitory.

This case again illustrates the need to test different CKs, and at different concentrations, especially in the lower range, before a true reflection of the actions of this class of compounds can be gained. Clearly, however, the root tip appears to play a role in preventing LRF in the region immediately proximal to it, and CKs appear to be strongly involved in this apical dominance.

Influences on Root Apical Dominance

It was proposed by Torrey (1962) that the declining gradient in CK activity from the root tip to the base of the primary root could play a controlling role in LRF. Wightman et al. (1980) reported total CK concentrations in the apical 5 mm segment of intact pea roots as high as 9.7×10^{-5} mol·l^{-1} declining to 8×10^{-6} mol·l^{-1} in the 20–25 mm region of the root. Böttger (1974) reported concentrations of 3.1×10^{-8} mol·l^{-1} and 6.7×10^{-9} mol·l^{-1} benzyladenine equivalents in 30 mm segments of intact and decapitated roots, respectively. Thus levels in the root tip can reach concentrations which would be inhibitory to lateral root production if applied exogenously (Böttger 1974, Biddington and Dearman 1982/83, Finnie and Van Staden 1985).

Wightman and Thimann (1980) confirmed that an acropetal sequence of lateral root development does occur in pea seedling roots. Primordia occurrence was nil in the apical 5 mm segment, but rose steeply to a maximum in the region 30–60 mm from the root tip. With a few exceptions (McCully 1975), this appears to be the general arrangement for roots of most plants. Böttger (1974) found that a 3 mm decapitation treatment resulted in a 50% increase (within 24 h) in lateral root primordia in the 30 mm segment proximal to the tip. Forsyth and Van Staden (1981), also working with pea roots decapitated 3 mm from the apex, detected significant increases in the initiation of lateral root primordia within 12 h of the treatment. Biddington and Dearman (1982/83) reported that removal of 3 mm of lettuce (*Lactuca sativa* L.) seedling root tips increased both the number and length of lateral roots. Wightman and Thimann (1980) confirmed that root tip removal results in a rapid increase in the number of lateral root primordia, but noted that this was a transitory effect, in that by the third day after decapitation, the number of primordia in treated roots was far less than in the controls. In the same study, 80% of lateral root primordia in decapitated roots had emerged after 48 h, whereas even after 72 h, only 3% emergence of primordia occurred in the controls.

Recently, Golaz and Pilet (1985) have examined the effects of decapitation on subsequent emergence and elongation of lateral roots of cultured apical maize root segments. Although decapitation had no effect on mean primordia number per cm of root, it did cause increased primordia initiation in the first apical 10 mm of root. Thus the main result of decapitation appears to be a shifting of the site of lateral root formation towards the root tip. Golaz and Pilet (1985) concluded that growth regulators produced or released by the root cap and diffusing basipetally may inhibit the initiation phase of LRF.

In contrast to these results, Goodwin and Morris (1979) found that LRF on pea roots was not affected, in terms of total lateral number, by decapitation. However, the response to tip removal depended on the length removed: as the length removed increased, so the site of LRF moved closer to the apex, until with 2 mm removed, lateral roots formed immediately adjacent to the root stump.

It is clear that CK production in the root tip probably has a profound effect on LRF, not only on the number of such roots formed but also on the position of these roots. Factors which therefore affect the rate of biosynthesis at the root tip or the export of these compounds from the site of production, as well as possible metabolism within different parts of the root itself, need to be considered more closely in future assessments of the role of CKs in root formation.

Applied Cytokinins and Root Growth

The exogenous application of CKs has often been shown to be inhibitory to root growth (Scott 1972, Svenson 1972, Torrey 1976, Goodwin 1978, Stenlid 1982). Although studies using applied growth regulators have contributed greatly to present knowledge, the results need not necessarily reflect the true status or involvement of these compounds within the plant (Van Staden and Davey 1979). In fact, stimulation of root growth by exogenous cytokinins has been reported, especially for roots formed in vitro (Lee 1959, Fridborg 1971, Nandi et al. 1977). The promotion of lateral root growth when applying low concentrations of zeatin (Biddington and Dearman 1982/83, Finnie and Van Staden 1985) and seaweed concentrate (Finnie and Van Staden 1985) have been mentioned.

CONCLUSION

The role of CKs in root formation is clearly a topic encompassing more questions than answers. What is equally apparent is that many of the "answers" derived from previous research have been misleading. This situation must in part be ascribed to a lack of knowledge concerning CK metabolism in plants, but recent advances in this field can to a great degree assist in integrating and elucidating some of the discrepancies recorded in the earlier literature with regard to ARF. Moreover, as it becomes clearer that CKs as a whole are more complex and varied than previously perceived, it becomes increasingly important to apply greater accuracy and definition when researching the actions of these phytohormones. We now know that different CK molecules have very different physiological effects; that these molecules are metabolized to forms which have varying activity; that a particular molecule and form of CK cause significantly different, even directly contrary responses at different concentrations; and that the relative strengths of other growth regulators present, in particular auxins, result in altered CK responses.

Furthermore, it is now apparent that CKs present or applied at different stages of rooting affect the eventual outcome of the process differently. Experimental material is also likely to contain higher and lower levels of differing forms of CK at the outset of a trial, depending on the prevailing environmental conditions. A good point to start experimentation on rooting would therefore be a determination of endogenous hormone levels, followed by sequential sampling to ascertain changes occurring during the various phases of root formation. Although biological materials may always remain relatively undefined media, the above precautions can at least contribute to eliminating further unnecessary discrepancies unrelated to the treatments applied.

Previous trials on ARF have concentrated on application of growth regulators to the basal ends of cuttings. This approach assumes that cutting material is relatively static physiologically, lacking the ability to transport or metabolize applied compounds. In fact, the greatest contribution of CKs to ARF is possibly gained by application to other parts of the plant segment and in particular to the leaves and buds. The few cases of rooting success involving "above-ground" applications indicate that CKs play a mobilizing or protective rather than a direct role in ARF. Rejuvenation and carbohydrate loading of cutting material through application of CKs are further aspects which deserve more attention. The whole question of CK transport within cuttings remains somewhat obscure, yet with labeled compounds now available, progress can certainly be made in this field.

On a cellular level, many questions remain unanswered. Why should a certain CK and auxin combination cause a cell or group of cells to begin differentiation into a root initial? Where do these molecules act within the cell? Are there specific receptor sites and why should bases be more active than ribosides or glucosides? These problems do not, of course, touch only on the process of ARF, but on all CK induced responses within the cell. Nevertheless, it may be that intensive study of a particular rooting system, such as the bean leaf system, could shed light on many other CK related problems. Just as studies on *Drosophila* or *E. coli* extended the boundaries of genetics and microbiology respectively, so the problem of ARF might best be approached by concentrating initially on a single species.

ACKNOWLEDGMENT

The financial support of the CSIR is gratefully acknowledged.

REFERENCES

Bachelard, E. P. and B. B. Stowe. 1963. Rooting of cuttings of *Acer rubrum* L. and *Eucalyptus camaldulensis* Dehn. *Aust. J. Biol. Sci.* 16:751–767.

Batten, D. J. and P. B. Goodwin. 1978. Phytohormones and growth and development of organs of the vegetative plant: Phytohormones and the induction of adventitious roots. In *Phytohormones and Related Compounds: A Comprehensive Treatise* (D. S. Letham, P. B. Goodwin and T. J. V. Higgins, eds). Elsevier/North-Holland, Amsterdam, The Netherlands. Vol. II. pp. 137–173. ISBN 0-444-80054-9.

Berry, J. B. 1984. Rooting hormone formulations: a chance for advancement. *Proc. Int. Plant Prop. Soc.* 34:486–491.

Biddington, N. L. and A. S. Dearman. 1982/83. The involvement of the root apex and cytokinins in the control of lateral root emergence in lettuce seedlings. *Plant Growth Regul.* 1:183–193.

Bollmark, M. and L. Eliasson. 1986. Effects of exogenous cytokinins on root formation in pea cuttings. *Physiol. Plant.* 68:662–666.

Bonnet, H. T. and J. G. Torrey. 1965. Chemical control of organ formation in root segments of *Convolvulus* cultured *in vitro*. *Plant Physiol.* 40:1228–1236.

Böttger, M. 1974. Apical dominance in roots of *Pisum sativum* L. *Planta* 121:253–261.

Bridglall, S. S. and J. Van Staden. 1984. Effect of auxin on rooting and endogenous cytokinin levels in leaf cuttings of *Phaseolus vulgaris* L. *J. Plant Physiol.* 117:287–292.

Carlson, W. C. and M. M. Larson. 1977. Changes in auxin and cytokinin activity in roots of red oak, *Quercus rubra*, seedlings during lateral root formation. *Physiol. Plant.* 41:162–166.

Chandra, G. R., J. F. Worley, L. E. Gregory and H. D. Clark. 1973. Effect of 6-benzyladenine on the initiation of adventitious roots on mung bean hypocotyl. *Plant Cell Physiol.* 14:1209–1212.

Chen, C. and B. Petschow. 1978. Cytokinin biosynthesis in cultured rootless tobacco plants. *Plant Physiol.* 62:861–865.

Danckwardt-Lillieström, C. 1957. Kinetin-induced shoot formation from isolated roots of *Isatis tinctoria*. *Physiol. Plant.* 10:794–797.

David, H., K. Isemukali and A. David. 1978. Obtention de plants de pin maritime (*Pinus pinaster* Sol) a pàrtir de brachyblastes ou d'apex caulinaires de très jeunes sujets cultivés *in vitro*. *C. R. Acad. Sci. Paris*, Ser. D 287:245–248.

Delegher-Langohr, V. 1974. Influence of five growth substances on root formation by cuttings of *Zebrina pendula*. *Bull. Soc. Royal Bot. Belg.* 107:73–89.

Dickinson, J. R., C. Forsyth and J. Van Staden. 1986. The role of adenine in the synthesis of cytokinins in tomato plants and in cell-free root extracts. *Plant Growth Regul.* 4:325–334.

Dirr, M. A. 1981. Rooting compounds and their use in plant propagation. *Proc. Int. Plant Prop. Soc.* 31:472–479.

Einset, J. W. 1977. Two effects of cytokinin on the auxin requirement of tobacco callus cultures. *Plant Physiol.* 59:45–47.

Engelbrecht, L. 1972. Cytokinins in leaf-cuttings of Phaseolus vulgaris L. during their development. Biochem. Physiol. Pflanzen 163:335–343.

Eriksen, E. N. 1974. Root formation in pea cuttings. III. The influence of cytokinin at different developmental stages. Physiol. Plant. 30:163–167.

Featonby-Smith, B. C. and J. Van Staden. 1981. Endogenous cytokinins and rooting of leaf cuttings of Phaseolus vulgaris L. Z. Pflanzenphysiol. 102:329–335.

_____ _____ 1984. Identification and seasonal variation of endogenous cytokinins in Ecklonia maxima (Osbeck) Papenf. Bot. Marina 27:527–531.

Feldman, L. J. 1975. Cytokinins and quiescent centre activity in roots of Zea. In The Development and Function of Roots (J. G. Torrey and D. T. Clarkson, eds). Academic Press, London, England. pp. 55–72. ISBN 0-12-695750-9.

Fellenberg, G. 1969. Der Einfluss von Gibberellinsäure und Kinetin auf die Auxininduzierte Wurzelneubidung und auf das Nucleoproteid von Erbsenepikotylen. Z. Pflanzenphysiol. 60:457–466.

Finnie, J. F. and J. Van Staden. 1985. Effect of seaweed concentrate and applied hormones on in vitro cultured tomato roots. J. Plant. Physiol. 120:215–222.

Forsyth, C. and J. Van Staden. 1981. The effects of root decapitation on lateral root formation and cytokinin production in Pisum sativum cultivar Kalvedon Wonder. Physiol. Plant. 51:375–379.

_____ _____ 1986. The metabolism and cell division activity of adenine derivatives in soybean callus. J. Plant Physiol. 124:275–287.

Fridborg, G. 1971. Growth and organogenesis in tissue cultures of Allium cepa var. proliferum. Physiol. Plant. 25:436–440.

Gaspar, T. and A. Xhaufflaire. 1967. Effect of kinetin on growth, auxin metabolism, peroxidase and catalase activities. Planta 72:252–257.

Golaz, F. W. and P. Pilet. 1985. Light and decapitation effects on in vitro rooting in maize root segments. Plant. Physiol. 79:377–380.

Goodwin, P. B. 1978. Phytohormones and growth and development of organs of the vegetative plant. In Phytohormones and Related Compounds: A Comprehensive Treatise (D. S. Letham, P. B. Goodwin and T. J. V. Higgins, eds). Elsevier/North-Holland, Amsterdam, The Netherlands. Vol. II. pp. 31–173. ISBN 0-444-80054-9.

_____ S. C. Morris. 1979. Applications of phytohormones to pea roots after removal of the apex: effect on lateral root production. Aust. J. Plant Physiol. 6:195–200.

Gresshoff, P. M. 1978. Phytohormones and growth and differentiation of cells and tissues cultured in vitro. In Phytohormones and Related Compounds: A Comprehensive Treatise (D. S. Letham, P. B. Goodwin and T. J. V. Higgins, eds). Elsevier/North-Holland, Amsterdam, The Netherlands. Vol. II. pp. 1–29. ISBN 0-444-8054-9.

Gur, A., A. Altman, R. Stern and B. Wolowitz. 1986. Improving rooting and survival of softwood peach cuttings. Scientia Hortic. 30:97–108.

Halperin, W. 1969. Morphogenesis in cell cultures. Ann. Rev. Plant Physiol. 20:395–418.

Hartmann, H. T. and D. E. Kester. 1983. Plant Propagation: Principles and Practices. Prentice Hall, Englewood Cliffs, NJ, USA. 4th ed. pp. 235–297. ISBN 0-13-680984-7.

Hendry, N. S., J. Van Staden and P. Allan. 1982a. Cytokinins in citrus. I. Fluctuations in the leaves during seasonal and developmental changes. Scientia Hortic. 16:9–16.

_____ _____ _____ 1982b. Cytokinins in citrus. II. Fluctuations during growth in juvenile and adult plants. Scientia Hortic. 17:247–256.

Hess, C. E. 1965. Phenolic compounds as stimulators of root initiation. Plant Physiol. (Suppl.) 40:XLV.

Heuser, C. W. and C. E. Hess. 1972. Isolation of three lipid root-initiating substances from juvenile Hedera helix shoot tissue. J. Amer. Soc. Hortic. Sci. 97:571–574.

Hewett, E. W. and P. F. Wareing. 1973a. Cytokinins in Populus × robusta: changes during chilling and bud burst. Physiol. Plant. 28:393–399.

_____ _____ 1973b. Cytokinins in Populus × robusta: changes during development. Physiol. Plant 29:386–389.

Humphries, E. C. 1960. Inhibition of root development on petioles and hypocotyls of dwarf bean (*Phaseolus vulgaris*) by kinetin. *Physiol. Plant.* 13:659–663.

Jain, M. L., P. G. Kadkade and P. Huysse. 1969. The effect of growth regulatory chemicals on abscission and IAA-oxidizing enzyme system of dwarf bean seedlings. *Physiol. Plant.* 51:399–401.

Kaminek, M. 1968. Dynamics of amino acids in pea stem sections during root formation and its inhibition by kinetin and ethionine. *Biol. Plant.* 10:462–471.

Kannangara, T. and A. Booth. 1974. Diffusable cytokinins in shoot apices of *Dahlia variabilis*. *J. Exp. Bot.* 25:459–467.

Kriesel, K. 1976. Activity of cytokinin-like substances in the development of buds, newly formed shoots and adventitious roots of willow cuttings. *Bull. Acad. Pol. Sci. Ser. Sci. Biol.* 14:299–302.

Lee, A. E. 1959. The effect of various substances on the comparative growth of excised tomato roots of clones carrying dwarf and normal alleles. *Amer. J. Bot.* 46:16–21.

Lee, T. T. 1971. Cytokinin controlled indoleacetic acid oxidase isoenzymes in tobacco callus cultures. *Plant Physiol.* 47:181–185.

_____ 1974. Cytokinin control in subcellular localization of IAA-oxidase and peroxidase. *Phytochemistry* 13:2445–2454.

Letham, D. S. 1978. Cytokinins. In *Phytohormones and Related Compounds: A Comprehensive Treatise* (D. S. Letham, P. B. Goodwin and T. J. V. Higgins, eds). Elsevier/North-Holland, Amsterdam, The Netherlands. Vol. I. pp. 205–263. ISBN 0-444-80053-0.

_____ L. M. S. Palni. 1983. The biosynthesis and metabolism of cytokinins. *Ann. Rev. Plant Physiol.* 34:163–197.

_____ _____ G. Tao, B. I. Gollnow and C. M. Bates. 1983. Regulators of cell division in plant tissues: XXIX. The activities of cytokinin glucosides and alanine conjugates in cytokinin bioassays. *J. Plant Growth Regul.* 2:103–115.

McCully, M. E. 1975. The development of lateral roots. In *The Development and Function of Roots* (J. G. Torrey and D. T. Clarkson, eds). Academic Press, London, England. pp. 105–124. ISBN 0-12-695750-9.

Maldiney, R., F. Pelese, G. Pilate, B. Sotta, L. Sossountzov and E. Miginiac. 1986. Endogenous levels of abscisic acid, indole-3-acetic acid, zeatin and zeatin-riboside during the course of adventitious root formation in cuttings of Craigella and Craigella lateral suppressor tomatoes. *Physiol. Plant.* 68:426–430.

Meredith, W. C., J. N. Joiner and R. H. Biggs. 1970. Influences of indole-3-acetic acid and kinetin on rooting and indole metabolism of *Feijoa sellowiana*. *J. Amer. Soc. Hortic. Sci.* 95:49–52.

Minocha, S. C. 1984. The role of benzyladenine in the differentiation of tracheary elements in Jerusalem artichoke tuber explants cultured *in vitro*. *J. Exp. Bot.* 35:1003–1015.

Mohammed, S. 1980. Root formation in pea cuttings: effects of combined applications of auxin and cytokinin at different development stages. *Biol. Plant.* 22:231–236.

Mothes, K. and L. Engelbrecht. 1956. Über den Stickstoffumsatz in Blattstecklingen. *Flora* 143:428–472.

Nakano, M., E. Yuda and S. Nakagawa. 1980. Studies on rooting of the hardwood cuttings of grapevine cultivar Delaware. II. Hormonal changes in the cuttings. *J. Jap. Soc. Hortic. Sci.* 48:385–394.

Nandi, S., G. Fridborg and T. Eriksson. 1977. The effects of 6-(3-methyl-2-buten-1-ylamino)-purine and α-naphthaleneacetic acid on root formation and cytology of root tips and callus in tissue cultures of *Allium cepa* var. *proliferum*. *Hereditas* 85:57–62.

Nemeth, G. 1979. Benzyladenine-stimulated rooting in fruit-tree rootstocks cultured *in vitro*. *Z. Pflanzenphysiol.* 95:389–396.

Noor-Saleh, A. 1981. The effect of kinetin on the indoleacetic acid level and indoleacetic acid oxidase activity in roots of young plants. *Physiol. Plant.* 51:399–401.

_____ T. Hemberg. 1980. The influence of kinetin on the endogenous content of indoleacetic acid in swelling seeds of *Phaseolus, Zea* and *Pinus* and young plants of *Phaseolus*. *Physiol. Plant.* 50:99–102.

Okoro, O. O. and J. Grace. 1978. The physiology of rooting *Populus* cuttings. II. Cytokinin activity in leafless hardwood cuttings. *Physiol. Plant.* 44:167–170.

Palmer, M. V., R. Horgan and P. F. Wareing. 1981. Cytokinin metabolism in *Phaseolus vulgaris* L. I. Variations in cytokinin levels in leaves of decapitated plants in relation to lateral bud outgrowth. *J. Exp. Bot.* 32:1231–1241.

Raviv, M. and O. Reuveni. 1984. Mode of leaf shedding from avocado cuttings and the effect of its delay on rooting. *HortScience* 19:529–531.

Reid, D. M. and W. J. Burrows. 1968. Cytokinin and gibberellin-like activity in the spring sap of trees. *Experientia* 14:189–190.

Reuveni, O. and M. Raviv. 1981. Importance of leaf retention to rooting of avocado cuttings. *J. Amer. Soc. Hortic. Sci.* 106:127–130.

Ryan, G. F., E. F. Frolich and T. P. Kinsella. 1958. Some factors influencing rooting of grafted cuttings. *Proc. Amer. Soc. Hortic. Sci.* 72:454–461.

Scott, T. K. 1972. Auxins and roots. *Ann. Rev. Plant Physiol.* 23:235–258.

Short, K. C. and J. G. Torrey. Cytokinins in seedling roots of pea. *Plant Physiol.* 49:155–160.

Skene, K. G. M. 1972. Cytokinins in the xylem sap of grape vine canes: changes in activity duṛ cold storage. *Planta* 104:89–92.

Skoog, F. and D. J. Armstrong. 1970. Cytokinins. *Ann. Rev. Plant Physiol.* 21:359–384.

_____ C. O. Miller. 1957. Chemical regulation of growth and organ formation in plant tissues cultured *in vitro*. *Symp. Soc. Exp. Biol.* 11:118–131.

_____ D. J. Armstrong, J. D. Cherayil, A. E. Hampel and R. M. Bock. 1966. Cytokinin activity: localization in transfer RNA preparations. *Science* 154:1354–1356.

Smith, D. R. and T. A. Thorpe. 1975. Root initiation in cuttings of *Pinus radiata* seedlings. I. Developmental sequence. *J. Exp. Bot.* 26:184–192.

Sriskandarajah, S., M. G. Mullins and Y. Nair. 1982. Induction of adventitious rooting *in vitro* in difficult-to-propagate cultivars of apple. *Plant Sci. Lett.* 24:1–9.

Stenlid, G. 1982. Cytokinins as inhibitors of root growth. *Physiol. Plant.* 56:500–506.

Svenson, S. 1972. A comparative study of the changes in root growth, induced by coumarin, auxin, ethylene, kinetin and gibberellic acid. *Physiol. Plant.* 26:115–135.

Syono, K. and T. Furaya. 1972. Effects of cytokinins on the auxin requirement and auxin content of tobacco calluses. *Plant Cell Physiol.* 13:843–856.

Takeno, K., J. S. Taylor, S. Sriskandarajah, R. P. Pharis and M. G. Mullins. 1982/83. Endogenous gibberellin- and cytokinin-like substances in cultured shoot tissues of apple, *Malus pumila* cv. Jonathan, in relation to adventitious root formation. *Plant Growth Regul.* 1:261–268.

Torrey, J. G. 1962. Auxin and purine interactions in lateral root initiation in isolated pea root segments. *Physiol. Plant.* 15:177–185.

_____ 1976. Root hormones and plant growth. *Ann. Rev. Plant Physiol.* 27:435–459.

Van Overbeek, J. and L. E. Gregory. 1945. A physical separation of two factors necessary for the formation of roots on cuttings. *Amer. J. Bot.* 32:336–341.

_____ S. A. Gordon and L. E. Gregory. 1946. An analysis of the function of the leaf in the process of root formation in cuttings. *Amer. J. Bot.* 33:100–107.

Van Staden, J. 1976a. Occurrence of a cytokinin glucoside in the leaves and in honeydew of *Salix babylonica*. *Physiol. Plant.* 36:225–228.

_____ 1976b. Seasonal changes in the cytokinin content of *Gingko biloba* leaves. *Physiol. Plant.* 38:1–5.

_____ 1978. The effect of centrifugation on the distribution of cytokinins in *Salix babylonica* cuttings. *Z. Pflanzenphysiol.* 90:279–284.

_____ N. A. C. Brown. 1977. The effect of ringing on cytokinin distribution in *Salix babylonica* L. *Physiol. Plant.* 39:266–270.

_____ N. A. Choveaux. 1980. Cytokinins in internodal stem segments of *Salix babylonica*. *Z. Pflanzenphysiol.* 96:153–161.

_____ J. E. Davey. 1979. The synthesis, transport and metabolism of endogenous cytokinins. *Plant Cell Environ.* 2:93–106.

_____ _____ 1981. Seasonal changes in the levels of endogenous cytokinins in the willow *Salix* babylonica L. Z. *Pflanzenphysiol.* 104:53–59.

_____ G. G. Dimalla. 1981. The production and utilisation of cytokinins in rootless dormant almond shoots maintained at low temperature. Z. *Pflanzenphysiol.* 103:121–129.

_____ C. Forsyth. 1984. Adenine incorporation into cytokinins in aseptically cultured tomato roots. *J. Plant Physiol.* 117:249–255.

_____ _____ L. Bergman and S. von Arnold. 1986. Metabolism of benzyladenine by excised embryos of *Picea abies. Physiol. Plant.* 66:427–434.

_____ H. Spiegelstein, N. Zieslin and A. H. Halevy. 1981a. Endogenous cytokinins and lateral bud growth in roses. *Bot. Gaz.* 142:177–182.

_____ N. Zieslin, H. Spiegelstein and A. H. Halevy. 1981b. The effect of light on the cytokinin content of developing rose shoots. *Ann. Bot.* 47:155–158.

Walker, K. A., P. C. Yu, S. J. Sato and E. G. Jaworski. 1978. The hormonal control of organ formation in callus of *Medicago sativa* L. cultured *in vitro. Amer. J. Bot.* 65:654–659.

Weiss, C. and Y. Vaadia. 1965. Kinetin-like activity in root apices of sunflower plants. *Life Sci.* 4:1323–1326.

Whitty, C. D. and R. H. Hall. 1974. A cytokinin oxidase in Zea mays. *Can. J. Biochem.* 52:789–799.

Wightman, F. and K. V. Thimann. 1980. Hormonal factors controlling the initiation and development of lateral roots. I. Sources of primordia-inducing substances in the primary root of pea seedlings. *Physiol. Plant.* 49:13–20.

_____ E. A. Schneider and K. V. Thimann. 1980. Hormonal factors controlling the initiation and development of lateral roots. II. Effects of exogenous growth factors on lateral root formation in pea roots. *Physiol. Plant* 49:304–314.

Wu, F. T. and M. F. Barnes. 1981. The hormone levels in stem cuttings of difficult-to-root and easy-to-root rhododendrons. *Biochem. Physiol. Pflanz.* 176:13–22.

Zimmerman, R. H. 1981. Micropropagation of fruit plants. *Acta Hortic.* 120:217–222.

CHAPTER 15

Polyamines and Adventitious Root Formation

Narendra Sankhla and Abha Upadhyaya

Department of Botany
University of Jodhpur
Jodhpur, INDIA 342001

INTRODUCTION..202
POLYAMINE METABOLISM..203
EXOGENOUS POLYAMINES AND ROOTING OF CUTTINGS 205
POLYAMINES AND ROOT GROWTH 206
POLYAMINES AND *IN VITRO* ROOTING 207
POSSIBLE MODES OF POLYAMINE ACTION IN ROOTING 207
CONCLUSION 208
ACKNOWLEDGMENTS 209
REFERENCES..209

Additional key words: growth regulators, putrescine, spermidine, spermine.
Abbreviations: ABA, abscisic acid; ADC, arginine decarboxylase (EC 4.1.1.19); DAO, diamine oxidase (EC 1.4.3.6); DCHA, dichlorohexyl amine; DFMA, difluoromethyl arginine; DFMO, difluoromethyl ornithine; GA_3, gibberellic acid; HEH, β-hydroxylethyl-hydrazine; IBA, indole-3-butyric acid; MGBG, methylglyoxal-bis(guanyl-hydrazone); ODC, ornithine decarboxylase (EC 4.1.1.17), PAO, polyamine oxidase (EC 1.4.3.4); SAM, S-adenosyl methionine.

INTRODUCTION

Intense mitotic activity, coupled to elaborate metabolic changes, appear to be essential prerequisites for root formation (Haissig 1986, Torrey 1986). A plethora of endogenous physiological factors have been shown to influence root formation (Jackson 1986; as well as this volume). Among the endogenous growth regulators, auxin is generally credited as the primary trigger for root initiation (Torrey 1986), although cytokinins may also be of importance. The influence of primary triggers may be considerably modified by interactions between other endogenously occurring factors, such as polyamines.

Polyamines are low molecular weight, aliphatic, nitrogenous polycations (Fig. 1). They occur ubiquitously in living organisms in free or conjugated form. The most commonly occurring polyamines are spermidine (a triamine), spermine (a tetramine), and putrescine (a diamine). Cadaverine, another diamine, is found in some plants; numerous other polyamines also occur in algae and higher plants (Slocum et al. 1984, Smith 1985a). In addition, conjugates of polyamines with hydroxycinnamic acids occur widely in plants, especially in flowers (Martin-Tanguy 1985).

Antonie van Leeuwenhoek first described the crystals of spermine phosphate in human semen over 300 yr ago. In comparison, the importance of polyamines in growth of *Helianthus tuberosus* L. explants was elucidated relatively recently (Bertossi et al. 1965, Bagni 1966). During the past 20 yr there has been a great surge of interest in polyamine research. Extensive studies indicate that these versatile polycations may play vital roles as modulators in a variety of biological processes (Bagni et

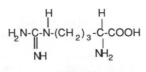

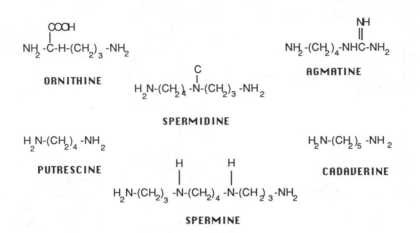

Figure 1. Chemical structures of some common polyamines.

al. 1982, Galston and Kaur-Sawhney 1982, 1987, Smith 1982, Galston 1983, Upadhyaya and Sankhla 1983a, Slocum et al. 1984, Flores et al. 1985, Galston and Smith 1985, Smith 1985ab, Sankhla et al. 1987). These processes include seed germination and dormancy (Bagni et al. 1980), cell division, elongation, and differentiation (Heimer et al. 1979, Palavan and Galston 1982, Schwartz et al. 1986), fruit growth and development (Cohen et al. 1982, Costa and Bagni 1983, Winer and Apelbaum 1986), and senescence of protoplasts and excised leaves (Altman et al. 1977, Galston et al. 1978, Kaur-Sawhney and Galston 1979). Conclusions based on the effect of exogenous polyamines, their precursors, analogues, and metabolic inhibitors, as well as correlations between endogenous titers of polyamines and ADC and ODC activities, strongly implicate these compounds as essential for cell growth and as intermediates in cellular responses to phytohormones and other external stimuli (Bagni et al. 1982, Galston and Kaur-Sawhney 1982).

Although little is known about the role of polyamines in the induction and development of adventitious roots in cuttings, it seemed appropriate to document and highlight the known effects of polyamines in rooting of plants because these substances have now been credited as representing a new group of endogenous growth factors (Bagni et al. 1982, Galston and Kaur-Sawhney 1987), and elucidation of polyamine action in plant growth is rapidly progressing (Smith 1985a, Sankhla et al. 1987). We hope that this chapter will stimulate interest and further research regarding polyamines and rooting.

POLYAMINE METABOLISM

Before examining the specific effects of polyamines on rooting it is important to understand how polyamines are synthesized and metabolized. In plants, two pathways are known for putrescine biosynthesis: Either arginine may be decarboxylated by ADC to form agmatine, which is then hydrolyzed to putrescine (Adiga and Prasad 1985, Smith 1985a), or ornithine may be formed from arginine by the loss of urea (Fig. 2). Ornithine is then decarboxylated by ODC to form putrescine. Both ADC and ODC have now been extensively studied and characterized from a wide range of plants. ADC is universally present in higher plants and is widespread in bacteria but not animals.

The diamine, cadaverine, may be formed via a separate pathway involving decarboxylation of

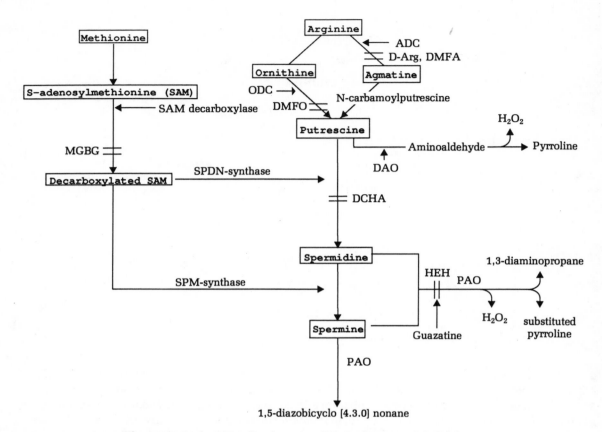

Figure 2. Pathways of polyamine biosynthesis and degradation.

lysine (Bachrach 1973, Smith 1977); the polyamines spermidine and spermine are derived from putrescine by the addition of aminopropyl groups donated by decarboxylated SAM.

The availability of the enzyme activated polyamine biosynthesis inhibitors DFMO (ODC inhibitor) and DFMA (ADC inhibitor) (Fig. 2) have greatly facilitated the elucidation of polyamine functions in plants (Pegg and McCann 1982, Fienberg et al. 1984, Rajam and Galston 1985, Smith et al. 1985a). Inhibitors are also available to block the decarboxylation of SAM which is required for formation of spermidine and spermine. The enzyme SAM decarboxylase (EC 4.1.1.50) can be inhibited by MGBG, but this inhibitor is not specific. Dichlorohexylamine, which suppresses spermidine levels in animals and bacteria (Bitonti et al. 1982), is an effective inhibitor of spermidine synthase in plants (Cohen et al. 1985, Biondi et al. 1986).

Some evidence implicates growth hormones and phytochrome in regulation of polyamine biosynthesis (Slocum et al. 1984). For instance, red light treatment decreased ADC activity and elongation in pea (*Pisum* sativum L.) internodes, while it increased ADC activity and growth in the terminal bud. Both responses were photoreversible, and ADC activity and polyamine titers corresponded to changes in the growth rates of stem tissues (Dai and Galston 1981, Goren et al. 1982).

Polyamine biosynthesis appears to be a prerequisite for auxin mediated growth of *Helianthus tuberosus* L. (Bagni et al. 1980), parthenocarpic development of tomato (*Lycopersicon esculentum* Mill.) fruit (Cohen et al. 1982), and adventitious root formation (Friedman et al. 1982, Jarvis et al. 1983a). In some systems, GA_3 has been reported to activate polyamine biosynthesis. During GA_3 promotion of elongation of dwarf pea internodes, a simultaneous increase was observed in polyamine titers and ADC activity (Dai et al. 1982). Putrescine synthesis also increased greatly following GA_3 treatment in lettuce (*Latuca sativa* L.) hypocotyls (Cho 1983). In germinating barley seedlings both GA_3 and IAA treatment brought about a 4-fold increase in ODC activity (Kyriakidis 1983).

Cytokinins have also been found to influence polyamine biosynthesis. Putrescine synthesis is enhanced in cotyledons of light grown lettuce seedlings by kinetin and benzyladenine (Cho 1983). In cucumber cotyledons both ADC activity and putrescine content increased in response to benzyladenine treatment (Suresh et al. 1978), whereas ABA proved inhibitory. However, in the latter test, ABA induced inhibition of polyamine synthesis was reversible by cytokinins.

The degradation of polyamines in plants may be accomplished by DAOs or PAOs (Smith 1985a) (Fig. 2). DAO is a copper containing enzyme with a broad substrate specificity, and occurs widely in the Leguminosae; PAO is highly specific for spermidine and spermine, and has been found only in the Gramineae (Smith 1985c).

Inhibitors of polyamine catabolism are also available. For instance, HEH inhibits PAO (Kaur-Sawhney et al. 1981, Shih et al. 1982) and suppresses conversion of spermidine or spermine into 1,3-diaminopropane. The synthetic fungicide guazatine has also recently been found to be a powerful inhibitor of PAO (Smith 1985c).

EXOGENOUS POLYAMINES AND ROOTING OF CUTTINGS

At present, the effect of exogenous application of polyamines on rooting has almost exclusively been studied using mung bean [Vigna radiata (L.) R. Wilcz.] hypocotyl cuttings (Friedman et al. 1982, 1985, Jarvis et al. 1983a, Shyr and Kao 1985). Problems and limitations associated with using mung bean rooting bioassays have been discussed previously (Blazich and Heuser 1979; see also Chapter 19 by Heuser, and Chapter 1 by Hackett). Nevertheless, studies with mung beans have provided some interesting information regarding polyamines and adventitious rooting.

Friedman et al. (1982) used Vigna radiata (Syn. Phaseolus aureus Roxb.) seedlings for their experiments, and found that exogenously applied polyamines (putrescine, spermidine, and spermine) did not promote adventitious rooting. In fact, putrescine and spermine slightly inhibited rooting, and at higher concentrations were toxic. These authors also studied the effect of polyamine precursors (L-arginine and L-ornithine), and their analogues (L-canavanine and L-canaline), as well as effects of MGBG on the rooting of mung bean cuttings in the presence of IBA. MGBG blocks spermidine and spermine biosynthesis (Corti et al. 1974). They observed that L-canavanine and L-canaline inhibited rooting, but the effect could be counteracted by applying L-arginine or L-ornithine, respectively. MGBG also significantly inhibited rooting. The endogenous titer of polyamines was found to be higher in IBA treated than in untreated cuttings. The changes in polyamines were dependent on IBA concentration and were organ-specific, which implicated biosynthesis of polyamines as a necessity for IBA induced rooting.

Jarvis et al. (1983a) evaluated the involvement of polyamines with adventitious root development in stem cuttings of mung bean (cv. Berkin). These authors reported for the first time that exogenously supplied spermine, alone or with IBA, had a stimulatory effect on adventitious rooting in the presence of boron. The spermine concentration $(7 \times 10^{-5} \text{ mol} \cdot \text{l}^{-1})$ most effective in inducing rooting also enhanced root growth. Spermidine, in contrast, did not influence root number or growth, except at high concentrations, where it decreased the number of roots. In addition, IBA treatment resulted in an enhancement in the contents of spermine, spermidine, and putrescine in the hypocotyl prior to development of any root primordia. Previously, Altman and Bachrach (1981) also found increases in putrescine and spermidine contents in hypocoytls of mung bean cuttings treated with IBA. However, these authors did not observe enhancement of rooting by exogenous application of spermine and spermidine. It appears that in light grown mung bean, boron is an essential requirement for IBA induced primordium initiation and subsequent root development (Middleton et al. 1978, Jarvis et al. 1983b), and the cuttings used by Altman and Bachrach (1981) possibly contained limiting levels of boron.

Studies with metabolic inhibitors such as MGBG lend further support to the contention that polyamines have an early involvement in adventitious rooting. MGBG not only inhibited rooting and root growth in the presence or absence of IBA, but reduced the levels of spermine and spermidine while increasing the level of putrescine in the hypocotyl prior to primordium formation (Altman and

Bachrach 1981). MGBG also prevented the IBA induced increase in spermine and markedly reduced spermidine content.

Jarvis et al. (1983a) observed that exogenously applied auxins were far superior to applied spermine in stimulating adventitious rooting in hypocotyl cuttings of mung bean. These authors suggested that auxin initiates rooting at an early stage and requires or is associated with increased polyamine levels. In mung bean, the level of spermine may be a key factor in determining the extent of development of adventitious roots. This hypothesis is further supported by the observation that ABA, which promoted rooting in mung bean cuttings, also enhanced the level of spermine (Jarvis et al. 1983a).

Shyr and Kao (1985) evaluated the effect of polyamines on rooting in hypocotyl cuttings of mung bean (cv. Tainan Select 3). In this cultivar, exogenous application of polyamines, as well as the polyamine precursors L-arginine and L-ornithine promoted rooting. Involvement of polyamines in rooting was further established from the fact that analogues of polyamines and MGBG also inhibited rooting. Even HEH, which inhibits PAO activity (Kaur-Sawhney et al. 1981, Shih et al. 1982), and hence conversion of spermidine or spermine into 1,3-diaminopropane, also effectively promoted rooting.

In light grown mung bean cuttings boron appears to be a prerequisite for rooting (Middleton et al. 1978) which, in the presence of IBA, is also stimulated by leaves on the cuttings (Middleton et al. 1980). In conformity with the above, Jarvis et al. (1983a) observed that spermine increased the number of roots in mung bean cuttings only in the presence of boron. In leafy cuttings of the mung bean cultivar 'Tainan Select 3' promotion of rooting by applied polyamines or IBA did not depend on the exogenous supply of boron (Shyr and Kao 1985), which suggested that the cuttings contained sufficient endogenous boron.

POLYAMINES AND ROOT GROWTH

Several lines of evidence indicate that there may be relations between root growth and polyamine content. In this regard, Walker et al. (1985) have demonstrated a requirement for polyamines during the cell division phase of radicle emergence. Polyamine titer and activity of associated biosynthetic enzymes have also been found to be highest in rapidly growing tissues such as root apices and germinating seeds, and during onset of growth (Bagni 1970, Villanueva et al. 1978, Galston and Kaur-Sawhney 1980, Serafini-Fracassini et al. 1980, Shih and Galston 1982). Shen and Galston (1985) recently reported that putrescine content increased as root elongation progressed; the putrescine/spermine ratio mirrored the sigmoidal growth curve until the initiation of laterals. In roots of maize (Zea mays L.), the gradient of polyamine content also paralleled the activity gradient of ADC and ODC (Dumortier et al. 1983).

Schwartz et al. (1986) studied the localization of ODC and changes in polyamine content in root meristems of maize. Lateral rooting was accompanied by an initial decline in putrescine content and by a significant increase in spermidine content. The root apex also contained high levels of spermidine and lower levels of putrescine. During lateral rooting a marked enhancement was also noticed in ADC and ODC activity. It has been postulated that ODC activity is correlated with cell division, whereas changes in ADC activity typically accompany a stress response (Cohen et al. 1982, Smith et al. 1985a, Young and Galston 1984). In contrast, the observations of Schwartz et al. (1986) showed that both pathways for putrescine biosynthesis may operate and respond to growth stimuli as shown earlier (Altman et al. 1983).

Comparative localization of active ODC and of meristematic activity in roots with α-[5^{14}C]-difluoromethyl ornithine or with [^{3}H]-thymidine has been accomplished using techniques adapted from animal systems (Henningsson et al. 1983, Zagon et al. 1983). Schwartz et al. (1986) have demonstrated a definite in situ correspondence between the site of ODC activity and the zone of cells that undergo mitosis in the lateral roots as well as in the apex of the primary root.

POLYAMINES AND *IN VITRO* ROOTING

Differentiation is an inherent process in the development of organized plant tissue. Organogenesis *in vitro* is a function of a delicate but complex interaction between exogenous and endogenous growth regulators in the cells at the time of organ formation (Chandler and Thorpe 1986). Recent evidence indicates that alterations in polyamine synthesis and titer are also involved in the differentiation of tissue cultures.

Malfatti et al. (1983) reported that in tobacco callus the rates of polyamine biosynthesis were extremely low, and that arginine and other amino acids accumulated. In contrast, when the tissue was induced to form roots by auxin supplementation, putrescine synthesis increased markedly.

Kaur-Sawhney et al. (1985) measured polyamine levels as related to growth, differentiation, and senescence in protoplast derived cultures. In *Vigna* spp. cultures, a dramatic increase in the level of free and bound putrescine and spermidine was correlated with cell division and progress of differentiation. The increase of free putrescine and spermidine titer was highest in root forming callus, whereas bound polyamines were highest in embryoid forming callus tissue. DMFA, DMFO, and DCHA, inhibitors of polyamine biosynthesis, not only inhibited cell division, but also arrested organogenesis. Low concentrations of exogenous polyamines restored these processes.

In *Picea abies* (L.) Karst. callus lines originating from megagametophytes, polyamines did not enhance growth (Simola and Honkanen 1983), but a specific combination of putrescine, spermidine, and spermine favored development of roots.

Desai and Mehta (1985) monitored the changes in polyamine levels during shoot formation, root formation, and callus induction in cultured *Passiflora* spp. leaf discs. Putrescine level increased severalfold before the appearance of shoots and roots, and also exhibited a sizeable promotion during callus induction. In contrast, increases in spermidine content were comparatively very slow, and no significant change was observed in spermine content during these events. Concurrently, an increase was also observed in the activity of agmatine aminohydrolase (EC 3.5.3.11), an enzyme involved in putrescine biosynthesis.

In "Hiproly" barley (*Hordeum vulgare* L.) callus, rooting after auxin withdrawal (Katoh et al. 1985) was associated with increased levels of arginine and methionine. Exogenous application of these amino acids also resulted in a strong stimulation of rooting. The levels of spermine and putrescine dramatically increased during root formation. Addition of polyamines to the medium stimulated root formation. These results indicated that in "Hiproly" callus arginine and methionine are probably closely associated with the cellular functions of polyamines. During rooting, the ADC activity remained almost constant, but ODC activity increased considerably, which likely accounted for the elevated putrescine levels. The enhancement of putrescine biosynthesis may, thus, be significant for rooting in "Hiproly" barley callus.

POSSIBLE MODES OF POLYAMINE ACTION IN ROOTING

With specific regard to adventitious rooting, little is known about the mode of action of polyamines. However, physiological and biochemical bases of polyamine action can be inferred from studies that did not involve adventitious rooting.

Many of the biological functions of polyamines have been ascribed to their roles in the regulation of macromolecular synthesis, structure and function of nucleic acids, properties of membranes, and modulation of enzyme activities (Slocum et al. 1984, Smith 1985a). By binding electrostatically with the anionic phosphate groups of nucleic acids, these versatile cations can function as structural reinforcement, and influence transcription and translation (Smith 1985a, Galston and Kaur-Sawhney 1987), both of which are presumably needed for adventitious rooting. In a variety of systems, a close correspondence has been demonstrated between increased polyamine biosynthesis and titers, and enhanced protein and nucleic acid synthesis (Slocum et al. 1984). In addition to macromolecular synthesis and mitosis, polyamines have been shown to interact with membranes by binding to negatively charged phospholipids or other anionic sites which results in an alteration of membrane stability and permeability (Slocum et al. 1984, Naik and Srivastava 1978, Sankhla et al. 1985).

Recently, the role of polyamines in modulating the activities of various enzymes, including membrane bound enzymes (Srivastava and Rajbabu 1983a, Upadhyaya and Sankhla 1983ab, Gaspar et al. 1984), has been further resolved (Slocum et al. 1984, Smith 1985a). The large accumulation of the diamine putrescine during potassium deficiency, and under a variety of stress conditions (Flores et al. 1985, Smith 1984, 1985a), implies that polyamines, at physiological pH, may serve as counterions and play an important role in maintaining cellular homeostasis. Whether these polyamine actions are involved in rooting is unclear.

Enhanced ethylene formation has been implicated in the flooding induced adventitious rooting of many intact plant species (Cohen and Kende 1986, see Chapter 11 by Mudge). Hence, it is interesting that polyamines and ethylene apparently have antagonistic effects on senescence and share the common intermediate, SAM, in their biosynthesis (Slocum et al. 1984). It has been suggested that ethylene and polyamine formation impose competitive demands on SAM; and allocation of this metabolite to either pathway may be a point of regulation in ethylene and polyamine biosynthesis (Even-Chen et al. 1982, Roberts et al. 1984). In fact, applied polyamines have inhibited ethylene formation in apple (*Malus pumila* L.) fruit tissue and protoplasts, bean and tobacco (*Nicotiana tabacum* L.) leaf discs (Apelbaum et al. 1981), *Tradescantia* spp. petals, and mung bean hypocotyls (Suttle 1981). Likewise, applied ethylene reduced ADC and SAMDC activities (Apelbaum et al. 1985c, Icekson et al. 1985). In this context, submergence of plants, which may lead to enhanced ethylene formation (Raskin and Kende 1984, Metraux and Kende 1983), may also lead to increased ADC and SAMDC activities, accumulation of polyamines, and adventitious rooting in the nodes.

In a recent study, gibberellin biosynthesis inhibitors have been shown to be associated with modulations of polyamine levels during rooting (Wang and Faust 1986). In apple seedlings, ancymidol, chlormequat chloride, and paclobutrazol reduced shoot extension, promoted root initiation, and increased root weight. The induction of rooting and increase in root weight were accompanied by a considerable increase in polyamine levels. Further, these authors suggested that, in apple, ODC may play a more important role than ADC in controlling the metabolic processes correlated with rooting.

Polyamines may also serve as a nitrogen source (Bagni 1979, Bagni et al. 1981) because both spermine and spermidine can be enzymatically oxidized to yield ammonium ions (White et al. 1968, Haissig 1986). Recently, however, Srivastava et al. (1985) have shown that during radish (*Raphanus sativus* L.) germination and growth, polyamines do not serve as a nitrogen source because a comparable increase in growth was obtained even in nitrate-grown seeds. Instead, in radish, the polyamines regulated growth of the embryo during the early period of germination by promoting reserve protein mobilization rather than the nitrogen assimilation. Whether such an effect plays a role in the rooting of cuttings is unclear.

It is generally accepted that arginine, ornithine, and methionine act as precursors for the synthesis of polyamines (Smith 1985a). It is known that these amino acids have either a stimulatory influence on formation of roots or accumulate in the rooting zone of cuttings (Suzuki and Kohno 1983, Katoh et al. 1985). Addition of polyamines in the medium stimulated root formation as well as arginine and methionine titer in "Hiproly" callus. This points towards a possible relationship between these amino acids and polyamines in influencing root formation.

CONCLUSION

Although polyamines have been shown to promote rooting from the hypocotyl cuttings of mung bean under at least some circumstances, and their requirement has been demonstrated in root ontogeny, our current knowledge relating to polyamine action in rooting is limited. This is partly due to the lack in knowledge about the physiological and biochemical roles of polyamines themselves (Slocum et al. 1984, Smith 1985ab). In plants very little is known about the subcellular compartmentalization of polyamines or the localization of enzymes related to their synthesis (Schwartz et al. 1986, Torrigiani et al. 1986). Thus far the evidence implicating polyamines in the regulation of rooting is largely correlational and indirect. Such evidence does not provide unambiguous proof of a regulatory role for polyamines in rooting. Nevertheless, the range and variety of correlations between polyamines and rooting response provide a strong impetus for future research. A glaring research deficiency on polyamines and rooting is that nearly all work done thus far has been with mung bean. Characterization of rooting response to polyamines in a wide array of plant species needs to be undertaken before further elucidation of polyamine action in adventitious rooting can be achieved.

Valuable information can also be obtained by resolving the nature of interaction between polyamines and other growth regulators during rooting. In view of the fact that polyamines and ethylene antagonize each other during senescence and both compete for the common precursor, SAM, studies relating to polyamine-ethylene interactions may be quite rewarding in providing useful information on the role of these substances in rooting. In addition, such studies may also help in the further characterization of an important developmental switch in plants.

ACKNOWLEDGMENTS

This work was supported by grants from University Grants Commission, New Delhi, India and Brigham Young University, Provo, USA. The authors thank Drs. Tim D. Davis and Bruce E. Haissig for helpful comments during preparation of the manuscript.

REFERENCES

Adiga, P. R. and G. L. Prasad. 1985. Biosynthesis and regulation of polyamines in higher plants. *Plant Growth Regul.* 3:205–226.

Altman, A. and U. Bachrach. 1981. Involvement of polyamines in plant growth and senescence. In *Advances in Polyamine Research* (C. M. Calderera, V. Zappia and U. Bachrach, eds). Raven Press, NY, USA. Vol. III. pp. 365–375. ISBN 0-89004-621-2.

_____ R. Friedman and N. Levin. 1983. Alternative metabolic pathways for polyamine biosynthesis in plant development. In *Advances in Polyamine Research* (U. Bachrach, A. Kaye and R. Chayen, eds). Raven Press, NY, USA. Vol IV. pp. 395–408. ISBN 0-89004-890-8.

_____ R. Kaur-Sawhney and A. W. Galston. 1977. Stabilization of oat leaf protoplasts through polyamine-mediated inhibition of senescence. *Plant Physiol.* 60:570–574.

Apelbaum, A., A. C. Burgoon, J. D. Anderson and M. Lieberman. 1981. Polyamines inhibit biosynthesis of ethylene in higher plant tissue and fruit protoplasts. *Plant Physiol.* 68:453–456.

_____ A. Goldlust and I. Icekson. 1985. Control by ethylene of arginine decarboxylase activity in pea seedlings and its implication for hormonal regulation of plant growth. *Plant Physiol.* 79:635–640.

Bachrach, U. 1973. *Function of Naturally Occurring Polyamines.* Academic Press, NY, USA. ISBN 0-12-070650-4.

Bagni, N. 1966. Aliphatic amines and a growth-factor of coconut milk as stimulating cellular proliferation of *Helianthus tuberosus* (Jerusalem artichoke) *in vitro*. *Experientia.* 22:732.

_____ 1970. Metabolic changes of polyamines during the germination of *Phaseolus vulgaris*. *New Phytol.* 69:159–164.

_____ 1979. Polyamines as sole nitrogen source for *Helianthus tuberosus*. In *Nitrogen Assimilation in Plants* (E. J. Hewitt and C. V. Cuttings, eds). Academic Press, NY, USA. pp. 585–587. ISBN 0-12-346360-2.

_____ B. Malucelli and P. Torrigiani. 1980. Polyamines, storage substances and abscisic acid-like inhibitors during dormancy and very early activation of *Helianthus tuberosus* tuber tissues. *Physiol. Plant.* 49:341–345.

_____ D. Serafini-Fracassini and P. Torrigiani. 1981. Polyamines and growth in higher plants. In *Advances in Polyamine Research* (C. M. Calderera, V. Zappia and U. Bachrach, eds). Raven, NY, USA. Vol. III. pp. 377–388. ISBN 0-89004-890-8.

_____ _____ _____ 1982. Polyamines and cellular growth processes in higher plants. In *Plant Growth Substances* (P. F. Wareing, ed). Academic Press, London, England. p. 683.

Bertossi, F., N. Bagni, G. Moryzi and C. M. Caldarera. 1965. Spermine as a new growth-promoting substance for *Helianthus tuberosum* (Jerusalem artichoke) *in vitro*. *Experientia*. 21:80–81.

Bitonti, A. J., P. P. McCann and A. Sjoerdsma. 1982. Restriction of bacterial growth by inhibition of polyamine biosynthesis by using monofluromethylornithine, difluoromethylornithine and dicycloxylamine sulphate. *Biochem. J.* 208:435–441.

Blazich, F. A. and C. W. Heuser. 1979. The mung bean rooting bioassay: A re-examination. *J. Amer. Soc. Hortic. Sci.* 104:117–120.

Chandler, S. F. and T. A. Thorpe. 1986. Hormonal regulation of organogenesis *in vitro*. In *Hormonal Regulation of Plant Growth and Development* (S. S. Purohit, ed). Agro Bot. Pub. Bikaner, India. Vol. 3. pp. 1–28.

Cho, S. C. 1983. Enhancement by putrescine of gibberellin-induced elongation in hypocotyls of lettuce seedlings. *Plant Cell Physiol.* 24:305–308.

Cohen, E. and H. Kende. 1982. The effect of submergence, ethylene, and gibberellin on polyamines and their biosynthetic enzymes in deep-water internodes. *Planta*. 169:498–504.

Cohen, E., Y. U. Heimer and Y. Mizrahi. 1982. Ornithine decarboxylase and arginine decarboxylase activities in meristematic tissues of tomato and potato plants. *Plant Physiol.* 70:544–546.

Cohen, S. S., R. K. Sindhu, M. Greenberg, B. Yamanoha, R. Balint and K. McCarthy. 1985. Cellular systems for the study of the biosynthesis of polyamines and ethylene, as well as of virus multiplication. *Plant Growth Regul.* 3:227–238.

Corti, A., C. Dave, H. G. Williams-Ashman, E. Mihich and A. Schenone. 1974. Specific inhibition of the enzymic decarboxylation of S-adenosylmethionine by methylglyoxal bis (guanzyhydrazone) and related substances. *Biochem. J.* 139:351–357.

Costa, G. and N. Bagni. 1983. Effects of polyamines on fruitset of apple. *HortScience*. 18:59–61.

Dai, Y. R. and A. W. Galston. 1981. Simultaneous phytochrome-controlled promotion and inhibition of arginine decarboxylase activity in buds and epicotyls of eliolated peas. *Plant Physiol.* 67:266–269.

Dai, Y. R., R. Kaur-Sawhney and A. W. Galston. 1982. Promotion by gibberellic acid of polyamine biosynthesis in internodes of light-grown dwarf pea. *Plant Physiol.* 69:103–105.

Desai, H. V. and A. R. Mehta. 1985. Changes in polyamine levels during shoot formation, root formation, and callus induction in cultured *Passiflora* leaf discs. *J. Plant Physiol.* 119:45–53.

Dumortier, F. M., H. E. Flores, N. S. Shekhawat and A. W. Galston. 1983. Gradients of polyamines and their biosynthetic enzymes in coleoptiles and roots of corn. *Plant Physiol.* 72:915–918.

Even-Chen, Z., A. K. Mattoo and R. Goren. 1982. Inhibition of ethylene biosynthesis by aminoethoxyvinylglycine and by polyamines shunts label from 3,4-[^{14}C]methionine into spermidine in aged orange peel discs. *Plant Physiol.* 69:385–388.

Fienberg, A. A., J. H. Choi, W. P. Lubich and Z. R. Sung. 1984. Developmental regulation of polyamine metabolism in growth and differentiation of carrot culture. *Planta* 162:532–539.

Flores, H. E., N. D. Young and A. W. Galston. 1985. Polyamine metabolism and plant stress. In *Cellular and Molecular Biology of Plant Stress*. (J. L. Key and T. Kosuge, eds). Alan R. Liss, NY, USA. pp. 93–114. ISBN 0-8451-2621-0.

Friedman, R., A. Altman and U. Bachrach. 1982. Polyamines and root formation in mung bean hypocotyl cuttings. I. Effects of exogenous compounds and changes in endogenous polyamine content. *Plant Physiol.* 70:844–848.

_____ _____ _____ 1985. Polyamines and root formation in mung bean hypocotyl cuttings. II. Incorporation of precursors into polyamines. *Plant Physiol.* 79:80–83.

Galston, A. W. 1983. Polyamines as modulators of plant development. *BioScience* 33:382–388.

Galston, A. W., A. Altman and R. Kaur-Sawhney. 1978. Polyamines, ribonuclease and the improvement of oat leaf protoplasts. *Plant Sci. Lett.* 11:69–79.

Galston, A. W. and R. Kaur-Sawhney. 1982. Polyamines: are they a new class of plant growth regulator? *In Plant Growth Substances* (P. F. Wareing, ed). Academic Press, London, England. pp. 451–461.

Galston, A. W. and R. Kaur-Sawhney. 1987. Polyamines as endogenous growth regulators. In *Plant Hormones and Their Role in Plant Growth and Development* (P. J. Davies, ed). Martinus Nijhoff Pub., Dordrecht, The Netherlands. pp. 280–295. ISBN 90-247-3497-5.

Galston, A. W. and T. A. Smith. 1985. *Polyamines in Plants.* Martinus Nijhoff/Dr. W. Junk Pub., Dordrecht, The Netherlands. ISBN 90-247-3245.

Gaspar, T., C. Kevers, M. Colemans, C. Panel and M. Greppin. 1984. Interaction of polyamines or their precursors with the calcium-controlled secretion of peroxidase by sugar beet cells. *Experientia* 40:696–697.

Goren, R., N. Palavan, H. Flores and A. W. Galston. 1982. Changes in polyamine titer in etiolated pea seedlings following red light treatment. *Plant Cell Physiol.* 23:19–26.

Haissig, B. E. 1986. Metabolic processes in adventitious rooting of cuttings. In *New Root Formation in Plants and Cuttings.* (M. B. Jackson, ed). Martinus Nijhoff Pub., Dordrecht/Boston/Lancaster. pp. 141–189. ISBN 90-247-3260-3.

Heimer, Y. M., Y. Mizrahi and U. Bachrach. 1979. Ornithine decarboxylase activity in rapidly proliferating plant cells. *FEBS Lett.* 104:146–148.

Henningsson, A., S. Henningsson, H. Tjalve, L. Hammer and G. Lowendahl. 1983. Whole body and microautoradiographic localization of ornithine decarboxylase in the kidneys of nandrolone-treated mice using tritium labelled difluoromethylornithine. In *Advances in Polyamine Research* (U. Bachrach, A. Kaye and R. Chayen, eds). Raven Press, NY, USA. Vol. 4. pp. 719–726.

Icekson, I., A. Goldlust and A. Apelbaum. 1985. Influence of ethylene on S-adenosylmethionine decarboxylase activity in etiolated pea seedlings. *J. Plant Physiol.* 119:335–345.

Jackson, M. B. 1986. *New Root Formation In Plants and Cuttings.* Martinus Nijhoff Pub., Dordrecht/Boston/Lancaster. ISBN 90-247-3260-3.

Jarvis, B. C., P. R. M. Shannon and S. Yasmin. 1983a. Involvement of polyamines with adventitious root development in stem cuttings of mung bean. *Plant Cell Physiol.* 24:677–683.

—————— ——— 1983b. Influence of IBA and cordycepin on rooting and RNA synthesis in stem cuttings of *Phaseolus aureus* Roxb. *Plant Cell Physiol.* 24:139–146.

Katoh, Y., T. Hasegawa, T. Suzuki and T. Fujii. 1985. Changes in the amounts of putrescine, spermidine and spermine in hiproly barley callus after auxin withdrawal. *Agric. Biol. Chem.* 49:1027–1032.

Kaur-Sawhney, R., M. E. Flores and A. W. Galston. 1981. Polyamine oxidase in oat leaves: A cell wall localized enzyme. *Plant Physiol.* 68:494–498.

—————— A. W. Galston. 1979. Interaction of polyamines and light on biochemical processes involved in leaf senescence. *Plant Cell Environ.* 2:189–196.

—————— N. S. Shekhawat and A. W. Galston. 1985. Polyamine levels as related to growth, differentiation and senescence in protoplast-derived cultures of *Vigna aconitifolia* and *Avena sativa. Plant Growth Regul.* 3:329–337.

Kyriakidis, D. A. 1983. Effect of plant growth hormones and polyamines on ornithine decarboxylase activity during the germination of barley seeds. *Physiol. Plant.* 57:499–504.

Malfatti, H., J. C. Vallee, E. Perdrizet, M. Carre and C. Martin. 1983. Acides amines libres d'explantes foliares de *Nicotiana tabacum,* cultives *in vitro* sur des milieux induisant la rhizogénèse ou la caulogénèse. *Physiol. Plant.* 57:492–498.

Metraux, J. P. and H. Kende. 1983. The role of ethylene in the growth response of submerged deep water rice. *Plant Physiol.* 72:441–446.

Martin-Tanguy, J. 1985. The occurrence and possible function of hydroxycinnamoyl acid amides in plants. *Plant Growth Regul.* 3:381–399.

Middleton, W., B. C. Jarvis and A. Booth. 1978. The boron requirement for root development in stem cuttings of *Phaseolus aureus* Roxb. *New Phytol.* 81:287–297.

—————— ——— 1980. The role of leaves in auxin and boron-dependent rooting of stem cuttings of *Phaseolus aureus* Roxb. *New Phytol.* 84:251–259.

Naik, B. I. and S. K. Srivastava. 1978. Effect of polyamines on tissue permeability. *Phytochemistry* 17:1885–1887.

Palavan, N. and A. W. Galston. 1982. Polyamine biosynthesis and titer during various developmental stages of *Phaseolus vulgaris*. *Physiol. Plant.* 55:438–444.

Pegg, A. E. and P. P. McCann. 1982. Polyamine metabolism and function. *Amer. J. Physiol.* 243(Cell Physiol. 12) C212–C221.

Rajam, M. V. and A. W. Galston. 1985. The effects of some polyamine biosynthetic inhibitors on growth and morphology of phytopathogenic fungi. *Plant Cell Physiol.* 26:683–692.

Raskin, I. and H. Kende. 1984. Regulation of growth in stem sections of deepwater rice. *Planta* 160:66–72.

Roberts, D. R., M. A. Walker, J. E. Thompson and E. B. Dumbroff. 1984. The effects of inhibitors of polyamine and ethylene biosynthesis on senescence, ethylene production and polyamine levels in cut carnation flowers. *Plant Cell Physiol.* 25:315–322.

Sankhla, D., T. D. Davis, N. Sankhla, A. Upadhyaya and B. N. Smith. 1985. Effect of 2-(3,4-dichlorophenoxy)-triethylamine on betacyanin efflux from beet root tissue. *Biochem. Physiol. Pflanzen.* 180:625–628.

Sankhla, N., A. Upadhyaya and T. D. Davis. 1987. Polyamines in Plant Growth and Development. In *Hormonal Regulation of Plant Growth and Development* (S. S. Purohit, ed), Agro Bot. Pub. Bikaner, India. Vol. IV. pp. 171–203.

_____ _____ D. Sankhla and T. D. Davis. 1985. Modulations of some salt-induced metabolic alterations by polyamines. In *12th Int. Conf. Plant Growth Sub.* (abst.). p. 128.

Schwartz, M., A. Altman, Y. Cohen and T. Arzee. 1986. Localization of ornithine decarboxylase and changes in polyamine content in root meristems of *Zea mays*. *Physiol. Plant.* 67:485–492.

Serafini-Francassini, D., P. Torrigiani and C. Branca. 1984. Polyamines bound to nucleic acids during dormancy and activation of tuber cells of *Helianthus tuberosus*. *Physiol. Plant.* 60:351–357.

_____ N. Bagni, P. G. Cionini and A. Bennici. 1980. Polyamines and nucleic acids during the first cell cycle of *Helianthus tuberosus* after the dormancy break. *Planta* 148:332–337.

Shen, H. and A. W. Galston. 1985. Correlations between polyamine ratios and growth patterns in seedling roots. *Plant Growth Regul.* 3:353–363.

Shih, L., R. Kaur-Sawhney, J. Fuhrer, S. Samanta and A. W. Galston. 1982. Effects of exogenous 1,3-diaminopropane and spermidine on senescence of oat leaves. I. Inhibition of protease activity, ethylene production, and chlorophyll loss as related to polyamine content. *Plant Physiol.* 70:1592–1596.

Shyr, Y. and C. Kao. 1985. Polyamines and root formation in mung bean hypocotyl cuttings. *Bot. Bull. Academia Sinica.* 26:179–184.

Simola, L. K. and J. Honkanen. 1983. Organogenesis and fine structure in megagametophyte callus lines of *Picea abies*. *Physiol. Plant.* 59:551–561.

Slocum, R. D., R. Kaur-Sawhney and A. W. Galston. 1984. The physiology and biochemistry of polyamines in plants. *Arch. Biochem. Biophys.* 235:283–303.

Smith, M. A., P. J. Davies and J. B. Reid. 1985. Role of polyamines in gibberellin-induced internode growth in peas. *Plant Physiol.* 78:92–99.

Smith, T. A. 1977. Recent advances in the biochemistry of plant amines. *Prog. Phytochem.* 4:27–81.

_____ 1982. The function and metabolism of polyamines in higher plants. In *Plant Growth Substances* (P. F. Wareing, ed). Academic Press, London, England. pp. 463–472. ISBN 0-12-735380-1.

_____ 1984. Putrescine and inorganic ions. *Adv. Phytochem.* 18:7–54.

_____ 1985a. Polyamines. *Ann. Rev. Plant Physiol.* 36:117–143.

_____ 1985b. Polyamines in plants. In *Polyamines in Plants* (A. W. Galston and T. A. Smith, eds). Martinus Nijhoff Pub., Dordrecht, The Netherlands. pp. vii–xxi.

_____ 1985c. The inhibition and activation of polyamine oxidase from oat seedlings. *Plant Growth Regul.* 3:269–275.

Srivastava, S. K., M. S. Kansara and S. M. Mungre. 1985. Effect of polyamines and guanidines on the growth, nitrogen assimilation and reserve mobilization in germinating radish seeds. *Plant Growth Regul.* 3:339–351.

_____ P. Rajbabu. 1983a. Effect of amines and guanidines on ATPase from maize and scutellum. *Phytochemistry* 22:2675–2679.

———— ———— 1983b. Effect of amines and guanidines on peroxidase from maize scutellum. *Phytochemistry* 22:2681–2686.

Suresh, M. R., S. Ramakrishna and P. R. Adiga. 1978. Regulation of arginine decarboxylase and putrescine levels in *Cucumis sativus* cotyledons. *Phytochemistry* 17:57–63.

Suttle, J. C. 1981. Effect of polyamines on ethylene production. *Biochemistry* 20:1477–1480.

Suzuki, T. and K. Kohno. 1983. Changes in nitrogen levels and free amino acids in rooting cuttings of mulberry (*Morus alba*). *Physiol. Plant.* 59:455–460.

Torrey, J. G. 1986. Endogenous and exogenous influences on the regulation of lateral root formation. In *New Root Formation in Plants and Cuttings* (M. B. Jackson, ed). Martinus Nijhoff Pub., Dordrecht/Boston/Lancaster. pp. 32–66. ISBN 90-247-3260-3.

Torrigiani, P., D. Serafini-Fracassini, S. Biondi and N. Bagni. 1986. Evidence for the subcellular localization of polyamines and their biosynthetic enzymes in plant cells. *J. Plant Physiol.* 124:23–29.

Upadhyaya, A. and N. Sankhla. 1983a. Responses of Indian desert plants to polyamines. In *Proc. Nat. Symp. Adv. Front. Plant. Sci.* Scientific Pub., Jodhpur, India. pp. 91–93.

———— ———— 1983b. Reversal of fluoride induced inhibition of nitrate reductase activity in wheat by polyamines. *Comp. Physiol. Ecol.* 8:138–140.

Villanueva, V. R., R. C. Adlakhla and A. M. Cantera-Soles. 1978. Changes in polyamine concentration during seed germination. *Phytochemistry* 17:1245–1249.

Walker, M. A., D. R. Roberts, C. Y. Shih and E. B. Dumbroff. 1985. A requirement for polyamines during cell division phase of radicle emergence in seeds of *Acer saccharum*. *Plant Cell Physiol.* 26:967–971.

Wang, S. Y. and M. Faust. 1986. Effect of growth retardants on root formation and polyamine content in apple seedlings. *J. Amer. Soc. Hortic. Sci.* 111:912–917.

White, A., P. Handler and E. L. Smith. 1968. *Principles of Biochemistry.* McGraw Hill, NY, USA.

Winer, L. and A. Apelbaum. 1986. Involvement of polyamines in the development and ripening of avocado fruits. *J. Plant Physiol.* 126:223–233.

Young, N. D. and A. W. Galston. 1984. Physiological control of arginine decarboxylase activity in K-deficient oat shoots. *Plant Physiol.* 76:331–335.

Zagon, I. S., J. E. Seely and A. E. Pegg. 1983. Autoradiographic localization of ornithine decarboxylase. *Methods Enzymol.* 94:169–176.

CHAPTER 16

Stock Plant Environment and Subsequent Adventitious Rooting

Roar Moe

Department of Horticulture
Agricultural University of Norway
P.O. Box 22, N-1432 Aas-NLH, Norway

and

Arne Skytt Andersen

Horticultural Institute
Royal Veterinary and Agricultural University
Rolighedsvej 23, DK-1958 Frederiksberg C, Denmark

INTRODUCTION..214
IRRADIANCE..215
    Seasonal Variation...215
    Irradiance Level...215
    Effect of Photoperiod..221
    Effect of Light Quality..222
    Artificial Lighting..223
TEMPERATURE EFFECTS..225
WATER STRESS..225
CARBON DIOXIDE..226
MINERAL NUTRITION...228
CONCLUSION..229
ACKNOWLEDGMENTS...230
REFERENCES..230

Additional key words: CO_2, irradiance, light quality, nutrition, photoperiod, temperature, water stress.
Abbreviations: ABA, abscisic acid; DNP, day neutral plants; FR, far-red light; GA, gibberellins; LCP, light compensation point; LD, long days; LDP, long day plants; ND, natural days; PEG, polyethylene glycol; R, red light; SD, short days; SDP, short day plants.

INTRODUCTION

In order to obtain high-quality young plants in the shortest possible time, cuttings must root quickly and abundantly. Cuttings must also be capable of good lateral branching and fast growth after rooting.

In this regard, Moe (1977) found that proper light and CO_2 enrichment of stock plants of *Campanula isophylla* Moretti reduced the propagation period by more than 50% (6–7 weeks) in the winter months. Similar positive after-effects of stock plant irradiation on rooting and subsequent growth of cuttings has been observed in *Chrysanthemum* spp. (Fischer and Hansen 1977, Borowski et al. 1981). Consequently, the effects of stock plant growing conditions on the successful rooting of cuttings has received increased attention in the last decade.

The physiological condition of the stock plant is the result of the interaction between genotype (species and cultivars) and environmental factors (light, temperature, water, CO_2, and nutrition). It is evident that the stock plant environment exerts a strong influence on root formation in stem cuttings (e.g. Hansen 1975), on root and bud formation in leaf cuttings (Heide 1964, 1965), and on regeneration in flower peduncle segments (Appelgren, 1976).

Although various environmental factors will be treated separately in the following discussion, it must be borne in mind that the "optimal" conditions for one factor are not always optimal if one or more of the other factors are changed. This raises a more general problem of environmental research: Do the optimal conditions remain unchanged throughout the experimental period? In particular, the question arises whether adaptation during stock plant growth to a particular environment imposes limitations on the environmental conditions that can best be utilized by the cuttings.

Although various types of vegetative propagation material, including leafy cuttings (stem or leaf cuttings) and cuttings without leaves (winter cuttings), have been reported in the literature (Hartmann and Kester 1983), we will, for several reasons, deal mainly with the influence of stock plant treatments on rooting of leafy cuttings because: 1) leafy cuttings are frequently used in practice, and 2) the reaction to variable growing conditions such as irradiance, photoperiod, CO_2, and water balance is reflected most strongly in the leaves and/or the buds.

IRRADIANCE

Seasonal Variation

Seasonal changes in adventitious rooting have been observed for many years (Hartmann and Kester 1975, 1983, and Fig. 1). This phenomenon seems to be related to seasonal variation in temperature, irradiance level, and/or an interaction between irradiance and photoperiod. When coniferous stock plants (e.g. *Picea, Pseudotsuga, Abies, Taxus* and *Juniperus* spp.) are grown under natural light conditions in the northern hemisphere, vegetative propagation is usually most successful with cuttings taken during autumn and early spring. The length of this period may vary from year to year (Roberts and Fuchigami 1973) and may differ between species (Thimann and Delisle 1942). It has also been noticed that the rooting ability of herbaceous cuttings fluctuates during the season even if they are taken from stock plants grown in greenhouses. For example, cuttings of *Pisum sativum* L. (pea) taken in spring and autumn had a high rooting capacity, whereas rooting was poor in summer-grown cuttings (Andersen et al. 1975). On the other hand, cuttings from stock plants of several glasshouse crops (e.g. *Begonia, Campanula, Euphorbia pulcherrima, Chrysanthemum*) responded positively to supplementary light to winter-grown stock plants (Fischer and Hansen 1977, Moe 1977, L. Bertram. 1985, M.S. Thesis, Royal Vet. and Agric. Univ., Copenhagen, Denmark). The seasonal variations in the rooting ability of cuttings indicate that the physiological status of the stock plant at the time the cutting are excised is of utmost importance for the rooting process.

Irradiance Level

The importance of irradiance level during stock plant growth has been investigated for several species (Table 1). The effect on subsequent rooting after exposing the stock plants to different irradiance levels is controversial (Andersen 1986). Increased irradiance may inhibit or delay rooting, promote rooting, or have no effect (Table 1). These results are based on research using different species and different experimental protocols which can be divided into four main groups.

In the first two groups the stock plants were grown under natural light conditions with either high irradiance during summer when irradiance was reduced by shading, or under low irradiance during winter when plants were given supplementary lighting. In the third group, stock plants were grown under variable irradiance levels in a controlled environment. The fourth group is the etiolation type where darkness is compared with light levels of various intensities.

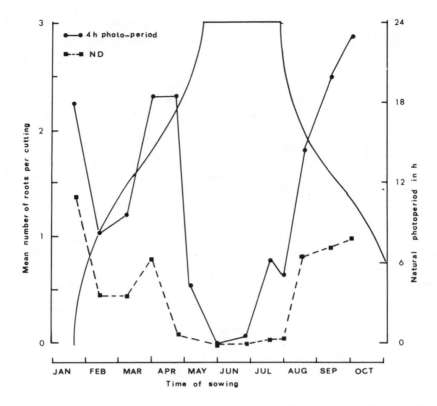

Figure 1. Effect of stock plant photoperiod on root formation in *Pinus sylvestris* L. hypocotyl cuttings. Stock plants were grown for six weeks under either natural photoperiods (ND) or 4 h photoperiod. Rooting occurred in tap water. Data are means of five replicates of 10 cuttings each. Analysis of variance showed significant effects of the main factors [season and photoperiod ($P<0.001$), and a significant interaction ($P<0.001$)]. The continuous solid curve represents the natural photoperiod throughout the experimental period. [From Hansen and Ernstsen (1982)]

Table 1. Species of plants where various stock plant irradiance experiments have been conducted. Experimental groups are: I = Stock plant shading, II = Low natural irradiance + supplementary lighting, III = Various irradiances in controlled environment rooms, and IV = Etiolation of the stock plants. The effect of increased irradiance on rooting is given as: + = positive effect, − = negative effect, 0 = no effect.

Plant Species	Expt. Group	Result of Increased Irradiance	References
Begonia × *hiemalis* Fortsch	II	+	Bertram (1986)
Begonia cheimantha Everett	III	+	Heide (1965)
Berberis thunbergii DC.	I	−	Knox and Hamilton (1982)
Chrysanthemum morifolium Ramat	III	+	Fischer and Hansen (1977)
	III	+	Borowski et al. (1981)
	III	+ or −	Weigel et al. (1984)

Plant Species	Expt. Group	Result of Increased Irradiance	References
Campanula isophylla Moretti	III	+	Moe (1976, 1977)
Dahlia variabilis Desf.	I	− or 0	Biran and Halevy (1973ab)
Euphorbia pulcherrima Willd.	II	+	Moe (1985b)
Hedera helix L.	III	−	Poulsen and Andersen (1980)
Hibiscus rosa-sinensis L.	I	−	Johnson and Hamilton (1977)
	II	0	Andersson (1985), Nielsen et al. (1984)
Juniperus sabina L.	II	−	Lin and Molnar (1980)
Juniperus squamata Buch.-Ham.	II	−	Lin and Molnar (1980)
Kalanchoe blossfeldiana V. Poelln	II	+	Andersson and Amsen (1985)
Ligustrum × *vicary*	I	0	Knox and Hamilton (1982)
Ligustrum obtusifolium var. *regelianum* Rehd.	I	−	Knox and Hamilton (1982)
Magnolia soulangiana Soul.-Bod.	II	+	Lin and Molnar (1980)
Malus spp.	III	0 or −	Christensen et al. (1980)
Malus (apple root stocks) Mill.	IV	− or +	Howard et al. (1985)
Pelargonium × *hortorum* L. H. Bailey	II	−	Nilsen (1976)
Pelargonium Zonale hybr.	II	− or +	Reuther and Forschner (1983)
	II	+	Tsujita and Harney (1978)
Phaseolus aureus Roxb.	III	+	Jarvis and Ali (1984)
Pinus sylvestris L.	III	−	Hansen et al. (1978)
	III	−	Hansen and Ernstsen (1982)
Pisum sativum L.	III	−	Leroux (1965)
	III	−	Hansen and Eriksen (1974)
	III	−	Andersen et al. (1975)
	III	−	Rajagopal and Andersen (1980a)
Populus tremula L.	III	−	Eliason and Brunes (1980)
Rhododendron L. spp.	II	− or +	Lin and Molnar (1980)
	I	−	Johnson and Roberts (1971)
Raphanus sativus L.	III	−	Lovell et al. (1972)
Sinapis alba L.	III	− or 0	Lovell et al. (1972)
			Lovell and Moore (1969)

Typical examples of the two main response groups to increasing stock plant irradiance on rooting are shown in Fig. 2 (Hansen and Eriksen 1974, Fischer and Hansen 1977). Plant species in the pea-response-group show a reduction in rooting while those of the chrysanthemum-response-group show an increase in rooting with increasing stock plant irradiance between about 0 and 100 W·m⁻². In contrast, rooting of some cuttings (e.g. apple, *Dahlia* spp.) has not been affected by irradiance pre-treatment (Christensen et al. 1980, Biran and Halevy 1973ab). The differences in response between species to increasing stock plant irradiance on rooting are not fully understood.

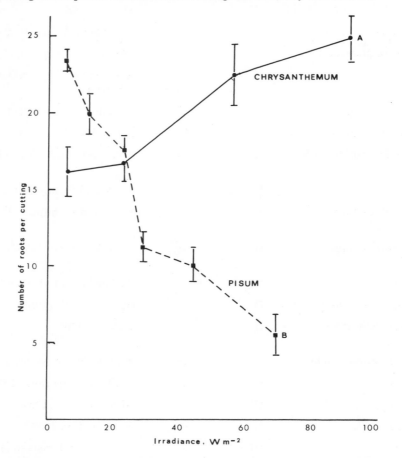

Figure 2. Root formation in cuttings as a function of irradiance to the stock plants: *Chrysanthemum morifolium* Ramat. (A) and *Pisum sativum* L. (B). Rooting took place under an irradiance of 16 W·m⁻² for *C. morifolium* and 15.5 W·m⁻² for *P. sativum* cuttings. Vertical lines denote 95% confidence limits. [From Fischer and Hansen (1977), and Hansen and Eriksen (1974)]

It is evident that stock plants and cuttings require a certain minimal level of light to root well, but that the optimal irradiance level for stock plants may vary from species to species and sometimes between cultivars. Therefore, more attention should be given to the influence of irradiance on the light acclimatization of the stock plants, and what consequences light acclimatization has on rooting. Normally, stock plants grown under high irradiance have a higher LCP and a higher respiration rate than those grown under low irradiance (Mortensen and Olsen 1987). If cuttings are taken from high-irradiance treated stock plants and rooted at irradiance below the LCP, the cuttings frequently root very poorly. It should be noted that the LCP drastically drops after taking the cutting due to a reduced respiration rate (Bjarke Veierskov, personal communication). This phenomenon might be a practical problem in shipping unrooted cuttings between the tropics and higher latitudes. Indeed, cuttings from *Pelargonium* spp. stock plants grown under high irradiance in the south rooted better when they were shipped to the north if they were given supplementary lighting and CO_2 enrichment during rooting in the winter (Reuther and Forschner 1983).

Davis and Potter (1981) contend that the current photosynthate supply to the base of pea cuttings is important for rooting. They found that root formation of leafy cuttings of *Pisum sativum* 'Alaska' was reduced by about 50% when net photosynthesis was adjusted to the compensation point by shading, by reducing the CO_2 concentration, or by blocking CO_2 exchange with an antitranspirant. A short (2–4 d) continuous dark treatment to *Chrysanthemum* spp. cuttings delayed the rooting process, and a longer period of darkness resulted in no roots even when the cuttings had been taken from high-irradiated stock plants (Borowski et al. 1981). From these observations it can be concluded that to fully evaluate the effect of stock plant irradiance on rooting, the light conditions during rooting also must be considered for each species.

Root emergence in *Chrysanthemum* spp. cuttings taken from medium-irradiance-treated stock plants (29 W·m^{-2}) was delayed by 1–2 d compared with low-irradiance-treated stock plants (7.2 W·m^{-2}), but a higher number of roots were formed on cuttings from medium-irradiance-treated stock plants (Borowski et al. 1981, Fig. 3). Veierskov et al. (1982ab) suggested that cuttings with the highest metabolic rate have the highest capacity for producing adventitious roots. Cuttings which have a high capacity for supplying the carbohydrate needed to maintain metabolic processes will form many roots. The ability of cuttings to supply carbohydrates, either from stored reserves (Haaland 1976) or through current photosynthesis during rooting, to the area where roots appear seems to be very

Figure 3. Effect of stock plant irradiance of *Chrysanthemum* 'Horim'.on rooting and subsequent growth of the young plants after planting. Upper: Rooted cuttings after two weeks at 18 °C air and 21 °C root temperature. Lower: Young plants grown for four weeks in an 18 °C greenhouse. [After Borowski et al. (1981)]

important for root development, but not for root initiation, which is hormonally controlled (Veierskov et al. 1982ab, Veierskov and Andersen 1982). The role of carbohydrates in rooting is discussed elsewhere (see Chapter 5 by Veierskov).

A suggestion for a response curve to irradiance and dark treatment on rooting is shown in Fig. 4. In many plant species a low stock plant irradiance resulted in poor rooting. With increasing stock plant irradiance rooting increased and reached a maximum at the optimal irradiance level, which differs for each plant species. When stock plant irradiance exceeds an optimal level, rooting will decrease. This type of stock plant response to different levels of irradiance has been shown in experiments with leafy pea cuttings (P. Fischer. 1981, Ph.D. Thesis, Royal Vet. and Agric. Univ., Copenhagen, Denmark). Fischer grew stock plants at three levels of irradiance (75, 100, or 300 μmol · $m^{-2} \cdot s^{-1}$) and cuttings from these irradiances were subsequently rooted under the various levels. With the various combinations maximum rooting was observed at the medium irradiance; the differences between stock plant treatments were similar to those given in Fig. 2. On the other hand, dark treatments of stock plants either as localized stem etiolation, which later forms the base of the cuttings, or as etiolation of the entire plant were favorable for rooting in many trees and shrubs (Howard et al. 1985). Increasing stock plant irradiance of these species types had a negative effect on rooting (see also Table 1). Fig. 4 also shows a generalized response curve of cutting irradiance on rooting which appears to be valid for most plant species. It is evident that leafy cuttings have a minimum light requirement for successful rooting (Loach and Whalley 1978) and that dark treatments of cuttings hindered rooting even when the stock plant irradiance had been optimal (Borowski et al. 1981). The optimal cutting irradiance for rooting appears to be strongly influenced by the previous stock plant irradiance history. Cuttings taken from stock plants under high irradiance have a high respiration rate and LCP. Rooting of these cuttings was enhanced by cutting irradiance somewhat above LCP. Too high a cutting irradiance will result in reduced rooting regardless of stock plant irradiance (Loach and Gay 1979). The effect of irradiance during rooting will be fully covered elsewhere (see Chapter 18 by Loach).

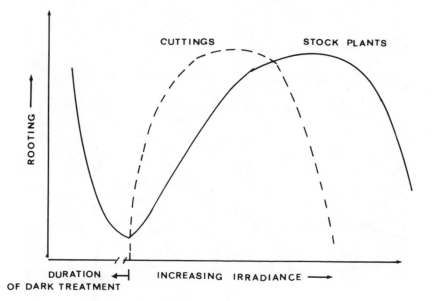

Figure 4. Schematic illustration of the generalized effect of irradiance to stock plants or to cuttings on rooting capacity of the cutting. The axes are relative and specific for species and/or cultivars. The main features of irradiance effects are that of optimum curves: Light is necessary but too little or too much hinders rooting. Dark treatments to stock plants either as localized etiolation of the stem, which later forms the base of the cuttings, or etiolation of the entire plant are both effective ways to increase rooting in many trees and shrubs (Howard et al. 1985).

The optimal irradiance level for rooting is also impacted by genotype and by the interaction between irradiance and the various environmental factors to which the stock plant is exposed. Leafy cuttings generally require a lower level of irradiance than stock plants to form good roots, but for some plant species etiolation of the stock plants stimulates rooting (Hartmann and Kester 1983, and Chapter 2 by Maynard and Bassuk).

At stock plant irradiance levels below optimum, rooting may be limited by a lack of carbohydrate and auxin supplies to the base of cuttings during rooting. This is evident for *Chrysanthemum* spp., in which Weigel et al. (1984) found more auxin present if stock plants had been grown under 40 $W \cdot m^{-2}$ than 4 $W \cdot m^{-2}$. Baadsmand and Andersen (1984) also found greater accumulation of externally applied auxin in the base of pea cuttings from high irradiance than from low irradiance (cf. Fig. 2 for rooting data). At supraoptimal levels of irradiance there is no lack of carbohydrate (but possibly too high a concentration), photo destruction of auxin, and possibly changes in water relations and rooting inhibitors and/or promotors.

High irradiance may also result in altered anatomical and morphological characteristics of the stock plant tissue where the roots later will appear. Anatomical studies of cuttings have shown that the sclerenchyma between phloem and cortex do not create a physical barrier to root formation (Beakbane 1961, 1969, Sachs et al. 1964, Edwards and Thomas 1980, Lovell and White 1986). It is our opinion that heavy sclerenchyma can cause some types of blockage for the translocation of root promotors to the origin cells where root primordia appear. There is, however, no clear connection between rooting ease and the amount of sclerenchyma tissue in *Hibiscus* spp. (Kachecheba 1976), and it is not known to what extent sclerenchyma development is related to irradiance. It might, therefore, be beneficial to study whether irradiance modifies the anatomical structure of stem tissue. Light modifications of differentiation should also be considered in this context. It has been shown that light conditions affect the wall thickness in xylem tissue as well as other parts of stem anatomy (Stafford 1948).

Effect of Photoperiod

The effect of stock plant photoperiod on rooting of cuttings seems to be related to the photoperiodic response to flowering, dormancy, and/or senescence. SD treatment of woody ornamental stock plants always inhibited rooting while LD promoted it (MacDonald 1969). Roberts and Fuchigami (1973) found that Douglas-fir (*Pseudotsuga menziesii* Franco) cuttings rooted poorly in September and October when bud dormancy was most pronounced. Cold treatment in October and November enhanced rooting and removed bud dormancy. Non-dormant cuttings also responded better to auxin application. With *Ficus infectoria* Willd., the rooting of cuttings correlated positively with cambial activity (Anand and Heberlein 1975). Smith and Wareing (1972) concluded that *Populus* spp. stock plants exposed to SD treatment showed a reduction in the number of roots in leafy stem cuttings compared to LD. Similarly, cuttings from seedlings of *Acer rubrum* L. grown under LD (16 h) rooted better than those from seedlings grown under SD (8 h) (Bachelard and Stowe 1963). In addition to auxin, leaves were essential for optimal rooting. In both SDP and LDP species, photoperiods that induce flowering appear generally to inhibit rooting (Heide 1965, Selim 1956, Moe 1976). Also in DNP such as roses (*Rosa* spp.) (Moe 1973), LD treatment was favorable for rooting. Hardwood cuttings of a blueberry species (*Vaccinium atrococcum* A. Heller) taken from shoots bearing flower buds did not root as well as those bearing only leaf buds (O'Rourke 1940). In the same way, *Dahlia* spp. cuttings from SD or LD treated stock plants bearing flower buds did not root as readily as cuttings with only vegetative buds (Biran and Halevy 1973ab). In the LDP *Campanula isophylla* Moretti, increased LD treatment durations to stock plants inhibited rooting of the cuttings (Moe 1976, Table 2). Auxin application enhanced rooting in cuttings from stock plants in a vegetative condition (SD), but not in a flowering condition (after 6–8 weeks of LD). Early removal of potential flower buds in *Rhododendron* spp. cuttings increased rooting, presumably through the elimination of a flower promoting stimulus which has adverse effects on rooting (C. R. Johnson. 1970, Ph.D. Thesis, Oregon State University, Corvallis, USA).

With the SDPs *Begonia* spp. (Heide 1965) and *Chrysanthemum* spp. (Heins et al. 1980), LD treated stock plants produced cuttings with better rooting ability than those exposed to SD treatment.

Table 2. The response of *Campanula isophylla* Moretti stock plants to 12 h short-day (SD) and 2, 4, 6, or 8 weeks of 24 h long-day (LD) treatments and indole-3-butyric acid (IBA) application as 5 s quick dip to cuttings on number of roots per cutting after three weeks of rooting at 18 °C. [From Moe (1976)]

Stock Plant Treatment	No. of Roots per Cutting IBA mg·l⁻¹	
	0	1800
Control (SD, 12 h)	6.6	24.5
2 weeks LD	2.1	22.2
4 weeks LD	0.5	2.3
6 weeks LD	0	0
8 weeks LD	0	0

Using *Begonia* spp. leaf cuttings, Heide (1968) found that seasonal changes in regeneration ability were due to a complex interaction of temperature and photoperiod that controlled the level of endogenous auxins and other growth regulators. Stock plants grown at high temperature and LD, a combination which is favorable for rooting, had a higher auxin activity than those from SD and low temperature (Heide 1968). In the facultative LDP *Begonia × tuberhybrida* Voss, LD to stock plants enhanced cutting production and rooting (Djurhuus 1984) presumably because LD maintained the plant in a non-dormant state.

In many herbaceous species, stock plants with the highest rooting ability are generally at the stage of active vegetative growth, while many woody ornamentals (e.g. *Rhododendron* spp., roses, *Picea pungens* Engelm.) show better rooting capacity when their shoots have entered a more mature stage of development (Moe 1973, Hartmann and Kester 1983). These changes in rooting ability with state of growth can be related to a change in the balance between rooting promotors and inhibitors (Hess 1969). In herbaceous cuttings of *Begonia* spp. and *Campanula* spp., GA seems to be an important rooting inhibitor when plants are entering the flowering stage. In some woody plant species, shoots actively growing show a high GA activity and poor rooting capacity. Application of auxins does not overcome GA inhibition of rooting in *Begonia* spp. cuttings (Heide 1968), but the possibility of using GA inhibitors or ABA to promote rooting should be pursued.

Effect of Light Quality

Very little attention has been paid to the influence of light quality on rooting. Light sources with a high R to FR ratio stimulated lateral branching (bud breaks) of stock plants and were favorable for cutting production in *Chrysanthemum* spp., roses, and poinsettia (Moe 1985a, Heins and Wilkins 1979, Hagen and Moe 1981). Irradiation of *Chrysanthemum* spp. cuttings during the rooting period with incandescent light, a light source with low R to FR ratio, increased rooting (Heins et al. 1980). FR irradiation of tomato shoots increased auxin content (Tucker 1977). Leaf discs from petunia plants exposed to R irradiation produced more shoots and fewer roots *in vitro* than discs treated with FR (cited in Wilkins 1982). This indicates that light quality may regulate regeneration and development of roots and shoots through its effect on the balance between cytokinin and auxin.

With few exceptions rooting is enhanced when cuttings are taken from the lower rather than from the upper portion of both herbaceous and woody stock plants (Loreti and Hartman 1964, Roulund 1973, Moe 1973, Poulsen and Andersen 1980, Leakey 1983, Hansen 1986). This effect might be attributed to the increased level of FR found down in the canopy due to the filtering of R radiation by the overlying leaves (Wilkins 1982). However, the effect of reduction in irradiance level by leaf shading on rooting cannot be excluded. Also the differences in ontogenetic age of the tissue from the top to the base may influence rooting of cuttings taken from different parts of the shoot (Roulund 1973).

In one test where stock plants were exposed to light sources of different quality for six weeks before the cuttings were taken, cuttings from plants exposed to blue light rooted most readily (Stoutemeyer 1947). Gupta et al. (1977) reported that the rooting of *Phaseolus mungo* L. [syn. *Vigna radiata* (L.) R. wilcz.] cuttings obtained from seedlings raised under FR light was completely inhibited, which could not be reversed by auxin application to the cuttings. Sucrose applications caused rooting under FR light and its effectiveness increased with the addition of auxin. In this system both the stock plants (seedlings) and the cuttings were kept in darkness or under white, R, or FR lights. FR light may retard or prevent sucrose and/or auxin translocation to the base of the cuttings. In this regard, Moe (1985b) recently reported that day-length extension using FR light inhibited bud breaks in 'Garnette' roses. Using $^{14}CO_2$ application to cuttings placed under FR light, Khayat and Moe (unpublished) observed that no radioactive material was translocated out of the leaves to the bud. Because there are conflicting results with regard to the influence of R and FR light on rooting, the distribution of light to various parts (leaf, bud, and basal part of the stem) of the cuttings must be considered. If the base of the cutting is not kept in the dark during rooting, root formation might be inhibited (Eliasson 1980).

Artificial Lighting

In recent years supplementary lighting and CO_2 enrichment of stock plants have become common practices in Scandinavia in the propagation of greenhouse plants to increase production and improve the "quality" of cuttings during the late autumn and winter months (Nilsen 1976, Moe 1977, Moe 1985a, Andersson and Amsen 1985). Supplementary high intensity irradiation of stock plants of *Begonia* spp. (Moe 1985a, Bertram 1986), *Chrysanthemum* spp. (Moe 1985b), *Campanula* spp. (Moe 1977, 1985a), *Hibiscus* spp. (Andersson 1985), *Kalanchoe* spp. (Andersson and Amsen 1985), *Pelargonium* spp. (Nilsen 1976, Tsujita and Harney 1978) and *Euphorbia* spp. (Moe 1985b) significantly increased the yield of cuttings when compared to non-irradiated or weakly irradiated stock plants (Tables 3 and 4). Increased irradiance of *Campanula* spp. stock plants resulted in better lateral branching and higher dry weight of cuttings (Table 4) and higher growth rate of cuttings after planting

Table 3. Some examples of stock plant responses to supplementary irradiance in terms of increased cutting yields. ND = natural daylight.

Species	Lighting Period	Supplementary Irradiance	Duration of Lighting (h)	Increase in Yield (%)	References
Pelargonium spp.	Nov.–Mar.	15 $W \cdot m^{-2}$	24	126	Nilsen (1976)
	Sept.–Jan.	100 $\mu mol \cdot m^{-2} \cdot s^{-1}$	16	30–48	Tsujita and Harney (1978)
Begonia × *hiemalis* Fotsch					
Tip cuttings	Nov.–Mar.	ND → 54–82 $\mu mol \cdot m^{-2} \cdot s^{-1}$	18	50–90	Bertram (1986)
Leaf cuttings	Oct.–Mar.	5–15 $W \cdot m^{-2}$	24	49	Moe (1985a)
Campanula spp.	Oct.–Feb.	10 $W \cdot m^{-2}$	24	60	Moe (1976)
Chrysanthemum	Oct.–Feb.	5–15 $W \cdot m^{-2}$	24	58	Moe (1985a)
Euphorbia spp. (Poinsettia)	Dec.–Apr.	13–21 $W \cdot m^{-2}$	24	47	Moe (1985b)
Hibiscus spp.	Oct.–Feb.	30–60 $\mu mol \cdot m^{-2} \cdot s^{-1}$	18	82	Andersson (1985)
	Oct.–Feb.	8 $W \cdot m^{-2}$	18	44	Nielsen et al. (1984)
Kalanchoe spp.	Nov.–Feb.	35–60 $\mu mol \cdot m^{-2} \cdot s^{-1}$	18	38–55	Andersson and Amsen (1985)
Kalanchoe 'Annette'	Nov.–Feb.	10 $W \cdot m^{-2}$	18	37	Nielsen et al. (1984)

Table 4. Effect of increasing irradiance to *Campanula isophylla* Moretti stock plants on yield of cuttings, number of lateral shoots per stock plant after five weeks, dry weight per cutting, and number of lateral shoots per rooted cutting after nine weeks of growth following rooting. [After Moe (1976, 1977)]

Irradiance 12 h · day⁻¹ (W · m⁻²)	Yield of Cuttings per Plant After 24 Weeks	No. of Lateral Shoots per Plant After 5 Weeks	Dry Weight per Cutting (mg)	No. of Roots per Cutting	No. of Lateral Shoots per Rooted Cutting After 9 Weeks
5.8	37	6.7	96	12.3	6.7
13.9	59	12.2	120	14.3	6.8
19.7	88	16.0	125	19.7	9.7
25.0	92	19.7	128	20.6	9.3

(Fig. 3). Further, better quality plants and reduced production periods resulted in more profitable propagation (Moe 1977, Borowski et al. 1981).

Even though the irradiance levels remain constant, there might be considerable differences in the yield of cuttings under different lamp types. For instance, a 12% increase in the yield of cuttings was obtained with *Chrysanthemum* spp. under fluorescent lamps as compared to metal halide or high pressure sodium lamps. With poinsettia plants, the yield of cuttings was respectively 8% and 40% less under high pressure sodium lamps (SON/T) and high pressure metal halide lamps (HPI/T), when compared to fluorescent lamps (TL33) (Moe 1985ab). In a comparison of different lamp types the efficiency order with respect to producing more and longer lateral breaks was: Fluorescent (Atlas warm white) > SON/T > HPI/T (Table 5). Energy ratio calculations between R (610–700 nm) and FR (700–750 nm) showed that HPI/T lamps have a low R:FR ratio (1.9) when compared to fluorescent lamps (5.2) (Grimstad 1981). Lamps with a low R:FR ratio seem to induce a low phytochrome photoequilibrium (Pfr/P total) and thus retard lateral branching and cutting production (Heins and Wilkins 1979, Wilkins 1982, Moe 1985ab, Hagen and Moe 1981). FR or incandescent light at the end of the daily light period will completely suppress axillary bud outgrowth in *Xanthium* spp. (Tucker and Mansfield 1972), tobacco (*Nicotiana tabacum* L.) (Kasperbauer 1971), tomato (*Lycopersicon esculentum* Mill.) (Tucker 1975, 1977), and *Fuchsia* spp. (Vince-Prue 1977). Healy et al. (1980) showed that R as a 4 h daily extension stimulated lateral bud activity and cutting production more effectively than FR in several green foliage plants (e.g. *Pilea involucrata* Urb.). Similar findings have been reported for roses (Moe 1985b). FR inhibition of lateral branching in rose plants was most pronounced when "end-of-day" lighting was used (Moe 1985b). This seems to be a general response to FR lighting in many plant species (Morgan and Smith 1981).

It can therefore be concluded that a lamp type with a rather high R:FR ratio appears to have beneficial effects on cutting production, and that end-of-day lighting with lamp types having low R:FR ratios should be avoided.

Table 5. Effect of lamp types on lateral branching and growth of laterals of unpinched poinsettia plants after three weeks of irradiance of 15 W · m⁻² (400–1000 nm) for 18 h per day. Significant differences denoted by different letters following means. (Temperature 21 °C) [From Hagen and Moe (1981)]

Lamp Type	Laterals (number)	Length of Lateral (mm)
Fluorescent (Atlas warm white)	7.1a	4.5a
Metal halide (HPI/T)	5.1c	2.7c
High pressure sodium (SON/T)	5.8b	3.4b

TEMPERATURE EFFECTS

Although it is difficult to separate the effect of temperature changes from a number of other environmental influences, temperature effects will be dealt with as if separation were possible. Temperature interactions with water relations are most important, but irradiance and CO_2 must also be considered.

The literature on this subject is rather limited: Heide (1964) and Hilding (1974) grew stock plants of *Begonia* spp. under different air temperatures (and day lengths) and later compared the rooting and adventitious bud formation of leaf cuttings from these plants (also at different temperatures). The differences between experiments carried out at 12 °C and 22 °C were minor and somewhat erratic. Also, there was no marked similarity in the responses of two cultivars. Bud initiation, however, was severely inhibited in cuttings from high temperature stock plants. Stem cuttings from stock plants of *Campanula isophylla* Moretti grown at 18 °C had a higher dry weight than those grown at 15 °C (Moe 1977), but rooting was not significantly influenced. Using a model system consisting of young pea plants, Kaminek (1962) found that stock plants grown at very low temperatures (3–4 °C) yielded cuttings that formed more roots than those grown at higher ambient temperature. Similar conclusions were reached for *Streptocarpus* spp. leaf cuttings (Appelgren and Heide 1972) although the optimal temperature was 12 °C.

A major problem in the above experiments is the effect that these temperature changes have on growth and development of the stock plant (Andersen 1986). Another uncertainty is the possible influence of an abrupt temperature change at the time of cutting propagation. Some of these problems have been dealt with in a series of experiments with pea cuttings (Fischer 1981 as cited in Andersen 1986). The principal conclusion of these studies was that cuttings exhibited no statistical difference in root number and days required for root emergence when stock plant temperatures were 15, 20, or 25 °C. On the other hand, higher rooting temperatures strongly reduced the number of roots as well as the time necessary for root emergence.

Thus the most detailed investigations—admittedly on very few species—point to a minor role of stock plant air temperature in subsequent rooting of cuttings. This generalization seems to apply to stock plants grown within a temperature range of about 12–27 °C.

WATER STRESS

Although drought is seldom applied as a stock plant treatment in nursery practice, it is likely to affect the abscisic acid and ethylene status of the cuttings (Rajagopal and Andersen 1980ab, Rasmussen and Andersen 1980, Ahmed and Andersen 1987). Methods for evaluating the effect of stock plant drought stress on rooting include growing the plants under various stress conditions or subjecting their root systems to PEG for varying lengths of time (Rajagopal and Andersen 1980ab). Cuttings from PEG stressed stock plants produced fewer roots only if the plants had been grown under low irradiance (16 W·m⁻² PAR). Under higher irradiance (38 W·m⁻²) there were more roots in cuttings from the moderately stressed stock plants (Fig. 5). To some extent, these data have been explained as a result of abscisic acid effects (Rasmussen and Andersen 1980).

The controversy over water stress effects is based first on the problem of the interaction of water stress with stock plant irradiance. That is, irradiance can directly influence rooting and thus confound water stress effects. Second, there is a problem of delay; cuttings treated with ABA form roots later than those that are untreated.

Too much water—i.e. waterlogging—often results in adventitious root initiation in intact stock plants. The ecological significance of adventitious root formation induced by waterlogging has been reviewed by Gill (1975). As far as we know, no practical use has been made of waterlogging in propagation. Kawase (1965) proposed that increased ethylene production in waterlogged stems was responsible for the induction of roots. However, the status of ethylene as a mediator of root formation in cuttings is inconclusive (Jarvis 1986, and Chapter 11 by Mudge).

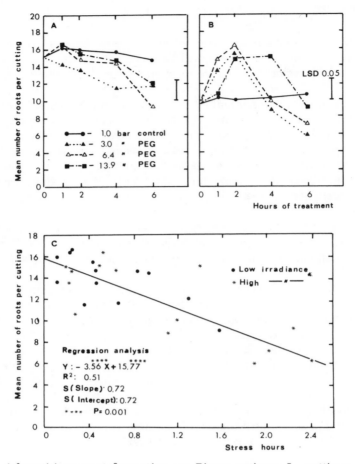

Figure 5. Adventitious root formation on *Pisum sativum* L. cuttings as a function of drought stress given to stock plants immediately prior to cutting. Drought stress was accomplished by immersing the roots in solutions of Polyethyleneglycol 6000 for 1, 2, 4, or 6 h. A: Stock plant irradiance 16 W·m⁻², B: 38 W·m⁻². Root numbers were counted after 13 d rooting in perlite under intermittent mist. Data are means of 80–120 cuttings per treatment. C: Data recalculated on basis of a "stress hour" factor. [From Rajagopal and Andersen (1980a)]

CARBON DIOXIDE

Carbon dioxide enrichment of the stock plant's environment increased the yield of cuttings in a number of species (Table 6). There is, however, considerable variation in the response to CO_2 application between different species and between different experiments on the same species. This difference may be due to the interaction between CO_2 and other environmental factors such as light intensity, temperature, relative humidity, and water status. *Campanula isophylla* responded to CO_2 enrichment by yielding more cuttings at low irradiance compared to the yield at irradiance close to the light saturation point (40.6 W·m⁻²). In this case the temperature and/or water supply might be the limiting factors for positive CO_2 response at high irradiance.

The principal reasons for increased cutting yields at enriched CO_2 concentrations are increased photosynthesis, higher relative growth rate, and greater lateral branching of the stock plants (Moe 1985ab, Moe and Mortensen 1986). Natural CO_2 concentrations (about 335 $\mu l \cdot l^{-1}$) are too low for maximum photosynthesis and growth. By increasing the CO_2 level to about 900 $\mu l \cdot l^{-1}$ the net photosynthetic rate is enhanced due to inhibition of photorespiration (Ehleringer 1979). Thus, carbon

Table 6. Effect of enriched CO_2 atmospheres on cutting yields from some typical greenhouse grown stock plants.

Plant Species	Increase (%) in Cutting Yield by CO_2	References
Campanula isophylla Moretti	(−6 to 123)	Moe (1976,1977)
Chrysanthemum morifolium Ramat	ca. 20	Walla and Kristoffersen (1974)
	ca. 40	Kobel (1965)
	ca. 69	Eng et al. (1983)
Euphorbia pulcherrima Willd. 'Paul Mikkelsen'	24	Walla and Kristoffersen (1974)
'Annette Hegg'	58	
Begonia × *tuberhybrida* Voss	23	Djurhuus (1984)
Dianthus spp. (cut carnation)	11	Goldsberry (1966)

Table 7. Effect of CO_2 (1000–2000 $\mu l \cdot l^{-1}$) enrichment during stock plant growth on lateral branching and rooting of some ornamental species. Symbols for lateral branching: $-$ = not recorded; + = 10–20 %; ++ = 20–30%; +++ = 30–50%; and ++++=50% increase of lateral breaks. Symbols for rooting: $-$ not recorded; 0 = no effect; + = positive effect.

Plant Species	No. of Lateral Breaks	Rooting	References
Aphelandra squarrosa Nees	++	−	Münch and Leinfelder (1968)
Begonia × *tuberhybrida* Voss	++	0	Djurhuus (1984)
Campanula isophylla Moretti	++	+	Moe (1976, 1977)
Chrysanthemum × *morifolium* Ramat	++	−	Walla and Kristoffersen (1974), Mortensen (1987)
Cissus rhombifolia Vahl	++	−	Anonymous (1980)
Dianthus spp.	−	+	Holley and Alstadt (1966)
Dieffenbachia maculata G. Don.	++	−	Saxe and Christensen (1984)
Fuchsia × *hybrida* Voss	++	+	Papenhagen (1983)
Euphorbia pulcherrima Willd.	+++	−	Mortensen (1985), Walla and Kristoffersen (1974)
Hedera helix L.	−	+	Daugaard (1981)
Hibiscus rosa-sinensis L.	+	−	Saxe and Christensen (1984)
Impatiens repens Moon	++++	−	Reimherr (1984)
Kalanchoe blossfeldiana V. Poelln.	++	−	Mortensen (1983), Papenhagen (1983)
Pelargonium × *hortorum* L. H. Bailey	−	+	Reuther and Forschner (1983)

dioxide enrichment of various greenhouse crops stimulated the relative growth rate by 10–25% (Moe and Mortensen 1986, Mortensen 1983, 1987) and improved cutting yield and quality (Walla and Kristoffersen 1974, Moe 1977). Carbon dioxide enrichment of the stock plant's environment also had a positive after-effect on rooting and lateral branching of some species (Table 7, Mortensen 1985). Optimal CO_2 concentration in these experiments was about 900 $\mu l \cdot l^{-1}$ for rooting and lateral branching whereas a higher CO_2 concentration (1800 $\mu l \cdot l^{-1}$) resulted in cuttings with a higher dry weight content as compared with lower CO_2 concentrations (Moe 1977, Table 8).

Photosynthetic rates of cuttings during the rooting period are relatively low (Davis and Potter 1981, Eliasson and Brunes 1980), and current photosynthate supply may limit adventitious root formation in some cuttings (Eliasson 1978, Davis and Potter 1981). Haaland (1976) reported that cuttings from CO_2 enriched Campanula isophylla stock plants showed a higher carbohydrate content, especially sucrose and the fructosyl sucroses, than plants not subjected to this treatment. In addition, cuttings from CO_2 enriched stock plants maintained favorable fructofuranosidase (EC 3.2.1.26) enzyme activity and high carbohydrate content during rooting and subsequent growth. These findings may explain the marked positive after-effect of CO_2 enrichment of stock plants on root formation and growth of the cuttings (Moe 1976, 1977, Table 8).

Table 8. Effects of atmospheric CO_2 enrichment of Campanula isophylla Moretti stock plants on subsequent rooting, lateral branching, and growth of young plants after rooting. [From Moe (1977)]

CO_2 Concentration ($\mu l \cdot l^{-1}$)	Roots per Cutting	Root Length (cm)	No. of Lateral Breaks per Plant After 6 Weeks	Young Plant Dry Weight (mg/plant)
330	11.0	2.1	5.4	275
900	21.1	2.8	8.2	405
1800	18.0	2.7	9.8	470

Apparently, the effect of CO_2 enrichment on root initiation is species-dependent (Lin and Molnar 1980). In some species roots emerged earlier on cuttings from stock plants grown in CO_2-enriched environments as compared to non-enriched stock plant environments (Moe 1976, 1977). But in Begonia × tuberhybrida Voss, Djurhuus (1984) reported that CO_2 enrichment of stock plants had no significant effect on rooting, but increased the yield of cuttings by 23%. Carbon dioxide enrichment during cutting propagation has shown similar variation between species in the rooting response. The effect of CO_2 application during rooting of cuttings will be discussed elsewhere (see Chapter 6 by Davis). From the present literature it is difficult to establish whether increased CO_2 concentrations to the stock plants or to cuttings influenced rooting directly or indirectly. Apparently, CO_2 enrichment can improve the water balance by inducing partial stomatal closure and can increase carbohydrate content of the plants, both of which would be favorable for rooting.

MINERAL NUTRITION

Stock plant nutrition has been found to influence quality and rooting of cuttings in several studies on a limited number of species (Pearse 1943, Knoblauch 1976, Hentig and Roeber 1976, Reuther and Roeber 1980, Roeber and Reuther 1982). Pearse (1943) grew Vitis spp. stock plants in nutrient solutions of different concentrations. The most diluted solution, which resulted in shoots showing deficiency symptoms for macronutrients, chiefly N, yielded cuttings which rooted better than those supplied with high nutrient concentrations where the shoots grew vigorously and had no deficiencies.

In a very thorough investigation, Roeber and Reuther (1982) claimed that they were able to predict the rooting (number and weight of roots) of Chrysanthemum spp. cuttings when the contents of N and carbohydrates in the cuttings were known. The N-content correlated negatively with both parameters, while the soluble carbohydrate concentration showed a positive correlation with rooting.

With *Pelargonium zonale* L'Herit. ex Ait. (Reuther and Roeber 1980) cuttings taken from stock plants grown hydroponically with different N supplies, no clear effect was found on root dry weight. The authors, however, reported that high N availability (160 mg $N \cdot l^{-1}$), through its effect on soluble carbohydrates (increase by 100% over 20 or 40 mg $N \cdot l^{-1}$), promoted rooting. The root mass produced was almost identical in all N concentrations. The composition of the N source for stock plants is also significant: too much NH_4 reduced the number of *Chrysanthemum* spp. cuttings and their "quality," but unfortunately no data on rooting were supplied (Roeber and Reuther 1982). For further discussion of the effects of carbohydrate/nitrogen contents on rooting, see Chapter 5 by Veierskov.

In an experiment with outdoor ornamental nursery stock (Knoblauch 1976) grown in rockwool ('Grodan'), the lowest nutrient solution concentration (0.025% of a standard 'Hornum' solution) yielded cuttings with the maximum number of roots on *Hypericum, Rosa,* and *Rhododendron* spp. In general, it appears that stock plants which are fertilized suboptimally for maximum shoot growth also result in cuttings that root best, but it is evident that much more detailed work is called for with regard to mineral nutrition optimization. For further information on mineral nutrition and rooting, see Chapter 4 by Blazich.

CONCLUSION

Irradiance levels under which stock plants are grown and cuttings are rooted influence cutting production and rooting capacities. The optimum irradiance level varies from species to species. The difference in response to irradiance can be attributed to an interaction between genotype and irradiance or other environmental factors such as photoperiod, temperature, CO_2, and relative humidity. When evaluating the effect of stock plant irradiance on rooting, the light conditions during rooting must also be considered.

Photoperiods that induce dormancy, promote adventitious bud formation, or promote flowering in stock plants generally inhibit or delay rooting of cuttings. The impact of stock plant irradiation with different light qualities has scarcely been investigated. Light with high R:FR ratio stimulated lateral branching and cutting production, but the effect of R and FR light on rooting has varied between species and research reports. Generally, supplementary lighting with high irradiance from lamps with high R:FR ratio increased cutting yields and stimulated subsequent rooting, lateral branching and the growth of the young plants.

CO_2 enrichment of stock plants had a beneficial effect on cutting production and reduced the light requirement by 30–40% during the winter due to a reduced light compensation point and inhibition of photorespiration. CO_2 enrichment had a positive after-effect on rooting and lateral branching of the rooted cutting. Increased CO_2 concentration resulted in higher carbohydrate content and dry weight, and improved the water status of the cuttings. Stock plant temperature seemed to have a minor effect on rooting, but low temperatures have generally promoted adventitious bud formation in leaf cuttings.

Water-stressing stock plants may inhibit or stimulate root formation, while waterlogged plants produce cuttings with better rooting ability in some species. Stock plants which are fertilized suboptimally for shoot growth yield cuttings which root best. Generally, rooting was negatively correlated with the content of N in the cuttings.

The influence of the abrupt change from the environment of the stock plant to the new propagation environment should be more carefully studied for plant species which belong to the pea- and chrysanthemum-response groups. The environmental factors of irradiance, light quality, and CO_2 deserve further attention. The influence of irradiance and light quality on anatomical structure of the base of the cuttings and on translocation of root promoters/inhibitors and assimilates to the rooting zone should be investigated during both stock plant growth and rooting.

ACKNOWLEDGMENTS

The Danish Agricultural and Veterinary Research Council supported part of the research for this chapter. The authors are grateful to Mrs. Solvi Olsen for typing our manuscript so carefully.

REFERENCES

Anon. 1980. Danish pot plants of top quality in all Europe (In Danish). *Gartner Tidende* 43:634–636.

Ahmed, A. and A. S. Andersen. 1987. Ethylene and rooting in *Pisum sativum* cuttings as affected by auxin. *Life Sci. Adv.* (in press).

Anand, V. K. and G. T. Heberlein. 1975. Seasonal changes in the effects of auxin on rooting in stem cuttings of *Ficus infectoria*. *Physiol. Plant.* 34:330–334.

Andersen, A. S. 1986. Environmental influences on adventitious rooting in cuttings of non-woody species. In *New Root Formation in Plants and Cuttings* (M. B. Jackson, ed). Martinus Nijhoff Pub., Dordrecht/Boston/Lancaster. pp. 223–254. ISBN 90-247-3260-3.

_____ E. N. Eriksen, J. Hansen and B. Veierskov. 1975. Stock plant conditions and root initiation on cuttings. *Acta Hortic.* 54:33–38.

Andersson, N. E. 1985. Kunstlys til Hibiscus. *Gartner Tidende* 101:1200–1201.

_____ M. G. Amsen. 1985. Kunstlys, stiklingeproduktion og kvalitet af Kalanchoe. *Gartner Tidende* 101:1203 and 1190.

Appelgren, M. 1976. Regeneration of *Begonia* × *hiemalis in vitro*. *Acta Hortic.* 64:31–38.

_____ O. M. Heide. 1972. Regeneration in *Streptocarpus* leaf discs and its regulation by temperature and growth substances. *Physiol. Plant.* 27:417–423.

Baadsmand, S. and A. S. Andersen. 1984. Transport and accumulation of indole-3-acetic acid in pea cuttings under two levels of irradiance. *Physiol. Plant.* 61:107–113.

Bachelard, E. P. and B. B. Stowe. 1963. Rooting of cuttings of *Acer rubrum* L. and *Eucalyptus camaldulensis* Dehn. *Aust. J. Biol. Sci.* 16:751–767.

Beakbane, A. B. 1961. Structure of the plant stem in relation to adventitious rooting. *Nature* 192:954–955.

_____ 1969. Relationships between structure and adventitious rooting. *Proc. Int. Plant Prop. Soc.* 19:191–201.

Bertram, L. 1986. Tilskudslys til moderplanter of *Begonia elatior*. *Gartner Tidende* 102:209–211.

Borowski, E., P. Hagen and R. Moe. 1981. Stock plant irradiation and rooting of chrysanthemum cuttings in light or dark. *Scientia Hortic.* 15:245–253.

Biran, I. and A. H. Halevy. 1973a. The relationship between rooting of dahlia cuttings and the presence and type of bud. *Physiol. Plant.* 28:244–247.

_____ _____ 1973b. Stock plant shading and rooting of dahlia cuttings. *Scientia Hortic.* 1:125–131.

Christensen, M. V., E. N. Eriksen and A. S. Andersen. 1980. Interaction of stock plant irradiance and auxin in the propagation of apple rootstocks by cuttings. *Scientia Hortic.* 12:11–17.

Daugaard, H. 1981. CO_2 to stock plants and cuttings of *Hedera* (In Danish). *Ugeskrift for Jordbruk* 28:403–405.

Davis, T. D. and J. R. Potter. 1981. Current photosynthate as a limiting factor in adventitious root formation on leafy pea cuttings. *J. Amer. Soc. Hortic. Sci.* 106:278–282.

_____ _____ 1983. High CO_2 applied to cuttings: effects on rooting and subsequent growth in ornamental species. *HortScience* 18:194–196.

Djurhuus, R. 1984. The effect of CO_2, daylength and light on the production and subsequent growth of *Begonia* × *tuberhybrida* cuttings. *Acta Hortic.* 162:65–74.

Edwards, R. A. and M. B. Thomas. 1980. Observations on physical barriers to root formation in cuttings. *Plant Prop.* 26(2):6–8.

Ehleringer, J. R. 1979. Photosynthesis and photorespiration: Biochemistry, physiology, and ecological implications. *HortScience* 14:217–222.

Eliasson, L. 1978. Effects of nutrients and light on growth and root formation in *Pisum sativum* cuttings. *Physiol. Plant.* 43:13–18.

——— 1980. Interaction of light and auxin in regulation of rooting in pea stem cuttings. *Physiol. Plant.* 48:78–82.

——— L. Brunes. 1980. Light effects on root formation in aspen and willow cuttings. *Physiol. Plant.* 48:261–265.

Eng, R. Y. N., M. J. Tsujita, B. Grodzinski, and R. G. Dutton. 1983. Production of chrysanthemum cuttings under supplementary lighting and CO_2 enrichment. *HortScience* 18:878–879.

Fischer, P. and J. Hansen. 1977. Rooting of chrysanthemum cuttings. Influence of irradiance during stock plant growth and of decapitation and disbudding of cuttings. *Scientia Hortic.* 7:171–178.

Gill, C. J. 1975. The ecological significance of adventitious rooting as a response to flooding in woody species with special reference to *Alnus glutinosa*. *Flora* 164:85–98.

Goldsberry, K. L. 1966. Effects of CO_2 on carnation stock plants. *Colorado Flower Growers Assoc. Bull.* 192:1–2.

Grimstad, S. 1981. Lampetyper og plantebestråling. 1. Virkning av lampetype og strålingsflukstetthet på vekst og utvikling hos tomat (*Lycopersicon esculentum* Mill.) dyrket i veksthus under ulike naturlige lysforhold. *Meld. Norg. LandbrHøgsk.* 60 (No. 22). 27 pp. English summary.

Gupta, S., V. K. Kochbar and K. K. Nanda. 1977. Effect of some metabolic inhibitors on rooting cuttings of *Phaseolus mungo* under varying light conditions and its relationship with auxin and nutrition. *Ann. Bot.* 41:507–515.

Haaland, E. 1976. The effect of light and CO_2 on the carbohydrates in stock plants and cuttings of *Campanula isophylla* Moretti. *Scientia Hortic.* 5:353–361.

Hagen, P. and R. Moe. 1981. Effect of temperature and light on lateral branching in poinsettia (*Euphorbia pulcherrima* Willd.). *Acta Hortic.* 128:47–54.

Hansen, J. 1975. *Adventitious Root Formation. Influence of Light During Stock Plant Growth.* Dept. Plant Physiology and Anatomy. *Internal review article.* The Royal Vet. and Agric. Univ., Copenhagen. 17 pp.

——— 1986. Influence of cutting position and stem length on rooting of leaf-bud cuttings of *Schefflera arboricola. Scientia Hortic.* 28:177–186.

——— E. N. Eriksen. 1974. Root formation of pea cuttings in relation to the irradiance of the stock plants. *Physiol. Plant.* 32:170–173.

——— A. Ernstsen. 1982. Seasonal changes in adventitious root formation in hypocotyl cuttings of *Pinus sylvestris* L.: Influence of photoperiod during stock plant growth and of indolebutyric acid treatment of cuttings. *Physiol. Plant.* 54:99–106.

——— L.-H. Strömquist and A. Ericsson. 1978. Influence of the irradiance on carbohydrate content and rooting of cuttings of pine seedlings (*Pinus sylvestris* L.). *Plant Physiol.* 61:975–979.

Hartmann, H. T. and D. E. Kester. 1975. *Plant Propagation: Principles and Practices.* Prentice-Hall, Inc., Englewood Cliffs, NJ, USA. 3rd Ed. 662 pp. ISBN 0-13-680991-X.

——— ——— 1983. *Plant Propagation: Principles and Practices.* Prentice-Hall, Englewood Cliffs, NJ. 4th ed. 727 pp. ISBN 0-13-681007-1.

Healy, W. E., R. D. Heins and H. F. Wilkins. 1980. Influence of photoperiod and light quality on lateral branching and flowering in selected vegetatively propagated plants. *J. Amer. Soc. Hortic. Sci.* 105:812–816.

Heide, O. M. 1964. Effects of light and temperature on the regeneration ability of *Begonia* leaf cuttings. *Physiol. Plant.* 17:789–804.

——— 1965. Photoperiodic effects on the regeneration ability of *Begonia* leaf cuttings. *Physiol. Plant.* 18:185–193.

——— 1968. Auxin level and regeneration of *Begonia* leaves. *Planta* 81:153–159.

Heins, R. and H. F. Wilkins. 1979. The influence of node number, light source, and time and irradiation during darkness on lateral branching and cutting production in 'Bright Golden Anne' chrysanthemum. *J. Amer. Soc. Hortic. Sci.* 104:265–270.

——— W. E. Healy and H. F. Wilkins. 1980. Influence of night lighting with red, far red and incandescent light on rooting of *Chrysanthemum* cuttings. *HortScience* 15:84–85.

Hentig, W.-U., von and R. Roeber. 1976. Der Einfluss von Zusatzbelichtung und N-Düngung auf die Stecklinge-produktion von *Euphorbia pulcherrima* 'Annette Hegg'. *Gartenbauwissenschaft* 41:275–279.

Hess, C. E. 1969. Internal and external factors regulating root initiation. In *Root Growth* (W. F. Whittington, ed) Butterworths, London, England. pp. 42–52.

Hilding, A. 1974. Inverkan av dagslängd och temperatur vid förökning av höstbegonia (*Begonia hiemalis*) med bladstecklinger. *Lantbrukshögskolans Med.* A209:1–15.

Holley, W. D. and R. A. Alstadt. 1966. Adding CO_2 to carnation stock plants improves performance of the cuttings. *Colorado Flower Growers Assoc. Bull.* 192:2–3.

Howard, B. H., R. S. Harrison-Murray and S. B. Arjyal. 1985. Responses of apple summer cuttings to severity of stock plant pruning and to stem blanching. *J. Hortic. Sci.* 60:145–152.

Jarvis, B. C. 1986. Endogenous control of adventitious rooting in non-woody cuttings. In *New Root Formation in Cuttings and Plants* (M. B. Jackson, ed). Martinus Nijhoff Pub., Dordrecht/Boston/Lancaster. pp. 191–222. ISBN 90-247-3260-3.

———— A. H. N. Ali. 1984. Irradiance and adventitious root formation in stem cuttings of *Phaseolus aureus* Roxb. *New Phytol.* 97:31–36.

Johnson, C. R. and A. N. Roberts. 1971. The effect of shading *Rhododendron* stock plants on flowering and rooting. *J. Amer. Soc. Hortic. Sci.* 96:166–168.

———— D. F. Hamilton. 1977. Rooting of *Hibiscus rosa sinensis* L. cuttings as influenced by light intensity and ethephon. *HortScience* 12:39–40.

Kachecheba, J. L. 1976. Seasonal effects of light and auxin on the rooting of *Hibiscus* cuttings. *Scientia Hortic.* 5:345–351.

Kaminek, M. 1962. Prspeek fyziologii otuzovani rostlin. Zakarenovani rostlinnych Rizkv. *Rostlin. Vyrd.* 8:959–978.

Kasperbauer, M. J. 1971. Spectral distribution of light in a tobacco canopy and the effects of end of day light quality on growth and development. *Plant Physiol.* 47:775–778.

Kawase, M. 1965. Etiolation and rooting in cuttings. *Physiol. Plant.* 18:1066–1076.

Knoblauch, F. 1976. Gødskning af *Hypericum Rosa* og *Rhododendron* i containere. Meddelelser fra Statens Forsøgsvirksomhed i Plantekultur 1275, 1276 and 1279.

Knox, G. W. and D. F. Hamilton. 1982. Rooting of *Berberis* and *Ligustrum* cuttings from stock plants grown at selected light intensities. *Scientia Hortic.* 16:85–90.

Kobel, F., Jr. 1965. Der Einfluss der CO_2 Anreicherung auf die Produktion von Chrysanthemumstecklingen. *Gartenbauwissenschaft* 30:549–556.

Leakey, R. R. B. 1983. Stockplant factors affecting root initiation in cuttings of *Triplochiton scleroxylon* K. Schum., an indigenous hardwood of West Africa. *J. Hortic. Sci.* 58:277–290.

Leroux, R. 1965. Etude de la rhizogénèse de segments d'épicotyles de Pois (*Pisum sativum* L.) en fonction de la lumière resçue par les plantules sur lesquelles ils ont été prélevés. *C.R. Acad. Sci. Paris* 261:5609–5611.

Lin, W. C. and J. M. Molnar. 1980. Carbonated mist and high intensity supplementary lighting for propagation of selected woody ornamentals. *Proc. Int. Plant Prop. Soc.* 30:104–109.

Loach, K. and A. P. Gay. 1979. The light requirement for propagating hardy ornamental species from leafy cuttings. *Scientia Hortic.* 10:217–230.

———— D. N. Whalley. 1978. Water and carbohydrate relationships during the rooting of cuttings. *Acta Hortic.* 79:161–167.

Loreti, F. and H. T. Hartmann. 1964. Propagation of olive trees by rooting leafy cuttings under mist. *Proc. Amer. Soc. Hortic. Sci.* 85:257–264.

Lovell, P. H. and K. G. Moore. 1969. The effects of light and cotyledon age on growth and root formation in excised cotyledons of *Sinapis alba* L. *Planta* 85:351–358.

———— J. White. 1986. Anatomical changes during root formation. In *New Root Formation in Plants and Cuttings* (M. B. Jackson ed). Martinus Nijhoff Pub., Dordrecht/Boston/Lancaster. pp. 111–140. ISBN 90-247-3260-3.

———— A. Illsley and K. G. Moore. 1972. Effects of light intensity and sucrose on root formation, photosynthetic ability and senescence in detached colyledons of *Sinapis alba* L. and *Raphanus sativus* L. *Ann. Bot.* 36:123–134.

MacDonald, A. B. 1969. Lighting—its effect on rooting and establishment of cuttings. *Proc. Int. Plant Prop. Soc.* 9:241–246.

Moe, R. 1973. Propagation, growth and flowering of potted roses. *Acta Hortic.* 31:35–40.

_____ 1976. Environmental effects of stock plants on rooting and further growth of the cuttings of *Campanula isophylla* Moretti. *Acta Hortic.* 64:71–81.

_____ 1977. Effect of light, temperature and CO_2 on the growth of *Campanula isophylla* stock plants and on the subsequent growth and development of their cuttings. *Scientia Hortic.* 6:129–141.

_____ 1985a. The effect of stock plant treatments on rooting and lateral branching in some greenhouse plants. *Acta Hortic.* 166:45–51.

_____ 1985b. Lysets innvirkning på knoppbryting hos morplanter og stiklinger. Vegetativ förökning. *NJF-seminarium,* Alnarp 29–31. Oktober, 1985. Trädgård 290:48–58.

_____ L. M. Mortensen. 1986. Carbon dioxide enrichment in Norway. In *Carbon Dioxide Enrichment of Greenhouse Crops.* (H. Z. Enoch and B. A. Kimball, eds). CRC Press, Boca Raton, FL, USA. Vol. 1. pp. 59–73. ISBN 0-8493-5611-3.

Morgan, D. C. and H. Smith. 1981. Non-photosynthetic responses to light quality. In *Encyc. Plant Physiol.* (P. Nobel, ed). Vol. 12. pp. 109–134. ISBN 0-387-12143-9.

Mortensen, L. M. 1983. Growth responses of some greenhouse plants to environment. XII. Effect of CO_2 on photosynthesis and growth of *Saintpaulia ionantha, Kalanchoe blossfeldiana* and *Nephrolepis exaltata. Meld. Norg. LandbrHøgsk.* 62(16):1–16.

_____ 1985. Effects of CO_2 enrichment and supplementary light on growth and flowering of poinsettia, *Euphorbia pulcherrima* Willd. *Meld. Norg. LandbrHøgsk.* 64(3):1–8.

_____ 1987. Review: CO_2 enrichment in greenhouses. Crop responses. *Scientia Hortic.* (In press).

_____ R. Olsen. 1987. Light acclimatization of some foliage plants. *Gartenbauwissenschaft* (In press).

Münch, J. and J. Leinfelder. 1968. Wirkung zusätzlicher Begasung mit CO_2 auf *Aphelandra squarrosa* 'Typ Königer'. *Gartenwelt* 68(11):239–241.

Nielsen, T. H., J. Hansen and M. G. Amsen. 1984. Forsøg med kunstlys i erhvervsgartnerier. *Gartner Tidende* 100(38):1201–1207.

Nilsen, J. 1976. Effects of irradiation of the mother plants on rooting of *Pelargonium* cutting. *Acta Hortic.* 64:65–69.

O'Rourke, F. L. 1940. The influence of blossom buds on rooting of hardwood cuttings of blueberry. *Proc. Amer. Soc. Hortic. Sci.* 40:332–334.

Papenhagen, A. 1983. Bessere Erträge durch CO_2—aber nicht überall. *Gärtnerbörse und Gartenwelt* 49:1244–1249.

Pearse, H. L. 1943. The effect of nutrition and phytohormones on the rooting of vine cuttings. *Ann. Bot.* 7:123–132.

Poulsen, A. and A. S. Andersen. 1980. Propagation of *Hedera helix:* Influence of irradiance to stock plants, length of internode and topophysis of cuttings. *Physiol. Plant.* 49:359–365.

Rajagopal, V. and A. S. Andersen. 1980a. Water stress and root formation in pea cuttings. I. Influence of the degree and duration of water stress on stock plants grown under two levels of irradiance. *Physiol. Plant.* 48:144–149.

_____ _____ 1980b. Water stress and root formation in pea cuttings. III. Changes in the endogenous level of abscisic acid and ethylene production in the stock plants under two levels of irradiance. *Physiol. Plant.* 48:155–160.

Rasmussen, S. and A. S. Andersen. 1980. Water stress and root formation in pea cuttings. II. Effect of abscisic acid treatment of cuttings from stock plants grown under two levels of irradiance. *Physiol. Plant.* 48:150–154.

Reimherr, P. 1984. CO_2 und Zusatzlicht bei *Impatiens repens* und *Pavonia multiflora. Deutscher Gartenbau* 51/52:2313–2314.

Reuther, G. and R. Roeber. 1980. Einfluss unterschiedlicher N-Versorgung auf Photosynthese und Ertrag von Pelargonienmutterpflanzen. *Gartenbauwissenschaft* 45:21–29.

_____ W. Forschner. 1983. Die Reaktion von Pelargonienstecklingen auf Licht und CO_2-Begasung in

234 Roar Moe and Arne Skytt Andersen

der Bewurzelungsphase. *Zierpflanzenbau* 21:948–954.

Roberts, A. N. and L. H. Fuchigami. 1973. Seasonal changes in auxin effect on rooting of Douglas fir stem cuttings as related to bud activity. *Physiol. Plant.* 28:215–221.

Roeber, R. and G. Reuther. 1982. Der Einfluss unterschiedlicher N-Formen und -Konzentrationen auf den Ertrag und die Qualitaet von Chrysanthemen Stecklingen. *Gartenbauwissenschaft* 47:182–188.

Roulund, H. 1973. The effect of cyclophysis and topophysis on the rooting ability of Norway spruce cuttings. *For. Tree Improv.* 5:21–41.

Sachs, R. M., E. Loreti and J. Debie. 1964. Plant rooting studies indicate schlerenchyma tissue is not a restricting factor. *Calif. Agric.* 18(9):4–5.

Saxe, H. and O. V. Christensen. 1984. Effects of carbon dioxide with and without nitric oxide pollution on growth, morphogenesis and production time of potted plants. *Acta Hortic.* 162:179–186.

Selim, H. H. A. 1956. The effect of flowering on adventitious root formation. *Meded. Landbouwhoogesch. Wageningen* 56(6):1–38.

Smith, N. G. and P. F. Wareing. 1972. The rooting of actively growing and dormant cuttings in relation to endogenous hormone levels and photoperiod. *New Phytol.* 71:483–500.

Stafford, H. A. 1948. Studies in the growth and xylary development of *Phleum pratense* seedlings in dark and in light. *Ann. J. Bot.* 35:706–715.

Stoutemeyer, V. T. 1947. Changes of rooting response in cuttings following exposure of the stock plants to light of different qualities. *Proc. Amer. Soc. Hortic. Sci.* 49:392–394.

Thimann, K. V. and A. L. Delisle. 1942. Notes on the rooting of some conifers from cuttings. *J. Arnold Arbor.* 23:103–109.

Tsujita, M. J. and P. M. Harney. 1978. The effects of Florel and supplemental lighting on the production and rooting of geranium cuttings. *J. Hortic. Sci.* 53:349–350.

Tucker, D. J. 1975. Far-red light as a supressor of side shoot growth in the tomato. *Plant Sci. Lett.* 5:127–130.

———— 1977. The effect of far-red light on lateral bud outgrowth in decapitated tomato plants and the associated changes in the levels of auxin and abscisic acid. *Plant. Sci. Lett.* 8:339–344.

———— T. A. Mansfield. 1972. Effects of light quality on apical dominance in *Xanthium strumarium* and the associated changes in endogenous levels of abscisic acid and cytokinins. *Planta* 102:140–151.

Veierskov, B. and A. S. Andersen. 1982. Dynamics of extractable carbohydrates in *Pisum sativum*. III. The effect of IAA and temperature on content and translocation of carbohydrates in pea cuttings during rooting. *Physiol. Plant.* 55:179–182.

———— ———— E. N. Eriksen. 1982a. Dynamics of extractable carbohydrates in *Pisum sativum*. I. Carbohydrate and nitrogen content in pea plants and cuttings grown at two different irradiances. *Physiol. Plant.* 55:167–173.

———— ———— B. M. Stummann and K. W. Henningsen. 1982b. Dynamics of extractable carbohydrates in *Pisum sativum*. II. Carbohydrate content and photosynthesis of pea cuttings in relation to irradiance and stock plant temperature and genotype. *Physiol. Plant.* 55:174–178.

Vince-Prue, D. 1977. Photocontrol of stem elongation in light-grown plants of *Fuchsia hybrida*. *Planta* 133:149–156.

Walla, J. and T. Kristoffersen. 1974. The effect of CO_2 application under various light and temperature conditions on growth and development of some florist crops. *Meld. Norg. LandbrHøgsk.* 53(27):1–46.

Weigel, V., W. Horn and B. Hock. 1984. Endogenous auxin levels in terminal cuttings of *Chrysanthemum*. *Physiol. Plant.* 61:422–428.

Wilkins, H. 1982. Influence of light on branching/cutting production and rooting. In *Proc. 21st Int. Hortic. Cong., Hamburg, W. Germany.* Vol II. pp. 894–903.

CHAPTER 17

Storage of Unrooted Cuttings

Volker Behrens

Institut für Obstbau und Baumschule
Universität Hannover
Am Steinberg 3
3203 Sarstedt, West Germany

INTRODUCTION...235
CONDITIONS FOR STORING CUTTINGS......................................236
    Relative Humidity 236
    Temperature...237
    Atmospheric Gases 237
DURATION OF STORAGE..238
PRETREATMENT OF CUTTINGS..241
PHYSIOLOGICAL CHANGES DURING STORAGE.............................242
    Food Reserves...242
    Breaking Dormancy..243
CONCLUSION 244
ACKNOWLEDGMENT 244
REFERENCES..245

Additional key words: adventitious root formation, atmospheric gases, controlled atmosphere storage, dormancy, low pressure storage, relative humidity, reserves, temperature, water status.
Abbreviations: CA, controlled atmosphere; LP, low pressure.

INTRODUCTION

It is generally accepted that the time between the excision of cuttings from the stock plant and insertion into the rooting medium should be as short as possible to ensure good rooting. While storage of cuttings is not generally of advantage to a propagation program, there may be several reasons for storing them:

Short term storage (up to a few weeks)
—long distance transport of cuttings (mainly herbaceous) produced in climatically favorable regions may be necessary.
—large numbers of cuttings (mainly softwood and herbaceous) harvested from stock plants or from prunings cannot always be inserted immediately due to space and time limitations.
—controlled chilling treatments of dormant cuttings (conifer and hardwood) in order to improve rooting may be desirable.
Long term storage (up to several months)
—herbaceous cuttings to be rooted during periods of low light intensity are sometimes stored after being produced during periods of high light intensity.
—propagation procedures may require collection and preparation of conifer cuttings before

severe frosts in early winter; storage during the winter and insertion into the propagation bed in early spring in order to reduce both energy costs and cutting losses.
—very long storage periods may be necessary for gene preservation programs (hardwood cuttings).

CONDITIONS FOR STORING CUTTINGS

For long periods of storage in particular, conditions must be chosen which insure the rooting potential of cuttings and which will not impair the development of a saleable liner. To achieve these conditions, four main factors should be taken into account:
—water stress should be minimized (Loach 1977, also see Chapter 8 by Loach).
—carbohydrate reserves should not be depleted to a level which will reduce disease resistance or rooting potential (Hansen et al. 1978, Loach and Whalley 1978, Behrens 1984, see Chapter 5 by Veierskov).
—spread of fungal diseases should be prevented (Doesburg 1962, Smith 1982).
—accumulation of gases (CO_2, ethylene) to concentrations noxious to plant material should be prevented.

Unfavorable storage conditions influence propagation success directly by killing cuttings or by reducing rooting potential. Cuttings can also be affected indirectly by adverse storage conditions. For instance, needles and stems may appear healthy after storage, but due to water and dry matter losses susceptibility to *Botrytis cinerea* (Grey Mold) and soil born fungi may be increased.

The potential storage duration is greatly influenced by relative humidity, temperature, the gaseous composition of the storage atmosphere, the species, pathogens, growing conditions of stock plants, and collection date. The three major environmental parameters influencing storage will be discussed in the following sections.

Relative Humidity

Biochemical activity of the protoplasm is not only influenced by the absolute water content, but also by the water potential. Loach (1977) showed that there is a connection between adventitious root growth and leaf water potential, optimum rooting requiring a high potential. Therefore, transpiration should be reduced as much as possible during storage. This can be done by maintaining the relative humidity at nearly 100%, as was shown by results with intact plants (de Haas and Wennemuth 1962, Nyholm 1975).

Water potential of plant tissue is influenced not only by water uptake and transpiration, but also by the anabolism and catabolism of osmotically active compounds, especially the low molecular weight carbohydrates. Thus, increasing or constant water potentials may occur during storage (Davis and Potter 1985), despite continuous transpiration.

In a forced-air cooled chamber high relative humidity is difficult to maintain at temperatures close to the freezing point and is not achievable below 0 °C (Bünemann 1961). Therefore, it is necessary to use a jacketed cold store or to wrap storage containers with plastic film, which is impermeable to water vapor, to maintain cutting water status. Packing of the cuttings in perforated polybags further lowers the risk of desiccation during long term storage. When cuttings are stored for long periods in tightly wrapped plastic film or poly bags, gas exchange must be maintained. This can be done by using a polyethylene film that is permeable to CO_2 and ethylene. Experience with storage of fruit indicates that polyethylene films of 0.03–0.05 mm in thickness are suitable (Duvekot 1961, Werminghausen 1964). Perforated plastic film or a wide ratio of indirectly cooled air to plant material is also effective. In any case, some water loss is unavoidable. Therefore, only completely turgid cutting material should be stored.

Freezing cuttings in ice blocks does not seem to be a viable means of reducing water loss during storage. For example, birch (*Betula* spp.) cuttings frozen at −18 °C for seven months exhibited normal shoot growth but did not root (Spethmann 1982), and conifer cuttings frozen at −4 °C fermented (Behrens and Bünemann 1985).

Temperature

Lowering storage temperature is an important treatment in retarding respiration and spread of fungal diseases during storage of plant material (de Haas and Wennemuth 1962, de Haas and Bünemann 1962, Kappen 1967, Edney 1973, Fidler 1973, Heursel and Kamoen 1976, Davis and Potter 1985). The extent to which temperatures may be lowered, however, depends on the chilling resistance or cold hardiness of the cuttings. Only a few trials on storing cuttings under various temperatures have been published and as yet only a rough grouping is possible:

— Cuttings of several plants, usually of tropical origin, cannot survive temperatures below 5 °C.
— Herbaceous cuttings of most indoor and greenhouse plants as well as softwood cuttings of hardy woody plants cannot be successfully stored at temperatures below −0.5 °C to +2 °C.
— Completely dormant conifer and broadleaved hardwood cuttings of hardy woody plants store best at temperatures below freezing. Such cuttings can tolerate −4 °C, if acclimated to cold on the stock plant (Behrens 1986).

A temperature of −10 °C was not suitable for storage of spruce cuttings (Jestaedt and Rapp 1977), and after 4–7 months of storage at −18 °C only very few cuttings of spruce rooted (Spethmann 1982). It is likely that leafless hardwood cuttings can be stored at temperatures down to −18 °C, but upon removal from storage they need to be thawed slowly.

Davis and Potter (1985) found no differences in the rooting potential of Rhododendron catawbiense Michx. 'Roseum Elegans' cuttings, if stored at 2 °C or at 21 °C for three weeks. In the case of scion-wood of Rhododendron cultivars stored for much longer periods (up to 34 weeks), however, Härig (1986) found a better graft take after 0 °C storage than after 2 °C storage. It appears that the longer the desired storage period the more important it is to use low temperatures (Heursel and Kamoen 1976).

Only frost resistant cuttings can be stored at temperatures below freezing (Ritchie 1982, Behrens 1986). Cuttings from stock plants which were exposed to several days of freezing temperatures were stored successfully at −4 °C whereas significant losses occurred from stock plants not completely acclimated to frost. Thus, the latter should be stored at 0 °C to −2 °C only.

It should be noted that within a batch of cuttings wrapped in plastic, the heat of respiration causes higher temperatures in the center of the bundle than in the surrounding environment. Depending upon how tightly the plant material is packed, temperatures within a bundle of cuttings could be 2 °C higher than an ambient storage temperature at 0 °C (Behrens 1986).

Atmospheric Gases

Manipulation of storage atmospheric gases is an additional treatment for reducing fungal attack, catabolism, and the formation of ethylene. Two methods are generally used:

— storage with controlled atmosphere (CA storage).
— storage at reduced pressure (hypobaric storage, LP storage).

As a rule, CA storage involves lowering the oxygen concentration and raising the CO_2 concentration depending on plant variety and to a lesser extent on storage duration. Considerable experience has been gained with the CA storage of fruit, vegetables, and cut flowers (Bünemann and Hansen 1973, Fidler 1973, Isenberg 1979), but very little is known about CA storage of cuttings (Behrens 1986). To maintain the composition of normal atmosphere (ca. 0.03% CO_2 and 21% O_2) constant during storage of conifer cuttings at −4 °C in air-tight containers, the continuous removal of CO_2 by an activated charcoal scrubber and enrichment with O_2 was necessary (Behrens 1986). Low temperatures slow down catabolism, nevertheless O_2 continues to decrease and CO_2 continues to accumulate. The O_2 content during CA storage should not be allowed to drop below 2–5%, otherwise fermentation may occur. The CO_2 content tolerated by cut flowers varies from 5–25%. Roses tolerate 13–21% CO_2, lilac 15% CO_2, and carnation for longer storage periods 4% (Jansen et al. 1984). How unrooted cuttings respond to raised concentrations of CO_2 is not yet known. According to Lipton (cited in Isenberg 1979) CO_2 toxicity appears primarily in tissue lacking chlorophyll.

Behrens (1986) stored cuttings of 10 different conifer genotypes at −2 °C for four months in an atmosphere of 3% O_2 and 3% CO_2. Only two genotypes showed a higher rooting percentage compared to storage without CA. Two further genotypes showed a lower rooting percentage with CA

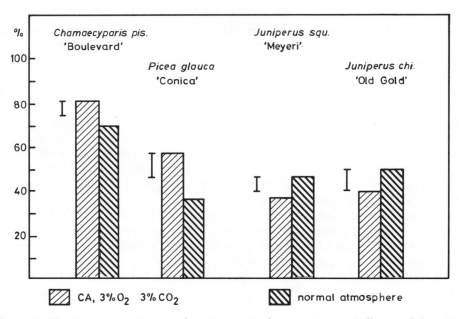

Figure 1. Rooting percentages of various conifer species as influenced by controlled atmosphere storage at −2 °C. Bars indicate L.S.D. 0.05. [From Behrens (1984)]

(Fig. 1). Visually detectable damage such as necrosis or leaf fall was not evident in either storage condition. CA during cold storage leads to a further reduction in the heat of respiration (Isenberg 1979), and the temperature within the plant material is closer to that of the storage unit. Thus, diminished rooting potential under CA conditions might be caused by frost damage.

LP storage entails the reduction of the atmospheric pressure to about 10% of the normal atmosphere, e.g. 10 kPa rather than 100 kPa (Dilley et al. 1975). Partial pressures of all atmospheric gasses are thereby also lowered to 10% of normal. Thus, an O_2 concentration of about 2% is achieved and respiration is diminished. Another important effect of low atmospheric pressure is an increased release of ethylene from the plant tissue. At an atmospheric pressure of about 10 kPa, ethylene escapes 10 times more rapidly than under normal atmosphere so the potential for ethylene induced injury is reduced. Few investigations have been carried out on the amount of ethylene produced by cuttings during cold storage, although experience with cut flowers (Jansen et al. 1984) and the results of Paton and Schwabe (1987) indicate that all vegetative plant tissue produces ethylene. Furthermore, storage under cold conditions does not ameliorate damage to vegetative plant parts by ethylene (Curtis and Rodney 1952).

Burg (1973) and Eisenberg et al. (1978) stored unrooted cuttings under LP conditions and in all cases LP storage was superior when compared with simple cold storage. One disadvantage of LP storage, however, is the lowering of water vapor concentration in the atmosphere. In order to prevent desiccation of the stored material at low pressure, continuous ventilation with water saturated air is necessary. Usually LP storage is combined with the lowest possible temperature tolerated by the plant material (Jensen and Bredmose 1979).

DURATION OF STORAGE

Table 1 lists various species and cultivars, and indicates how long and under what conditions cuttings have been successfully stored. Only the best storage conditions as found by the respective author are included, i.e. the longest possible duration without significant reduction of rooting potential or cutting quality in comparison to a control. Hardwood cuttings are not included because the winter storage of completely dormant, leafless, hardy cuttings is a well established propagation practice without any major storage problems. Very long storage periods are necessary for gene preservation programs, but are not considered here.

Table 1. Maximum duration of storage period of unrooted cuttings from a variety of species. Storage conditions and references are also shown.

Species	Excision Date	Maximum Storage Duration (d)	Temperature (°C)	Humidity Control[1]	Reference
Begonia Elatior cvs.	—	10	10	PF	Röber (1977)
	—	35	15	LP	Kirk and Andersen (1986)
Schlumbergera cv.	—	35	15	LP,MB	Kirk and Andersen (1986)
Hedera helix L. cv.	—	35	15	LP,MB	Kirk and Andersen (1986)
Dieffenbachia maculata cv.	—	35	15	LP,MB	Kirk and Andersen (1986)
Epipremnum pinnatum	—	35	15	LP,MB	Kirk and Andersen (1986)
Kalanchoe blossfeldiana v. Poellnitz cv.	—	35	15	LP,MB	Kirk and Andersen (1986)
Nemathanthus radicans	—	35	15	LP	Kirk and Andersen (1986)
Fuchsia triphylla L.	—	35	15	LP	Kirk and Andersen (1986)
Cissus rhombifolia Vahl.	—	35	15	LP	Kirk and Andersen (1986)
Hydrangea macrophylla Ser. cvs.	—	20	3–6	PF	Röber (1977)
Euphorbia pulcherrima Willd. cvs.	—	10–12	10	PF	Röber (1977)
	—	28	12–13	MB	von Hentig and Knösel (1986)
	—	7	5	LP	Eisenberg et al. (1978)
Pelargonium × *hortorum* Bailey cv.	—	14–21	5	LP	Eisenberg et al. (1978)

[1]PF: plastic film or poly bags; LP: low pressure storage; S: sealed in plastic bags or jars; J: jacketed cold store and perforated poly bags; CA: controlled atmosphere storage; MB: moistened burlap or humidified air

Species	Excision date	Maximum Storage Duration (d)	Temperature (°C)	Humidity Control[1]	Reference
Pelargonium × hortorum Bailey cv.	—	35	4	S	Paton and Schwabe (1987)
Pelargonium spp.	—	35	−0.5	S	Baudendistel (1957)
Hibiscus rosa-sinensis L. cv.	—	21	15	LP	Andersen and Kirk (1986)
Impatiens spp.	July	7–14	5	S	Judd (1976)
Chrysanthemum cv.	—	42–56	−0.5	S	Langhans (1954)
Chrysanthemum vestitum cv.	—	35	15	LP	Kirk and Andersen (1986)
Chrysanthemum indicus L. cvs.	—	30	2–3	PF	Röber (1977)
Chrysanthemum (tender cvs.)	—	42	0–2	LP	Burg (1973)
Chrysanthemum (hardy cvs.)	—	84	0–2	LP	Burg (1973)
Dianthus cv.	—	240	0–2	LP	Burg (1973)
	May/June	150–180	0.5	—	Holley (1961)
Dianthus caryophyllos L. cvs. (carnation)	—	90–120	0.5–1	PF	Röber (1977)
Rhododendron simsii Planch. cvs.	—	56	3–4	S	von Hentig and Bohländer (1972)
	—	70	1	PF	Heursel and Kamoen (1976)
Rhododendron obtusum Planch. cvs.	—	70	1	PF	Heursel and Kamoen (1976)
Rhododendron ponticum L. cvs.	—	70	1	PF	Heursel and Kamoen (1976)
Rhododendron (Kurume)	—	70	−0.5–4	S	Pryor and Stewart (1963)
Rhododendron catawbiense Michx. cv.	Aug.	21	2–21	MB	Davis and Potter (1985)
Rhamnus frangula L. cv.	—	42	1–2	LP	Eisenberg et al. (1978)
Ligustrum obtusifolium Sieb. & Zuci. cv.	—	42–63	1–2	LP	Eisenberg et al. (1978)

Species	Excision date	Maximum Storage Duration (d)	Temperature (°C)	Humidity Control[1]	Reference
Viburnum opulus L. cv.	—	42	1–2	LP	Eisenberg et al. (1978)
Ilex crenata Thunb. cvs.	Feb.	80	1–2	S	Parkerson (1980)
Taxus × media Rehd. cv.	Jan.	70	1	—	van Elk (1976,1977)
Juniperus communis L. cv.	Jan.	70	1	—	van Elk (1976,1977)
Thuja occidentalis L. cv.	Dec.	90	1	—	van Elk (1976,1977)
	Nov.	118	−2–0	J	Behrens (1986)
Chamaecyparis lawsoniana Parl. cvs.	Nov.	118	−2–0	J	Behrens (1986)
Chamaecyparis pisifera Sieb. & Zucc. cvs.	Nov.	118	−2–0	J	Behrens (1986)
Juniperus chinensis L. cvs.	Nov.	118	−2–0	J	Behrens (1986)
Juniperus squamata Lamb. cv.	Nov.	118	−2–0	J	Behrens (1986)
Picea glauca (Moench) Voss	Nov.	118	−2–0	CA,J	Behrens (1986)
Taxus × media Rehd. cv.	Nov.	118	−2–0	J	Behrens (1986)
Picea abies (L.) Karst.	Jan.	138	0	PF	Jestaedt and Rapp (1977)
Picea sitchensis Carr.	Nov.	70	2	—	Van Den Driessche (1983)
Pinus radiata D. Don	Nov/Dec.	90	3	PF	Libby and Conkle (1966)
Pseudotsuga menziesii Mirb (Franco)	Oct/Nov.	60	0	PF	Roberts and Fuchigami (1973)
Abies fraseri Poir.	Oct.	56	4	S	Wise et al. (1985)

PRETREATMENT OF CUTTINGS

Only cuttings of good quality should be stored. This means that pretreatment begins with the stock plant which should be well supplied with all growth factors. In order to obtain completely turgid cuttings on collection, the plants must be well watered before cutting harvest. Other factors should conform to those usually required for optimum rooting (see Chapter 16 by Moe and Andersen). Treatments to promote food reserve accumulation should be undertaken before cutting if the cuttings are scheduled for long term storage. In particular, plants should be exposed to high light intensities while high nitrogen feeding of the stock plants should be avoided. Excessive nitrogen fertilization may lead to decreased rooting potential and increased rotting of stored cuttings (Röber 1977).

Lignified, hardened, or quiescent cuttings store better than those growing actively. For this reason, the date of collecting cuttings should be taken into account as much as possible.

For successful storage, stock plants should be disease free. According to Heursel and Kamoen (1976) a prophylactic treatment of *Rhododendron* spp. mother stock with a systemic fungicide 10 d before harvest was more effective than treating cuttings directly prior to storage. Herbaceous and softwood cuttings should be surface dry (Langhans 1954, Baudendistel 1957) or allowed to dry after fungicide treatments (Eisenberg et al. 1978) before storing. The latter is labor intensive and hazardous on a commercial scale. Behrens and Bünemann (1985) found that applying fungicide powder only to the base of conifer cuttings was as effective as a liquid or powder treatment of the entire cutting. Species susceptible to fungal attack need particular attention to good hygiene. If rooting hormones are necessary they should be combined with the fungicide because post-storage application of rooting hormone is often impossible due to the fungicide film. Investigations by van Elk (1976, 1977) did not give definite results on whether hormone treatments should be done prior to or after storage. Most published storage procedures report that hormone treatments after storage result in adequate rooting (e.g. Jestaedt and Rapp 1977, Eisenberg et al. 1978, Davis and Potter 1985).

Many investigators have stored cuttings that are ready for planting, but most have wounded or recut the basal ends prior to insertion into the rooting medium (Libby and Conkle 1966, Judd 1976, Eisenberg et al. 1978, Davis and Potter 1985, Wise et al. 1985). The need for this additional operation has not yet been documented. Von Hentig and Bohländer (1972) found improved propagation results for *Rhododendron simsii* Planch if intact shoots (10 cm) were stored instead of cuttings. Stripping of leaves or needles at the basal ends of cuttings can be done prior to storage without detrimental effects on rooting (Roberts and Fuchigami 1973, Behrens 1986).

In general, cuttings should not be bundled or packed tightly during storage because higher temperatures and reduced gas exchange may lead to fungal attack and diminished rooting (Behrens 1985). Refrigeration for at least several hours at higher temperatures than the final storage temperature might be advantageous for acclimatizing cuttings harvested from a heated greenhouse (Heursel and Kamoen 1976, Eisenberg et al. 1978).

PHYSIOLOGICAL CHANGES DURING STORAGE

Three main endogenous factors are influenced during storage of cuttings: water potential, food reserves, and growth substances. By using optimum storage conditions, changes in water potential can be kept to a minimum as discussed earlier. Some studies have followed changes in food reserves during storage, but the significance of these changes regarding rooting potential is still unclear. Few studies have demonstrated changes in endogenous growth substances during storage, but such changes are presumed to occur because dormancy can often be broken during storage.

Food Reserves

In general, cuttings are not storage organs, so none of the typical plant food reserves such as sugars, starch, lipids, nitrogenous compounds, hemicelluloses, and fructosans (Kramer and Kozlowski 1979) can be expected to be found in extensive quantities. Low temperatures and CA storage slow down the metabolic processes which deplete food reserves. Nevertheless, there is a constant need for energy and organic compounds. Therefore, a decline in carbohydrate content of cuttings should be expected in a dark storage room during long term storage. Such a decline was reported by Ritchie (1982) for Douglas-fir and by McCracken (1979) for pine seedlings. In contrast, Kappen (1967) analyzed four sugars of Douglas-fir seedlings separately and found a constant decline in raffinose only. Sucrose, fructose, and galactose content remained about the same before and after storage, but showed erratic decreases and increases of up to 100% during storage. Behrens (1984, Diss. Univ. Hannover, W. Germany) found that changes in the carbohydrate content of conifer cuttings depended on storage temperature and species (Fig. 2). Only *Juniperus squamata* Lamb. 'Meyeri' and *Taxus* × *media* Rehd. 'Hicksii' showed a more or less constant decline in carbohydrates, if stored

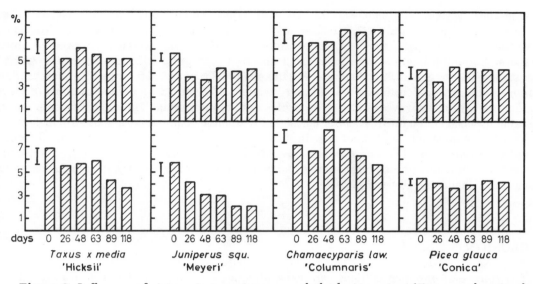

Figure 2. Influence of storage temperature on carbohydrate content (% sugar plus starch in dry matter) of various conifer species during a 118 d storage period at −2 °C (top row) or +2 °C (bottom row). Bars indicate L.S.D. 0.05. [From Behrens (1984)]

at +2 °C. If stored at −2 °C the reduction was limited to the initial storage weeks only. *Chamaecyparis lawsoniana* Parl. 'Columnaris' and *Picea glauca* (Moench) Voss 'Conica' showed erratic carbohydrate increases and decreases at both temperatures during storage. The carbohydrate contents before and after the four month storage period, however, were about the same. In all cases starch was degraded rapidly in the first weeks of storage. Likewise, Robinson and Schwabe (1977) failed to find a continuous decline in carbohydrate content of stored apple root cuttings, neither for total soluble sugars nor for total extractable polysaccharides. These results suggest that food reserves, other than the carbohydrates investigated, must have been used during storage.

After storing *Rhododendron catawbiense* Michx. 'Roseum Elegans' cuttings for 21 d, Davis and Potter (1985) found that changes in carbohydrate concentrations were not large enough to influence rooting. At the end of a 118 d storage period, however, a strong correlation was found between the content of the investigated carbohydrates and the rooting percentages of conifer cuttings (V. Behrens. 1984, Diss., Univ. Hannover, W. Germany). The sucrose content in particular was strongly correlated to rooting though this in itself may not have had a regulatory role in rooting. The role of carbohydrates in adventitious root formation is discussed more fully in Chapter 5 by Veierskov.

Low lipid levels prior to storage and the minor changes by the end of a four month storage period indicated that lipid was not an important food reserve for stored conifer cuttings (V. Behrens. 1984, Diss., Univ. Hannover, W. Germany). In the same study, neither protein nor other soluble nitrogenous compounds appeared to be used as reserves. The significance of hemicelluloses as food reserves is still unclear. These polysaccharides are often important components of the cell wall (Beck and Wieczorek 1977), but whether or not they influence rooting is not clear.

Breaking Dormancy

Cuttings of *Taxus, Juniperus,* and *Thuja* spp. root much better if stock plants have been subjected to a period of low temperature (Wells 1979). Completely dormant hardwood cuttings collected in the fall may require a cold treatment for optimum rooting (Zsuffa and Saul 1977, Hare and Land 1982). Because weather conditions may either prevent sufficient chilling of stock plants, prevent harvesting of cuttings, or cause winter damage, the chilling requirements of conifer and hardwood cuttings are more reliably achieved in controlled cold storage (Shelton and Moore 1981, Cram and Lindquist 1982). In fact, several researchers have stored conifer cuttings (Table 1) with the objective of improving rooting by chilling (Libby and Conkle 1966, Roberts and Fuchigami 1973, van den

Driessche 1983, Wise et al. 1985). Rooting seems to require less chilling than does vegetative bud break (Roberts and Fuchigami 1973, Hinesley and Blazich 1980). Investigations with hardwood cuttings of *Castanea sativa* Mill. showed that cold storage influenced the content of endogenous growth substances. Rooting inhibitors were inactivated and root promoting cofactors seemed to increase during storage (Gesto et al. 1981). Fadl and Hartmann (1967) reported similar results for pear (*Pyrus communis* L.) hardwood cuttings. Iwasaki and Weaver (1977) stored hardwood cuttings of grapes (*Vitis vinifera* L.) for eight weeks at 0 °C to achieve optimum rooting. Their data indicate that the extractable abscisic acid content was not correlated with either bud break or root growth.

Very active buds may start shoot growth shortly after inserting cuttings and thus hinder rooting and further development (Behrens 1986). Such undesirable shoot growth resulting from collection after breaking dormancy or long term storage can be retarded by short day treatments during rooting (Snyder 1955, Bhella and Roberts 1979).

CONCLUSION

The storage of unrooted cuttings can be an effective element in a propagation program. The duration of the storage period depends on storage conditions, state of the cuttings, and species. Desiccation, dry matter losses, and fungal diseases must be minimized during storage. Within the storage unit the relative humidity should be kept as close as possible to 100%, and temperature must be maintained as low as can be tolerated by cuttings of a given species and their physiological condition. Reduced O_2 and ethylene levels in the storage atmosphere are also helpful in maintaining rooting capacity.

The amount of food reserves, and the degree of lignification, frost hardiness, and dormancy depend on the species and the collection date, and have an influence on storage duration. For these reasons, the potential storage duration varies from some days up to several months.

Future research regarding storage of cuttings is needed in two general areas. First the practical side of storing cuttings on a commercial scale:

—how long can cuttings of various species actually be stored in a forced-air cooled chamber with simple wrapping in plastic film?

—how tightly can storage containers be sealed for various storage durations?

—do the advantages of CA, LP, or jacketed cold storage more than compensate for the additional costs?

—can intact branches be stored more successfully than prepared cuttings?

—what are the minimum temperatures and the maximum storage durations for various species and varieties?

Second, the scientific aspects need to be addressed:

—which growth substances change during storage and how is this influenced by storage conditions (temperature, photoperiod, atmospheric gas composition, humidity)?

—what amount of ethylene is produced by cuttings and what is its significance?

—in addition to sugars and starch, do cuttings use other materials as food reserves?

—can other CO_2/O_2 combinations improve maintenance of vigor and food reserve status?

—is there an interaction between food reserves and growth substances that influences rooting potential?

ACKNOWLEDGMENTS

The author wishes to thankfully acknowledge Prof. Dr. G. Bünemann and Dr. D. MacCarthaigh, University of Hannover, West Germany, and Dr. W. Spethmann, Lower Saxony Forest Research Station, West Germany for reviewing this paper.

REFERENCES

Andersen, A. S. and H. G. Kirk. 1986. Low pressure storage of herbaceous cuttings. *Acta Hortic.* 181:305–312.

Baudendistel, R. 1957. Geranium storage. *NY State Flower Growers Bull.* 140:4.

Beck, E. and J. Wieczorek. 1977. Carbohydrate metabolism. In *Progress in Botany 39* (H. Ellenberg, K. Esser, H. Merxmüller, E. Schnopf and H. Ziegler eds.). Springer Verlag, NY, Heidelberg, Berlin. pp. 62–82. ISBN 0-387-08501-7.

Behrens, V. 1984. Storage of unrooted coniferous cuttings. *Proc. Int. Plant Prop. Soc.* 34:274–286.

_____ 1986. Kühllagerung von unbewurzelten Koniferenstecklingen II. Lagertemperatur und atmosphäre. *Gartenbauwissenschaft* 51:118–125.

_____ G. Bünemann. 1985. Kühllagerung von unbewurzelten Koniferenstecklingen I. Vorbehandlungen und Verpackung. *Gartenbauwissenschaft* 50:161–169.

Bhella, H. S. and A. N. Roberts. 1974. The influence of photoperiod and rooting temperature on rooting of Douglas-fir (*Pseudotsuga menziesii*). *J. Amer. Soc. Hortic. Sci.* 99:551–555.

Bünemann, G. 1961. Die Regelung der relativen Luftfeuchtigkeit im Lagerraum durch die Mantelkühlung. Dt. *Gartenbauwirtschaft* 9:92–94.

_____ H. Hansen. 1973. *Frucht- und Gemüselagerung.* Eugen Ulmer, Stuttgart, W. Germany. pp. 23–86. ISBN 3-8002-5256-8.

Burg, S. P. 1973. Hypobaric storage of cut flowers. *HortScience* 8:202–205.

Cram, W. H. and C. H. Lindquist. 1982. Overwinter storage of Walker poplar cuttings. *For. Chron.* 58:167–177.

Curtis, O. F., Jr. and D. R. Rodney. 1952. Ethylene injury to nursery trees in cold storage. *Proc. Amer. Soc. Hortic. Sci.* 60:104–108.

Davis, T. D. and J. R. Potter. 1985. Carbohydrates, water potential, and rooting of stored *Rhododendron* cuttings. *HortScience* 20:292–293.

Dilley, D. R., W. J. Carpenter and S. P. Burg. 1975. Principles and application of hypobaric storage of cut flowers. *Acta Hortic.* 41:249–268.

Doesburg, J. van. 1962. Use of fungicides with vegetative propagation. In *Proc. 16th Int. Hortic. Cong. Brussels.* Vol. IV. pp. 365–372.

Driessche, R. van den. 1983. Rooting of sitka spruce cuttings from hedges, and after chilling. *Plant and Soil* 71:495–499.

Duvekot, W. S. 1961. *Storage of Apples and Pears Under Plastic.* Annexe 1961-1 au Bull. de l'institut intern. du Froid.

Edney, K. L. 1973. Fungal disorders. In *The Biology of Apple and Pear Storage* (J. C. Fidler, B. G. Wilkinson, K. L. Edney and R. O. Sharples, eds). Commonwealth Agric. Bur. Res. Rev. 3. pp. 133–172. ISBN 85198-256-5.

Eisenberg, B. A., G. L. Staby and T. A. Fretz. 1978. Low pressure and refrigerated storage of rooted and unrooted ornamental cuttings. *J. Amer. Soc. Hortic. Sci.* 103:732–737.

Elk, B. C. M. van. 1976, 1977. Het koelen van coniferenstek. In *Jaarboek Proefstation voor de Boomkwekerij, Boskoop.* pp. 30–31, 31–32.

Fadl, M. S. and H. T. Hartmann. 1967. Relationship between seasonal changes in endogenous promoters and inhibitors in pear buds and cutting bases and the rooting of pear hardwood cuttings. *Proc. Amer. Soc. Hortic. Sci.* 91:96–112.

Fidler, J. C. 1973. Conditions of storage. In *The Biology of Apple and Pear Storage.* (J. C. Fidler, B. G. Wilkinson, K. L. Edney and R. L. Sharples, eds). Commonwealth Agric. Bur. Res. Rev. 3. pp. 17–44. ISBN 85189-256-5.

Gesto, M. D. V., A. Vazquez and E. Vieitez. 1981. Changes in the rooting inhibitory effect of chestnut extracts during cold storage of the cuttings. *Physiol. Plant.* 51:365–367.

Haas, P. G. de and G. Bünemann. 1962. Kühllagerung von Baumschulgenhölzen, IV. Die Mantelkühlung bei −2 °C im Vergleich zur Direktkühlung bei +1.5 °C und zur herkömmlichen Einschlagmethode. *Gartenbauwissenschaft* 27:243–246.

_____ G. Wennemuth. 1962. Kühllagerung von Baumschulgehölzen, II. Pflanzenbauliche und physiologische Probleme. III. *Botrytis*- und *Fusarium*befall an Gehölzen im Kühllager. *Gartenbauwissenschaft* 27:243–246.

Hansen, J., L. H. Strömquist and A. Ericsson. 1978. Influence of the irradiance on carbohydrate content and rooting of cuttings of pine seedlings (*Pinus sylvestris* L.). *Plant Physiol.* 61:975–979.

Hare, R. C. and S. B. Land, Jr. 1982. Effect of cold storage and chemical treatment on rooting of hardwood sycamore cuttings. *Can. J. For. Res.* 12:417–419.

Härig, R. 1986. Ergebnisse zur Kühllagerung von *Rhododendron*reisern. *Baumschulpraxis* 16(8):304–305.

Hentig, W.-U., von and H. Bohländer. 1972. Mutterpflanzenbehandlung und Stecklingslagerung bei Azaleen. *Gartenwelt* 72:365–367.

_____ K. Knoesel 1986. Storage of poinsettia cuttings. *Acta Hortic.* 181:313–322.

Heursel, J. and O. Kamoen. 1976. Preservation of cuttings of *Rhododendron obtusum* Planch., *R. ponticum* L. and *R. simsii* Planch. *Scientia Hortic.* 4:87–90.

Hinesley, L. E. and F. A. Blazich. 1980. Influence of postseverance treatments on the rooting capacity of Fraser fir stem cuttings. *Can. J. For. Res.* 11:316–323.

Holley, W. D. 1961. Effects of long term storage on carnation cuttings. *Colorado Flower Growers Assoc. Bull.* 133:1–2.

Isenberg, F. M. R. 1979. Controlled atmosphere storage of vegetables. *Hortic. Rev.* 1:337–394.

Iwasaki, K. and R. J. Weaver. 1977. Effects of chilling, calcium cyanamide, and bud scale removal on bud break, rooting, and inhibitor content of buds of 'Zinfandel' grape (*Vitis vinifera* L.). *J. Amer. Soc. Hortic. Sci.* 102:548–587.

Jansen, H., E. Bachthaler, E. Fölster and H.-C Scharpf. 1984. *Gärtnerischer Pflanzenbau.* Ulmer Verlag, Stuttgart, W. Germany. pp. 414–426. ISBN 3-8001-2524-2.

Jensen, H. E. K. and N. Bredmose. 1979. Planter under lavtryk. *Gardner Tidende* 95:512–513.

Jestaedt, M. and H.-J. Rapp. 1977. Versuche zur Lagerung von Fichtenstecklingen. *Forstarchiv* 48:10–13.

Judd, R. W., Jr. 1976. Storage and rooting of *Impatiens* cuttings. *Conn. Greenhouse Newsl.* 71:8–9.

Kappen, L. 1967. Physiologische Einflüsse der Kühlhauslagerung auf die Douglasie (*Pseudotsuga menziesii* (Mirb.) Franco). *Allg. Forst- u. Jagdztg.* 138:181–185.

Kirk, H. G. and A. S. Andersen. 1986. Influence of low pressure storage on stomatal opening and rooting of cuttings. *Acta Hortic.* 181:393–397.

Kramer, P. J. and T. T. Kozlowski. 1979. *Physiology of Woody Plants.* Academic Press, NY, USA. pp. 268–277. ISBN 0-12-425050-5.

Langhans, R. W. 1954. *Chrysanthemum* cutting storage. *NY State Flower Growers Bull.* 106:1.

Libby, W. J. and M. T. Conkle. 1966. Effects of auxin treatment, tree age, tree vigor, and cold storage on rooting young Monterey pine. *For. Sci.* 12:484–502.

Loach, K. 1977. Leaf water potential and the rooting of cuttings under mist and polythene. *Physiol. Plant.* 40:191–197.

_____ D. N. Whalley. 1978. Water and carbohydrate relationships during the rooting of cuttings. *Acta Hortic.* 79:161–168.

McCracken, I. J. 1979. Changes in the carbohydrate concentration of pine seedlings after cool storage. *N. Z. J. For. Sci.* 9:34–43.

Nyholm, I. 1975. Cold-storage of plants. *Acta Hortic.* 54:143–145.

Parkerson, C. H. 1980. Cold storage pretreatment of cuttings. *Proc. Int. Plant Prop. Soc.* 30:483–484.

Paton, F. and W. W. Schwabe. 1987. Storage of cuttings of *Pelargonium* × *hortorum* Bailey. *J. Hortic. Sci.* 62:9–87.

Pryor, R. L. and R. N. Stewart. 1963. Storage of unrooted softwood *Azalea* cuttings. *Proc. Amer. Soc. Hortic. Sci.* 82:483–484.

Ritchie, G. A. 1982. Carbohydrate reserves and growth potential in Douglas-fir seedlings before and after cold storage. *Can. J. For. Res.* 12:905–912.

Roberts, A. N. and L. H. Fuchigami. 1973. Seasonal changes in auxin effect on rooting of Douglas-fir stem cuttings as related to bud activity. *Physiol. Plant.* 28:215–221.

Robinson, J. C. and W. W. Schwabe. 1977. Studies on the regeneration of apple cultivars from root cuttings. II. Carbohydrate and auxin relations. *J. Hortic. Sci.* 52:221–233.

Röber, R. 1977. Lange Lagerzeiten möglich. *Deutsche Gärtnerbörse* 77:140–142.

Shelton, L. L. and J. N. Moore. 1981. Field chilling vs. cold storage of highbush blueberry cuttings. *HortScience* 16:316–317.

Smith, P. M. 1982. Diseases during propagation of woody ornamentals. In *Proc. 21st Int. Hortic. Cong.* Vol. 2. pp. 884–893.

Snyder, W. E. 1955. Effect of photoperiod on cuttings of *Taxus cuspidata* while in the propagation bench and during the first growing season. *Proc. Amer. Soc. Hortic. Sci.* 66:397–402.

Spethmann, W. 1982. Einlagerungsversuch mit Fichten-, Birken- und Kirschenreisern. Tätigkeitsber. Nieders. Forstl. Vers. Anst., Abt. Forstpfl. Zücht., Germany. pp. 39.

Wells, J. S. 1979. *Plant Propagation Practices.* MacMillan Pub., NY, USA. pp. 243–304.

Werminghausen, B. 1964. Erfahrungen über Kunststoffe im Obstbau. *Erwerbsobstbau* 6:146–148.

Wise, F. C., F. A. Blazich and L. E. Hinesley. 1985. Propagation of *Abies fraseri* by semidormant hardwood stem cuttings. *HortScience* 20:1065–1067.

Zsuffa, L. and G. H. Saul. 1976. *The Rooting Ability of Winter Stored and Freshly Planted Cuttings of Leuce Duby (Aspen) Hybrid Clones.* For. Res. Note No. 4, Maple, Ontario, Canada.

CHAPTER 18

Controlling Environmental Conditions to Improve Adventitious Rooting

K. Loach

Institute of Horticultural Research, Worthing Road,
Littlehampton, West Sussex, BN17 6LP, England

INTRODUCTION..249
ENCLOSURES...249
   Outdoor Propagation Under Low Tunnels............................249
      The environment 249
      Rooting results 250
   Enclosures in a Greenhouse (Non-Misted)...........................251
      Shading..251
      Tent and contact systems...................................252
      Wet tents..254
MIST..255
   The Environment Under Mist..255
      Temperature..255
      Leaf wetness and ambient humidity 255
      Irradiance...256
      Heating requirements under mist propagation 257
      Outdoor mist 259
   Enclosed Mist 259
      Temperature in misted enclosures...........................259
      Humidity, leaf wetness, and turgor 259
      Rooting results and practical guidelines...................259
FOGGING...261
   Principles 261
   Development 261
   Fogging Equipment 262
   Fog Controllers 262
   Capacity 263
   Distribution 265
   Conclusion 265
BASAL HEAT..266
   Equipment 266
   Basal Heat Settings 266
ROOTING MEDIA...267
   Practice 267
   Chemical and Biological Aspects...................................268
SANITATION..269
CONCLUSION..270
ACKNOWLEDGMENTS...270
REFERENCES..271

Additional key words: basal heat, disease, environment, fog, mist, propagation, rooting media, shading.

Abbreviations: ACR, air change rate; LAVPG, leaf-to-air vapor pressure gradient; PE, polyethylene; r.h., relative humidity; s.d. saturation deficit; V_{air}, vapor pressure of water in the air; V_{leaf}, vapor pressure in the leaf; ψ, water potential; π, osmotic potential.

INTRODUCTION

Without proper environmental control, the rooting of many types of cuttings is very difficult. Tremendous variability in rooting, or in some cases complete failure to root occur if environmental conditions are not adequately controlled. The conditions required for propagation from leafy cuttings fall under three main headings. First and foremost, propagation systems aim to maintain an atmosphere with a low evaporative demand, so that transpiration from the cuttings is minimized and any substantial tissue water deficits are avoided. This is because cuttings without roots have no effective organs for replacing transpired water, and cells at the site of root initiation must maintain an adequate level of turgor to foster the regenerative processes (see Chapter 8 by Loach for more information). A second requirement is the maintenance of a suitable temperature to stimulate metabolism in the base of the cutting. However, high leaf temperatures are to be avoided because they can increase water loss and promote bud growth at the expense of rooting. The third need is to maintain the irradiance within a range sufficient to generate photosynthetic production of carbohydrates for root growth, yet avoid excessive leaf temperatures.

The aim of this chapter is to describe and assess the methods which have been developed to achieve these conditions.

ENCLOSURES

The driving force for loss of water from leafy cuttings is the LAVPG, i.e. $V_{leaf}—V_{air}$. This gradient can be minimized either by reducing V_{leaf} through controlling leaf temperature, or by increasing the V_{air} around the cuttings. Enclosures are used primarily to maintain V_{air} at near-saturation by preventing the escape of water vapor which has been transpired by the cuttings or evaporated from wet surfaces beneath the covers. The major advantages of enclosures are simplicity and low cost. The main disadvantage is that because the system is closed, heat is trapped below the covers if irradiance is high. In this case, the s.d. of the air increases as does leaf temperature, and in consequence the LAVPG rises thus increasing the rate of water loss and making shading essential.

In traditional cold-frame propagation the covers used were glass held in wooden frames but these were relatively leaky to water vapor and the most commonly used material at present is PE film. This material has a low permeability to water vapor (Dubois 1978), is inexpensive, and is produced in a wide range of convenient widths. PE-covered enclosures are used both outdoors and in the greenhouse.

Outdoor Propagation Under Low Tunnels

The environment

Low, PE-covered frames have been used over outdoor nursery beds to root cuttings since the 1950s (Templeton 1953, Buckley 1955, van Hoff 1958, Rowe-Dutton 1959). Left unshaded, however, soil and air temperatures beneath the covers are frequently excessive. For example, on a clear August day in England (outdoor air temperature = 27 °C), Deen (1971) measured air temperatures of up to 43 °C under a clear PE tunnel and 30 °C under white, translucent film. Even with two layers of white PE (56% shade), air temperatures > 40 °C were reported at Efford Experimental Horticulture Station on six days out of 19 in July (Anon. 1985). In a Maryland (USA) summer, Lewandowski and Gouin (1982) measured very high temperatures under a low, white PE tunnel; when the air temperature outdoors was 35.5 °C, the temperature under the tunnel was reportedly 65.5 °C. Hence, for climates with high irradiance in summer, heavy shading is essential. For this purpose, the use of microfoam thermo-blankets has proven useful. A layer of microfoam 63.5 mm thick, covered with 100 μm white polyethylene, gave 81% shade (Lewandowski and Gouin 1985), and the maximum air temperature below the cover was a relatively moderate 43.3 °C, when outdoor ambient was 35.5 °C. Over 42 d in the months of July and August, when the mean daily temperature outdoors was approximately 28 °C, the corresponding temperatures beneath the thermo-blanket cover averaged only 2.3 °C higher (Lewandowski and Gouin 1985).

While many cuttings will survive temperatures $> 40\,°C$, sustained exposure may cause cell injury and tissue water deficits sufficient to retard metabolic activity. Wong et al. (1971) found that soil temperatures in excess of 35–$40\,°C$ inhibited root growth of woody species for several days after exposure. Nevertheless, there are reports of cuttings surviving short exposures to quite extreme temperatures (e.g. $52\,°C$, Templeton 1953; $57.5\,°C$, Buckley 1955).

Deen (1971) pointed out that for low outdoor tunnels, the shading which is necessary for bright days, prevents soil-warming and causes generally sub-optimal temperatures in the rooting medium when the irradiance is lower. To counter this, Lewandowski and Gouin used rubberized heating pads on the ground under their tunnels after late October. However in March, cuttings of *Euonymus kiautshovica* Loes. rooted equally well, with or without this basal heating (Lewandowski and Gouin 1982).

PE films are now available with a range of additives (vinyl acetate, alumina, or magnesium silicates) which substantially increase their opacity to long-wave, infrared radiation (Nisen et al. 1984). Incoming short-wave solar radiation absorbed by the cuttings and rooting medium beneath the cover and re-radiated at longer wavelengths, is trapped by these "modified" films to a greater extent than occurs with ordinary PE. The rise in temperature should therefore be greater, but because this so-called "greenhouse effect" owes more to a reduction in turbulent air movement within the enclosed space than to the transmission characteristics of the cover (Maher and O'Flaherty 1973), there is relatively little difference in daytime temperatures whichever film is used. For example, comparative measurements in unheated greenhouses covered with modified and ordinary PEs showed a $2.5\,°C$ difference at an outdoor irradiance of $200\ W\cdot m^{-2}$ and $3.4\,°C$ difference at $400\ W\cdot m^{-2}$ (Bernaud et al. 1984). Night-time temperatures should also be greater under the modified PE because outgoing long-wave radiation from the cuttings and medium will be retained more effectively. Again, the measurements by Bernaud et al. showed that the difference was small ($<1\,°C$) for unheated houses. However, if base heating were used, the energy savings from the use of modified PE should be substantial and worthwhile. Comparative tests of ordinary and modified PEs for low, propagation tunnels have not been reported but, on the basis of the above observations, we should not expect large differences in rooting.

Provision of irrigation, either as mist or applied directly to the rooting medium, has often been introduced to reduce the s.d. which develops in bright conditions. Mist under outdoor, PE-covered, low tunnels was the basis of the "Phytotektor Method" developed by Templeton (1953). With appropriate management, this system has proven adaptable to a wide range of conditions.

Rooting results

Trials at Efford Experimental Horticultural Station showed that in the British climate, mist proved more effective in summer than did the use of an additional lightweight ($20\ \mu m$), clear PE film, laid directly over the cuttings (Burgess 1984). Mist was also superior to ground irrigation supplied by seephose for 26 out of 29 propagations of woody species in summer (mean rooting percentages were 82 vs. 54% respectively; Anon. 1985). However, for propagations started in March and November mist was not used. At these times, an additional layer of PE supported just above the cuttings gave reasonably good results with many species (*Rhododendron* spp. excepted), and excellent results for some of the more easily rooted conifers (e.g. × *Cupressocyparis leylandii* (Dallim. and Jack.) Dallim., *Prunus laurocerasus* L. 'Otto Luyken', *Skimmia* and *Euonymus* cvs.). The mean percentage rooting in non-misted tunnels was 60% for 30 varieties inserted in autumn and 53% for 21 insertions in spring. Burgess (1984) lists 61 species and varieties which rooted successfully in low PE-covered tunnels and this list has subsequently been extended (Anon. 1985).

In Maryland, Lewandowski and Gouin (1985) compared the effectiveness of thermo-blankets covered with white polyethylene vs. a glasshouse mist unit. Cuttings of *Euonymus kiautschovica* Loes. inserted in July rooted equally well in both systems (100%), though misted cuttings showed the greatest gain in root dry weight. Cuttings of both green and variegated cultivars of *Pachysandra terminalis* Sieb. & Zucc. rooted better under glasshouse mist versus thermoblankets (97 vs. 84% for the variegated and 84 vs. 62% for the green cultivar). Nevertheless, it was found that a wide range of easy-rooting species would tolerate the summer conditions under microfoam covers and root well.

These included *Forsythia × intermedia* Zab., *Viburnum rhytidophyllum* Hemsl., *Hedera helix* L., *Juniperus horizontalis* Moench., *Ilex crenata* Lindl., *Pyracantha crenulata* (D. Don) Roem., *Jasminum nudiflorum* Lindl., *Rubus* spp., and others (Gouin 1980). The system proved effective throughout the year, if base-heating was provided in winter.

In summary, outdoor propagation under low tunnels is undoubtedly a cost-effective method for many species. For summer propagations, substantial shading is essential for good results. If mist is provided under the covers, the quality of the plants is often improved and the range of successful species increased. Even with mist, low tunnels outdoors remain inexpensive systems relative to greenhouse propagation. They present a particularly attractive proposition for the direct insertion of cuttings into containers, which would otherwise occupy large areas of comparatively expensive glasshouse space. Use of such methods is widespread and variations in technique are more extensive than can be covered in this review.

Enclosures in a Greenhouse (Non-Misted)

Propagation under covers in a greenhouse presents the opportunity for closer control of the environment than is possible outdoors; consequently more attention has been devoted to characterizing optimal conditions. More difficult species can be rooted successfully under this system and the speed of rooting is often improved compared to outdoor propagation.

Shading

Loach and Whalley (1978) propagated a number of species in the glasshouse under PE covers with varied levels of shading. When the cuttings received an average daily radiation < 1.5 $MJ \cdot m^{-2} \cdot d^{-1}$ (total short-wave), the percentage rooting declined sharply (Fig. 1), probably because the supply of photosynthate was inadequate to support rooting. These data are helpful in devising shading schemes, but can provide no more than a guideline. Cuttings have different reserves of carbohydrate at insertion, different rates of photosynthesis and respiration, and different rates of utilization of carbohydrate at growing points within the cutting, all of which influence the light requirement during propagation.

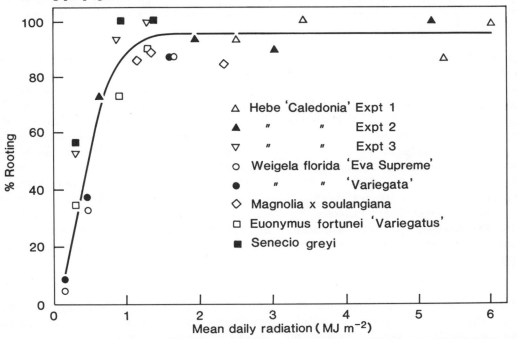

Figure 1. The relationship between rooting and the mean daily radiation (total short-wave) received by cuttings under a polyethylene cover in a glasshouse. Glasshouse heating was initiated at 10 °C and venting at 15 °C. With < 1.5 $MJ \cdot m^{-2} \cdot d^{-1}$, the rooting percentage decreased substantially. [From Loach and Whalley (1978)]

Loach (1980) devised shading schemes for glasshouse propagation under PE covers in U.K. conditions on the assumption that the optimal radiation integral would be no more than twice the minimum value, i.e. 3 MJ·m^{-2}·d^{-1}. This should allow sufficient photosynthesis but not risk the development of tissue water deficits in the cuttings. The shading scheme calls for different levels of shade on bright and dull days, and in different months of the year, calculated from average radiation data at a particular location (e.g. West Sussex, U.K. in the example shown in Fig. 2).

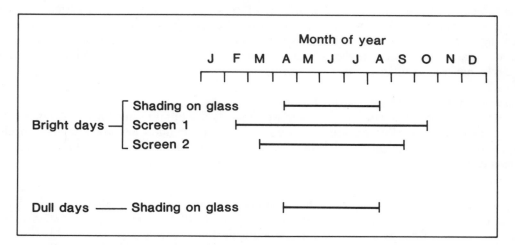

Figure 2. Example of a suggested shading scheme for propagation of hardy ornamental species under polyethylene covers in West Sussex, U.K. The scheme is based on average radiation data for the site and aims to achieve 3 MJ·m^{-2} · d^{-1} (total short-wave) at the cuttings. For purposes of calculation, a bright day is defined as one with radiation 50% greater than average; an overcast (dull) day is one with 50% less mean daily radiation. The operator judges the appropriate shade requirements using these guidelines. Shading of 40% is assumed for each layer. The scheme requires modification for different climates.

Automatically operated, light regulated shading curtains are available to maintain irradiance within a pre-set range in the greenhouse. A single layer of shade cloth is insufficient for full summer sunlight. For example, to reduce an irradiance of 1000 W·m^{-2} to a desirable 100 W·m^{-2}, requires 86% shading (assuming 70% transmission by the greenhouse structure itself); in contrast, use of a single screen of this density would give too much shade for other times of year. The solution to this dilemma is to provide additional layers of shade in summer—either as a shading compound applied externally to the house (Mastalerz 1977, p. 18), or as further layers of shading fabric mounted above or within the house. This extra shading should be removed for winter. Internal shade screens reduce the leaf temperatures (e.g. by 5.3 °C for a 50% shade cloth according to Gray, quoted in Mastalerz 1977), but air temperatures are not greatly altered because the amount of solar energy entering through the greenhouse cover is unaffected. In warm climates, shading alone may prove inadequate for avoiding excessive air temperatures under a PE enclosure within a greenhouse. Additional means of reducing the greenhouse air temperature are then necessary, e.g. fan and pad cooling or ventilated fogging (see section on Fogging).

Tent and contact systems

Polyethylene covers used for propagation can either be laid directly on the cuttings ("contact" PE), or supported above them in the form of a low tent. These two methods can give different results, depending upon the time of year. In summer, the contact system gave better results than a tent but in winter the results were reversed (Table 1, Loach unpublished data). Changes in the water content of cuttings were measured by harvesting, drying, and weighing samples at insertion and again five weeks later, prior to root emergence. It was shown that cuttings in the contact system gained water

Table 1. Contrasted performance of polyethylene (PE) propagation systems in summer and winter: (1) Contact-PE laid directly on the cuttings and (2) Tent-PE supported above the cuttings. All differences between systems statistically significant at P<0.05. Cuttings of *Santolina* wilted in both systems but most severely in the tent.

| Season | Species | | Percentage Rooting | |
			Contact	Tent
Summer	*Cornus alba* L. 'Spaethii'		83	70
	Forsythia × intermedia Zab.		97	68
	Olearia × scilloniensis Dorrien-Smith		92	78
	Santolina chamaecyparissus L.		26	8
		Means	75	56
Winter	*Cupressus glabra* Sudw.		35	71
	Cryptomeria japonica D. Don 'Elegans'		53	95
	Rhododendron 'Joseph Whitworth'		42	67
	Rhododendron 'Fastuosum Flore Pleno'		76	50
		Means	52	71

during propagation (expressed relative to their water content at the time of insertion, Table 2) but those in the tent often lost water. In summer conditions, when maintaining the turgor of the cuttings is a prime concern, the contact system is preferred. In winter when irradiance is lower and transpiration from the cuttings is correspondingly reduced, the tissues at the base of the cutting can become waterlogged and rooting may be poor with the contact system. The drier, tent system gives best results under these conditions. For further details see Chapter 8 by Loach.

Grange and Loach (1983) examined reasons for differences in the water status of cuttings under contact and tent PE systems. The vapor pressure of the air surrounding the cuttings was similar in both systems but the leaves were cooler under contact PE. This occurred partly because in the contact system, many of the leaves were in direct contact with the PE sheet whose temperature was intermediate between glasshouse and enclosure air temperatures, but also because of evaporative cooling of the leaves wetted by condensate from the cover. In contrast, leaves under the tent are usually dry. Leaf wetting has an additional significance to which we shall return when mist is discussed; water lost to the surrounding air from externally wetted leaves does not originate entirely from within the tissues of the cutting. Consequently, they suffer less internal water stress than dry leaves, when exposed to the same LAVPG.

From regression equations relating measured leaf and air temperatures and saturation deficits to irradiance (Grange and Loach 1983), LAVPGs for tent and contact systems were calculated in Table 3. There was little difference between systems at $100 \text{ W} \cdot \text{m}^{-2}$ (total short-wave) but at higher irradiances, the LAVPG was smaller under contact PE. This, combined with the advantage of leaf wetting in the latter system, accounts for the difference in water status of the cuttings observed in Table 2.

Table 2. Changes in water content of cuttings under contact and tent polyethylene systems, prior to root emergence. All differences between systems significant at P<0.05. The cuttings were inserted in October (first three species) and December (last species). Glasshouse heating was initiated at 10 °C and venting at 15 °C.

| Species | | Change in Water Content $(\text{mg} \cdot \text{g}^{-1} \cdot \text{d}^{-1})$ | |
		Contact	Tent
Cupressus glabra Sudw.		+ 4.7	− 3.3
Cryptomeria japonica D. Don 'Elegans'		+ 7.0	+ 1.2
Rhododendron 'Fastuosum Flore Pleno'		+ 0.7	− 2.3
Rhododendron 'Joseph Whitworth'		+ 1.5	− 1.9
	Means	+ 3.5	− 1.6

Table 3. Leaf (V_{leaf}) and air (V_{air}) vapor pressures for cuttings of *Forsythia × intermedia* Zab. under contact and tent polyethylene covers at three irradiances. Data calculated for a glasshouse temperature of 20 °C, from regression equations derived by Grange and Loach (1983, Figs. 2 and 3, and Table 1).

Irradiance at the Cutting ($W \cdot m^{-2}$ total shortwave)	Vapor Pressures (kPa) Under					
	Contact			Tent		
	V_{leaf}	V_{air}	V_{leaf}-V_{air}	V_{leaf}	V_{air}	V_{leaf}-V_{air}
100	3.0	2.4	0.6	3.2	2.4	0.8
200	3.9	2.7	1.2	4.4	2.7	1.7
300	5.1	3.2	1.9	6.0	3.2	2.8

Wet tents

Quite a different kind of enclosure for greenhouse propagation has been developed at Oklahoma State University, following an idea used by Franclet in France. A tent of polyester fabric is suspended from a water-filled PVC pipe, centrally positioned above the propagation bed. The fabric is inserted into a longitudinal slit in the pipe and thus receives a small but continuous flow of water through it (Whitcomb 1982). Antecedents of this kind of system include the "Burlap Cloud" method (Hancock 1959) which was used outdoors and consisted of light, portable wooden frames with burlap covers and sprinklers to wet the burlap. For propagating cuttings of cacao (*Theobroma cacao* L.), Evans (1952) used a white domestic cloth supported over the cuttings and wetted by a continuous mist spray from above. This system was not successful in tropical conditions and was abandoned. In a glasshouse, Loach (1977) used a fabric cover supported on a frame over the cuttings and misted intermittently from above. This "mist plus mesh" method gave greater leaf water potentials (i.e. the foliage was less water-stressed) and better rooting than did conventional intermittent mist or nonmisted white or clear PE enclosures (Table 4).

Table 4. Mean leaf water potentials and rooting parameters for cuttings of *Hebe* 'Caledonia' under four propagation systems. Root number was estimated on the basis of a 0–5 visual score; all differences were significant at $P<0.05$. Percentage rooting and root length means followed by the same letter are not significantly different. Water potentials are means of daily measurements between 21 April and 14 May.

Treatment	Water Potential (MPa)		Rooting Parameters		
	am	pm	%	Number	Length (mm)
Mist	−1.30	−1.61	99a	4.0	55.5a
Mist + mesh	−0.96	−1.36	100a	4.4	59.1a
White polyethylene	−1.28	−1.63	93a	3.2	41.8b
Clear polyethylene	−1.17	−1.83	87a	2.7	43.1b

A wet tent serves to maintain a consistently high ambient humidity (V_{air}) within, while providing shade to reduce leaf temperatures and hence V_{leaf}. Convective air movements ensure some passage of air through the fabric, which helps prevent extreme air temperatures from developing within the tent. The Oklahoma State University system has been used effectively on a wide range of cuttings, e.g. Whitcomb et al. (1982) reported that 3 of 11 varieties rooted better or gave less leaf drop in a wet tent than under intermittent mist. Practical disadvantages associated with use of wet tents include the inconvenience of access to the cuttings through a wet fabric and algal contamination of the fabric in some circumstances. No physical measurements of the environment within the tent to facilitate a direct comparison of its effectiveness with other systems have yet been reported.

A recent refinement of the wet tent method includes the use of two mist lines, one in the tent and one above the tent (Whitcomb et al. 1984). The inner mist line operates on a reduced frequency and serves to wet the foliage, a feature found necessary on very warm days. The upper mist line moistens the tent. Advantages over the original system are: less reliance on the wicking quality of the fabric so that a lighter material with good light transmission can be used, and elimination of the need to insert and hold the fabric into the slit pipe (Whitcomb et al. 1984). The major disadvantage of this system is its relative complexity.

MIST

Mist propagation originated in the late 1930s and early 1940s, was promoted commercially in the 1950s, and probably remains the most widely used propagation system for leafy cuttings. Its basic principle, as outlined in the studies of Hess and Snyder (1955), is that leaves which are sprayed intermittently are effectively cooled through evaporation of this applied film of water. The V_{leaf} and LAVPG thus remain low and water loss is restricted, which maintains the cuttings in a turgid condition. Unlike the enclosure systems already described, mist seeks to minimize V_{leaf} rather than maximize V_{air}. A favored advantage of open mist over enclosed systems is that cuttings are immediately accessible and the propagator can easily view their condition and attend to any obvious problems.

The Environment Under Mist

Temperature

The effects of mist on leaf and air temperatures are illustrated in Table 5 which compares misted and non-misted sections of a propagation bench over a 2 h period in a shaded glasshouse (Loach 1979). Leaf temperatures were measured by 10 differential leaf/air thermocouples on each 1 m^2 section of bench, and air temperatures and vapor pressures by a miniature aspirated psychrometer, shielded from the mist spray. The average irradiance (total short-wave) over the 2 h was 88 W·m^{-2}. Leaf temperatures of the misted cuttings averaged 7.2 °C lower than for non-misted cuttings. Air temperatures were 4.8 °C lower under mist, presumably because falling droplets cooled the air through evaporation. Movement of cooled air over the cuttings on the open bench (advection) makes a significant, additional contribution to the moderation of leaf temperatures. Advective cooling does not occur to any great extent in the enclosed systems previously described.

Misted cuttings were only 0.3 °C cooler than the adjacent air but dry leaves on the non-misted section were 2.1 °C above adjacent air temperature. Assuming that the internal leaf air is saturated at these leaf temperatures, and using the measured air vapor pressures, the calculated LAVPGs are shown in Table 5. The LAVPG was three times greater for non-misted vs. misted cuttings. If the leaf resistances to water loss (primarily stomatal) were the same for misted and non-misted cuttings, mist would reduce the water loss correspondingly by a factor of three.

Table 5. Effects of mist on leaf and air temperatures, vapor pressures, and the leaf-to-air vapor pressure gradient (LAVPG) for cuttings of *Forsythia* × *intermedia* Zab. Data are the means on misted and non-misted sections of a propagation bench, over a 2 h period with an average irradiance of 88 W·m^{-2} (total short-wave) at the cuttings (Loach 1979).

Parameter	Mist	No Mist
Leaf temperature (°C)	21.7	28.9
Air temperature (°C)	22.0	26.8
Leaf vapor pressure (kPa)	2.6	4.0
Air vapor pressure (kPa)	2.0	2.2
Leaf-to-air vapor pressure gradient (kPa)	0.6	1.8

Leaf wetness and ambient humidity

A further important feature of wetted foliage under mist, which has already been mentioned in the comparison of tent and contact PE systems, is that much of the water lost to the surroundings is not internal, tissue water but is externally applied water. Therefore, for misted cuttings the water deficit developed in the tissues should be appreciably less than that expected from the magnitude of the LAVPG. Moreover in some species, absorption of water can occur through the foliar surfaces (see Chapter 8 by Loach for further information).

On a misted bench, the s.d. of the air around the cuttings increases towards the ambient greenhouse value between mist bursts (Fig. 3). The film of water deposited on the foliage is inevitably imperfect because of different foliar inclinations and differences in the wettability of the leaf surfaces. For example, the spray seldom reaches the leaf under-surfaces where a majority of the stomata may be located. Loss of some tissue water from exposed portions of the foliage is thus unavoidable. Increasing the mist frequency to wet the foliage more thoroughly can increase the leaching of nutrients from the leaves (see Chapter 4 by Blazich) and also over-wet the rooting medium, thus reducing oxygen diffusion to developing root initials. More frequent misting also reduces evaporative cooling of the leaves and increases V_{leaf}, so that the expected reduction in LAVPG is not fully realized.

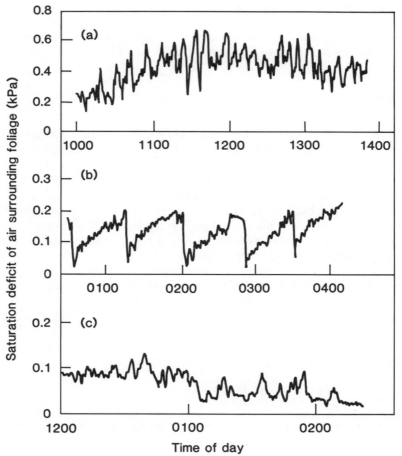

Figure 3. Changes in the saturation deficit of the air around cuttings on an open mist bench, with misting frequency controlled by an electronic leaf: (a) a bright March day; (b) a dry, cold winter night; and (c) a humid summer night. [From Grange and Loach (1983)]

Irradiance

Hess and Snyder (1955) argued that an important advantage of mist, in comparison to propagation under double-glass covered frames, was the higher levels of irradiance which could be tolerated, thus increasing photosynthesis in the cuttings. However, there is evidence that photosynthesis may be limited by the restricted "sink-capacity" for carbohydrates in unrooted cuttings (Loach and Gay 1979) and by stomatal closure, so that very little light is needed to saturate photosynthesis (Davis and Potter 1987). Shading experiments suggest that reduced light can be beneficial for rooting of some species under mist. Grange and Loach (1985) propagated three species under mist, with shading to produce four radiation regimes averaging from 1.2 to 3.4 MJ·m⁻²·d⁻¹ at the cuttings. With increasing irradiance the osmotic potentials decreased, i.e. solutes accumulated, either through concentration of existing solutes following water loss or through photosynthetic production of new solutes. This served to offset declining water potentials so that in two of the species, leaf turgor differed by little more than 0.2 MPa between treatments. In *Hibiscus syriacus* L. 'Woodbridge', water stress caused the leaves to roll so that they intercepted mist less effectively and leaf turgor was low, especially at high irradiance (Fig. 4a).

Rooting did not generally relate to leaf turgor (Fig. 4b). Cuttings of *Viburnum* × *bodnantense* Stearn 'Dawn' rooted best in the lowest radiation treatment, 1.2 MJ·m⁻²·d⁻¹, *Hibiscus syriacus* L. 'Woodbridge' in the second lowest treatment, 1.6 MJ·m⁻²·d⁻¹, while *Weigela florida* A.D.C. 'Eva Supreme' behaved irregularly relative to radiation (Fig. 4b). Poor rooting in the first two species appeared to be associated with higher solute accumulation. The practical implication is that shading to achieve a radiation integral of around 1.5 MJ·m⁻²·d⁻¹ will improve rooting of some species under mist. Subsequent growth of initiated roots requires higher radiation, an important consideration in formulating a shading policy.

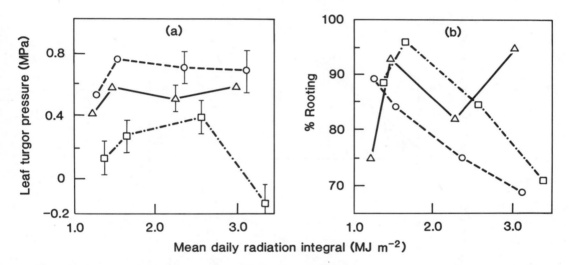

Figure 4. Leaf turgor (a) in cuttings under open-bench mist, shaded to vary the daily radiation integral (total short-wave). Rooting percentage (b) in relation to daily radiation integral. Species were *Weigela florida* (Bge.) A.D.C. 'Eva Supreme' (△), *Viburnum* × *bodnantense* Stearn 'Dawn' (○), and *Hibiscus syriacus* L. 'Woodbridge' (□). In the latter two species, there was a sharp decrease in rooting with increasing radiation, which was unrelated to changes in leaf turgor. Bars show ± one S.E. if > 0.05 MPa. [From Grange and Loach (1985)]

Heating requirements under mist propagation

As already pointed out, open-bench mist in summer maintains cooler temperatures around the cuttings than do enclosed systems. However, at other times of the year these lower temperatures prevail and can be disadvantageous. Table 6 shows the relative usage of electricity for basal heating set at 20 °C, with glasshouse heating initiated at 13 °C and venting at 17.5 °C, during the winter months of December and January. Open-bench misting used around twice as much electricity as the two enclosed systems. Similar measurements at Luddington Experimental Horticultural Station (Cooper 1977) showed that even in late summer, a shaded, PE-covered enclosure in a glasshouse without mist, used only 35% of the electricity required on an open mist bench, with base heat settings of 20 °C in both systems. The question of optimal temperature settings for rooting is considered in a separate section later in the chapter.

Table 6. Relative electricity usage expressed as a percentage of usage in open-bench mist system for basal heating in different propagation systems during winter. Basal heat setting 20 °C, glasshouse heating initiated at 13 °C and venting at 17.5 °C.

| | Relative Electricity Usage | | |
Month/Year	Open-Bench Mist	Enclosed Mist	Contact Polyethylene
December, 1980	100	75	52
January, 1981	100	56	31
Means	100	66	42

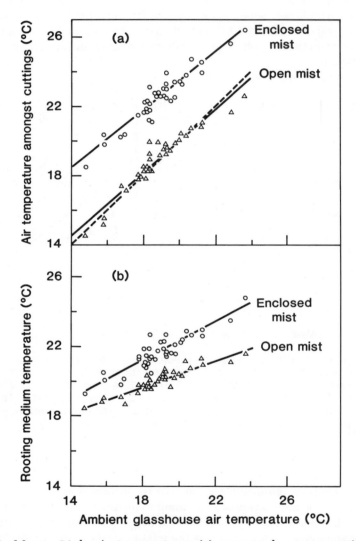

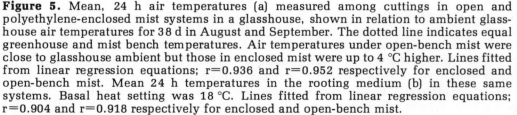

Figure 5. Mean, 24 h air temperatures (a) measured among cuttings in open and polyethylene-enclosed mist systems in a glasshouse, shown in relation to ambient glasshouse air temperatures for 38 d in August and September. The dotted line indicates equal greenhouse and mist bench temperatures. Air temperatures under open-bench mist were close to glasshouse ambient but those in enclosed mist were up to 4 °C higher. Lines fitted from linear regression equations; r=0.936 and r=0.952 respectively for enclosed and open-bench mist. Mean 24 h temperatures in the rooting medium (b) in these same systems. Basal heat setting was 18 °C. Lines fitted from linear regression equations; r=0.904 and r=0.918 respectively for enclosed and open-bench mist.

Outdoor mist

Mist is often used outdoors in warm climates (Rowe-Dutton 1959, Weller 1959). Wind protection, such as a PE or shade netting screen, is essential to prevent drift of the mist spray. A coarser nozzle, to give a heavier droplet than is customary for greenhouse mist is preferred. Shading for bright weather and provision of a sand bed with good drainage are also beneficial. Basal heating is not usually supplied so outdoor mist is most often used only in the summer.

Enclosed Mist

Mist has been used within PE enclosures since the 1950s (Templeton 1953). Enclosures inevitably change the environment, with important consequences for the cuttings. These changes will be discussed in the following sections.

Temperature in misted enclosures

In temperate climates, temperatures on an open mist bench are frequently suboptimal, in which circumstances a low PE tent mounted above the bench at a height sufficient to clear the mist spray can give a useful temperature rise. This is illustrated in Fig. 5a which reports the 24 h mean air temperatures for cuttings on an open mist bench and for a similar misted bench enclosed with clear, 38 μm PE, in the same glasshouse over 38 consecutive days in August and September in the U.K. Temperatures in the rooting medium (Fig. 5b) and ambient glasshouse air temperatures were recorded by a computer scanning at 20 s intervals. Control settings were as follows: glasshouse heating initiated at 10 °C, base-heating at 18 °C, venting at 15 °C, and automatic shading provided by two shade screens, activated at 100 W·m^{-2} and giving 15% of outdoor radiation at the cuttings with both in place. Misting frequency was controlled by an electronic leaf on the open bench but by time-clock under the enclosure—a 2 s burst every 9 min.

Not surprisingly, mean (24 h) air temperatures measured on the bench amongst the cuttings were closely related to glasshouse air temperatures (r=0.952 and 0.936 for open and enclosed mist, resp.; Fig. 5a). However, air temperatures within the enclosure were 3.6 °C greater than glasshouse ambient. Differences in the temperature of the rooting medium were greatest on warm days but averaged 1.6 °C higher in the enclosure than on the open bench (Fig. 5b). Air temperatures were also recorded during the warmest part of the day at 1300 h. Differences between systems were, of course, greater than for the 24 h means, and averaged 4.6 °C warmer in the enclosure. Maximum temperatures recorded were 27.4 °C on the open bench and 33.7 °C in the enclosure, neither of which is excessive. Providing that enclosed mist is able to maintain the cuttings in a turgid condition, temperatures within the tent should be closer to the optimum for rooting than those on the open mist bench (though this will depend greatly on climate).

Humidity, leaf wetness, and turgor

The PE and mist combination reduce the fluctuations in ambient humidity that are a feature of open-bench mist. Typically, the s.d. in enclosed mist is about one-quarter of that in the latter system (Grange and Loach 1983) and leaves are at least partially wetted, which provides further "insurance" against desiccation. We should, therefore, expect that leaf ψ of cuttings under enclosed mist would be higher than in open mist. Measurements (at 1400 h on bright days) confirmed this expectation (see Table 1 of Chapter 8 by Loach) but π was also greater (i.e. less solute), possibly because higher temperatures increased the rate of respiration. In consequence, cell turgor (π-ψ) was similar in both systems. Nevertheless, cuttings under enclosed mist showed a greater gain in water content in the first few weeks following insertion, than did those in open mist (see Fig. 4 of Chapter 8 by Loach).

Rooting results and practical guidelines

Comparative trials of open-bench and PE-enclosed mist systems in glasshouses have been carried out at the New Zealand Nursery Research Centre and in the U.K. (Grange and Loach 1984, Richards 1984). Gains in percentage rooting due to enclosure averaged 15% (Table 7). Several important practical observations have been made from the foregoing trials. First, the enclosure should be no larger than is needed to clear the mist spray. Second, experience has shown that over-frequent misting

Table 7. Early rooting trial results comparing open-bench and polyethylene-enclosed misting systems in New Zealand and the U.K. [The New Zealand results are taken from two separate trials in the Annual Report of the New Zealand Nursery Research Centre (Anon. 1982); U.K. results from Loach (1987)]

Location	Year	Number of Species	% Rooting in Mist		Gain Through Enclosure
			Open	Enclosed	
New Zealand	1982a	7	67	81	+14
	1982b	6	62	76	+14
U.K.	1983	6	44	67	+23
	1984	9	70	81	+11
	Means		61	76	+15

must be avoided. For some species, especially conifers, over-misting can entirely prevent root initiation, though maintaining the cuttings in a conspicuously fresh condition. Results of an experiment to examine the influence of the misting frequency on rooting of *Weigela florida* A.D.C. 'Variegata', *Berberis thunbergii* DC. 'Pink Queen', and *Thuja plicata* D. Don in summer are summarized in Fig. 6. Best rooting occurred with mist applied for 3 s at 15 min intervals during daylight hours. Two rooting media were used, mixtures of peat and fine pine bark in 3/1 and 1/3 volume proportions. Neither of these two media greatly influenced rooting, except at 60 min misting frequency, when the more open medium with a higher proportion of bark may have been too dry.

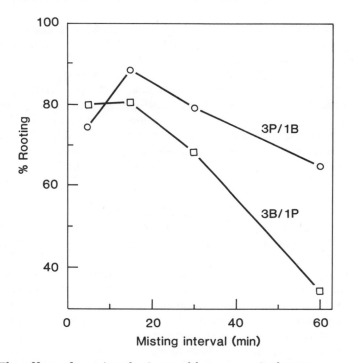

Figure 6. The effect of varying the interval between mist bursts on percent rooting in a polyethylene-enclosed mist system. Bursts were 3 s in length, over 18 h per day. Rooting percentages are means for three species, *Weigela florida* (Bge.) A.D.C. 'Variegata'. *Berberis thunbergii* DC. 'Pink Queen', and *Thuja plicata* D. Don. Two rooting media were used: 3/1 (3P/1B) and 1/3 (3B/1P) volume proportions of peat/pine bark. The S.E.D. (P=0.05, n=6) for the interaction, misting interval × medium = 11.3. The species were inserted respectively, on 28, 29, and 30 August, 1985; glasshouse heating was initiated at 10 °C, venting at 15 °C.

Table 8. Summary of guidelines for propagation under polyethylene-enclosed mist in a glasshouse (Loach 1987). These are based on U.K. conditions and may need modification when used elsewhere.

1. Mist frequency (controlled by interval timer; OFF 1 h after sunset, ON 1 h before dawn) 　　　　Summer— 2 s burst each 15 min 　　　　Winter—　2 s burst each 30 min
2. Shading 　　　　Summer—two layers 50% shade cloth 　　　　Winter—no shade 　　　　Intermediate times and overcast summer weather—one layer

Finally, temperature within a PE enclosure depends not only on the irradiance but also on glasshouse and outdoor air temperatures. In warm climates, the temperatures within the enclosure may become super-optimal in summer unless forced ventilation or evaporative cooling is employed. In these conditions, shading is also advisable to reduce incoming radiation to around 2–$3\,MJ\cdot m^{-2}\cdot d^{-1}$.

Recommendations based on current experience are summarized in Table 8 but may need modification in different climates.

FOGGING

Principles

While mist systems aim to reduce V_{leaf} through cooling the leaves, fog systems operate at the other end of the LAVPG, and aim to maximize V_{air}, i.e. raise the ambient humidity. In the case of ventilated fog (see below), there may be some cooling of the leaves through conductive transfer of heat from the leaves to the cooled air passing around them, but this is not the primary aim of fogging.

To achieve maximum humidity, fog generators produce very fine droplets ($<20\,\mu m$) which have two helpful characteristics. First they remain airborne for a relatively long period, which increases the likelihood of evaporation; and, second, the surface/volume ratio (which doubles for each halving of the droplet diameter) is high, so that finely divided water presents a very large surface over which evaporation can occur. Because the water passes into the air as vapor rather than falling on to the surfaces of the leaves and rooting medium, leaching of nutrients and metabolites from the foliage, and over-wetting of the medium, are avoided. Also, because less water is evaporating from the surface of the rooting medium than occurs with mist, the medium is not being cooled to the same extent so less basal heating is required to maintain a favorable base temperature.

Development

Fogging systems were being used for propagation at about the same time that mist was being developed (Stoutemyer 1942). Because the fog generating equipment used at that time had a low output (1–$4\,l\cdot h^{-1}$) it did not readily produce a visible fog and was therefore referred to as humidification rather than fogging. Another consequence of the low output was that the propagation house had to be tightly closed to maintain the humidity, hence it was difficult to avoid excessive temperatures. Heavy shading was used so disease in the cuttings was difficult to prevent. Humidification was tested extensively at the Boskoop Experimental Station in Holland in the 1950s and was used in some commercial nurseries (Stroombeek 1958). However, it was still handicapped by the low output of the equipment used and gradually fell out of favor in comparison with mist. Equipment with a larger output was available even at that time, e.g. units with up to $45\,l\cdot h^{-1}$ or assemblies of smaller nozzles (each $2.4\,l\cdot h^{-1}$) (Chadwick 1951), but these either produced relatively large droplets or were expensive.

In recent years a wide range of new equipment has been developed, mainly for purposes other than propagation, e.g. in the textile, pulp, and paper industries, and for frost protection in orchards. Availability of this equipment has re-awakened interest in use of fog for propagation of cuttings.

Further impetus has resulted from the advent of micropropagation. Micropropagules removed from the humid protection of their culture vessels are prone to wilting when transferred to the greenhouse. Moreover, they are easily damaged if water droplets from misting adhere to them, thus making fog advantageous over mist in this context.

Fogging equipment

The range of equipment available is extensive but all employ one of three basic operational principles:

1. *Centrifugal foggers,* which use centrifugal force to atomize water. Their output ranges from 1–4 $l \cdot h^{-1}$ for the models used in the 1940s to a mammoth 190 $l \cdot h^{-1}$ for equipment that was developed initially for ventilated cooling of greenhouses. Generally speaking, the models with a large output produce a rather mixed range of droplet sizes but with a preponderance of diameters $< 25 \mu m$. Centrifugal foggers are produced in a variety of models in many countries. Electric motors from 1/40 to 3/4 horsepower are used to drive the units and in many models an additional fan is employed to disperse the fog.

2. *High pressure water nozzles,* similar in principle to mist nozzles but operating at much higher water pressures (up to 6.9 MPa = 1000 p.s.i.), have a very fine orifice and a carefully engineered impact pin in place of the "anvil" or impact plate of a mist nozzle. Other designs more akin to conventional swirl jet nozzles are also produced. Originally these nozzles were used for frost protection of outdoor crops, which called for very large pumps and filtering systems, as ultrafine orifices are easily blocked. However, smaller pump and filter systems more appropriate for propagation houses have recently become available. Individual nozzles typically output 5–8 $l \cdot h^{-1}$ and are inserted at around 2 m spacing along thick-walled PVC or copper pipes running the length of the greenhouse.

3. *Pneumatic nozzles* use compressed air and water. Nozzles in which the air, in effect, tears the water from an air-water interface within the nozzle, have been available for many years and typically output up to 7 $l \cdot h^{-1}$ of water droplets. More recently, "sonic" nozzles have been developed, in which the water is disrupted by passage through a field of high-frequency sound waves generated by the air at the tip of the nozzle. These have a greater output than the older type; those most commonly used for propagation produce up to 20 $l \cdot h^{-1}$ of uniformly fine droplets but larger sizes producing up to 55 $l \cdot h^{-1}$ are also available. Another nozzle uses a piezoceramic element to produce mechanical oscillations which, when transmitted to the atomizer, produce fine capillary waves from which droplets are detached. Outputs of 50 $l \cdot h^{-1}$ and more, of droplets ranging from 30–60 μm, are available from this type of equipment.

Fog controllers

Accurate control of high humidities presents problems. Time clocks have been used but a fixed rate of fog input is not entirely satisfactory because changes in irradiance in a fog enclosure elicit rapid changes in saturation deficit. If a high rate of input is set, to satisfy the most demanding, bright conditions, then excessive fogging and fall-out on the cuttings and medium occurs when the irradiance declines.

The simplest automatic controllers use hygroscopic threads which contract on drying to operate a microswitch and initiate fogging. There can be an appreciable lag in response to changing conditions, especially when a fall in humidity follows a prolonged period of high humidity. The "deadband" of operation for this type of controller is therefore unsatisfactorily large. Other controllers use thin films of hygroscopic materials whose resistance or capacitance varies with changing humidity. These have the advantage of very fast response times but unfortunately are sensitive to contaminants in the water in very humid conditions so their humidity sensing characteristics change with time.

Aspirated psychrometers using closely matched wet- and dry-bulb temperature sensors have proved satisfactory for relative humidities up to about 98%. It is not easy to prevent the dry bulb from getting wet, and thus overestimating the r.h., if over-fogging is permitted. However, more recent designs have provided the most effective controllers currently available at reasonable cost.

Most controllers operate to maintain a fixed r.h. This is the simplest and least expensive option in practice, but, theoretically, is less than perfect. For example, a r.h. of 90% at 30 °C represents a saturation deficit 2.5 times greater than the same r.h. at 15 °C (0.42 vs. 0.17 kPa). Cuttings would therefore lose water 2.5 times faster at the higher temperature, even though r.h. is the same (neglecting any further effects of air temperature on leaf temperature or leaf resistances). Nevertheless, r.h. controllers are most often used because controllers based on measurement of the saturation deficit or dewpoint are impractically expensive at present.

Capacity

In any fogging installation it is important to provide sufficient but not excessive fogging capacity. The fogging requirement hinges principally on the ventilation which is essential to avoid extreme temperatures on warm days, but must be considered alongside the need to maintain a high humidity. Because of the central importance of ACR in relation to temperature, a brief examination of this subject is necessary.

Even in cool climates, greenhouse temperatures rise rapidly on bright days if ventilation is restricted. Fogged greenhouses are usually lined with clear PE to prevent leakage of humid air and reduce condensation on the glass walls. A shade screen is often mounted above the fogged area to reduce the irradiance and further enclose the fog. The maximum opening of the roof ventilators may also be restricted for the same purpose. These measures have the effect of reducing the ACR and increasing the temperature within the house. Some idea of the magnitude of this rise in temperature can be gained by reference to computer models of temperature relationships in glasshouses, based on energy balance equations (Maher and O'Flaherty 1973, Landsberg et al. 1979). Natural (unforced) ventilation rates in glasshouses are determined primarily by the wind velocity outdoors and the ventilator apertures; equations expressing this relationship for glasshouses with roof ventilators have been derived from empirical measurements, e.g. Goedhart et al. 1984. In Table 9, the temperature rise in an unshaded glasshouse, relative to outdoor temperature, is predicted for different wind conditions and vent openings using data from Maher's and O'Flaherty's Fig. 1 and Goedhart's equation.

On calm days with high irradiance the temperature rise can be substantial. Note that in calm conditions, opening the vents further has little effect, because the ACR remains very low. To avoid excessive temperature under these conditions there are two options: 1) use external shading to reduce the radiation entering the house; and 2) introduce forced ventilation. Reducing irradiance from 850 to 450 $W \cdot m^{-2}$ halved the temperature rise (Table 9); heavier shading than this is commonly used for propagation houses.

The alternative measure of using forced ventilation is acceptable for propagation only if the incoming air is humidified on entry to near-saturation levels. This is the basis of so-called "ventilated fogging," which provides the capability of reducing air temperatures in the glasshouse below those outdoors (Milbocker and Wilson 1979), i.e. theoretically close to the wet-bulb temperature outdoors.

Table 9. Calculations of air temperature rise (above outdoor ambient) in an unshaded glasshouse with restricted ventilation, at 450 and 850 $W \cdot m^{-2}$ outdoor irradiance and four wind speeds. [Air change rates (ACR) calculated from Goedhart et al. (1984, p. 397), assuming K=0.067 and 0.5% leakage; temperature rise from Fig. 1 of Maher and O'Flaherty (1983). Glasshouse volume 100 m^3, floor area 35 m^2.]

Wind Characteristics			Ventilator Opening (% of maximum)					
			10			20		
			ACR (volumes· h^{-1})	Temp. Rise (°C) at ($W \cdot m^{-2}$)		ACR (volumes· h^{-1})	Temp. Rise (°C) at ($W \cdot m^{-2}$)	
Beaufort Designation		Velocity ($m \cdot s^{-1}$)		450	850		450	850
Calm	0	0.1	0.07	17	31	0.14	16	30
Gentle breeze	3	5	3.5	12	20	6.9	8	16
Fresh breeze	5	10	7.0	8	15	13.7	5	10
Gale	8	20	14.1	6	10	27.5	3	6

Milbocker (1980) developed a centrifugal fogger suitable for ca. 200 m³ of propagation space, with an air movement rate of 6,800 m³·h⁻¹ (or 34 air changes·hr⁻¹ in 200 m³) and 190 l·h⁻¹ of fog output. High pressure water nozzles also offer a flexible means of varying the fogging capacity and are customarily positioned below the roof ventilators in a glasshouse, so that natural ventilation currents distribute the fog throughout the house. However, as already discussed, in calm conditions ventilation may be inadequate for ventilated cooling by this means alone, so forced ventilation (e.g. by PE convection tubing, see Mastalerz 1977, pp. 15–17) is helpful.

Sonic nozzles are individually expensive and the provision of compressed air adds appreciably to their cost. Consequently, it is usual to install systems with a relatively low fogging capacity, operate with restricted ventilator settings (around 20% of maximum), and rely on heavy shading for temperature control. Even so, high temperatures remain a problem in summer. If the nozzles are enclosed within PE enclosures over the individual propagation benches or beds, then unrestricted house venting can be used. The temperature problem is reduced, but the physical arrangement of the fogging equipment can be problematical and the cuttings are less accessible.

The calculation of fogging requirements for particular installations and specific conditions requires a knowledge of:

1. Outdoor temperature and humidity.
2. Air change rate.
3. Volume of the greenhouse, and air temperature and humidity before fog is introduced.
4. Operational "on-time" for the fogging equipment (full-time operation is often inadvisable).
5. Losses through fall-out and condensation (most often estimated).
6. Output per nozzle (or unit).

With this information the number of nozzles required to generate a particular humidity at any specified greenhouse temperature can be calculated. A microcomputer is useful but the calculations can easily be performed on a hand-calculator if tables of saturation vapor density at different temperatures are available. Illustrative results using these calculations for differing weather conditions are shown in Table 10. The strong dependence of the fog input requirement on the ventilation rate is evident, as is the need for a flexible and preferably automatic control system to accommodate variations in weather. The criteria applied to assess the capacity required will vary according to the particular application envisaged and are inevitably somewhat arbitrary. For example, for glasshouse propagation in the U.K. I have used the following criteria for calculating the fogging requirement: the minimum input required is that necessary to raise the r.h. within the enclosure to 98% at 30 °C, starting with 75% r.h. inside and 7 changes per h of outdoor air at 20 °C and 50% r.h. Use of shading to avoid excessive temperatures is essential with such a low ACR.

Table 10. Calculated fog input required per 100 m³ of greenhouse volume, at different air change rates in: (1) demanding summer conditions, and (2) cooler weather. Assumed requirements were: (1) achievement of 98% r.h. at a greenhouse temperature of 35 °C with outdoor conditions of 50% r.h. at 25 °C, and (2) achievement of 98% r.h. at a greenhouse temperature of 15 °C with 50% r.h. at 10 °C outdoors. Other assumptions: r.h. in greenhouse before fogging = 75%, loss through fall-out or condensation = 10% of input.

Air Change Rate (volumes·h⁻¹)	Fog Input (l·h⁻¹) Required to Meet the Set Criteria	
	(1)	(2)
0	1.0	0.3
0.5	2.5	0.8
1	4.0	1.2
2	7.1	2.1
5	16.2	4.7
10	31.3	9.1
20	61.7	17.8
50	152.6	44.0

Distribution

I have stressed the need for adequate fogging capacity but it is also important that over-fogging be avoided. If fog is introduced into an already saturated atmosphere, fall-out of liquid water on cuttings and rooting medium is inevitable. Some leaf wetting is beneficial (except perhaps for micropropagules) but it is easily overdone for reasons already discussed. Moreover, fall-out is always greatest in the area immediately adjacent to the nozzles so wetting is therefore spatially uneven. For this reason, it is often useful to install ancillary ventilation fans to move the fog gently around the greenhouse, with pneumatic and high-pressure systems. The importance of achieving even distribution is illustrated in the experiment shown in Fig. 5 of Chapter 8 by Loach (see also Loach 1985). Subsequent experiments confirmed that cuttings positioned at regular intervals along a fogged PE tunnel at stations ranging from 0–8.5 m from a single, sonic nozzle (r.h. 100% at the nozzle, 90% at the remote end) show striking variations in rooting (unpublished data). Evidently, even small, local variations in either humidity or fall-out have a substantial effect on rooting. For this reason, a "distributed" system with a large number of small nozzles is advantageous in comparison with one relying on fewer fog sources.

Conclusion

Fogging is still in its infancy and technical developments are occurring at a rapid pace. Similarly, there are few "standard" approaches to the installation of equipment. Fog will eventually offer a more closely regulated propagation environment than has previously been possible and will likely displace mist in many applications.

However, further research is needed in several areas relating to the use of fogging. Van Bavel et al. (1981) used a computer simulation to analyze the effectiveness of evaporative cooling, and concluded that while air temperatures were successfully reduced in an evaporatively-cooled greenhouse, plant and soil temperatures and daytime water stress were not substantially affected. This occurred because leaf temperatures are governed more by irradiance and transpiration rate than by the temperature of the surrounding air. Shading (in this case with a fluid roof system) was necessary to reduce water stress. The demonstrated effectiveness of ventilated fogging for cuttings may owe as much to the wetting of the foliar surfaces that occurs as to any reduction in leaf temperatures. A computer simulation of the energetics associated with cuttings in propagation environments would clarify the principles and guide future practice.

Currently, fogging systems are controlled by interval timers or r.h. sensors. The inflexibility of the former leads easily to over- or under-fogging with changes in weather. But control on the basis of r.h. (or even dewpoint or s.d.) is also not entirely satisfactory. This control system neglects an important time component in the development of water stress in the cuttings; changes in setting according to length of day and seasonal variations in irradiance would seem sensible if the aim is to maintain a standard, low range of water stress in the cuttings. Again, computer simulations should provide guidance, and are currently being developed. "Intelligent" control of the propagation environment using microprocessors should eventually be possible.

Table 11. A subjective comparison of different fogging systems (***=highest rating). The relative merits must be considered in relation to the particular application intended.

	Sonic Nozzles	Centrifugal Generators	High-Pressure Water Nozzles
Low cost	*	***	**
Capacity	*	**	***
Effectiveness for ventilated cooling	*	***	**
Spatially uniform distribution	**	*	***
Uniformly fine droplets (minimal fall-out)	***	*	**
Even air flow	**	*	***
Maintenance requirement	***	**	**

The relative merits of different systems can only be considered in the light of the particular application, e.g. requirements for micropropagated material differ from those for direct-stuck, easily rooted, summer cuttings. Furthermore, cost-effectiveness is as important a consideration as technical efficiency. A subjective comparison of the strengths and weaknesses of different fogging systems is given in Table 11 but must be considered with these qualifications in mind.

BASAL HEAT

Temperatures in the rooting medium may sometimes be below the optimum required for rooting. In this circumstance, bud development may occur in advance of root formation thus forming a competitive sink for assimilates or causing desiccation. For these reasons it is often beneficial to provide basal heating in the rooting medium.

Equipment

Electrical heating cables offer easy installation and temperature control in the rooting medium but running costs can be high. In some countries, electricity is supplied at lower cost during off-peak hours which reduces costs substantially (see below). Incorporation of heating cables in flat-sheet "blankets" of plastic or metal foil simplifies installation further and can improve the uniformity of heat dispersal, but can present drainage problems. Recently, electrical resistance heating using solid metal "cables" or welded metal mesh has been used. Costs of operation are similar to conventional electrical heating though installation is simpler.

Alternatively, warm water heating can be used. The pipework and more complex control systems which are necessary result in higher installation costs but lower operating costs than with electricity. Warm water lends itself particularly well to solar heating systems (Whitcomb et al. 1984). Warm water heating systems are now available in proprietary "units" complete with base insulation. The reader is referred to more specialized texts (e.g. Mundy 1981) or extension and advisory services for further information on the different systems.

Whalley and Randall (1976) stressed the importance of using accurate temperature controllers. Electronic sensors with fast response times, stable calibration characteristics, and ready compatibility with digital or computer based monitoring systems are gradually replacing older rod thermostat controllers.

Basal heat settings

Much effort has been expended in determining the optimal settings for base heating. The consensus view is that temperatures of 18–25 °C are optimal for most cool-season species and up to 7 °C greater for those from warm climates (Kester 1970, Dykeman 1976). Several factors interact to determine the likely optimal setting for a particular species which makes it unlikely that optima can be precisely determined. First, the base heat setting only controls minimum temperature in the rooting medium; actual daytime temperatures are a function of irradiance and therefore dependent on weather. Second, high basal temperatures are usually accompanied by drier conditions because in practice basal and foliar temperatures are unavoidably linked. The more supportive the aerial environment from a water relations viewpoint (i.e. the higher the humidity or the greater the leaf wetting), the greater the tolerance of, and potential benefit from, higher base temperatures. Thus experiments in controlled conditions at relatively low irradiance (when evaporative demand is low) have suggested relatively high optima, e.g. 25–30 °C for rooting of *Humulus lupulus* L. (Howard 1965), *Forsythia* × *intermedia* Zab. and *Chrysanthemum* × 'Bright Golden Anne' (Dykeman 1976). Optimal settings are likely to be system-dependent. Third, the effects of temperature interact with the disease status of the cuttings. Whalley and Loach (1977) and Smith (1982) found that cuttings of *Rhododendron* varieties which are particularly prone to basal rots, rooted less well at 25 and 20 °C basal stem temperatures than at 15 °C due to disease. However, disease-free cuttings of these same varieties rooted best at the higher temperatures. Kelly and Lamb (1979) propagated six varieties of conifer under contact PE in autumn, both in a glasshouse and in an outdoor frame. The rooting percentages were generally low and many of the cuttings decayed at the base of the stem. However, best rooting occurred in the frame without base heat and in the glasshouse with a reduced base-heating regime;

these two treatments were associated with less basal decay than when base heat was supplied throughout the propagation period. Such damage is not unusual in winter propagations of conifers (see also section on Enclosed Mist and Chapter 8 by Loach., Fig. 4).

Basal temperature also interacts with hormone treatment. Whalley (unpublished) found that cuttings of *Chamaecyparis lawsoniana* Parl. 'Ellwoodii' and *C. pisifera* Endl. 'Boulevard' were indifferent to base temperature (15, 20, and 25 °C) if the cuttings were treated with hormone at insertion; untreated cuttings responded positively to each temperature increment. Conversely, Dykeman (1976) observed that cuttings were much more responsive to auxin at higher temperatures.

Following studies by Burholt and Vant Hoff (1970) which examined the influence of temperature on cell division and growth kinetics of sunflower (*Helianthus annuus* L.), Dykeman (1976) studied the different temperature responses of the two component processes of rooting, i.e. root initiation and root growth. Root initiation is controlled by the rate of cell division and the number of cells dividing, which in *Chrysanthemum* and *Forsythia* spp. had a temperature optimum around 30 °C. Root development depends on the supply of photosynthate and was best at lower temperatures of 22 and 25 °C, probably because respiration rates were reduced. Best rooting results with *Chrysanthemum* spp. were obtained by using 30 °C for root initiation followed by a reduction to 25 °C after root emergence. Use of a two-temperature system may be of limited practicality in a propagation house containing a wide range of ornamental species at different stages of propagation. However, these findings support the contention that optimal base heat settings cannot be rigidly defined.

With economy in mind, researchers have tested the need for base heat for a full 24 h per day. Work at both Kinsealy Research Centre in Eire and Efford Experimental Horticultural Station in the U.K. suggested that for winter propagations rooting percentages were not substantially affected if the base heating was switched off during the night or day as compared with heating available (i.e. "on-call") all of the time. Scott (1980) reported rooting percentages of 91, 88, and 89% for continuous, night-only, and day-only base heating regimes, respectively, averaged over the seven species used in the trial. Using fully insulated beds and PE covers, the total electricity used did not differ greatly between treatments (night- and day-only heating used 88 and 80%, respectively, of that required in the continuous heat treatment), but because off-peak electricity at night was available at 43% of normal cost, the night-time treatment was appreciably more economical.

In a similar trial but using mist, Cooper (1977) compared continuously available and day-only base heating. Two easy-rooting species, *Hebe* 'Blue Gem' and *Helianthemum nummularium* (L.) Mill., gave similar results in both treatments. *Juniperus communis* L. 'Hibernica' and *Choisya ternata* H.B.K., rooted much faster with continuous heat but to a similar final percentage in both treatments. Under mist, daytime-only heating used one-third of the electricity required for continuous heat, so if speed of rooting is not a major concern day-only heating is more cost-effective.

Other studies have suggested that for winter propagation, base heat can actually be detrimental in rooting of some species. In these cases disease was often involved, as previously discussed. Scott (1980) propagated seven species under a PE tent in winter, using two air temperatures, 5 and 10 °C, and two base heat temperature settings, 15 and 21 °C. Differences in rooting percentages were small; averaged across species they were 91, 83, 86, and 81% for the 5/15, 5/21, 10/15, and 10/21 air/base temperature treatment combinations, respectively. The small advantage in favor of the lower base temperature is unexpected; and some interaction with other features of the propagation environment must be suspected. For example, conditions may have been slightly "drier" and cutting turgor marginally lower with 21 °C base heat.

ROOTING MEDIA

Practice

Rooting media must be considered as an integral part of the propagation system. Mixes of an organic component such as peat and an inert ingredient such as grit, pumice, or perlite are most commonly used. Peat is included primarily because of its large total pore space and its ability to hold water; the mineral ingredient increases the proportion of large air-filled pores and improves drainage.

Other constituents are often used according to local availability and cost; thus bark, sawdust, leaf mould, sphagnum moss, or rice hulls have been used as the organic component along with scoria, vermiculite, polystyrene, clay granules, or rockwool as the inorganic component.

The proportions of the organic and mineral components should be varied according to the wetness of the propagation system employed, the softness of the cutting material, and the likely irradiance levels (i.e. time of year), e.g. one volume of peat to three volumes of grit for wetter systems, winter conditions, and mature cuttings; reversed proportions for drier systems, summer conditions, and soft cuttings. Thus for enclosed mist, contact PE or densely fogged systems, especially in autumn and winter, the former mix (with a low proportion of peat) is preferable, whereas spring and summer propagations of less mature cuttings under open-bench mist or shaded tent PE benefit from the inclusion of relatively more peat. These same general principles apply whatever components are used in the mix.

To test this approach, nine species of woody ornamentals were propagated within a fogged enclosure in two media with differing water contents; i.e. peat alone, with 35% water by volume and a 2/1 grit/peat mix with 21% water. Four species were inserted in October–November when the mean daily radiation integral was < 5 MJ$\cdot$m$^{-2}\cdot$d^{-1} through the propagation period and five species in May–July when the radiation integral was > 15 MJ$\cdot$m$^{-2}\cdot$d^{-1}. For the October–November insertions, all four species rooted better in the drier, grit/peat medium; 69% rooting vs. 43% in peat alone (Loach unpublished). For May–July insertions the wetter medium, peat, gave better results than grit/peat for three of the five species and the mean rooting percentages were 74 and 66%, respectively. It is often found that the effect of rooting medium is least evident in summer, probably because at high irradiances tissue water deficits occur daily, whatever the medium.

It has commonly been supposed that the optimal requirements for rooting cuttings can be specified in terms of the physical characteristics of the medium, in particular, the relative proportions by volume of air and water. However, this is not strongly supported by the differing results reported in the literature, where air-filled porosities ranging from 1–20% have been found to be optimal. Comparative trials of different rooting media show that, while the nature of the medium can substantially influence successful rooting, results differ between species and are not necessarily reproducible in repeated propagations. The literature and the possible reasons for these diverse results were discussed by Loach (1985). Until our understanding improves, the suggestions outlined above provide a sensible basis for practice.

Chemical and Biological Aspects

In spite of the approach recommended above, it is unlikely that the relative effectiveness of different rooting media can entirely be explained in terms of their physical properties. Both chemical and biological influences have been observed. For example, toxins occur in some sources of bark which may either reduce rooting or, through their influence on the microflora of the cuttings and medium, may improve rooting.

Beneficial effects of incorporating inoculum of mycorrhizal fungi in the rooting medium have been observed. Linderman and Call (1977) reported enhanced rooting of Arctostaphylos uva-ursi L. Spreng. and Vaccinium ovatum Parsh. by as many as 13 different ectomycorrhizal fungi. This enhancement occurred before or in the absence of any obvious mycorrhizal association with the tissues of the cuttings, suggesting that the fungi may produce growth substances, or perhaps may inhibit the activities of damaging elements of the microflora. Navratil and Rochon (1981) similarly reported improved root development in Populus spp. cuttings when the medium was treated with a vermiculite-based inoculum of Pisolithus tinctorius (Pers.) Coker & Couch. Nelson (1987) inoculated 16 different woody ornamental species wih Glomus intraradices Schenck & Smith to form an endomycorrhizal association during propagation. In some species this delayed early root development, perhaps due to a parasitic phase in the early stages of endomycorrhizal infection. However, subsequent root development of infected plants approached or surpassed that in non-mycorrhizal plants.

These more subtle influences of the rooting medium, both chemical and biological, are attracting more attention, but their basis is poorly understood.

SANITATION

The warm, humid conditions of a propagation unit are conducive to growth and spread of many fungal pathogens, especially those with water-borne zoospores, e.g. *Pythium* and *Phytophthora* spp.; with spores spread by water splash, e.g. *Rhizoctonia, Pestalotiopsis, and Glomerella* spp.; or others particularly virulent in a humid environment, e.g. *Botrytis* and *Peronospora* spp. Similarly, pests such as sciarid flies or aphids will on occasion multiply to near-epidemic proportions during propagation unless suitable precautions are taken.

Geard (1979) pointed out that the presence of a host plant and a pathogen does not necessarily mean that disease will result. Two further determining influences are the environment and management practices. Disease problems during propagation are always worse if the host's tissues are damaged in some way. If cuttings have been allowed to wilt then leaf or bud tissues can be irreparably affected even though the cutting apparently recovers. Fungal infection of the damaged tissue can occur and spread to healthy tissue. In summary, conditions which best favor turgid cuttings not only improve rooting but prevent disease.

Orndorff (1982), Moody (1983), and many others stress the central importance of preventive rather than curative management practices for successful propagation. These measures start with the stock plants. Some nurseries practice routine spraying with fungicides and pesticides on the stock ground (Rumbal 1977, Moody 1983). Typically these involve broad-spectrum chemicals because the target organisms are largely unknown. Others have preferred to restrict spraying to varieties which regularly present disease problems during propagation, e.g. certain hardy hybrid *Rhododendron* varieties such as 'Cynthia'; *Cotinus coggygria* Scop. 'Royal Purple', which is particularly prone to attack by *Botrytis cinerea* Pers.; and soft, summer cuttings of *Euonymus* cultivars. Restricted spraying of stock plants has the advantage of economy as few varieties regularly present pest or disease problems, which reduces the risk of developing strains of organisms resistant to the applied chemicals and is more acceptable in relation to the increasing public sensitivity to environmental contamination.

Spraying is particularly important if cuttings are to be cold-stored before insertion. Heursel and Kamoen (1976) found that a benomyl treatment applied to stock plants of *Rhododendron obtusum* Planch., *R. ponticum* L., and *R. simsii* Planch. varieties was very effective in reducing disease during subsequent cold-storage and was better than sprays applied at the time of storage. (For further information, see Chapter 17 by Behrens).

Sanitation measures in the cutting preparation area are described by Rumbal (1977), and Hartmann and Kester (1983); suitable disinfectants are listed by McCain (1977). Routine treatment of cuttings with fungicides or with insecticides, if insect damage is evident prior to insertion, is particularly important. Smith (1982) isolated and identified several species of pathogenic fungi on cuttings of *Rhododendron, Juniperus,* and *Camellia* spp. immediately after removal from the stock plant, and identified additional pathogens during subsequent propagation. Best protection was obtained if, after immersing the cuttings in fungicidal solution, the rooting medium was also drenched with it. Regular inspection of cuttings during propagation and prompt removal of infected cuttings or leaves is a further important routine measure for disease and pest control.

Currently, there are very few crop protection chemicals with label recommendations for use during propagation. Phytotoxicity is frequently reported for many of the commonly used fungicides, but there have been reports of enhanced rooting perhaps due to hormonal properties of the fungicides (refs. in Morgan and Colbaugh 1983). When damage has occurred, plant species have shown differing susceptibilities. For example, Morgan and Colbaugh (1983) found that of 21 soak or drench treatments applied to cuttings of *Buxus microphylla* Sieb. & Zucc. propagated under mist, only one reduced the rooting percentage. Again, drenching the medium was more effective for controlling disease than simply soaking the cuttings. However, of 34 sanitation treatments used for rooting of *Peperomia caperata* Yuncker, 13 treatments reduced rooting. Smith (1982) applied eight fungicidal treatments to four *Rhododendron* hybrids. Of these, six increased the rooting percentage relative to the untreated control, primarily by controlling rot, and none were detrimental. She pointed out that in addition to the intrinsic fungitoxicity of the fungicide, many other factors influence its effectiveness, e.g. its formulation; method and frequency of application (mist presumably washes off the fungicide, necessitating more frequent applications); temperature and composition of the rooting medium; and the interaction of the fungicide with any applied root promoting compounds. Disparate results and occasional phytotoxicity would seem to be inevitable, though clear benefits are common. The addition of fungicides to rooting hormone formulations is discussed in Chapter 10 by Blazich.

As previously stated, broad-spectrum fungicides are usually used during propagation because the target organisms are unknown. Where particular problems are expected, a more specifically active compound should be included, e.g. iprodine for control of *Rhizoctonia solani* Kuhn during propagation of *Ericas* (Hutchinson 1982). Guidance on the spectra of the commonly used fungicides is given by Ormrod (1975), and Scopes and Ledieu (1983, Tables 13.2 and 13.3), and is available from extension or advisory services. Dip and drench treatments are effective for prolonged periods under PE but sprays at two week intervals may be necessary under mist. Insecticidal sprays are best applied as needed, rather than routinely.

The rooting medium can be the source of pathogens in some instances. For example, *Pythium* and *Penicillium* spp. have been isolated from peat. Pasteurization of the medium with aerated steam (Coyier 1978) has been advocated in these cases. However, in addition to making additional cost, even mild treatment can have a detrimental effect on the structure of the peat. Inoculation of rooting and growing media with organisms antagonistic to pathogens may prove possible in the future (Baker 1971), but is not yet a reality.

CONCLUSION

The propagation methods described in this review vary in complexity from simple outdoor enclosures to glasshouse systems with provision of basal heating, automatic shading, misting or fogging, and relatively complex controlling devices. The benefits accruing from increasing complexity are a greater assurance of success and the ability to root a wider range of species. The simpler methods need not necessarily give poorer results, but because they operate within a less well regulated environment, they call for more intensive management and, probably, more skill on the part of the nurseryman.

The requirements for rooting leafy cuttings are not easily arrived at by the exercise of intuition. Rather there is a need for a decision framework based on a sound appreciation of the environmental principles involved. This certainly applies to the simpler systems where, in many cases, the principle means of instruction is best described as "learning through disaster." It may teach the nurseryman how to avoid the worst situations but not necessarily how to achieve the best.

Our understanding of how the cutting operates in its environment is increasing rapidly, fortunately at the same time as the means for controlling that environment have substantially improved, thanks to computer control, automatic shading, and fogging. In glasshouse-based systems we can look forward to propagation environments controlled by a computer to maintain conditions within an optimal range, taking into account changes in weather and season. The experience gained in this way will feed back into the less sophisticated systems in the form of improved operational guidelines. These simpler systems will no doubt also remain with us due to their cost-effectiveness.

ACKNOWLEDGMENTS

Dr. R. I. Grange, Dr. R. L. Jinks, and Dr. F. A. Langton kindly reviewed this chapter, and I thank them for their helpful comments and suggestions. Numerous colleagues have assisted with formerly unpublished work which is included, and I gratefully acknowledge their help.

REFERENCES

Anon. 1982. *Evaluation of Propagation Systems for Leafy Cuttings.* Ann. Rept. N. Z. Nursery Res. Centre. pp. 24–28 and 49–57.

Anon. 1985. *Field Grown Nursery Stock. Sun Tunnel Propagation—Reports on Trials.* Min. Agric., Fish and Food, Efford E.H.S., Lymington, Hampshire, England.

Baker, K. F. 1971. Disease-free propagation in relation to standardization of nursery stock. *Proc. Int. Plant Prop. Soc.* 21:191–198.

Bernaud, P., J. Y. Champagne, G. Le Palec, Ph. Bournot, B. de Muynck and R. Vandevelde, 1984. Effect of plastic cover properties on the thermal efficiency of a greenhouse. *Acta Hortic.* 154:141–147.

Buckley, A. R. 1955. Mist and polyethylene tents for rooting softwood cuttings. In *Proc. Int. Plant Prop. Soc.* 5th Ann. Mtg. pp. 136–137.

Burgess, C. M. 1984. *Sun Tunnel Propagation—a Review of Work 1981–84.* Rev. Efford Exp. Hortic. Sta. for 1984 (1985). pp. 22–35.

Burholt, D. R. and J. Vant Hoff. 1970. The influence of temperature on the relationships between cell population and growth kinetics of *Helianthus annuus* roots. *Amer. J. Bot.* 96:80–82.

Chadwick, L. C. 1951. Controlled humidification as an aid to vegetative propagation. In *Proc. Int. Plant Prop. Soc.* 1st Ann. Mtg. pp. 38–39.

Cooper, P. D. E. 1977. *Effect of Bottom Heat in a Mist Propagation Unit.* Rept. Luddington Exp. Hortic. Sta. for 1977 (1978). pp. 180–182.

Coyier, D. L. 1978. Pathogens association with peat moss used for propagation. *Proc. Int. Plant Prop. Soc.* 28:70–72.

Davis, T. D. and J. R. Potter. 1987. Physiological response of *Rhododendron* cuttings to different light levels during rooting. *J. Amer. Soc. Hortic. Sci.* 112:256–259.

Deen, J. L. W. 1971. Rooting cuttings under polythene tunnels. *Proc. Int. Plant Prop. Soc.* 21:248–252.

Dubois, P. 1978. (Transl. by C. A. Brighton). *Plastics in Agriculture.* Applied Science, Barking, England. ISBN 0-85334-776-X.

Dykeman, B. 1976. Temperature relationship in root initiation and development of cuttings. *Proc. Int. Plant Prop. Soc.* 26:201–207.

Evans, H. 1952. Physiological aspects of the propagation of cacao from cuttings. In *Proc. 13th Int. Hortic. Cong.* 2:1179–90.

Geard, I. D. 1979. Fungal diseases in plant propagation. *Proc. Int. Plant Prop. Soc.* 29:589–94.

Goedhart, M., E. M. Nederhoff, A. J. Udink ten Cate and G. P. A. Bot. 1984. Methods and instruments for ventilation rate measurements. *Acta Hortic.* 148:393–400.

Gouin, F. R. 1980. Vegetative propagation under thermoblankets. *Proc. Int. Plant Prop. Soc.* 30:301–305.

Grange, R. I. and K. Loach. 1983. Environmental factors affecting water loss from leafy cuttings in different propagation systems. *J. Hortic. Sci.* 58:1–7.

_____ _____ 1984. Comparative rooting of eighty-one species of leafy cuttings in open and polyethylene-enclosed mist systems. *J. Hortic. Sci.* 59:15–22.

_____ _____ 1985. The effect of light on the rooting of leafy cuttings. *Scientia Hortic.* 27:105–111.

Hancock, L. 1959. The burlap cloud method of rooting softwood summer cuttings. In *Proc. Int. Plant Prop. Soc.* 9th Ann. Mtg. pp. 165–168.

Hartmann, H. T. and D. E. Kester. 1983. *Plant Propagation: Principles and Practices.* Prentice-Hall, NJ, USA. 4th ed. ISBN 0-13-681007-1.

Hess, C. E. and W. E. Snyder. 1955. A physiological comparison of the use of mist with other propagation procedures used in rooting cuttings. In *Proc. 14th Int. Hortic. Cong.* 2:1125–1139.

Heursel, J. and O. Kamoen. 1976. Preservation of cuttings of *Rhododendron obtusum* Planch., *R. ponticum* L. and *R. simsii* Planch. *Scientia Hortic.* 4:87–90.

Howard, B. H. 1965. Regeneration of the hop plant (*Humulus lupulus* L.) from softwood cuttings. I. The cutting and its rooting environment. *J. Hortic. Sci.* 40:181–91.

Hutchinson, D. 1982. Disease control in *Ericas* and *Callunas*. *Proc. Int. Plant Prop. Soc.* 32:228–230.

Kelly, J. C. and J. G. D. Lamb. 1979. *Nursery Stock.* An Foras Taluntais Ann. Rept. pp. 21–22.

Kester, D. E. 1970. Temperature and plant propagation. *Proc. Int. Plant Prop. Soc.* 20:153–163.

Landsberg, J. J., B. White and M. R. Thorpe. 1979. Computer analysis of the efficacy of evaporative cooling for glasshouses in high energy environments. *J. Agric. Eng. Res.* 24:29–39.

Lewandowski, R. J. and F. R. Gouin. 1982. Rooting Euonymus cuttings outdoors under thermo-blankets or under greenhouse intermittent mist using propagating media with and without composted sewage sludge. *Proc. Int. Plant Prop. Soc.* 32:525–534.

_____ _____ 1985. Rooting of *Pachysandra terminalis* and *Euonymus kiautschovica* stem cuttings under intermittent mist and outdoor thermo-blanket tents. *J. Environ. Hortic.* 3:162–165.

Linderman, R. G. and C. A. Call. 1977. Enhanced rooting of woody plant cuttings by mycorrhizal fungi. *J. Amer. Soc. Hortic. Sci.* 102:629–632.

Loach, K. 1977. Leaf water potential and the rooting of cuttings under mist and polythene. *Physiol. Plant.* 40:191–197.

_____ 1979. Mist propagation—past, present and future. *Proc. Int. Plant Prop. Soc.* 29:216–229.

_____ 1980. Shading success. *Gardeners Chronicle and HTJ.* Oct. 10. pp. 21–24.

_____ 1985. Rooting of cuttings in relation to the propagation medium. *Proc. Int. Plant Prop. Soc.* 35:472–485.

_____ 1987. Mist and fruitfulness. *Hortic. Week* Apr. 10. pp. 28–29.

_____ A. P. Gay. 1979. The light requirement for propagating hardy ornamental species from cuttings. *Scientia Hortic.* 10:217–230.

_____ D. N. Whalley. 1978. Water and carbohydrate relationships during the rooting of cuttings. *Acta Hortic.* 79:161–168.

Maher, M. J. and T. O'Flaherty. 1973. An analysis of greenhouse climate. *J. Agric. Eng. Res.* 18:197–203.

Mastalerz, J. W. 1977. *The Greenhouse Environment—The Effect of Environmental Factors on the Growth and Development of Flower Crops.* John Wiley and Sons, NY, USA. ISBN 0-471-57606-9.

McCain, A. H. 1977. Sanitation in plant propagation. *Proc. Int. Plant Prop. Soc.* 27:91–93.

Milbocker, D. C. 1980. Ventilated high humidity propagation. *Proc. Int. Plant Prop. Soc.* 30:480–482.

_____ R. Wilson. 1979. Temperature control during high humidity propagation. *J. Amer. Soc. Hortic. Sci.* 104:123–126.

Moody, E. H. 1983. Sanitation: A deliberate, essential exercise in plant disease control. *Proc. Int. Plant Prop. Soc.* 33:608–613.

Morgan, D. L. and P. F. Colbaugh. 1983. Influence of chemical sanitation treatments on propagation of *Buxus microphylla* and *Peperomia caperata*. *Proc. Int. Plant Prop. Soc.* 33:600–607.

Mundy, M. 1981. *Heated Propagating Beds.* Min. Agric. Fish. and Food. Technical Note 67. pp. 2–3.

Navratil, S. and G. C. Rochon. 1981. Enhanced root and shoot development of poplar cuttings induced by *Pisolithus* inoculum. *Can. J. Forest Res.* 11:844–848.

Nelson, S. D. 1987. Rooting and subsequent growth of woody ornamental softwood cuttings treated with endomycorrhizal inoculum. *J. Amer. Soc. Hortic. Sci.* 112:263–366.

Nisen, A., J. Nijskens, J. Deltour and S. Coutisse. 1984. Détermination des properties radiométriques des matériaux plastiques utilisés en couverture des serres. *Acta Hortic.* 154:19–32.

Ormrod, D. J. 1975. Fungicides and their spectra. *Proc. Int. Plant Prop. Soc.* 25:112–115.

Orndorff, C. 1982. Constructing and maintaining disease-free propagation structures. *Proc. Int. Plant Prop. Soc.* 32:599–605.

Richards, M. 1984. Enclosed mist system for propagation of broad-leaved evergreens. *Proc. Int. Plant Prop. Soc.* 34:50–53.

Rowe-Dutton, P. 1959. *Mist Propagation of Cuttings.* Comm. Bur. Hortic. and Plant. Crops. Digest No. 2.

Rumbal, J. M. 1977. Aspects of propagation hygiene. *Proc. Int. Plant Prop. Soc.* 27:323–24.

Scopes, N. and M. Ledieu. 1983. *Pest and Disease Control of Hardy Nursery Stock, Bedding Plants and Turf.* Pest and Disease Control Handbook. BCPC Publications, Croydon. 2nd ed. ISBN 0-901436-83-6.

Scott, M. A. 1980. *Hardy Nursery Stock—Fuel Economy in the Propagation Bench.* Rev. Efford Exp. Hortic. Sta. for 1980 (1981). pp. 14–27.

Smith, P. M. 1982. Diseases during propagation of woody ornamentals. In *Proc. 21st Int. Hortic. Cong.* 2:884–893.

Stoutemyer, V. T. 1942. Humidification and the rooting of greenwood cuttings of difficult plants. *Proc. Amer. Hortic. Sci.* 40:301–304.

Stroombeek, F. 1958. The propagation of softwood cuttings in the foghouse. In *Proc. Plant Prop. Soc. 8th Ann. Mtg.* pp. 47–53.

Templeton, H. M. 1953. The phytotektor method of rooting cuttings. In *Proc. Plant Prop. Soc. 3rd Ann. Mtg.* pp. 51–56.

van Bavel, C. H. M., E. J. Sadler and J. Damagnez. 1981. *Cooling greenhouse crops in a Mediterranean summer climate. Acta Hortic.* 115:527–536.

van Hoff, M. 1958. Rooting under plastic. In *Proc. 8th Ann. Mtg. Plant Prop. Soc.* pp. 168–169.

Weller, H. A. 1959. Outdoor mist propagation. In *Proc. Plant Prop. Soc. 9th Ann. Mtg.* pp. 168–170.

Whalley, D. N. and K. Loach. 1977. Effects of basal temperature on the rooting of hardy hybrid *Rhododendrons. Scientia Hortic.* 6:83–89.

_____ R. E. Randall. 1976. Temperature control in the rooting medium during propagation. *Ann. Appl. Biol.* 83:305–309.

Whitcomb, C. E. 1982. Rooting cuttings under a wet tent. *Proc. Int. Plant Prop. Soc.* 32:450–455.

_____ C. Gray and W. Cavanaugh. 1982. Propagating under a wet tent. *Aust. Hortic.* May:97–98.

_____ _____ W. Cavanaugh. 1984. *The "Ideal" Greenhouse for Propagation?* OK Agric. Exp. Sta. Res. Rept. P-855. pp. 4–5.

Wong, T. L., R. W. Harris and R. E. Fissell. 1971. Influence of high soil temperatures on five woody-plant species. *J. Amer. Soc. Hortic. Sci.* 96:80–82.

CHAPTER 19

Bioassay, Immunoassay, and Verification of Adventitious Root Promoting Substances

Charles W. Heuser

The Pennsylvania State University
Department of Horticulture
303 Tyson Building
University Park, PA 16802

INTRODUCTION. 274
HIGHER PLANT BIOASSAYS. 275
    Light Grown Mung Beans 275
        Hess test. 275
        Heuser and Norcini variation 275
        Bassuk and Howard variation. 276
    Pinto Beans 276
    Shoot Apices of English Ivy 277
    Tomato Leaf Discs. 277
    Summary 278
FUTURE ASSAYS 278
    Immunochemical Assays 279
    Physicochemical Verification 280
CONCLUSION 281
ACKNOWLEDGMENTS 282
REFERENCES. 282

Additional key words: endogenous factors, mung bean cuttings, pinto bean cuttings, tomato leaf discs, English ivy apices, radioimmunoassay, monoclonal.
Abbreviations: BHT, butylated hydroxytoluene; BSA, bovine serum albumin; CMC, carboxymethyl cellulose; ELISA, enzyme-linked immunosorbent assay; ETOH, ethanol; GC-MS, gas chromatography-mass spectrometry; HCl, hyrochloric acid; HPLC, high performance liquid chromatography; IAA, indole-3-acetic acid; IAAsp, IAA-aspartate; IBA, indole-3-butyric acid; LPI, leaf plastochron index; MEOH, methanol; NAA, naphthaleneacetic acid; PAR, photosynthetically active radiation; PVPP, polyvinylpolypyrrolidone; RIA, radioimmunoassay; TLC, thin layer chromatography.

INTRODUCTION

Investigations of endogenous factors that influence rooting in cuttings required the development of bioassay systems. The term bioassay describes the use of living material to test effects of known and putative biologically active substances. The bioassays described here are based on the initiation and development of adventitious roots.

Bioassays for phytohormonal-like activities are based on varied physiological responses, such as cell enlargement and division. However, all bioassays must have certain minimal requirements to be effective (Devlin and Witham 1983): 1) specificity; 2) sensitivity; 3) ease in measuring a detectable and relatively fast response; 4) relative ease in setup and control; and 5) absence in the bioassay plant material of the chemical(s) being tested.

Following the early rooting research of Went (1934) with etiolated pea (*Pisum sativum* L.) cuttings and of Hemberg (1951) with hypocotyls of bean (*Phaseolus vulgaris* L.), Hess developed the mung bean bioassay (C. E. Hess. 1957, Ph.D. Thesis, Cornell University, Ithaca, NY, USA). Although many bioassays for root promoting substances have been devised, only the Hess mung bean bioassay and variations thereof have been generally used. The following four higher plant bioassays will be presented here: light grown mung bean [*Vigna radiata* (L.) R. Wilcz.], tomato (*Lycopersicon esculentum* Mill.) leaf disc, aseptically cultured shoot apices of English ivy (*Hedera helix* L.), and pinto bean.

Any critical assessment of the significance of a compound in adventitious root initiation and/or development depends on the unequivocal identification of that compound. Unequivocal identification can only be made by using physicochemical methods. Future research therefore should incorporate a physicochemical method, such as GC-MS, or RIA validated by a physicochemical method. Thus, I have included information on such methods.

HIGHER PLANT BIOASSAYS

Light Grown Mung Beans

Hess test

The mung bean test was developed to detect naturally occurring substances that stimulate rooting in the presence of IAA. The original test developed by Hess utilized 5-d-old etiolated seedlings (C. E. Hess. 1957, Ph.D. Thesis, Cornell Univ., Ithaca, NY, USA). Etiolated seedlings were subsequently replaced by light grown seedlings (Hess 1962, 1964). Later research established the presence of preformed root primordia in etiolated mung bean cuttings, which could confound the etiolated test (C. W. Heuser unpublished). Hess (1962, 1964) used the new test to detect four rooting cofactors from English ivy, chrysanthemum (*Chrysanthemum* spp.), and hibiscus (*Hibiscus rosa-sinensis* L.) that stimulated rooting in the presence of auxin.

The procedure for the light grown Hess test is:

1) Dry mung bean seeds are surface sterilized in 10% sodium hypochlorite solution for 3 min, rinsed thrice in water, and then soaked in room temperature, running water for 24 h.

2) Seeds are germinated and grown in a growth chamber at 27 °C, with ca. 425 μmol·m^{-2}·s^{-1} supplied by fluorescent and incandescent lamps for a 16 h photoperiod.

3) Cuttings are harvested from 9- to 10-d-old seedlings. A cutting consists of 3 cm of hypocotyl and the epicotyl containing primary leaves plus unexpanded trifoliolate leaf bud.

4) Ten cuttings are placed in shell vials (19 × 65 mm) containing 4 ml of a 5 μmol·l^{-1} solution of IAA per vial. The IAA is dissolved in a minimal amount of ETOH and diluted with distilled water. The incubation solution is taken up by the cuttings within 18–24 h; whenever required, distilled water is added during the 6- to 7-d-long assay. Chemical substances on paper chromatogram strips are added to the basal solution for testing.

Heuser and Norcini variation

We developed a modified mung bean bioassay because of inconsistent rooting results in the Hess test, which were caused by bacterial contamination of the distilled water (Blazich and Heuser 1978) and some unknown component of vermiculite (J. Norcini. 1986, Ph.D. Thesis, The Pennsylvania State Univ., Univ. Park, USA).

The procedure for the Heuser and Norcini test is:

1) Seeds are surface sterilized in 0.5% sodium hypochlorite solution, aerated in tap water for 24

h, and then sown below 1 cm of perlite.

2) The mung beans are germinated and grown in a growth chamber under a 16 h photoperiod, quantum flux density of 54 μmol·m^{-2}·s^{-1}, and a temperature of 27.5 ± 1 °C. The bioassay is similarly conducted.

3) Cuttings are made as previously described in the Hess procedure.

4) Seedlings are watered daily with tap water, except on the fifth day when a full strength Hoagland's nutrient solution is used.

5) The bioassay is terminated on day 7 by placing the leafy cuttings in 95% ETOH for 24 h, followed by water containing Safranin O. This stain and the clearing effect of the alcohol facilitate counting of root primordia.

6) All glassware and water are sterilized before use in the actual rooting test.

Bassuk and Howard variation

Bassuk and Howard (1981) modified the Hess light grown test to reduce variability in rooting response. The presence of cotyledons and leaves promoted rooting only during specific stages of growth—the cotyledons promoted rooting in young seedlings and the leaves did so with 5-d-old and older cuttings. Thus Bassuk and Howard (1981) removed cotyledons from young seedlings, and leaves from older seedlings, to reduce rooting of untreated cuttings. Reduced rooting of untreated cuttings increased sensitivity to applied chemicals.

The procedure for the Bassuk and Howard test is:

1) Seeds are sown over a 3-cm-deep bed of vermiculite and lightly covered with vermiculite. A typical sowing might use 25 ml of dry seeds spread in a 25 × 20 × 4 cm tray. Trays are watered to saturation, covered with a glass plate for 3 d, and then placed under warm-white fluorescent lamps at a quantum flux density of 99 μmol·m^{-2}·s^{-1} (21.2 W m^{-2}), 16-h photoperiod at 23 °C.

2) The cotyledons are removed 4 d after sowing. At this stage, the hypocotyls have just straightened and the seedlings are about 8 cm tall. The cuttings are prepared by severing uniform seedlings 4 cm below the cotyledonary node. The epicotyl should be no longer than 0.5 cm, with nearly vertically positioned, unexpanded primary leaves.

3) Four cuttings are placed in a 5 ml vial containing 4–5 ml of test solution. Test solution is replenished with distilled water as needed.

4) Visible roots are counted after 7 d.

Pinto Beans

The pinto bean assay was originally devised by Hemberg (1951) and involved placing cuttings in 100 ml brown glass jars containing water. The jars were placed in boxes containing a layer of moist sand, and covered with glass to increase humidity.

The pinto bean test presented here is a variation by Haissig (1979):

1) Pinto bean seeds are germinated and grown in perlite in a growth chamber, under an 18 h photoperiod from a mixture of fluorescent (Sylvania F96T12/CWX/VHO) and incandescent (Sylvania 1950L/P25/8) lamps (ca. 350 μmol·m^{-2}·s^{-1}), and a temperature of 24 °C during the day and 22 °C at night.

2) Cuttings are prepared about 9 d after planting, from seedlings with primary leaves 3.5–4.5 cm long. Hypocotyls are severed 5 cm below the cotyledonary node.

3) Treatment solutions are prepared in 10 mol^{-1} MES-NaOH buffer, pH 6.0. Auxins are dissolved in some of the NaOH used to adjust pH of the MES, which precludes influences of organic solvents such as ETOH on rooting.

4) Five cuttings are placed in a vial containing 15 ml of treatment solution. Vials with cuttings are randomly placed in shallow plastic pans in an incubator. After 24 h, treatment solutions are replaced with double-deionized water, which is then replaced each 24 h during the experiment.

5) The cuttings are incubated at 25 °C in a chamber with an 18 h photoperiod from fluorescent lamps (Sylvania F14T12, ca. 30 μmol·m^{-2}·s^{-1}).

6) Rooting is evaluated by counting the number of primordia (day 3) and number of elongated roots (day 5, > 1 mm).

Shoot Apices of English Ivy

The aseptically cultured shoot apices bioassay using English ivy was developed by Hackett (1970). He used it to determine the root promoting activity of fractionated extracts from shoots of juvenile and adult English ivy plants. The juvenile phase of English ivy forms adventitious roots on the stem of intact plants, but the adult phase does not, and is difficult-to-root from leafy stem cuttings, even under the most favorable conditions. Shoot apices from juvenile plants exhibit concentration dependent responses to several auxins. Adult apices do not form roots in the presence of from 0–50 mg·l⁻¹ of IAA, IBA, or NAA. This test is a model for testing factors that promote adventitious rooting of both easy- and difficult-to-root cuttings of identical genotype.

The procedure for the aseptically cultured shoot apices test is:

1) Rooting tests are performed aseptically in 6 dr glass vials using basal culture medium (Table 1). A folded strip of Whatman No. 3MM filter paper serves as a wick and platform for the apices. The basal medium plus substances to be tested are filter sterilized, and vials and support papers are autoclaved at 1.05 kg·cm^2 (15 lb·in^2) for 20 min. Paper strips from chromatographed extracts are also autoclaved (which limits testing of heat-labile compounds).

2) Shoot apices are obtained from juvenile and mature plants that originated from the same stock plant. Apices of 2–3 mm length are aseptically excised from sterilized shoots and placed on the wicks in the culture bottles.

3) Bioassays are conducted at 21 °C and a quantum flux density of ca. 70 μmol·m^{-2}·s^{-1} from daylight type fluorescent lamps, 16 h photoperiod. Visible roots are counted after 28 d.

Table 1. Composition of basal medium for rooting bioassay using aseptically cultured shoot apices of English ivy (Hackett 1970).

Components	mg·l⁻¹
KH_2PO_4	170.2
KCl	149.2
NaCl	2.3
$MgSO_4 \cdot 7H_2O$	123.3
Na_4Fe EDTA	25.3
$Ca(NO_3)_2 \cdot 4H_2O$	472.4
$MnSO_4 \cdot H_2O$	1.7
KI	0.17
$ZnSO_4 \cdot 7H_2O$	0.29
H_3BO_3	0.12
$CuSO_4 \cdot 5H_2O$	0.25
$NaMoO_4 \cdot 2H_2O$	0.24
Myo-inositol	90.1
Thiamin HCl	0.17
Urea	300.0
Sucrose	20,000.0
pH	5.8

Tomato Leaf Discs

The Coleman and Greyson (1977) tomato leaf bioassay uses discs cultured *in vitro* on a chemically defined nutrient medium. In their bioassay, discs are cut from the third true leaf; the developmental age of the leaf (expressed as LPI for leaf 3) is between LPI_3 2.0–4.0. Rooting in this test theoretically depends only on external factors such as organic and inorganic nutrients and auxins or similar hormones. Applied gibberellic acid inhibits rooting. Root primordia originate from two distinct tissues: phloem parenchyma of the primary and secondary veins, and sheath parenchyma of minor veins. By 24 h, profuse mitotic activity is evident in the vascular parenchyma and cortex of major veins, and in the sheath parenchyma of minor veins. After 36 h, small meristematic centers are present in minor and major veins; root primordia have differentiated by 48 h, and have elongated and begun to emerge by 132 h.

The procedure for the tomato leaf disc test is:

1) True leaf number 3 from tomato cv. 'Farthest North' is severed at the petiole base, and the cut surface sealed with liquid Paraplast wax.

2) The leaves are surface sterilized for 7 min with freshly prepared aqueous 5% calcium hypochlorite solution and then rinsed four times with sterile distilled water.

3) Five mm discs are cut from leaves placed with their adaxial sides down on a glass plate. Two discs are removed from the basal half of each leaflet such that the midrib is medially located within each disc.

4) Discs are separately placed abaxial side down on 10 ml of 0.8% agar medium in screw-top vials containing Murashige and Skoog (1962) mineral elements, 100 μmol $\cdot$ l^{-1} thiamine, and 2% sucrose. This is termed the basal medium, to which IAA or other putative root promoting substances are added.

5) Discs are incubated for up to 7 d at 25 $\pm$ 1 °C under continuous illumination from cool-white fluorescent (Sylvania) lamps (29–36 μmol $\cdot$ m^{-2} $\cdot$ s^{-1}). Discs are then cleared in 70% ETOH and roots are counted.

Summary

Rooting bioassays are suitable for detecting physiological activity and monitoring the purification of root promoting substances. The simplest rooting test is the mung bean bioassay and it is also the only bioassay that can easily provide the considerable degree of replication for statistically satisfactory results. With the mung bean bioassay, the timing and location of cell division activity is also known, which is important in understanding the multitude of factors involved in a complex process such as rooting. The timing and location of cell division activity is similarly known for the tomato leaf disc bioassay, which theoretically depends only on external factors for stimulation of rooting. The English ivy bioassay is, as is the tomato leaf disc assay, not suitable for a large number of replications or extensive testing of crude extracts. However, the English ivy bioassay is ideal for testing substances on apices with different rooting abilities but identical genotypes. The pinto bean bioassay has been used mainly with pure rooting substances.

A major underlying premise of growth regulator research, of which root promoting substances are a part, is that the endogenous levels of plant growth regulators control the growth and development of plants. There are major problems associated with the use of bioassays for the quantification and identification of root promoting substances. Bioassays measure a developmental response in contrast to the amount of a chemical or substance, and must be strictly controlled to obtain definitive results. Biological variation exists in all rooting bioassay plant material used, which requires a considerable number of replications that are often difficult to obtain. Bioassays are properly used only if restricted in their use to individual species, and so are required for each species because there is no certainty that a root promoting substance from one species is the same in another. Therefore, care is required when classifying an unknown compound as a rooting substance on the basis of activity discovered in a single species. Lastly, in bioassays the physiological response to a substance in an impure chemical fraction is the sum of promotive and inhibitory responses that result from the various chemical constituents; such interactions can confound the interpretation of results.

FUTURE ASSAYS

Critical assessment of the importance of putative endogenous hormones or exogenous growth regulators to adventitious rooting depends on unequivocal demonstration of the compound(s) in the plant. It is also necessary to determine the amount of compound(s) produced or contained in a particular organ, tissue, or cell, if demonstrating correlative roles is important. Such unequivocal determination can only be made with, for example, GC-MS or assays using monoclonal antibodies (immunoassay) that are physicochemically validated. Future critical determinations of endogenous root promoting compounds and their correlation with developmental processes will require using techniques such as GC-MS and immunoassay.

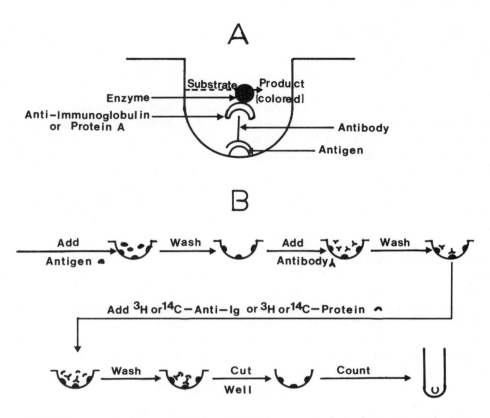

Figure 1. Diagrammatic representations of (A) the enzyme-linked immunosorbant assay (ELISA) and (B) solid-phase radioimmunoassay (RIA).

Immunochemical Assays

RIA and ELISA are general methods for routinely measuring physiological concentrations of hormones (Fig. 1) (Jaffe and Behrman 1974). A RIA for IAA is sensitive in the range from 0.2–12 ng. The assay depends upon the competition of IAA and ^{3}H-IAA for binding sites on rabbit anti-IAA antibody molecules. Low molecular weight compounds such as IAA are not usually antigenic and must be covalently attached to a suitable carrier molecule, e.g. BSA, to elicit antibodies. Antibodies are proteins manufactured by the immune system and present in the bloodstream. However, RIA may not unequivocally identify a compound due to interfering compounds. RIA must be validated by one or more physicochemical methods which reveal the chemical structure of unknown compound(s). For example, GC-MS analysis can provide unequivocal identification. Therefore, combined RIA and GC-MS can provide unequivocal identification. RIA and GC-MS for measuring IAA in extracts of maize (*Zea mays* L.) have been compared. As a result, a model for validation was developed (Pengelly et al. 1981).

The RIA described below is based on research by Pengelly and Meins (1977) for IAA, but could be adapted for other root promoting substances. The procedure for extraction and quantification of IAA from *Nicotiana tabacum* L. is:

1) Tissue containing 10–200 ng IAA is homogenized and suspended in four volumes of 100% MEOH (0 °C) containing 1,250 Bq ^{3}H-IAA. The homogenate is filtered and the filter pad is washed thrice with cold MEOH (0 °C).

2) MEOH is removed from the combined filtrates *in vacuo* at 37 °C, leaving an aqueous solution. The aqueous solution is mixed with 20 ml of 0.5 mol·l^{-1} K_2HPO_4 at pH 8.5 and purified by acid-base partitioning with diethyl ether.

3) Antibodies are prepared by repeated injection of rabbits with IAA-BSA conjugate. When a high antibody titre has been reached, the rabbits are bled and the serum used for assay.

The standard assay mixture consists of 10 μl ^{3}H-IAA (166 Bq), 80 μl sample in phosphate-buffered saline, and 10 μl anti-IAA-globulin solution. After agitation, the tubes are incubated in the dark for 24 h. The IAA-antibody complex that forms is precipitated. The amount of nonradioactive IAA in the sample is then calculated by comparing the fraction of total ^{3}H-IAA in the bound form with a standard curve.

Hybridomas can provide another method for the production of antibodies. In 1975, Kohler and Milstein developed a technique for producing monoclonal antibodies by hybridomas. Hybridomas are cells that are formed by combining the nuclei of normal antibody forming cells (spleen) with those of their malignant counterparts. The hybridomas formed as a result of polyethylene glycol fusion can be screened for their ability to produce specific antibodies by standard assays, such as ELISA (Engvall and Pesce 1978) or RIA (Catt and Treger 1967). Once a desirable hybridoma is obtained, it is relatively inexpensive to produce hundreds of mg of specific antibody. The hybridoma technique supports assay of 200–500 plant samples per day because it is simpler and less time consuming than standard bioassays.

Physicochemical Verification

MS is one of the most discerning methods for identifying many chemical compounds. MS identifies a compound by its molecular weight or characteristic fragmentation pattern, or both. Direct quantification is also possible. Biochemicals are usually purified by GC before MS, which resulted in the development of combined GC-MS instruments.

The method presented below (Fig. 2) can be used to extract and analyze IAA and IAAsp from small amounts of pigmented plant material, such as mung bean [adapted from Laws and Hamilton (1982) and Norcini and Heuser (1985)].

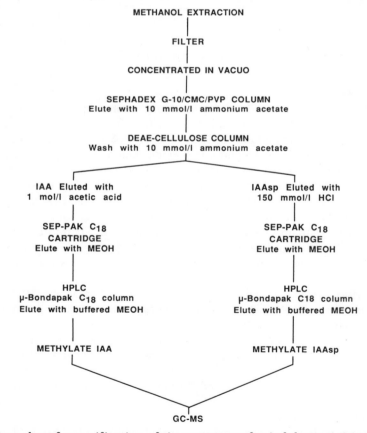

Figure 2. Flow chart for purification of tissue extracts for indole-3-acetic acid (IAA) and IAA-aspartate (IAAsp) analysis by gas chromatography-mass spectrometry (GC-MS).

1) Mung bean stems (ca. 35 g fresh wt., 1.4 g dry wt.) are harvested and immediately placed into 210 ml of boiling 100% MEOH. About 370 Bq of both 1-^{14}C-IAA and 1-^{14}C-IAAsp are added as an internal standard to monitor possible loss. About 21 mg BHT is also added. The tissue is ground in a mortar and reextracted with agitation in the dark for 20 min. After filtration (a Buchner funnel fitted with Whatman #1 paper), the residue is rinsed with 140 ml 80% MEOH and all filtrates are combined. An aliquot of combined filtrate, equivalent to 3 g fresh wt. (120 mg dry wt.), is used for analysis.

2) For IAA isolation the extract can be concentrated *in vacuo* at 40 °C to the aqueous phase, acidified (pH 2–3 with 3 mol·l^{-1} HCl), and partitioned four times with equal volumes of anhydrous diethyl ether. After evaporating the ether, the sample is resuspended by washing with 1 ml water, 0.5 ml 100% MEOH, and 2.5 ml 10 mmol·l^{-1} ammonium acetate. Alternatively, the initial sample extraction can be concentrated to 1 ml and the flask washed with 0.5 ml 100% MEOH and 2.5 ml 10 mmol·l^{-1} ammonium acetate.

3) Following preliminary purification, the extract is further purified by column chromatography. The combined fractions are applied to a 2.2 × 15 cm column containing Sephadex G-10 overlaid with CMC and PVPP in the respective proportions of 10:5:10 (v/v/v). The column is eluted with 10 mmol·l^{-1} ammonium acetate and the labeled fractions are pooled. A 2 × 10 cm column filled with 10 ml DEAE-cellulose is prepared by washing with 20 ml of 6 mol·l^{-1} acetic acid followed by 30 ml of 10 mmol·l^{-1} ammonium acetate. The labeled fraction is applied and washed with 25 ml of 10 mmol·l^{-1} ammonium acetate. IAA is eluted with 1 mol·l^{-1} acetic acid; and IAAsp with 150 mmol·l^{-1} HCl. The IAA fraction is applied to a Sep-Pac C$_{18}$ cartridge and eluted with 1 ml 100% MEOH. IAAsp is similarly treated except that a few drops of 6 mol·l^{-1} HCl are added to the Sep-Pak C$_{18}$ cartridge for better retention just prior to application of the IAAsp fraction.

4) TLC is an optional method sometimes used before HPLC and GC-MS. If silica gel is used, the recovery of IAA must be done within 2 h, before the IAA is degraded. In the best method, samples are streaked on polyamide TLC plates (Iino and Carr 1982) and chromatographed with benzene:ethyl acetate:acetic acid (70:25:5, v/v/v, ascending).

Putative active substances are removed, after location by methods such as short wavelength UV light and Salkowski reagent, by scraping off the medium containing the putative active substance, which is then eluted.

5) HPLC can be used for additional purification. The HPLC is equipped with a 30 cm × 7.8 mm μ-Bondapak C$_{18}$ column and glassy carbon electrochemical detector. The elution solvent is 100% MEOH:buffer (1:4, v/v) with a buffer consisting of 50 mmol·l^{-1} perchloric acid, 5 mmol·l^{-1} acetic acid, and 1 mmol·l^{-1} EDTA adjusted with NaOH to pH 3.5 for IAAsp or pH 5.0 for IAA. Fractions that co-chromatograph with authentic IAA and IAAsp are collected, concentrated to dryness, and dissolved in 100% MEOH.

6) Determination of IAA by GC-MS is carried out using a selected ion monitoring technique (Caruso et al. 1978, Little et al. 1978, Mugnus et al. 1980). Deuterated IAA is used as an internal standard, and the mixture of deuterated and endogenous IAA is derivatized before GC-MS. IAA and IAAsp are dissolved in MEOH for methylation with ethereal diazomethane (Fales et al. 1973) prior to GC-MS. Samples are injected into a 182.9 cm × 2 mm column containing 3% SP2250 on 80/100 Supelcoport. The He flow rate is about 23 ml·min^{-1} with the injector temperature 325 °C. A temperature program of 250–350 °C at 10 °C·min^{-1} is used. Mass spectra are recorded at an ionizing voltage of 70 eV with a source temperature of 250 °C. One ng of sample can be readily detected.

CONCLUSION

Higher plant bioassays have contributed to our understanding of the initiation and development of adventitious roots. Historically such bioassays were a necessary first step in demonstrating the presence and importance of root promoting substances. Higher plant bioassays, however, measure a plant developmental response as contrasted to the measure of a chemical substance. A particular chemical is detected and measured only by inference in a bioassay. Unequivocal determination can only be made with physicochemical methods. There will be less future acceptance of higher plant bioassays, and more reliance on physicochemical analyses and immunochemical assays that have been physicochemically validated.

ACKNOWLEDGMENTS

I thank Drs. Richard N. Arteca, Bruce E. Haissig, and Robert H. Hamilton for reviewing the manuscript.

REFERENCES

Blazich, F. A. and C. W. Heuser. 1978. The mung bean rooting bioassay: a re-examination. *J. Amer. Soc. Hortic. Sci.* 104:117–120.

Bassuk, N. L. and B. H. Howard. 1981. Factors affecting the use of mung bean (*Vigna radiata* L. Wilczek) cuttings as a bioassay for root-promoting substances. *J. Hortic. Sci.* 56:295–300.

Caruso, J. L., R. G. Smith, L. M. Smith and T.-Y. Cheng. 1978. Determination of indole-3-acetic acid in Douglas fir using a deuterated analog and selected ion monitoring. Comparison of micro-quantities in seedling and adult tree. *Plant Physiol.* 62:841–845.

Catt, K. J. and G. Tregear. 1967. Solid-phase radioimmunoassay in antibody-coated tubes. *Science* 158:1570–1572.

Coleman, W. K. and R. I. Greyson. 1977. Analysis of root formation in leaf discs of *Lycopersicon esculentum* Mill. cultured *in vitro. Ann. Bot.* 41:307–320.

Devlin, R. M. and F. H. Witham. 1983. *Plant Physiology.* Willard Grant Press, Boston, USA. ISBN 0-87150-765-X.

Engvall, E. and A. J. Pesce. 1978. *Quantitative Enzyme Immunoassay. Scand. J. Immunol.* Suppl. No. 7. Vol. 8. Blackwell Pub., Oxford.

Fales, H. M. and T. M. Jaouni. 1973. Simple device for preparing ethereal diazomethane without resorting to codistillation. *Anal. Chem.* 45:2301–2303.

Hackett, W. P. 1970. The influence of auxin, catechol, and methanolic tissue extracts on root initiation in aseptically cultured shoot apices of the juvenile and adult forms for *Hedera helix. J. Amer. Soc. Hortic. Sci.* 95:398–402.

Haissig, B. E. 1979. Influence of aryl esters of indole-3-acetic and indole-3-butyric acids on adventitious root primordium initiation and development. *Physiol. Plant.* 47:29–33.

Hemberg, T. 1951. Rooting experiments with hypocotyls of *Phaseolus vulgaris* L. *Physiol. Plant.* 47:358–369.

Hess, C. E. 1962. A physiological analysis of root initiation in easy and difficult-to-root cuttings. In *Proc. 16th Int. Hortic. Cong.* pp. 375–381.

———— 1964. Naturally-occurring substances which stimulate root initiation. In *Col. Int. du Centre Nat. Recherche Sci. Paris.* No. 123. pp. 517–527.

Iino, M. and D. J. Carr. 1982. Estimation of free, conjugated, and diffusible indole-3-acetic acid in etiolated maize shoots by the indolo-α-pyrone fluorescence method. *Plant Physiol.* 69:950–956.

Jaffe, B. M. and H. R. Behrman. 1974. *Methods of Hormone Radioimmunoassay.* Academic Press, NY, USA. ISBN 0123792509.

Kohler, G. and C. Milstein. 1975. Continuous culture of fused cells secreting antibody of predefined specificity. *Nature* 256:495–497.

Laws, D. M. and R. H. Hamilton. 1982. A rapid isotope dilution method for analysis of indole-3-acetic acid and indoleacetyl aspartic acid from small amounts of plant tissue. *Biochem. Biophys. Res. Commun.* 106:1035–1041.

Little, C. H. A., J. K. Heald and G. Browning. 1978. Identification of indoleacetic and abscisic acids in the cambial region of *Picea sitchensis* (Bong) Carr. by combined gas chromatograph-mass spectrometry. *Planta* 139:133–138.

Mugnus, V., R. S. Bandurski and A. Shulze. 1980. Synthesis of 4,5,6,7 deuterium-labeled indole-3-acetic acid for use in mass spectrometric assays. *Plant Physiol.* 66:775–781.

Murashige, T. and F. Skoog. 1962. A revised medium for rapid growth and bio assays with tobacco tissue cultures. *Physiol. Plant.* 15:473–497.

Norcini, J. G. and C. W. Heuser. 1985. Changes in free and conjugated indole-3-acetic acid during initiation and early development of adventitious roots in mung bean. *J. Amer. Soc. Hortic. Sci.* 110:528–533.

Pengelly, W. and R. Meins, Jr. 1977. A specific radioimmunoassay for nanogram quantities of the auxin, indole-3-acetic acid. *Planta* 136:173–180.

_____ R. S. Bandurski and A. Schultz. 1981. Validation of a radio-immunoassay for indole-3-acetic acid using gas chromatography-selected ion monitoring-mass spectrometry. *Plant Physiol.* 68:96–98.

Went, F. H. 1934. A test method for rhizocaline, the root-forming substance. *Proc. K. Ned. Akad. Wetensch. Amst.* 37:445–55.

CHAPTER 20

Agrobacterium rhizogenes: **A Root Inducing Bacterium**

Gary A. Strobel

Department of Plant Pathology
Montana State University
Bozeman, Montana 59717

and

Avi Nachmias

Agricultural Research Organization
Gilat Regional Experiment Station
Mobile Post Negev
Israel

INTRODUCTION. 284
BIOLOGY OF *AGROBACTERIUM RHIZOGENES* 285
ROOTING OF BARE ROOT STOCK MATERIAL 285
ROOTING OF CUTTINGS 286
CONCLUSION 287
REFERENCES. 288

Additional key words: adventitious root formation, lateral root formation, seedlings, cuttings, genetic engineering.
Abbreviations: IBA, indole-3-butyric acid; TBZ, thiabendazole.

INTRODUCTION

Agrobacterium rhizogenes, a soil inhabiting microorganism, occasionally enters a plant root through a wound or natural opening eventually causing proliferation of secondary roots. This condition is known as "hairy root." *A. rhizogenes* has a wide host range including many dicotyledonous but no poly- or monocotyledonous plants (DeCleene and Delay 1981). The phenomenon of "hairy root" was first described by Riker (1930) who also showed that the causal organism was distinct from the species causing tumors (*A. tumefaciens*). The bacterium was isolated from "hairy-rooted" apple and other roseaceous plants. Plants in which the syndrome occurred appeared, at maturity, to possess less top growth than those not infected, but no significant differences in fruit yield were noted. Riker (1930) also suggested that the hairy root organism may be used to stimulate root production in certain plants.

After a flurry of reports in the early 1930s, *A. rhizogenes* and the "hairy root" syndrome were given little or no attention. Since no dramatic crop loss, nor pathological symptoms exist in plants transformed by *A. rhizogenes,* it is probable that the syndrome escaped the attention of plant pathologists—and the time was not ripe for molecular biologists to study the phenomenon until the 1970s and 1980s. In the early 1970s, several reports suggesting that *A. rhizogenes* produced "super rooting" hormones in liquid culture media were published. These hormones seemed to be associated with the aqueous phase after extraction of liquid culture media with various organic solvents. However, other workers showed that the intact bacterium was required to induce rooting (Moore et al. 1979). The foregoing results indicated that "super rooting" was apparently induced by living bacteria remaining in the aqueous phase, as opposed to a previously produced hormone. As few as 100 *A. rhizogenes* cells could induce rooting on a carrot (*Daucus carota* L.) root disc, the method of choice for a bioassay (Moore et al. 1979).

What then was responsible for rooting? It seemed that the phytohormones commonly involved in rooting were somehow involved in the rooting of carrot discs, but not totally responsible for the phenomenon. Eventually, Moore et al. (1979) demonstrated that a large bacterial plasmid (1.2×10^8 Daltons or ca. 200,000 base pairs) called the Ri (also Hr) plasmid was inextricably linked with the root inducing potential of *A. rhizogenes*. This research was confirmed by the efforts of White and Nester (1980).

BIOLOGY OF *AGROBACTERIUM RHIZOGENES*

Moore et al. (1979) demonstrated the involvement of the Ri plasmid in the rooting of dicotyledonous plants in the following ways: 1) bacteria lacking the Ri plasmid did not induce rooting; 2) Ri plasmid transconjugants, prepared by *in vitro* mating experiments between *A. rhizogenes* and *A. radiobacter*, demonstrated root promoting activity; and 3) mutants prepared by ethidium bromide treatment were unable to cause rooting, apparently because they had sustained one or more lesions in the Ri plasmid.

The entire process of rooting in plants induced by *A. rhizogenes* is a natural example of genetic engineering in which the agent is the Ri plasmid. Chilton et al. (1982) were the first to observe that about 10% of DNA (the T-DNA) from the Ri plasmid is inserted into the genome of the higher plant cell and transcribed. Translation of the resulting mRNA results in biosynthesis of the novel amino acid mannopine (an opine), which *A. rhizogenes* can metabolize (Lippincott et al. 1973, Chilton et al. 1982, Petit et al. 1983). Opine synthesis continues when the affected plant tissue is cultivated *in vitro* in the absence of *A. rhizogenes* (Chilton et al. 1982).

Following these discoveries, many more molecular details, such as the base sequences of T-DNA, have been described (Durant-Tardif et al. 1986). However, the mechanisms of T-DNA transfer and insertion, and an explanation of how these events lead to more rooting remain a mystery.

ROOTING OF BARE ROOT STOCK MATERIAL

As the molecular processes involved in plant genetic transformation have become better understood, the overall biological effects of *A. rhizogenes* are being more closely examined for their potential usefulness in agriculture, horticulture, and forestry. In this connection, one of the first species to be studied was almond (*Prunus amygdalus* L.). A "super rooting" mutant of *A. rhizogenes* was selected from cultures of this organism (Fig. 1). The ability of this mutant to cause rooting in the standard carrot root assay was considerably greater than the wild type *A. rhizogenes* (Strobel and Nachmias 1985). In one test, this strain (MT232) was inoculated onto bare root stock almond seedlings (Israeli root stock var. 14 with scion stock no. 51). Autoclaved MT232, filter sterilized MT232, *A. radiobacter* NTI (without an Ri plasmid), *A. radiobacter* (autoclaved), and medium alone were used as control inoculation treatments. After 90 d, new roots developing on the original bare roots were counted, harvested, and weighed. The *A. rhizogenes* MT232 (living cells) treatment yielded greater new root number and mass then any of the controls. Furthermore, at 90 and 150 d compared to plants inoculated with medium alone, plants inoculated with *A. rhizogenes* MT232 had a significantly larger number of branches, length of lateral branches, and stem diameter. However, after several years the growth of control (medium alone) trees equaled that of MT232 treated trees (Strobel and Nachmias unpublished). The development of large lateral roots was lower in the treated vs. the control trees.

Figure 1. Rooting of parsnip root discs three weeks after treatment with wild-type strain of *Agrobacterium rhizogenes* TR105 (top two discs), the mutant strain 232 (lower right), and nothing (lower left).

Interestingly, treated almond trees did not lose their leaves in the fall and winter seasons, suggesting a variation in endogenous phytohormones. The phenomenon of leaf retention may be related to the relatively mild winters usually experienced in the upper Negev Desert where the trees were planted.

At the time of planting of the almond trees, a second experiment using bare root stock olive trees was also undertaken. Growth and development of *A. rhizogenes* MT232 treated vs. control olive trees within the first year paralleled that of almond (Strobel and Nachmias 1985). The treated trees have maintained an impressive and significant lead in growth and flowering vs. untreated controls during the first 4 yr after planting. Many other fruit and forest species, including those used for reclamation, are presently being tested for root induction, growth effects, and drought tolerance in various parts of the world.

ROOTING OF CUTTINGS

Seemingly unlike *A. tumefaciens,* which transforms most dicotyledonous plant tissues into tumors, *A. rhizogenes* apparently restricts itself to the pericycle (Jaynes and Strobel 1981). This is evidenced by the appearance of new roots in a circular pattern paralleling that of the pericycle in bacterial treated, mature carrot root discs (Fig. 1), or in a patchy pattern from the pericycle located in the vascular bundles scattered throughout the cortical tissue of red beet or sugar beet (*Beta* spp.) root pieces. Thus, *A. rhizogenes* appears to require competent target cells such as those in the pericycle to effectively transform root tissue.

The potential for developing *A. rhizogenes* as a treatment to enhance the formation of adventitious roots in stem cuttings may depend on its ability to produce rooting hormones in culture. Other compounds exuded by the bacteria as they colonize the stem tissue may be required to properly condition the rooting response. Thus, the processes of transformation may not necessarily be involved. Although the process is not clear a few observations have been made which indicate that *A. rhizogenes* promotes adventitious rooting in some plant species.

In preliminary studies with stem cuttings of several fruit trees and vine species we used *A. rhizogenes* MT232 as one dip treatment and as a pretreatment (control) a dip treatment in 5% TBZ with 3 mg·l⁻¹ IBA. The phytohormone was used as a control since it is the accepted treatment for the vegetative propagation of the plant species tested. The cuttings (10–12 cm in length) were placed in a rooting mixture of 60% peat and 40% perlite, and misted with a water cloud once every 10 min. The results were evaluated 4–6 weeks after the beginning of the experiment (Table 1). There was an observable positive effect only in the rooting of mango cuttings with the bacterial treatment, in contrast to the phytohormone treatment. However, in other tests, a positive effect on the rooting of violet cuttings after *A. rhizogenes* treatment has been noted (Strobel unpublished). In addition, cuttings of avocado, rooted after *A. rhizogenes* treatment, were much more tolerant to drought than controls (Strobel and Nachmias unpublished).

Table 1. Promotion of root formation in various species of plant stem cuttings with *Agrobacterium rhizogenes* 232.[1]

Plant Species (Variety)[2]	Mean Number of Rooted Cuttings by Treatment[3]	
	Phytohormone	*A. rhizogenes*
Avocado (hard Western Indian 1)	4.6 ± 3.2	5.2 ± 5.0
Avocado (Mexican)	58.0 ± 10.4	61.9 ± 20.7
Grape (root stock 16/13)	80.0 ± 2.0	71.0 ± 14.0
Jojoba (wild type)	85.0 ± 5.0	85.0 ± 5.0
Mango (13/1 salt resistant)	41.0 ± 7.5	79.0 ± 7.0

[1]Cuttings, four replications with at least 40 cuttings, were treated with a mixture of synthetic phytohormones (normal method of rooting) or with a dip in 10^8 cells/ml of *A. rhizogenes*.
[2]Avocado, *Persea americana* Mill.; Grape, *Vitis vinifera* L.; Jojoba, *Simmondsia chinensis* (Link) Schneider; Mango, *Mangifera indica* L.
[3]Average number of cuttings that rooted per treatment plus the standard deviation. Only mango, phytohormone treatment (IBA) vs. bacterial treatment was statistically different ($P < 0.05$) using the Student's t-test.

CONCLUSION

A. rhizogenes, including all of its genetically diverse types around the world, may have potential in promoting the rooting of bare root stock material, vegetable transplants, and stem cuttings. More trials on a variety of plant species using various isolates and mutants of *A. rhizogenes* should be tested in the future. Further, the mechanism(s) involved in root initiation may differ for cuttings (adventitious rooting) than for initiation of lateral roots on primary roots, and so must be tested.

It is apparent that strain differences occur in *A. rhizogenes,* for example, Lam et al. (1984) constructed TN5 mutants of *A. rhizogenes* which were host-restricted. That is, these mutants caused rooting in members of the family Chenopodiaceae (beets) but virtually no others. Thus, it is conceivable that wild type strains of *A. rhizogenes* have evolved that are more or less host selective. From this it follows that more than one strain of *A. rhizogenes* should be tested for root inducing ability.

Additionally, the best means of applying the organism to plants—by dipping, injection, or smearing directly on the cut surface—requires assessment.

A. rhizogenes should also be examined for its ability to induce rooting in callus tissue culture, on meristems, and in embryo culture. The organism does not destroy or otherwise degrade tissue and seems to be compatible with plant cells in culture (Strobel unpublished). Some preliminary success has been realized in several hardwood tissue culture systems. Additional work along this line is needed. It must be remembered, however, that plants arising from tissue culture should be carefully examined by biochemical means (e.g. presence of opines) in order to determine whether root initiation results from transformation or by direct hormonal action. If the former, it is then possible that the phenotype of the plant may have been permanently altered which may be a positive or a negative event depending upon the ultimate growth pattern of the plant.

Ultimately, *A. rhizogenes* may provide a much needed vector of genetic information. For instance, the genetic modification of the T-DNA region of the Ri plasmid with special genes that are expressed in the transformed host plant may result in insect or disease resistance.

There are many other options for study with this amazing soil microbe. For example, the Ri plasmid has recently been placed into *Rhizobium* spp. with the resulting transconjugants being able to produce more nodules in the case of *R. meliloti,* or no nodules at all in the case of *R. trifolii* (Strobel et al. 1985, 1986). However, the *R. trifolii* transconjugants behaved like *A. rhizogenes* in that they were root inducers.

The organism may also find use in the treatment of root systems diseased or impaired by various pathogenic soil microbes, i.e. *Phytophthora* spp. nematodes. When used in conjunction with a proper fungicide or nematacide, new rooting from the healthy regions of the root system may be the salvation of the diseased plant.

Release of *A. rhizogenes* into the soil environment should be done only with great caution and under carefully controlled circumstances. Questions have been raised about its effects on weed species, its ability to transform already established species, and its long term effect on the growth and yield of plants. What is needed at this time is enlightened consideration of *A. rhizogenes* by plant scientists. It should not be considered only as a pathogen, with the connotation of "pest," but rather as a potential biological mediator to achieve improved plant reproduction, growth, and health through specific genetic changes.

REFERENCES

Chilton, M. D., D. A. Tepfer, A. Petit, C. David, F. Casse-Delbart and J. Tempe. 1982. *Agrobacterium rhizogenes* inserts T-DNA into the host plant root cells. *Nature* 295:432–434.

Constantino, P., P. J. J. Hooykaas, H. denDulk-Ras and R. A. Schilperoot. 1980. Tumor formation and rhizogenicity of *Agrobacterium rhizogenes* carrying Ti plasmids. *Gene* 11:79–87.

Decleene, M. and J. Deley. 1981. The host range of infectious hairy root. *Bot. Rev.* 47:147–193.

Durand-Tardif, M., R. Broglie, J. Slightom and D. Tepfer. 1986. Structure and expression of Ri T-DNA from *Agrobacterium rhizogenes* in *Nicotina tabacum.* Organ and phenotypic specificity. *J. Mol. Biol.* 186:557–564.

Jaynes, J. and G. A. Strobel. 1981. The position of *Agrobacterium rhizogenes. Int. Rev. Cytol.* (Suppl.) 13:105–125.

Lam, S., B. Lam, L. Harrison and G. A. Strobel. 1984. Genetic information on the Ri plasmid of *Agrobacterium rhizogenes* determines host specificity. *Plant Sci. Lett.* 34:345–352.

Lippincott, J. A., R. Beiderbeck and B. A. Lippincott. 1973. Utilization of octopine and nopaline by *Agrobacterium. J. Bacteriol.* 116:378–383.

Moore, L. W., G. Warren and G. Strobel. 1979. Involvement of a plasmid in the hairy root disease of plants caused by *Agrobacterium rhizogenes. Plasmid* 2:617–626.

Petit, A., C. David, G. A. Dahl, J. G. Ellis, P. Guyon, F. Casse-Delbart and J. Tempe. 1983. Further extension of the opine concept: Plasmids in *Agrobacterium rhizogenes* cooperate for opine degradation. *Mol. Gen. Genet.* 190:204–214.

Riker, A. J. 1930. Studies on infectious hairy root of nursery apple trees. *J. Agric. Res.* 4:507–540.

Strobel, G. A. and A. Nachmias. 1985. *Agrobacterium rhizogenes* promotes the initial growth of bare root stock almond. *J. Gen. Microbiol.* 131:1245–1249.

_____ B. Lam, L. Harrison, B. M. Hess and S. Lam. 1985. Introduction of the hairy root plasmid results in increased nodulation on its host. *J. Gen. Microbiol.* 131:355–361.

_____ M. Heide and B. M. Hess. 1986. Biology of *Rhizobium trifolii* bearing the hairy root plasmid. *J. Gen. Microbiol.* 132:653–660.

White, F. and E. W. Nester. 1980. Hairy root plasmid encodes virulence traits in *Agrobacterium. J. Bacteriol.* 144:710–720.

CHAPTER 21

Adventitious Rooting of Tissue Cultured Plants

Brent H. McCown

Department of Horticulture
University of Wisconsin-Madison
Madison, WI 53706

INTRODUCTION..289
CULTURE ESTABLISHMENT AND MICROCUTTING QUALITY 290
ROOTING AND ACCLIMATION OF MICROCUTTINGS........................294
COMBINING MICROPROPAGATION AND CLASSICAL
    PROPAGATION...296
MICROCULTURE AS A RESEARCH TOOL...................................296
CONCLUSION 299
ACKNOWLEDGMENTS 299
REFERENCES...299

Additional key words: in vitro culture, micropropagation, microcuttings.
Abbreviations: BA, benzyladenine; IAA, indole-3-acetic acid; NAA, naphthaleneacetic acid; IBA, indole-3-butryic acid; GA, gibberellin A_3.

INTRODUCTION

The use of in vitro culture for the clonal propagation of plants has gained acceptance in a number of areas of commercial agriculture. The earliest use was with the herbaceous ornamentals (Murashige 1974). Only recently has significant application to woody crops (fruits, plantation crops, landscape ornamentals, forest trees) been advanced (Zimmerman et al. 1986). Woody plants have not only been more difficult to establish in culture, but problems with the rooting and acclimation of in vitro produced propagules have complicated removal from culture and subsequent commercialization.

The factors previously discussed in this book for the formation of adventitious roots on cuttings generally apply to the rooting of tissue cultured stock. However, two differences are apparent. Because the "stock plant" in a tissue culture system is contained in a sterile, highly defined environment, the degree of control of factors predisposing the cuttings to rooting is much greater than with the stock used for the more classical propagation. Thus the "quality" of in vitro produced cuttings can be managed to a high degree so that the formation of adventitious roots can be optimized. Much less emphasis is placed on manipulative treatments to induce rooting of cuttings after removal from the in vitro stock cultures than on producing the best "stock culture" which will yield cuttings with a high biological potential for rooting (Maene and DeBergh 1983). Secondly, cuttings taken from tissue culture are usually in a very active state of growth so should be considered "softwood" in quality. Thus observations made with normal hardwood cuttings may not be readily applicable to tissue cultured stock.

A number of special terms have developed that refer to various aspects of the tissue culture system; the terms utilized in this chapter are defined as follows:

Tissue culture: Refers to *in vitro* culture in general. More properly, tissue culture is the growth *in vitro* of organized tissue systems such as the epidermis or vascular elements.

Microculture: *In vitro* culture; suggested as a replacement for the term "tissue culture."

Micropropagation: Clonal propagation using *in vitro* techniques to generate the initial propagules.

Shoot culture: Multiplication based principally on the stimulation of axillary buds of shoots grown in microculture, although adventitious shoot meristem development may also be involved.

Micropropagule (microplant): A plant derived from microculture; also termed plantlet.

Microcutting: A shoot derived from microculture, usually shoot culture, that is to be rooted.

Macrocutting: Conventional cuttings taken from stock plants grown *ex vitro*.

This chapter emphasizes microcuttings obtained from shoot cultures because this is by far the most utilized methodology and where most of my experience with adventitious rooting has occurred. Information about microcutting responses will also be the most relevant to the discussions in this book concerning conventional cutting propagation because of the general morphological and anatomical similarities between these two types of cuttings.

This chapter reviews the factors which have been shown to influence the rooting of microcuttings from myriad plant species. However, because rooting of microcuttings of woody perennial species is more challenging than rooting of herbaceous species, considerably more research has been directed toward defining the conditions important in the rooting of woody perennials; thus this review emphasizes woody species. In addition to reviewing the rooting of microcuttings, this chapter discusses the use of *in vitro* techniques as a research tool for adventitious rooting studies.

CULTURE ESTABLISHMENT AND MICROCUTTING QUALITY

Manipulation of the shoot culture (the "stock plant" for micropropagation, Fig. 1) is of paramount importance to the rooting of microcuttings. Thus an understanding of how such cultures are generated is important and is the topic of most of the published research on micropropagation including many books and symposia (e.g. George and Sherrington 1984, Zimmerman et al. 1986). For this discussion, it is particularly important to note that the development of a shoot culture consists of three phases (McCown and McCown 1986):

Figure 1. A stabilized shoot culture of *Populus* spp. (left); a microcutting newly harvested from such a culture (center); and a micropropagule rooted and partially acclimated in a plug (right).

Isolation phase: This phase is usually of short duration and consists of the expression (growth) of the buds isolated with the original explant. Growth is highly dependent on the explant itself; such factors as the condition of the source plant, injury during sterilization, and endemic bacterial contamination are important.

Stabilization phase: During this period, growth of shoots will change in character and physiology from that typical of the original source plant to growth typical of actively growing shoot cultures. Such changes usually involve a reduction in size of the organs (leaf size, stem diameter) (Smith et al. 1986) and the assumption of a continuous (non-flushing or non-episodic) growth habit. A completely stabilized shoot culture is characterized by uninterrupted and uniform growth, and no detectable change in character of the shoots after repeated subculturing.

Production phase: The cultures are fully stabilized, and the rate of shoot multiplication and the individual shoot quality is now optimized. Microcuttings are harvested for rooting at the conclusion of this phase.

The latter two phases are of paramount importance for the adventitious rooting of microcuttings. In the stabilization phase, changes in the physiological nature of the shoots occur that usually lead to an increased ability to form adventitious roots. In the production phase, physical characteristics (overall size, leaf development) of the harvested microcuttings that are important in subsequent rooting are determined.

The exact nature of the changes that occur in the stabilization phase are not known. There are two lines of evidence that indicate that "rejuvenation" is probably a major factor. Explants originating from juvenile stock plants (Bonga 1987, Franclet et al. 1987) or from herbaceous annuals progress through this phase most rapidly and can often be stabilized after a few subcultures; explants from very mature (adult) stock plants may never stabilize or may require long periods of culture involving a year or more.

More direct evidence that rejuvenation is involved has been obtained in studies where the ability of microcuttings to form adventitious roots or the change in visual characteristics associated with phase change (leaf shape) are used to monitor the cultures. Such partial rejuvenation has been recorded with herbaceous crops such as potato (*Solanum tuberosum* L.) (Denton et al. 1977), but more often with woody plants such as *Populus* spp. (Whitehead and Giles 1977), *Vaccinium* spp. (Lyrene 1980, 1981), *Rubus* spp. (Snir 1981), *Eucalyptus* spp. (McComb and Bennett 1982, Standardi 1982), and *Prunus amygdalus* Batsch. (Rugini and Verma 1983). Recently, Mullins (1987) reviewed the research on pome and stone fruits, in which gradual improvement in the rootability of microcuttings with subculturing has been commonly observed. For example, with apple (*Malus* spp.) shoot cultures grown at 26 °C under continuous light, microcutting rootability improved from 8% in the first subculture to 70% after five subcultures and to 95% after nine subcultures. Associated with this change in rootability was a visible change in the form of the shoots, from thick stems and large leaves (nonrooting form) to small leaves and thin stems (rooting form). The rooting form accumulated anthocyanin after treatment with auxins and, in addition, contained higher levels of phloridzin and lower levels of GAs and abscisic acid than the nonrooting form. Similar correlations with anthocyanin development (Bachelard and Stowe 1962), phenolics (Jarvis 1985, Haissig 1986), and hormones (Jarvis 1985) have been noted with macrocutting rootability.

When adventitious shoots are generated, then the immediate appearance of the juvenile form has been universally observed in the resultant shoots (Banks 1979, Bilkey and McCown 1979, Hackett 1987).

The stimulation of rooting potential in mature stock by continuous *in vitro* culture has not always been successful. Problems with tree genera such as *Celtis, Quercus, Sassafras, Magnolia,* and many of the conifers have been apparent in commercial laboratories. The rooting ability of shoots of avocado (*Persea americana* Mill.) rootstocks was not improved after nine subcultures (Barcelo-Munoz and Pliego-Alfaro 1986). However, a variant of a technique used successfully for macrocuttings was developed which included successive grafting to seedling rootstocks. By grafting microshoots to seedlings that had been germinated *in vitro* (*in vitro* grafting or micrografting), microcutting rooting percentages increased to almost 50% after four successive grafts. Whether this character remained stable in subsequent shoot cultures was not determined.

Possible correlations between rootability of different genotypes using macrocuttings or microcuttings has not been intensively studied. Observations in commercial laboratories and in our research with *Rhododendron* spp. (McCown and Lloyd 1985), *Betula* spp., and *Populus* spp. (J. C. Sellmer. 1987, M.S. Thesis, Univ. of Wisconsin, Madison, WI, USA and Table 1) indicated that there was not a strong correlation between macrocutting and microcutting rootability when the stock cultures were well stabilized and maintained.

Table 1. Some factors (other than stock plant influences and time in culture) reported to be especially important in the stimulation of adventitious rooting of plant organs *in vitro*.

Factor	Exemplary References	Comments
MEDIUM COMPONENTS **Generally stimulatory factors**		
Auxins (IAA, NAA, IBA)	Many references; for a listing, see George and Sherrington 1984. Also note Amerson et al. 1985.	Dosage effect (conc. $\times$ time) important, 0.05–10 mg·l^{-1} for prolonged periods (days to weeks) or 50–100 mg·l^{-1} for short periods (s to h).
High sugar/nitrogen ratio	Hyndman et al. 1982b.	Effect and optimum conc. varies with mineral medium used and light levels.
Phenols		
Phlorglucinol, phloridzin	Jones and Hatfield 1976, Jones and Hopgood 1979, Welander and Huntrieser 1981, Zimmerman and Broome 1981, Pontikis and Sapoutzaki 1984.	Appears to be synergistic with auxins; response dependent on genotype and culture preconditioning; can also be inhibitory.
Chlorogenic acid	Hammerschlag 1982.	IAA synergist when in lighted conditions.
Charcoal, activated	Constantin et al. 1977, Defossard et al. 1978, Bressan et al. 1982, Khoshkhui and Sink 1982, George and Sherrington 1984 (pp. 365–366), Mission et al. 1983.	Effect dependent on medium and genotype; may reduce light levels in medium and/or absorb medium components such as inhibitory phenolics and excess auxins.
Generally inhibitory Cytokinins	Many references, see George and Sherrington 1984.	A very common observation; usually eliminated during rooting phase.
Gibberellins	Heide 1969.	Strong interaction with other hormones.
High ionic strength of medium	Sriskandarajah and Mullins 1981, Hyndman et al. 1982a, McComb and Bennett 1982.	Confounded by effects of individual nutrients.
Agar	Harris and Stevenson 1979, Werner and Boe 1980, Von	Exact causes of the effect are not known since agar is

Factor	Exemplary References	Comments
	Arnold and Erickson 1984, Lee et al. 1986.	impure and variable in content.
ENVIRONMENTAL FACTORS Etiolation/darkness, low light	Amerson et al. 1985, Lovell et al. 1972, Druart et al. 1982, 1983, Hammerschlag 1982.	Interacts with medium sugar levels, auxins, and phenols, and endogenous enzymes (peroxidases particularly).
WOUNDING	Sriskandarajah and Mullins 1981.	May produce wound factors and/or increase the absorption of medium components.

In vitro induced rejuvenation appears to be only a partial reversal of phase change. The best evidence is based on observations of micropropagated plants that, after removal from culture, were planted in comparative growth situations with seedlings. Micropropagated plants often flower sooner than seedlings of the same species (Banks 1979). Mullins (1986) reported that the increased rootability of apples developed in culture is lost after microcultured plants become well established in the field. Such apple plants also flowered in 2 yr after removal from culture. However Amerson and coworkers (1985) reported that Pinus taeda L. plants derived from adventitious buds of cotyledons showed more morphologically mature characteristics than seedlings planted in comparable field plots. Thus the precise effect on phase state of the in vitro culture environment is unclear.

Based on the foregoing discussion, a major factor affecting the rootability of microcuttings is the stabilization of shoot cultures. Once the stabilized state has been reached, the biological potential for rooting may be increased dramatically over that of the original source tissue. The expression of this potential in rooting then becomes a function of both the quality of the microcuttings, as determined in the shoot production environment prior to harvest, and of the rooting environment itself.

The factors determining the quality of microcuttings in relation to adventitious rooting have not been thoroughly defined. Overall size of the microcutting has been important (Maene and Debergh 1983). Microcuttings less than a minimum size, usually about 1 cm in length, do not root with uniform and high percentages. Leaf development is also important because microcuttings with fully developed leaves root the most consistently. A major effect of microcutting size is merely a survival factor—smaller cuttings have a high mortality rate in the rooting environment. However, leaves have been indicated as sources of factors important in rooting of macrocuttings (Rappaport 1940, Reuveni and Raviv 1981) and thus fully developed leaves on microcuttings may be metabolically important for rooting. In addition, stem elongation has been recognized as beneficial to rooting (Miller et al. 1982, Maene and Debergh 1985), possibly as an important part of the inductive phase (Druart et al. 1983). The ability to remain in active growth and thus to continue to elongate is a phenomenon influenced by exogenous GA treatment and thus accounts for some of the favorable reports of GA treatments on rooting in vitro (Hammerschlag 1982).

Another measure of quality is the degree of vitrification of the microcuttings. Under some conditions such as high cytokinin levels in the medium, tightly sealed vessels, and liquid media (Gaspar et al. 1987), and with some genotypes of plants, shoot growth will become watery (as evidenced by a translucent character and a high fresh wt./dry wt. ratio); such "vitreous" microcuttings have a high mortality rate and do not root vigorously.

Microcuttings taken from actively growing shoot cultures generally root more quickly and with greater uniformity than those harvested from quiescent or senescent cultures. Such observations have also been made with macrocuttings (see Hartmann and Kester 1983) and are at least partly attributed to the activity of the stem cambium (Davies 1984) and buds (Lanphear and Meahl 1961).

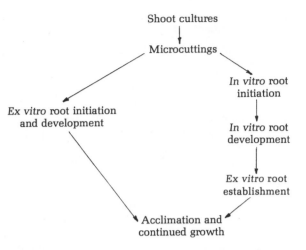

Shoot cultures

↓

Microcuttings

In vitro root
initiation

↓

In vitro root
development

↓

Ex vitro root
establishment

Ex vitro root initiation
and development

Acclimation and
continued growth

Figure 2. A flowchart showing the alternative strategies for rooting microcuttings.

ROOTING AND ACCLIMATION OF MICROCUTTINGS

The rooting of microcuttings can be accomplished *in vitro* or *ex vitro* (Fig. 2). Although *in vitro* rooting can often be accomplished more readily than rooting *ex vitro*, there are distinct advantages to rooting outside of culture (Debergh and Maene 1981):

1. *In vitro* rooting is expensive and can account for more than 50% of the cost of the propagule. The expense arises from the need for an additional culture step (using a medium different than that used for shoot culture), increased manual manipulation of the microcutting, and inefficient use of the culture space (the microcuttings are widely spaced in large rooting vessels).

2. Roots initiated *in vitro* may not be well formed. Cultured roots have few root hairs and are often easily separated from the stem where they arise. A common observation in our studies as well as that of P. L. Debergh (personal communication) is that roots formed *in vitro* will often die when the cutting is removed from culture and potted, although a new root system may rapidly be formed by subsequent adventitious rooting.

3. *In vitro* initiated roots are very easily damaged by handling of the rooted micropropagule.

Even with these disadvantages, there is still more literature describing the adventitious root development of microcuttings *in vitro* than *ex vitro*. Such *in vitro* rooting systems lend themselves to the analysis of various factors, media components as well as environmental, that might influence rooting. Table 1 gives a partial list of some of the more important factors that have been explored in this way. It is interesting to note that many of the same factors that have been found influential in the rooting of macrocuttings have also been observed to affect microcuttings.

As noted above, *in vitro* rooting is usually accomplished by subculturing the shoot clumps or the microcuttings to a medium different from that used for multiplication of the cultures. Characteristics common to media used for rooting include no cytokinin and reduced levels of mineral salts, both of which may inhibit rooting. A typical response is shown in Table 2 where the rooting of two *Populus* clones is detailed. In both clones, increasing the level of the cytokinin BA in the rooting medium progressively decreased rooting response until complete inhibition was observed. However, a high mineral salt medium (MS) only inhibited rooting significantly in one clone (*Populus tremula* L. 'erecta'). *Populus tremula* 'erecta' is a clone that also shows shoot growth inhibition when grown on high salt media whereas the other clone (NC 5339) grows equally well on both media (McCown and Sellmer 1987). Thus although cytokinins appear to be strong inhibitors of adventitious rooting, high salts may have indirect inhibitory effects by reducing microcutting quality and vigor. High auxin/cytokinin ratios appear to be necessary for root initiation (Skoog and Tsui 1948, Haissig 1965); in high quality shoot cultures, the actively growing shoot tips appear capable of supplying the stimulatory levels of auxin if cytokinin levels in the medium are not overwhelmingly inhibitory.

Table 2. The *in vitro* rooting response of two clones of *Populus* when grown on two standard mineral salt media and a range of BA (benzyladenine) levels. NC5339 is a moderately difficult-to-root clone and *P. tremula* is very difficult-to-root using standard macrocuttings. WPM (Woody Plant Medium) is a relatively low salt medium (42 μmol·l^{-1} in total salts) and MS (Murashige and Skoog medium) is a high salt (94 μmol·l^{-1}) medium (McCown and Sellmer 1987). The data are from an experiment where each vessel contained five, 2-node stem pieces originating from stabilized shoot cultures. Five replicates per treatment. Data presented as mean (S.E.)

Clone	BA Conc. (μmol·l^{-1})	Average Number of Roots per Culture Vessel	
		WPM	MS
NC5339	0	5.0 (0.3)	3.0 (0.5)
	0.04	3.6 (0.5)	3.0 (0.5)
	0.10	3.6 (0.4)	3.6 (0.6)
	0.40	0.8 (0.2)	1.8 (0.6)
	1.0	1.2 (0.3)	2.6 (0.5)
	4.0	0	0
P. tremula	0	16.4 (1.9)	3.1 (1.0)
'erecta'	0.04	14.7 (2.7)	4.9 (1.0)
	0.10	8.7 (1.2)	1.1 (0.6)
	0.40	3.1 (2.1)	0.1 (0.3)
	1.0	3.0 (1.6)	0
	4.0	0	0

Activated charcoal has also been commonly included in rooting media (Table 1). The effect involves at least two stimulatory aspects: 1) reducing the light at the base of the shoots, thus providing an environment conducive to the accumulation of auxins and/or cofactors (Druart et al. 1983); and 2) absorbing inhibitory components (Misson et al. 1983). The latter may include the removal of any cytokinins that are carried-over from the previous media.

The factors influencing *ex vitro* rooting are much less well established. Microcutting quality is probably more important in *ex vitro* rooting than with *in vitro* rooting as the microcutting is no longer being "cultured" with complex media. There have been some reports of "pretreating" the microcuttings before sticking in order to enhance their rooting response. Such *in vitro* pretreatments usually involve sugar and/or auxin solutions (Maene and Debergh 1985, Zimmerman and Fordham 1985) and may be accompanied by a dark period.

The treatment of microcuttings directly with rooting hormone preparations has in general not been as advantageous as with macrocuttings. Microcuttings are extremely succulent and in a very active state of growth and thus are easily overtreated with auxins. The application of rooting hormones to such small cuttings is tedious and difficult to standardize in a commercial setting. In general, such treatments are usually not necessary for rooting to occur, but may be useful for increasing the speed and uniformity of rooting as well as the number of adventitious roots. In addition, some genotypes of a species may benefit from auxin treatments, especially if rootability was not markedly increased by continuous microculture (Zimmerman 1984).

For *ex vitro* rooting, humidity control in the rooting environment is extremely important. Microcuttings will desiccate and die within minutes on exposure to low humidity after harvest from the culture vessel. When placed in a rooting environment with a uniform and high humidity (greater than 80%), the tissues formed in the culture environment can be maintained until rooting occurs. With most crops, immediately upon root emergence, very active shoot growth resumes and by the time the micropropagule is sufficiently rooted to remove from the rooting environment, the size of the cutting has more than doubled (Fig. 1). The new tissues thus formed are much less subject to desiccation and can be acclimated to low humidities by the gradual lowering of the humidity, usually in concert with increased light exposure. The leaves originally formed in culture usually desiccate or senesce by this time. Thus an important aspect of microcutting rooting and acclimation is the maintenance of shoot growth during root system development.

Although the photosynthetic capacity of microcuttings is low, sometimes less than 50% that of seedlings, full capacity is restored upon rooting of the microcutting (Smith et al. 1986).

COMBINING MICROPROPAGATION AND CLASSICAL PROPAGATION

Many commercial organizations are finding that a close interaction between classic clonal propagation and micropropagation allows them to develop a hybrid system that can capitalize on the benefits of both approaches. For example, we are seeing an increasing number of cases where micropropagation is used to generate stock plants that are then used for macrocuttings. Stock plants are generated on demand and used for a series of macrocutting harvests, then destroyed. The advantages of this approach include guaranteeing disease-free stock, generating actively growing and relatively juvenile stock at any time of the year, eliminating the need to maintain stock plant blocks throughout the year, and being able to rapidly respond to changes in market demand by adjustment of stock plant numbers. As important, however, is the reliance on the macrocutting technology which tends to produce propagules at less cost than micropropagation.

A second integration of micropropagation with classic propagation is in the use of modern transplant plug technology. Micropropagules are similar to seedlings, thus the technology used to handle large quantities of seedlings, especially in the vegetable and bedding plant industries, is applicable if microcuttings are rooted and acclimated in the plug systems used for seedling transplants (Fig. 1). This approach has proven successful in some applications (McCown 1986), however the physical character of many of the plug systems presents problems to the uniform production of adventitious roots on some crops, especially woody plants. The obvious next step is the automation of the transfer of the micropropagules to the plugs, but research on this aspect has only begun.

MICROCULTURE AS A RESEARCH TOOL

Microculture presents an excellent opportunity to intensively study some of the factors that influence adventitious rooting. As Table 1 indicates, many of the factors that influence adventitious rooting in vitro are similar to those that have been found with softwood macrocuttings. However, some of the experimental problems encountered when utilizing macrosystems may be overcome by employing microculture-based systems. Of major concern is the repeatability and reproducibility of experiments. With both macrocuttings and microcuttings, the previous history of the stock plant has a paramount influence on the responses observed with the cuttings obtained from these plants. Indeed, the inductive phase may be already completed by the time the cuttings are taken from the stock (Gaspar 1981). Lack of full appreciation of this phenomenon has undoubtedly led to some of the problems in interpretation and reproducibility of research in propagation. Precise and reproducible control of the stock plants is considerably easier with microcultured stocks. Microcultured stocks can be kept in continuously active growth throughout the year and thus a consistency in physiological states of the microcuttings can be obtained. Thousands of such stocks, virtually identical to each other and free of the variables created by pathogens or pest infection, can be generated on short time cycles (20–40 d subculture periods) and independent of season.

In addition to standardizing the biological variables, the environment in which stock cultures are grown and in which in vitro experiments are conducted is more readily standardized using microculture. The culture vessel can be specified and reproduced around the world from commercially available components. Since these culture vessels are small, many can be grouped in a minimum of shelf space, thus maximizing the use of environmentally controlled areas. As important, if each shoot culture/culture vessel is viewed as an experimental unit, and each culture shelf as analogous to a field plot, then statistically relevant experiments using standard plot/subplot designs can be planned to further treat known or suspected environmental gradients and external variation.

The types of variables that can best be studied using microculture have not been well analyzed. There are a number of basic tissue systems that seem appropriate. Organ cultures (leaf and internodes derived from shoot cultures; microcuttings) can be used in experiments analogous to those performed with softwood macrocuttings. In addition, the ability to maintain high viability of detached organs in culture is useful in experiments designed to determine the importance of interactions between organs on a cutting (e.g. the essentiality of hormones/cofactors derived from leaves/buds on the initiation of adventitious roots from stem tissues). Such culture systems are useful in studies of genetic differences in rooting potential (Haissig 1986); the environment and source tissue physiology can be highly standardized in microculture, thus making the genetic component the major variable being analyzed.

A second tissue system that appears relevant to studies on adventitious rooting is microcalli. Many hundreds of small (1–3 mm dia.) microcalli can be generated by plating of suspension culture cells or protoplasts (Fig. 3). Such calli have the capacity to regenerate roots (Fig. 4), the root primordia often being as large as the original microcalli source. The minimum size of calli that can be induced to

Figure 3. Microcalli developed from protoplasts of *Populus* that were isolated from leaf tissue and plated on a floating polyester screen.

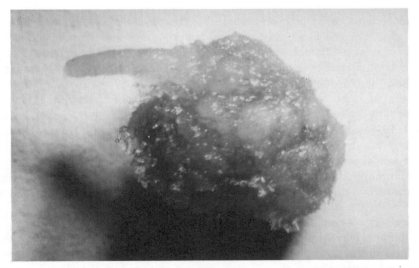

Figure 4. A microcallus of *Populus* spp. differentiating a root. The root was initiated by moving the web containing the callus to a medium containing 0.1 mg·1⁻¹ of IBA (indole-3-butyric acid).

regenerate roots has not been determined, however their usefulness for research lies in their small, relatively uniform cell mass and unorganized nature. Thus experiments such as the effect of specific exogenously applied compounds on adventitious root primordia initiation and development can be studied near or at the cellular level, and without the complexities of correlative interactions between tissue and organ systems. Such microcalli are reproducible and highly responsive because their biological history can be standardized, and they should be amenable to infusion of agents on both a long term or pulse dosage basis. Such microcalli may be particularly useful in studies of the progenitor cells of root initials (Nougarède and Rondet 1983) and on the inductive phase of rooting, both of which are areas needing intensive investigation. Additionally, microcalli experiments may be appropriate in determining the exact cellular effects of some environmental variables. For example, water stress has been implicated in the inductive phase of rooting (Haissig 1986); water stress can be predictably controlled in in vitro culture by addition of such osmotic components as polyethylene glycol to the basic support medium.

Microcultured tissue is highly applicable to studies requiring analysis of endogenous components. Tissue from microculture is very succulent (low amounts of dead support tissue) and, because of its rapid growth, is metabolically active. Such tissues are responsive to treatments and readily macerated and extracted. Experiments requiring precise timing of treatments and rapid extraction of labile, rapidly cycling, or radiolabelled components are facilitated by use of microculture. As important, many multiples of a test tissue system can be grown repeatedly in relatively short time periods and in a minimum of space, thus facilitating the capture of significant quantities of tissue of the desired physiological responsiveness. Such research systems may be useful in studies on the balance of endogenous factors known to be of potential importance in adventitious rooting such as oligopeptide/cytokinin/auxin (Klämbt 1983), phenolics/peroxidase (Druart et al. 1983), and hormone balances (Jarvis 1985).

Even though microculture may be a valuable tool for studying various phases of adventitious rooting, researchers should realize that problems unique to microculture may develop. Probably only softwood and primarily juvenile tissues are readily grown in culture. Although mature and juvenile callus has been maintained in culture for extended periods (Stoutemeyer and Britt 1965), adult (mature phase) tissues are generally difficult to grow uniformly and in quantity. In addition, the stability of the adult phase in long term culture is unpredictable, especially for tissues with organized meristems (Hackett 1987). Even though shoot cultures can be induced into dormancy, mimicking a seasonal cycle and the development of the hardwood character, such cultures are commonly non-uniform in individual shoot responses within a culture and thus do not appear to be highly useful for precise research.

Microcuttings have unique morphological and physiological characteristics, especially "abnormal" leaf structure and function. Common abnormalities include poor stomatal functioning (Brainerd and Fuchigami 1982), lack of an effective cuticle (Sutter and Langhans 1982, Wardle et al. 1983), reduced mesophyll and/or vascular structure (Grout and Aston 1977a, Wetzstein and Sommer 1982, Donnelly and Vidaver 1984, Smith et al. 1986) and reduced or modified photosynthetic/respiratory function (Grout and Aston 1977a, Smith et al. 1986). After rooting and removal from the culture environment, the new leaves and some of the leaves developed in vitro resume normal function (Grout and Aston 1977ab, Donnelly and Vidaver 1984, Smith et al. 1986). However, newly harvested microcuttings may respond to treatments in ways not readily transferrable to macrocuttings.

The culture vessel environment itself is unique and usually contains accumulations of gaseous components not normally found in significant quantities in most plant growth environments (Dunwell 1979). Some of these, particularly ethylene, have been implicated in the rooting response itself (Jarvis 1985).

Finally, especially when the undifferentiated tissue systems such as calli are used, genetic variants may be produced by way of somaclonal variation (Vasil 1986, Ahuja 1987). Although I do not know of any report documenting somaclonal variation in the capacity to form adventitious roots, there is no reason to assume that such variation can not be found. Although the generation of such mutants may be useful for comparative research studies, it presents yet another source of uncontrolled experimental error in other types of experiments.

CONCLUSION

The *in vitro* formation of adventitious roots is not well understood. As summarized by Halperin (1986), a diversity of factors control morphogenesis *in vitro* and no clear theory invariably explains the responses observed. Since similar uncertainty surrounds adventitious rooting *ex vitro*, with both microcuttings and macrocuttings, our present approaches to rooting most cuttings remain highly empirical.

One approach to deciphering some of the complex factors affecting rooting would be for physiologists to capitalize on the genetic differences in capacity to root that exist between closely related genotypes. This permits meaningful comparative studies leading to the physiological and genetic bases of root morphogenesis.

I believe that we will see an increasing trend toward the development and utilization of cutting propagation strategies which employ both microculture and conventional techniques. The advantages of microculture coupled with the well tested and trusted macrocutting techniques make a powerful and flexible approach for clonal propagation.

Microculture should be useful in helping to decipher the biochemical, genetic, and physiological bases of rooting. Intelligently designed research incorporating both whole plant and microculture approaches will be useful in testing hypotheses and in differentiating between primary and secondary effects. As importantly, the use of microculture can help to make repeatability and reproducibility of the results a standard instead of only a hope.

ACKNOWLEDGMENTS

Aspects of this work were performed in projects supported by the Agricultural Experiment Station and the Graduate School, University of Wisconsin-Madison, and by the Horticultural Research Institute. This paper benefitted from the comments and contributions of Jim Sellmer, Julie Russell, and Bernie Fourrier.

REFERENCES

Ahuja, M. R. 1987. Somaclonal variation. In *Cell and Tissue Culture in Forestry, General Principles and Biotechnology* (J. M. Bonga and D. J. Durzan, eds). Martinus Nijhoff Pub., Boston, USA. Vol. 1. pp. 272–285. ISBN 90-247-3430-4.

Amerson, H. V., L. J. Frampton, Jr., S. E. McKeand, R. L. Mott and R. J. Weir. 1985. Loblolly pine tissue culture: Laboratory, greenhouse, and field studies. In *Tissue Culture in Forestry and Agriculture* (R. R. Henke, ed). Plenum Press, NY, USA. pp. 271–287. ISBN 0-306-41919-X.

Bachelard, E. P. and B. B. Stowe. 1962. A possible link between root initiation and anthocyanin formation. *Nature* 194:209–210.

Banks, M. S. 1979. Plant regeneration from callus from two growth phases of English ivy, *Hedera helix* L. *Z. Pflanzenphysiol.* 92:349–353.

Barcelo-Munoz, A. and F. Pliego-Alfara. 1986. *In vitro* propagation of avocado rootstock G.A.-13. In *Abst. VI Int. Cong. Plant Tissue and Cell Cult.* (D. A. Somers, ed). University of Minnesota, Minneapolis, USA. pp. 278.

Bilkey, P. C. and B. H. McCown. 1979. *In vitro* culture and propagation of *Episcia* sp. (Flame Violet). *J. Amer. Soc. Hortic. Sci.* 104:109–114.

Bonga, J. M. 1987. Clonal propagation of mature trees: Problems and possible solutions. In *Cell and Tissue Culture in Forestry, General Principles and Biotechnology* (J. M. Bonga and D. J. Durzan, eds). Martinus Nijhoff Pub., Boston, USA. Vol. 1. pp. 249–271. ISBN 90-247-3430-4.

Brainerd, K. E. and L. H. Fuchigami. 1982. Stomatal functioning of *in vitro* and greenhouse apple leaves in darkness. *J. Exp. Bot.* 33:388–392.

Bressan, P. A., Y.-J. Kim, S. E. Hyndman, P. M. Hasegawa and R. A. Bressan. 1982. Factors affecting *in vitro* propagation of rose. *J. Amer. Soc. Hortic. Sci.* 107:979–990.

Constantin, M. J., R. R. Henke and M. A. Manson. 1977. Effect of activated charcoal on callus growth and shoot organogenesis in tobacco. *In Vitro* 13:293–296.

Davies, F. T. 1984. Shoot RNA, cambial activity and indolebutyric acid effectivity in seasonal rooting of juvenile and mature *Ficus pumila* cuttings. *Physiol. Plant.* 62:571–575.

Debergh, P. L. and L. J. Maene. 1981. A scheme for commercial propagation of ornamental plants by tissue culture. *Scientia Hortic.* 14:335–345.

Defossard, R. A., M. T. Bennett, R. Gorst and R. A. Bourne. 1978. Tissue culture propagation of *Eucalyptus ficifolia* F. Muell. *Proc. Int. Plant Prop. Soc.* 28:427–434.

Denton, I. R., R. J. Westcott and B. V. Ford-Lloyd. 1977. Variation in potato plants after tissue culture. *Potato Res.* 20:131–136.

Donnelly, D. J. and W. E. Vidaver. 1984. Leaf anatomy of red raspberry transferred from culture to soil. *J. Amer. Soc. Hortic. Sci.* 109:172–176.

Druart, P. H., C. Kevers, P. H. Boxus and H. Gaspar. 1982. *In vitro* promotion of root formation by apple shoots through darkness: Effect on endogenous phenols and peroxidases. *Z. Pflanzenphysiol.* 108:429–436.

Dunwell, J. M. 1979. Anther culture of *Nicotiana tabacum*: the role of culture vessel atmosphere in pollen embryo induction growth. *J. Exp. Bot.* 30:419–428.

Franclet, A., M. Boulay, F. Bekkaoui, Y. Fouret, B. Verschoore-Martouzet and N. Walker. 1987. Rejuvenation. In *Cell and Tissue Culture in Forestry, General Principles and Biotechnology* (J. M. Bonga and D. J. Durzan, eds). Martinus Nijhoff Pub., Boston, USA. Vol. 1. pp. 232–248. ISBN 90-247-3430-4.

Gaspar, Th. 1981. Rooting and flowering, two antagonistic phenomena from a hormonal point of view. In *Aspects and Prospects of Plant Growth Regulators* (B. Jeffcoat, ed). *British Plant Growth Regulator Group,* Wantage, England. Monograph 6. pp. 39–49.

——— C. Keevers, P. DeBergh, L. Maene, M. Paques and Ph. Boxus. 1987. Vitrification: Morphological, physiological, and ecological aspects. In *Cell and Tissue Culture in Forestry, General Principles and Biotechnology* (J. M. Bonga and D. J. Durzan, eds). Martinus Nijhoff Pub., Boston, USA. Vol. 1. pp. 152–166. ISBN 90-247-3430-4.

George, E. F. and P. D. Sherrington. 1984. *Plant Propagation by Tissue Culture.* Exegetics, Hants, England. 709 pp. ISBN 0-9509325-0-7.

Grout, B. and M. J. Aston. 1977a. Transplanting of cauliflower plants regenerated from meristem culture, I. Water loss and water transfer related to changes in leaf wax and to xylem regeneration. *Hortic. Res.* 17:107.

——————— 1977b. Transplanting of cauliflower plants regenerated from meristem culture, II. Carbon dioxide fixation and the development of photosynthetic ability. *Hortic. Res.* 17:65–71.

Hackett, W. P. 1987. Juvenility and Maturity. In *Cell and Tissue Culture in Forestry, General Principles and Biotechnology* (J. M. Bonga and D. J. Durzan, eds). Martinus Nijhoff Pub., Boston, USA. Vol. 1. pp. 216–231. ISBN 90-247-3430-4.

Haissig, B. E. 1965. Organ formation *in vitro* as applicable to forest tree propagation. *Bot. Rev.* 31:607–626.

——— 1986. Metabolic processes in adventitious rooting. In *New Root Formation in Plants and Cuttings* (M. B. Jackson, ed). Martinus Nijhoff Pub., Dordrecht/Boston/Lancaster. pp. 141–189. ISBN 90-247-3260-3.

Halperin, W. 1986. Attainment and retention of morphogenetic capacity *in vitro*. In *Cell Culture and Somatic Cell Genetics of Plants, Plant Regeneration and Genetic Variability* (I. K. Vasil, ed). Academic Press, Orlando, FL, USA. Vol. 3. pp. 3–47. ISBN 0-12-715003-X.

Hammerschlag, F. 1982. Factors influencing *in vitro* multiplication and rooting of the plum rootstock myrobalan (*Prunus cerasifera* Ehrh.). *J. Amer. Soc. Hortic. Sci.* 107:44–47.

Hartmann, H. T. and D. E. Kester. 1983. *Plant Propagation: Principles and Practices.* Prentice Hall, NJ, USA. 4th ed. pp. 265–268. ISBN 0-13-681007-1.

Heide, O. M. 1969. Non-reversibility of gibberellin-induced inhibition of regeneration in *Begonia* leaves. *Physiol. Plant.* 22:671–679.

Hyndman, S. E., P. M. Hasegawa and R. A. Bressan. 1982a. Stimulation of root initiation from cultured rose shoots through the use of reduced concentration of mineral salts. *HortScience* 17:82–83.

_____ _____ _____ 1982b. The role of sucrose and nitrogen in adventitious root formation on cultured rose shoots. *Plant Cell Tissue Organ Cult.* 1:229–238.

Jarvis, B. C. 1985. Endogenous control of adventitious rooting in non-woody cuttings. In *New Root Formation in Plants and Cuttings* (M. B. Jackson, ed). Martinus Nijhoff Pub., Dordrecht/ Boston/Lancaster. pp. 191–222. ISBN 90-247-3260-3.

Jones, O. P. and S. G. S. Hatfield. 1976. Root initiation in apple shoots cultured *in vitro* with auxins and phenolic compounds. *J. Hortic. Sci.* 51:495–499.

_____ M. E. Hopgood. 1979. The successful propagation *in vitro* of two rootstocks of *Prunus:* The plum rootstock Pixy (*P. insititia*) and the cherry rootstock F12/1 (*P. avium*). *J. Hortic. Sci.* 54:63–66.

Khosh-Khui, M. and K. C. Sink. 1982. Rooting-enhancement of *Rosa hybrida* for tissue culture propagation. *Scientia Hortic.* 17:371–376.

Klämbt, D. 1983. Oligopeptides and plant morphogenesis: a working hypothesis. *J. Theor. Biol.* 100:411–435.

Lanphear, F. O. and R. P. Meahl. 1961. The effect of various photoperiods on rooting and subsequent growth of selected woody ornamental plants. *Proc. Amer. Soc. Hortic. Sci.* 77:620–634.

Lee, N., H. Y. Wetzstein and H. E. Sommers. 1986. The effect of agar vs. liquid medium on rooting in tissue-cultured sweetgum. *HortScience* 21:317–318.

Lovell, P. H., A. Illsley and K. G. Moore. 1972. The effect of light intensity and sucrose on root formation, photosynthetic ability, and senescence in detached cotyledons of *Sinapis alba* L. and *Raphanus sativa* L. *Ann. Bot.* 36:123–134.

Lyrene, P. M. 1980. Micropropagation of rabbiteye blueberries. *HortScience* 15:80–81.

_____ 1981. Juvenility and production of fast-rooting cuttings from blueberry shoot cultures. *J. Amer. Soc. Hortic. Sci.* 106:396–398.

Maene, L. and P. Debergh. 1983. Rooting of tissue cultured plants under *in vivo* conditions. *Acta Hortic.* 131:201–208.

_____ _____ 1985. Liquid medium additions to established tissue cultures to improve elongation and rooting *in vitro. Plant Cell Tissue Organ Cult.* 5:23–34.

McComb, J. A. and I. J. Bennett. 1982. Vegetative propagation of *Eucalyptus* using tissue culture and its application to forest improvement in Western Australia. In *Plant Tissue Culture 1982. Proc. V. Int. Cong. Plant Tissue and Cell Cult.* (A. Fujiwara, ed). Jap. Assoc. Plant Tissue Cult., Tokyo, Japan. pp. 721–722.

_____ _____ 1985. Micropropagation of *Eucalyptus*. In *Biotechnology in Agriculture and Forestry, Trees* (Y. P. S. Bajaj, ed). Springer Verlag, Berlin, W. Germany. Vol. I. pp. 340–362. ISBN 3-540-15581-3.

McCown, B. H. and G. B. Lloyd. 1985. A survey of the response of *Rhododendron* to *in vitro* culture. *Plant Cell Tissue Organ Cult.* 2:77–85.

_____ J. C. Sellmer. 1987. Media and Physical Environment. In *Cell and Tissue Culture in Forestry, General Principles and Biotechnology* (J. M. Bonga and D. J. Durzan, eds). Martinus Nijhoff Pub., Boston, USA. Vol. 1. pp. 1–16. ISBN 90-247-3430-4.

McCown, D. D. 1986. Plug systems for micropropagules. In *Tissue Culture as a Plant Production System for Horticultural Crops* (R. H. Zimmerman, R. J. Griesbach, F. A. Hammerschlag and R. H. Lawson, eds). Kluwer-Nijhoff Pubs., Norwell, MA, USA. pp. 53–60. ISBN 90-247-3378-2.

_____ B. H. McCown. 1987. North American Hardwoods. In *Cell and Tissue Culture in Forestry, Case Histories: Gymnosperms, Angiosperms and Palms.* (J. M. Bonga and D. J. Durzan, eds). Kluwer-Nijhoff Pubs., Norwell, MA, USA. Vol. 3. pp. 247–260. ISBN 90-247-3432-0.

Miller, N. F., L. E. Hinesly and F. A. Blazich, 1982. Propagation of Fraser fir by stem cuttings: effects of type of cutting, length of cutting, and genotype. *HortScience* 17:827–829.

Misson, J. P., P. Boxus, M. Coumans, P. Giot-Wirgot and Th. Gaspar. 1983. Rôle du charbon de bois dans les milieux de culture de tissus végétaux. *Med. Fac. Landbouww. Rijksuniv Gent.* 48:1151–1157.

Mullins, M. G. 1987. Propagation and genetic improvement of temperate fruits: the role of tissue culture. In *Plant Tissue and Cell Culture, Proc. VI Int. Cong. of Plant Tissue and Cell Cult.* (D. A. Somers, ed), University of Minnesota, Minneapolis, USA. pp. 395–406.

Murashige, T. 1974. Plant propagation through tissue cultures. *Ann. Rev. Plant Physiol.* 25:135–166.

Nougarède, A. and P. Rondet. 1983. Bases cytophysiologiques de l'induction rhizogène en reponse à un traitement auzinique dans l'épicotyle du Pois nain. *Ann. Sci. Nat. Bot. Paris.* 13e:121–149.

Pontikis, C. A. and E. Sapoutzaki. 1984. Effect of phloroglucinol on successful propagation *in vitro* of troyer Citrange. *Plant Prop.* 30(4):3–5.

Reuveni, O. and M. Raviv. 1980. Importance of leaf retention to rooting of avocado cuttings. *J. Amer. Soc. Hortic. Sci.* 106:127–130.

Rugini, E. and D. C. Verma. 1983. Micropropagation of difficult-to-propagate almond (*Prunus amygdalus*, Batsch.) cultivars. *Plant Sci. Lett.* 28:273–281.

Rappaport, J. 1940. The influence of leaves and growth substances on the rooting response of cuttings. *Natuurw Tijdschr.* 21:356–359.

Skoog, F. and C. Tsui. 1948. Chemical control of growth and bud formation in tobacco stem and callus. *Amer. J. Bot.* 35:782–787.

Smith, M. A. L., J. P. Palta and B. H. McCown. 1986. Comparative anatomy and physiology of microcultured, seedling, and greenhouse-grown Asian white birch. *J. Amer. Soc. Hortic. Sci.* 111:437–442.

Snir, I. 1981. Micropropagation of red raspberry. *Scientia Hortic.* 14:139–143.

Standardi, A. 1982. Effects of repeated subcultures in shoots of *Actinidia chinensis* (Pl). In *Plant Tissue Culture 1982, Proc. V. Int. Cong. Plant Tissue and Cult.* (A. Fujiwara, ed). Japanese Assoc. Plant Tissue Cult., Tokyo, Japan. pp. 737–738.

Stoutemyer, V. T. and O. K. Britt. 1965. The behavior of tissue cultures from English and Algerian ivy in different growth phases. *Amer. J. Bot.* 52:805–810.

Sriskandarajah, C. and M. G. Mullins. 1981. Micropropagation of Granny Smith apple: Factors affecting root formation *in vitro*. *J. Hortic. Sci.* 56:71–76.

Sutter, E. and R. W. Langhans. 1982. Formation of epicuticular wax and its affect on water loss in cabbage plants regenerated from shoot-tip culture. *Can. J. Bot.* 60:2896–2902.

Vasil, M. R. 1986. *Cell Culture and Somatic Cell Genetics of Plants, Plant Regeneration and Genetic Variability.* Academic Press, Orlando, FL, USA. Vol. 3. 460 p. ISBN 0-12-715003-X.

von Arnold, S. and T. Erickson. 1984. Effect of agar concentration on growth and anatomy of adventitious shoots of *Picea abies* (L.) Karst. *Plant Cell Tissue Organ Cult.* 3:257–264.

Wardle, K., E. B. Dobbs and K. C. Short. 1983. *In vitro* acclimatization of aseptically cultured plantlets to humidity. *J. Amer. Soc. Hortic. Sci.* 108:386–389.

Welander, M. and I. Huntrieser. 1981. The rooting ability of shoots raised *in vitro* from apple rootstock A2 in juvenile and in adult growth phases. *Physiol. Plant.* 53:301–306.

Werner, E. M. and A. A. Boe. 1980. *In vitro* propagation of Malling 7 apple rootstock. *HortScience* 15:509–510.

Wetzstein, H. and H. Sommer. 1982. Leaf anatomy of tissue cultured *Liquidamber styraciflua* (Hammamelidaceae) during acclimation. *Amer. J. Bot.* 69:1579–1586.

Whitehead, H. C. M. and K. L. Giles. 1977. Rapid propagation of poplar by tissue culture methods. *N. Z. J. For. Sci.* 7:40–43.

Zimmerman, R. H. 1984. Apple. In *Handbook of Plant Cell Culture, Crop Species* (W. R. Sharp, D. A. Evans, P. V. Ammirato and Y. Yamada, eds). Macmillan Pub. Co., NY, USA. Vol. 2. pp. 369–395. ISBN 0-02-949230-0.

_____ O. L. Broome. 1981. Phloroglucinol and *in vitro* rooting of apple cultivar cuttings. *J. Amer. Soc. Hortic. Sci.* 106:648–652.

_____ I. Fordham. 1985. Simplified method for rooting apple cultivars *in vitro*. *J. Amer. Soc. Hortic. Sci.* 110:34–38.

_____ R. J. Griesbach, F. A. Hammerschlag and R. H. Lawson. 1986. *Tissue Culture as a Plant Production System for Horticultural Crops.* Kluwer-Nijhoff Pub., Norwell, MA, USA. 371 pp. ISBN 90-247-3378-2.

CHAPTER 22

Future Directions in Adventitious Rooting Research

Bruce E. Haissig

USDA-Forest Service
North Central Forest Experiment Station
Forestry Sciences Laboratory
P.O. Box 898, Rhinelander, WI 54501

INTRODUCTION. 303
QUESTIONS. 304
APPROACHES. 305
MEASUREMENTS 308
CONCLUSION 308
ACKNOWLEDGMENTS 309
REFERENCES. 309

Additional key words: vegetative propagation, organogenesis, genetic engineering, micropropagation.

INTRODUCTION

It is a profound mistake to think that everything has been discovered; as well think the horizon the boundary of the world. —Lemierre

We do not know how, when, or where the need for vegetative propagation of plants arose during the 10,000 yr development of agriculture (Weaver 1985). At least within written history, selection of special genotypes and their vegetative propagation has been a prime basis for crop improvement that has not been completely replaced by breeding and planting of seeds. Unfortunately, whereas some plants propagate vegetatively with ease, others do not, which is the primary factor upon which hundreds of written communications, and the present book, have been based.

Thus, there has been a long and fairly orderly progression in the acquisition of knowledge of vegetative propagation. Deficiencies in practice have led to research and development. Most recently, fundamental investigations in genetics, biochemistry, and physiology have been undertaken, all targeted at finding better procedures for rooting of cuttings *per se* or an understanding of the process of rooting itself, or both.

Many researchers have collected numerical data, based on the hypothesis that enough numbers collected and properly compared will yield repeatable, interrelated sequences that can be used to decipher the most basic processes underlying rooting. However, that hypothesis remains unproven. We have not been able to identify any underlying law(s) that our data and, therefore, rooting obeys. Thus, we do not grasp the causes and controls that would allow us to root what we wish, when we wish, as has been our aim for centuries.

Nonetheless, based on the increasing progress of the last 25 yr, as recounted and discussed in this book and another (Jackson 1986), there is hope for better enumeration and understanding of the causes and controls of rooting by cuttings, probably within this century and certainly by early in the next. Therefore, I will briefly describe in this chapter how progress in our understanding of rooting may be accelerated. I have not relied directly on many specific articles from the scientific literature, although my ideas have been conditioned during the past 27 yr by the written and spoken ideas of others in many disciplines, to whom I am grateful. In sum, I cannot point to any indisputable basis for my evaluation, which to some degree makes the present chapter one of personal opinion, albeit conditioned by Dryden's caution: "Stiff in opinions; always in the wrong." Therefore, I present here mild and flexible opinions, those interpretations and suggestions that constitute the proverbial "food for thought." For brevity, I have used the term *rooting* to mean *adventitious rooting*.

QUESTIONS

Examine well your own thoughts.—Chaucer

A colleague recently suggested that he would rather have poor measuring equipment but a good question to ask (hypothesis) than the reverse. His thought again brought to mind that we must formulate meaningful questions concerning rooting before worrying about testing. Rooting research on cuttings during this century has pursued diverse answers, mostly based on two broad questions: 1) How do cells form primordia that will develop into functional roots?; and 2) How does whole plant biology affect rooting of cuttings?

Future progress in understanding rooting may depend upon how these two questions are interrelated in terms of actual experimentation. The tendency has been to simplify rooting research by recognizing and then largely ignoring aspects of one or the other question. That penchant is understandable because rooting appears to be incredibly complex. How can we study processes of primordium initiation and development apart from influences of the donor (stock) plant? How can influences of the donor plant be studied when we do not understand primordium initiation and development? Overall, one question confuses and obscures the other. It may be more profitable to combine elements of each question into a single, guiding question that is neither oversimplified nor unassailable. From a good single question, very specific individual hypotheses might be developed and rationally tested. In doing so, we might learn that rooting of cuttings is not as complex as it presently appears. But, first, the essential elements of these two questions and each question's merits and deficiencies require brief examination.

The first question aims at discovering the basic genetics, molecular biology, biochemistry, biophysics, and physiology of rooting. All the problems of rooting cuttings would probably be solved if the first question could be fully answered. However, a complete answer may be very long in coming because the first question does not offer sufficient experimental guidance. A wide variety of studies can legitimately be targeted at answering the first question. However, as evidenced in many chapters of the present book, we have been unable to synthesize disparate answers into any meaningful rooting hypothesis. The first question is deficient because it does not provide guidance on how to choose species, donor plants (clones), donor plant environments, propagation environments, etc. For example, do we presently have any basis for assuming that an understanding of root primordial cell physiology in mung bean will apply to giant sequoia? Would an understanding of primordial cell responses in giant sequoia apply equally to clones separated in age by 500 yr? Probably, yes, but only if the questions that we are asking about rooting go far beyond what is occurring exclusively in the primordial cells of cuttings.

The second question aims at defining the fundamental processes of rooting by cuttings in the context of influences imposed by the donor plant's life cycle (i.e. whole plant biology). Thus, the second question affords important guidelines to aid in fully answering the first question. For example, the second question can be used to define the specific constraints within which an answer to the first question is sought. Initially, an investigation might target the understanding of rooting only during a very limited portion of the donor plant's life cycle. The answers obtained in this fashion would still be constrained but the constraints would have been defined by the investigator, not by the results. If constraints are properly defined, constrained answers may be very useful for improving practical propagation and advancing our basic knowledge of rooting.

As another example, the second question may tell us much about rooting via negation. Rooting competes, often ineffectively, with other processes that occur during the donor plant's life cycle, including the obscure processes of maturation. Whereas we may never directly obtain a complete answer to the first question by studying cuttings alone, much can probably be learned about the process of rooting by understanding competing processes that for some reason may be more experimentally accessible. This negation approach may also aid in formulating specific, discerning tests about rooting. As an example, rooting (an early life trait) and flowering (a later trait) somehow compete, in some species nearly to the point of being mutually exclusive (e.g. Gaspar 1980 and Chapter 9 by Gaspar and Hofinger). Therefore, what is learned about flowering may also yield new information directly about rooting or at least about how to study rooting.

Study of only the second question might, however, subvert study of the first by a substitution of indirect approaches for direct, leading to the definition of influential as contrasted with causal factors in rooting. In that event, the rate of accumulating knowledge would quickly wane. But, that situation need not arise.

Studies of the first and second questions can be related if genetics is used as the basis for studies of biochemistry, molecular biology, biophysics, and physiology. Genetics is suggested as a common basis because it is the only discernible constant between the donor plant and its cuttings (ramets) once the cuttings are severed. The act of making the cutting abruptly starts a new life cycle and dramatically changes genotype-environment interactions, resulting in profound changes in all but genetics, which is the least transient and variable aspect of physiology. The life cycle of the donor plant is the product of genotype and is conditioned by genotype-environment interactions. Rooting of a cutting, which initiates the new life cycle, seemingly results from the donor plant's genetics per se and also from a quantitatively and qualitatively different expression of the donor plant's genetics. Whether genetic expression can change in the cutting, causing rooting, seems strongly related to the genotype-environment interaction in the donor plant's life cycle.

The foregoing formulation has a basis in studies of root system development. For example, Zobel (1986) has indicated that genes which control the root system can act directly or indirectly, and that of the 30% of the genome which modifies rooting characteristics about one-third of the genes affect only the root system. Therefore, future research on the rooting of cuttings may become more holistic, asking the single question: What quantitative (structural) and qualitative (regulatory) genetic characters of the stock plant and its cuttings, interacting with environment, predispose initiation of adventitious root primordia and their development into functional root systems?

Future research should address this question because it embodies every element of genetics, and every direct and indirect manifestation of genetics such as developmental morphology and anatomy, molecular biology, biochemistry, biophysics, and physiology, which in total comprehend all the underlying causes and controls of rooting in cuttings.

APPROACHES

It is a bad plan that admits of no modification. —Publilius

Genetics is the only identifiable link between the donor plant and the cutting, as discussed above. Therefore, genetic approaches appear to be the most useful for future explorations of rooting, whether these studies concern interpositions of donor plant influences or the direct processes of primordium initiation and development in the cutting. There is almost no factual basis to support the foregoing statement because, of all possible approaches to studying rooting, the genetic approach has either been virtually unused, or inefficiently or improperly used. This state of affairs has not been true over the last 10 yr in other areas of vegetative regeneration research such as adventitious bud initiation and somatic embryogenesis by in vitro cultures of herbaceous agronomic crop plants (see Chapter 3 by Haissig and Riemenschneider); nor has it been true in studies of root system development (e.g. Zobel 1986).

This difference in approaches has resulted in a profound consequence: More knowledge has been gained during the last 10 yr in areas of vegetative regeneration other than rooting of cuttings. To increase effectiveness, future rooting studies might employ, for example, mutants of agronomic and other plants that have little genetic variation (e.g. Veierskov et al. 1982), and full-sib families of species that have wild-type genetic variation (see Chapter 3 by Haissig and Riemenschneider). Heterogeneous species could also, and possibly most effectively, be studied by using self-pollinated (inbred) lines.

One type of genetic approach, comparison of easy- and difficult-to-root plants, has been used in rooting research but never with complete propriety. Given an otherwise suitable experimental design, it is completely proper to perform comparative studies of, for example, two closely related varieties, only one of which roots easily and profusely, with the goal of determining the genetic basis. Such research assumes that most of the genotypic differences between the two varieties relate directly or indirectly to rooting because most of the genomes are homologous. The assumption of genomic homology is acceptable in a working hypothesis but needs validation before rooting tests are conducted. Unfortunately, such genomic comparisons have never preceded rooting tests, which makes the "varietal approach" suspect as it has been used. There have been, however, far more presumptive and, therefore, far less valid uses of the easy- vs. difficult-to-root approach. These less trustworthy tests have been "species approaches" wherein species, even in different genera, were compared. The species approach is of little potential value because it presumes that indirect genetic effects which influence rooting (e.g. Zobel 1986) are similar and similarly expressed between species. This assumption could only be true if there were broad genomic homology between species, which would contradict their identification as separate species. Based on the foregoing, the species approach may never be useful as a genetic approach in rooting tests. However, specific valid genetic approaches have been fully discussed and illustrated elsewhere in this book (see Chapter 3 by Haissig and Riemenschneider).

A focus that includes genetics helps in selecting specific experimental plants for study. At present, it seems impossible to transfer knowledge of rooting between species, which apparently countermands most "model species" approaches. The difficulty in transferring knowledge of rooting between species is probably due to pronounced interspecific genetic differences. Thus, it has generally been most reliable to predict rooting of a species based on known rooting characteristics of a close relative. As a consequence, fundamental knowledge of rooting should be obtained with the species for which propagation by rooting of cuttings is needed, if improved practical propagation is the goal. However, that may be possible only as we learn more about woody plant genomes and how to quickly manipulate them to produce special study plants. At present, a genetic focus indicates that suitable test species should have little, perhaps less than 20%, genetic redundancy, in order to allow study of recessive traits (Zobel 1986). Small genome size would facilitate genomic mapping. These criteria are not satisfied by most woody species that have been studied to date. Thus, the most discerning genetically based rooting research may now be restricted to herbaceous model species. However, there are untried but promising genetic approaches that are applicable to many difficult-to-root woody species (see Chapter 3 by Haissig and Riemenschneider).

Future investigations may not quickly decipher rooting in cuttings only by studying cuttings. Certain important limitations in studying the metabolism of rooting in cuttings were recently discussed (Haissig 1986). Certain of these limitations, when more generally stated, apply equally to the overall study of rooting:

- —Genetic, biochemical, and physiological attributes of rooting by cuttings are seemingly not unique enough to be quickly, easily, and unambiguously distinguished from attributes of "normal" whole plant development.
- —There may be no generally occurring, single status of whole plant or organ development that can be unequivocally linked to the peculiarities of rooting by cuttings.
- —The developmental anatomy of the donor plant and cuttings may be so variable and complex as to obscure studies of rooting by cuttings.
- —The biomass of the cutting is many times greater than those cells directly involved in the initiation and development of root primordia, which may obscure observation of the specifics of primordium initiation.

These limitations might be eliminated or substantially alleviated by use of *in vitro* culture. That idea is not new (e.g. Haissig 1965) but may be more useful now thanks to our increased knowledge of the last 10–15 yr in how to culture various types of organs and tissues, particularly of angiosperms. Important advances have been made in formulating suitable inorganic growth media and hormone supplements. A complete discussion of the potential uses of *in vitro* cultures in studying the rooting of cuttings is not within the scope of this chapter. Thus, I will limit discussion to one example.

With some angiospermous woody species, continuous shoot cultures may offer a means to easily study how maturation influences rooting. Establishment of continuous shoot cultures of certain species leads to a "stabilized," "juvenile," "rooting-prone" physiology, even though it may not have existed in the donor plant (McCown 1985 and see Chapter 21 by McCown). Therefore, the influence of maturation during the donor plant's life cycle on rooting of its cuttings may be assailable through the aid of *in vitro* cultures.

In the previous example, *in vitro* cultures can be used to anatomically and physiologically simplify the experimental system for precise study, including genotypic and genotype-environment influences. However, *in vitro* studies alone may not explain rooting in cuttings, which suggests simultaneous, comparative studies of rooting of cuttings from the same species and donors (clones) from which the *in vitro* cultures were originated. Previously, *in vitro* cultures have mostly been studied without the element of comparative testing of cuttings. Again, a more holistic technological approach may be beneficial.

There are new and potentially powerful approaches that might quickly and significantly increase knowledge of rooting. Molecular biology and the related technologies of recombinant DNA and genetic engineering offer remarkable promise in understanding diverse plant developmental processes, including rooting (e.g. Perani et al. 1986, Wilke-Douglas et al. 1986). The potential of *Agrobacterium* mediated genetic transformation has already been demonstrated by the natural genetic engineer of rooting, *Agrobacterium rhizogenes* (e.g. Cardarelli et al. 1985, Ryder et al. 1985, Boulanger et al. 1986, see also Chapter 20 by Strobel and Nachmias). Genetic engineering of herbaceous plants commonly appears in various newsletters and scientific journals; genetic engineering of a forest tree has recently been achieved (Fillatti et al. 1987); and genetic engineering of horticultural trees is underway (Anon. 1986). In addition to *Agrobacterium* mediated transformation there are methods to modify the genome without the need for additional biological agents (Perani et al. 1986, Zachrisson and Bornman 1986), such as microinjection, electroinjection (e.g. Hashimoto et al. 1985, Morikawa et al. 1986), and electroporation (e.g. Fromm et al. 1985, Langridge et al. 1985, Nishiguchi et al. 1986). Somatic cellular hybridization offers another powerful means for unique genomic modification (Zachrisson and Bornman 1984). Therefore, the ability to attain specific genetic modification of diverse plant species exists and most probably can be used to study the exact genes influencing rooting. We have found, for example, that introduction of the oncogenes from a strain of *Agrobacterium tumefaciens* into a hybrid poplar nearly eliminates rooting *in vitro* (Fillatti et al. 1987). This hybrid poplar roots readily even when genetically transformed by the same bacterial strain but without the oncogenes. The reduction in rooting ability is apparently due to higher endogenous cytokinin biosynthesis due to expression of cytokinin-coding genes, but there may also be other reasons. Such specific genetic engineering for studies of development will progress only as quickly as gene identification and isolation permit. But, gene identification methods constantly improve. Transposable elements, for example, offer remarkable promise for use in such probing, identifying activities (Fedoroff 1984, Freeling 1984, McClintock 1984, Doring and Starlinger 1986). Overall, applications of molecular biology via genetic engineering should allow study of quantitative and qualitative differences in gene activities on rooting. Previously, only breeding was available for this purpose, but breeding is too slow to be useful for developmental investigations with species that have long sexual generation times. These are often the very species that we wish to learn how to propagate.

MEASUREMENTS

When you can measure what you are speaking about, and express it in numbers, you know something about it. —Lord Kelvin

Gaining a better understanding of rooting in cuttings seems related to making more precise measurements, and making them earlier and with greater frequency during primordium initiation and early development. In these respects, future investigations should not perpetuate past discrepancies. Current analytical methods, properly chosen and applied, are adequate for measuring responses of donor plants and cuttings, basically organ-level measurements. However, the major problem is that most methods destroy the sample and, therefore, stop its development. Thus, studying the development of preprimordial and primordial cells entails discontinuous sampling and analyses. Analytical discontinuity causes uncertain or even grossly improper interpretation. What is needed seems to be presently impossible: Cellular group measurements of intact regions of the cutting during the dynamic preprimordial and primordial phases of rooting.

Cellular-level measurements, especially those related to preprimordial and primordial cells, are beyond the capabilities of most existing methods. Adventitious rooting in cuttings involves so few and often obscure primordial cells that they cannot readily be removed for microanalysis. Yet, future studies should be much concerned with preprimordial and primordial cells because we know so little about them beyond cytological observations. One needed improvement is better methods to easily and accurately identify preprimordial cells. In addition, cellular level analyses might be improved by the development of sophisticated microchemistry (e.g. Crossway and Houck 1985) and use of *in vitro* systems to increase the biomass of preprimordial and primordial cells, compared to nonregenerating tissues. Success in future tests may require combining both approaches, assuming the potential insufficiency of either alone. Microautoradiography offers promise for studying some cellular phenomena such as nucleic acid synthesis and translocation of various radioactively labeled compounds. However, microautoradiography has been little used in rooting research (e.g. Haissig 1970, 1971, Tripepi et al. 1983), which makes its general value uncertain. Further tests of microautoradiography are needed in the future, but other micro methods must also be devised or adapted for rooting research (e.g. Molnar and LaCroix 1972ab). Microimmunochemical tests (e.g. Hughes and Dunn 1985, Sagee et al. 1986) might be developed and prove useful, especially if such methods are combined with *in vitro* cultures as stated above. However, there is great overall uncertainty about the amount and rapidity of future progress in understanding rooting through direct measurements of primordial cells. Micromeasurement, especially nondestructive methods, would make a valuable individual area of study within vegetative and sexual regeneration research.

CONCLUSION

The longest way round is the shortest way home—Anon.

Rooting research should be redirected if we are to quickly grasp the underlying biology and benefit therefrom in practical propagation of higher plants. In the process, past research should be used positively as a basis for redirection, rather than negatively as an example of need for redirection. The greatest force driving redirection of rooting research is the body of knowledge about rooting that has developed in the last 25 yr, coupled with advanced knowledge primarily in genetics and molecular biology, and their vehicles of implementation, genetic engineering and plant breeding. Redirection of rooting research implies that new opportunities exist for advancing our knowledge, but opportunities are mostly achieved only during times of change, whether personal or societal. Redirection of rooting research now seems possible because of changing individual and societal perceptions, as described by Jackson (1986, p. XI):

The study of plant roots is currently enjoying something of a resurgence. Perhaps we are coming to realize more keenly than before the benefits to agriculture and horticulture that

can flow from such work. Furthermore, the essential nature of root research for any under-
standing of plants as whole organisms that interact in complex ways with their environ-
ment is self evident.

Even though I have attempted to do so in this chapter, I cannot recommend with certainty the
"best" redirection for rooting research. Possibly no one person can now or ever will be able to do so.
However, I feel that the community of researchers dedicated to this scientifically and commercially
important, interesting aspect of higher plant biology can effectively redirect efforts by considering
changes in only three critical aspects of experimentation: How to question, approach, and measure
the subject. Addressing each of these from historical perspective will result in the needed redirection.
The perspective should be broader than has been common, including not only our knowledge of
adventitious rooting but also knowledge of root system development in general and genetic impacts
thereon (e.g. Jackson 1986, Zobel 1986).

ACKNOWLEDGMENTS

I thank Drs. T. D. Davis, W. P. Hackett, and D. E. Riemenschneider for critical review of this
chapter, and Ms. Sandra Haissig for clerical assistance.

REFERENCES

Anon. 1986. Transformation of trees by *Agrobacterium* inoculation. *Agricell Rept.* 7(5):36.

Boulanger, F., A. Berkaloff and F. Richaud. 1986. Identification of hairy root loci in the T-regions of
Agrobacterium rhizogenes Ri plasmids. *Plant Mol. Biol.* 6:271–279.

Cardarelli, M., L. Spano, A. De Paolis, M. L. Mauro, G. Vitali and P. Costantino. 1985. Identification of the genetic locus responsible for non-polar root induction by *Agrobacterium rhizogenes*
1855. *Plant Mol. Biol.* 5:385–391.

Crossway, A. and C. M. Houck. 1985. A microassay for detection of DNA and RNA in small
numbers of plant cells. *Plant Mol. Biol.* 5:183–190.

Doring, H.-P. and P. Starlinger. 1986. Molecular genetics of transposable elements in plants. *Ann.
Rev. Genet.* 20:175–200.

Fedoroff, N. V. 1984. Transposable genetic elements in maize. *Sci. Amer.* 250:84–90, 95–98.

Fillatti, J. J., J. Sellmer, B. McCown, B. Haissig and L. Comai. 1987. *Agrobacterium* mediated transformation and regeneration of *Populus. Mol. Gen. Genet.* 206:192–199.

Freeling, M. 1984. Plant transposable elements and insertion sequences. *Ann. Rev. Plant Physiol.*
35:277–298.

Fromm, M., L. P. Taylor and V. Walbot. 1985. Expression of genes transferred into monocot and
dicot plant cells by electroporation. *Proc. Nat. Acad. Sci. USA* 82:5824–5828.

Gaspar, T. 1980. Rooting and flowering, two antagonistic phenomena from a hormonal point of view.
In *Aspects and Prospects of Plant Growth Regulators* (B. Jeffcoate, ed). British Plant Growth
Regulator Group, Wantage, England. Monograph 6. pp. 39–49.

Haissig, B. E. 1965. Organ formation *in vitro* as applicable to forest tree propagation. *Bot. Rev.*
31:607–626.

———— 1970. Influence of indole-3-acetic acid on adventitious root primordia of brittle willow. *Planta*
95:27–35.

———— 1971. Influence of indole-3-acetic acid on incorporation of ^{14}C-uridine by adventitious root
primordia of brittle willow. *Bot. Gaz.* 132:263–267.

———— 1986. Metabolic processes in adventitious rooting of cuttings. In *New Root Formation in
Plants and Cuttings* (M. B. Jackson, ed). Martinus Nijhoff Pub., Dordrecht/Boston/Lancaster.
pp. 141–189. ISBN 90-247-3260-3.

Hashimoto, H., H. Morikawa, Y. Yamada and A. Kimura. 1985. A novel method for transformation of intact yeast cells by electroinjection and plasmid DNA. *Appl. Microbiol. Biotechnol.* 21:336–339.

Hughes, M. A. and M. A. Dunn. 1985. The use of immunochemical techniques to study plant proteins. *Plant Mol. Biol. Reporter* 3:17–23.

Jackson, M. B. (ed). 1986. *New Root Formation in Plants and Cuttings.* Martinus Nijhoff Pub., Dordrecht/Boston/Lancaster. p. xi. ISBN 90-247-3260-3.

Langridge, W. H. R., B. J. Li and A. A. Szalay. 1985. Electric field mediated stable transformation of carrot protoplasts with naked DNA. *Plant Cell Repts.* 4:355–359.

McClintock, B. 1984. The significance of responses of the genome to challenge. *Science* 226:792–801.

McCown, B. H. 1985. From gene manipulation to forest establishment: shoot cultures of woody plants can be a central tool. *TAPPI J.* 68:116–119.

Molnar, J. M. and L. J. LaCroix. 1972a. Studies on the rooting of cuttings of *Hydrangea macrophylla:* Enzyme changes. *Can. J. Bot.* 50:315–322.

_____ _____ 1972b. Studies on the rooting of cuttings of *Hydrangea macrophylla:* DNA and protein changes. *Can. J. Bot.* 50:387–392.

Morikawa, H., A. Iida, C. Matsui, M. Ikegami and Y. Yamada. 1986. Gene transfer into intact plant cells by electroinjection through cell walls and membranes. *Gene* 41:121–124.

Nishiguchi, M., W. H. R. Langridge, A. A. Szalay and M. Zaitlin. 1986. Electroporation-mediated infection of tobacco leaf protoplasts with tobacco mosaic virus RNA and cucumber mosaic virus RNA. *Plant Cell Repts.* 5:57–60.

Perani, L., S. Radke, M. Wilke-Douglas and M. Bossert, 1986. Gene transfer methods for crop improvement: Introduction of foreign DNA into plants. *Physiol. Plant.* 68:566–570.

Ryder, M. H., M. E. Tate and A. Kerr. 1985. Virulence properties of strains of *Agrobacterium* on the apical and basal surfaces of carrot root discs. *Plant Physiol.* 77:215–221.

Sagee, O., A. Maoz, R. Mertens, R. Goren, and J. Riov. 1986. Comparison of different enzyme immunoassays for measuring indole-3-acetic acid in vegetative citrus tissue. *Physiol. Plant.* 68:265–270.

Tripepi, R. R., C. W. Heuser and J. C. Shannon. 1983. Incorporation of tritiated thymidine and uridine into adventitious-root initial cells of *Vigna radiata. J. Amer. Soc. Hortic. Sci.* 108:469–474.

Veierskov, B., A. S. Andersen, B. M. Stummann and K. W. Henningsen. 1982. Dynamics of extractable carbohydrates in *Pisum sativum.* II. Carbohydrate content and photosynthesis of pea cuttings in relation to irradiance and stock plant temperature and genotype. *Physiol. Plant.* 55:174–178.

Weaver, K. F. 1985. The search for our ancestors. *Nat. Geog.* 168:561–623.

Wilke-Douglas, M., L. Perani, S. Radke and M. Bossert. 1986. The application of recombinant DNA technology toward crop improvement. *Physiol. Plant.* 68:560–565.

Zachrisson, A. and C. H. Bornman. 1984. Application of electric field fusion in plant tissue culture. *Physiol. Plant.* 61:314–320.

_____ _____ 1986. Electromanipulation of plant protoplasts. *Physiol. Plant.* 67:507–516.

Zobel, R. W. 1986. Rhizogenetics (root genetics) of vegetable crops. *HortScience* 21:956–959.

Index

Abscisic acid 19, 113–114, 168, 175–176, 194, 205, 225, 244
Acclimation 34, 294–296
Acclimatization 218–219, 242
Activated charcoal 292, 295
Adventitious buds 17
Aging 11–25
Agrobacterium rhizogenes 284–288
  biology of 285
  effect on rooting of bare root stock 285–286
  effect on rooting of cuttings 286–287
Air change rates 263–264
Alcohol 138–140
Aminocyclopropane-1-carboxylic acid 154, 156
Amino acids 73, 207–208
Amino ethoxyvinylglycine 154–157
Amylase 93
Anatomy 13–15, 22–24, 37–39, 189, 193–194, 221
Ancymidol 169, 177, 180
Anthocyanin 291
Antigibberellins 175–180, 208
Antitranspirant 66, 83
Arginine decarboxylase 203–208
Artificial lighting 223–224
Assimilate supply 168
Assimilate transport 79, 176, 181, 192
Atmospheric gases 237–238
Auxin metabolism 117–125
Auxins 14–16, 18, 23–24, 35, 39–42, 62–63, 65, 82, 84, 90–95, 114, 117–125, 132–146, 150, 152, 154–158, 162, 168–169, 174–175, 177, 179, 181, 186–189, 191, 193–194, 202, 204, 221–223, 267, 279–281, 286–287, 292–295
  biosynthesis of 118–120
  catabolism of 120–125
  cofactors 41–42
  conjugated 120, 123
  endogenous levels of 122–123
  factors influencing response to 139–141
  relative efficacy 134–136
  sensitivity of etiolated tissues to 40–41
  synthesis of 39
  techniques for applying 136–139
  transport of 40, 72, 181
  uptake of 40

Bacteria 284–288
Banding 29–43
  comparison versus etiolation 35–37
  effect on anatomy 37–39
  history 31–33
  location of band 34–35
  materials for 31–33, 35
  physiological effects 39–42
  responsive species 30–31, 36–37
  width of band 35
Basal heating 140, 251, 257–258, 266–267
Benzylamino purine 32
Benzylisothiocyanate 155, 157
Bioassay 39, 274–282
  English ivy 277
  limitations 278
  mung bean 275
  pinto bean 276
  tomato leaf discs 277–278
Blanching 30, 34–35
Boron 41, 64–65, 205–206
Botrytis 32, 34
Bound auxins 120
  amide conjugates 120
  ester conjugates 120
Buds 167–168, 191–192

C-effects 48, 53
  M-effects 53
  m-effects 53
Cadaverine 202–203
Calcium 62–66
Carbon/nitrogen ratios 41, 70, 73, 228–229
Carbohydrates 63, 70–76, 79, 82–84, 93–95, 113, 169, 181, 219, 221, 226, 228, 236, 242–243, 251, 256–257
  accumulation during rooting 73–74
  exogenous 74–75
  metabolism 169
  phosphorylated 73, 75–76
  roles in rooting 75–76
  stock plant 71–73
Carbon dioxide 79–85, 112, 214–215, 219, 223, 226–228, 237–238
  enrichment 214–215, 219, 223, 226–228
  exchange 79–85
Cell division 114, 169, 179, 181
Cell expansion 114

Cell wall synthesis 114
Cellular solvent capacity 76
Cellulase 158
Centrifugal foggers 262, 265
Chemical rooting agents 132–146
Chlorflurenol 177, 179
Chlormequat chloride 169, 176–177
4-Chloroindolyl-acetic acid 119
Chlorophyll 33
Chlorphonium chloride 169, 177, 180
Clone 48
Computers 266, 270
Conjugated auxins 120, 123
    amide conjugates 120
    ester conjugates 120
Controlled atmosphere storage 237–238
Cotyledons 276
Current photosynthesis 79–85
Cyclohexamide 91
Cytochrome oxidase 89
Cytokinins 81, 175, 177, 185–197, 205, 292,
        294–295, 307
    basal applications 188–189
    binding 186
    biosynthesis 194–195
    endogenous 189–193
    exogenous 186–189
    foliar applications 188–189
    in root apical dominance 195–196
    metabolism 186
    root growth 194–196
    seasonal activities 189, 191–192

Daminozide 177, 180
DCMU 75, 83
Decapitation 168
Defoliation 33
DFMA 204, 207
DFMO 204, 207
Diamine oxidase 205
Differentiation 34, 38–39, 41
Disease 236, 242, 266, 269–270
DNA 285
DNase 95–96
Dominance 48–49
Donor Plant 11–25
Dormancy 221–222, 243–244
Dry matter accumulation 80–81

Enclosures 249–255
    contact systems 252–255
    for mist 258–261
    in greenhouse 251–255

    low tunnels 249–251
    tents 252–255
Environmental conditions 214–229, 248–270
Enzyme-linked immunosorbent assay 279–
        280
Enzymes 88–97, 178–179, 203–205, 207–208
    hydrolytic 93–96
    oxidases 89–93
    respiratory 89
Epistasis 48–49
Ethephon 153–157
Ethylene 19, 82, 113–114, 134, 150–158, 194,
        208, 225, 237–238
    analogues of 152
    endogenous 154–155
    exogenous 151–153
    factors affecting response to 153–154
    inhibitors of 155–156
    interactions with auxin 156–157
    mode of action 157–158
    sensitivity of cuttings to 158
Etiolation 29–43, 141, 166, 220–221, 275
    comparison versus banding 35–37
    effect on anatomy 37–39
    history 31–33
    physiological effects 39–42
    responsive species 30–31, 36–37
    timing 33
    transmissable effect 38
Experimental approaches 303–309

Factorial mating 56
Family mean correlations 55
Flooding 155
Flowering 11–12, 16, 124–125, 221–222, 305
Flurprimidol 177, 180
Fogging 106, 261–266
    distribution 265
    equipment 262–265
    general principles 261
    output 263–264
    requirements 263–264
    system development 261–262
Foliar fertilization 66
Food reserves 242–243
Freezing of cuttings 237
Full-sib families 54–56
Fungicides 137, 242, 269–270

G compounds 17, 21–22
Gas chromatography-mass spectrometry
        280–281
Genetic engineering 307

Genetics 47–57, 305–307
  biochemical and physiological measur-
    ments related to 56–57
  effects on organogenesis and somatic
    embryogenesis 49–51
  effects on rooting of cuttings 51–52
  experimental approaches 52–56
  interaction with environment 53
  models 54–56
Gibberellin antagonists 169, 176–180
Gibberellins 13, 19–20, 32, 162–170, 174–180,
    190, 194, 204, 208, 222, 291, 292
  effects on rooting of various species 164
  endogenous content of 163
  exogenous application of 163
  interactions with buds 167–168
  interactions with light 166–167
  interactions with other growth regulators
    168–169
  mode of action in rooting 169
  sensitivity of rooting to 163–165
  shoot growth 167–168
  timing of application 163–165
Girdling 33, 35, 141
Glutamic dehydrogenase 96
Glycolytic-citric acid cycle 73, 76
Glycosylation 75
Grafting 13, 17, 22–24, 291
Grandinol 21–22
Gravimetric studies 106–108
Growth inhibitor 174–182
Growth regulators 117–209, 243–244
Growth retardants 84, 169, 174–182
Growth substances 117–209, 243–244

Hairy root syndrome 284–285
Half-sib families 54
Hedging 15, 16, 33
Heritability 47–57
Human chorionic gonadotropin 180
Humidity 33, 81, 83, 112, 236, 238, 249–266
Hydrolytic enzymes 93–96, 169
Hypobaric storage 237–238
Hypotheses for studying rooting 304–305

Immunochemical assays 279–280
IAA oxidase 42, 90–92, 120–121, 123–124,
    194
Indolyl-3-acetic acid 118, 119, 120–123
Indolyl-3-acrylic acid 120
Inheritance 47–57
Internode length 38
Invertase 94

In vitro culture 13–14, 17–18, 19, 20, 24, 47,
    48, 49–51, 163, 166, 168, 207, 262, 289–
    299, 307
Irradiance 33–34, 38, 40–42, 103–106, 111,
    112, 113, 166, 170, 215–226, 248–270
  artificial 223–224
  light quality 222–223
  photoperiod 221–222
  seasonal variations 215
Isoenzymes 89, 90, 92, 93, 96, 189, 194
Isoperoxidases 120–121, 124–125

Juvenility 11–25, 42, 189–190, 291, 293

Lateral rooting 195–196, 206
Layering 30
Leaching of mineral nutrients 65–67
Leaf cuttings 190–191
Leaf retention/shedding 188–189
Light 29–43, 71–72, 74, 79, 80, 81, 82, 83, 90,
    166–167, 215–226, 248–270
  artificial 223–224
  quality 222–223
Lignification 38, 42, 153
Lignotubers 14
Lipids 242–243

Magnesium 62, 64, 65–66
Malate dehydrogenase 89
Manganese 63, 66
Maleic hydrazide 90
Mannopine 285
Mass spectrometry 279, 280, 281
Maturation 11–25, 49, 189–190, 291, 293, 305
  factors associated with reduced rooting
    18–24
  overcoming adverse effects of 16–18
Membranes 207–208
Meristematic activity 41
Methodology 308
Microcalli 297–298
Microculture 13–14, 17–18, 19, 20, 24, 47, 48,
    49–51, 163, 166, 168, 207, 262, 289–299,
    307
Microcutting 289–299
Microfoam 249, 250
Micropropagation 13–14, 17–18, 19, 20, 24,
    207, 262, 289–299, 307
Microtechnique 308
Mineral nutrition 61–67, 82, 228–229
  effect on post-propagation vigor 65
  in root initiation 62–63
  leaching 65–67

root growth and development 63–65
Mineral salts 294
Mist 65–67, 103, 104, 106, 107, 108, 110, 113
    enclosed 258–261
    environment 255–259
    frequency 256, 260–261
    outdoor 259
Mobilization of mineral nutrients 62, 63, 64, 65
Monophenols 42
Morphactins 90, 177, 179
Mycorrhizae 268

Nitrogen 62–64, 65, 70, 73, 208, 228–229, 241
Norbornadiene 155
Nozzles 262, 265
Nucleases 95–96
Nucleic acids 95–96, 207
Nucleotides 94, 95–96
Nutrient diversion 168

Opines 285
Organogenesis 49–51
Ornithine decarboxylase 203, 204, 206, 207
Ortet 48
Osmoregulation 76
Osmotic potential 104–106
Oxidative phosphorylation 89
Oxygen 237–238

Paclobutrazol 89, 93, 177–179
Pasteurization 270
Pentose phosphate pathway 73, 75–76
Pericycle 286
Peroxidases 42, 89–90, 91, 120–121, 124–125, 189
Pesticides 269–270
Phase change 11–25, 189–190, 291, 293
Phases of rooting 89–90, 118, 124
Phenolase 92
Phenolics 20, 41–42, 84, 89–93, 120–121, 291–292
Phenoxy compounds 134, 136
Phenylalanine ammonia-lyase 42
Phloridzin 92
Phosphatases 95
Phosphorous 62, 63, 64, 65, 66
Phosphorylated carbohydrates 73, 75–76
Photo-oxidation 80
Photoperiod 83, 160–167, 216, 221–222
Photosynthesis 71, 72, 73, 79–85, 114, 218–219, 226, 229, 249, 251–252, 256–257, 267, 296

effect of auxins on 82
effects of carbon dioxide on 81–82
effect of leaf number on 82
effect of sucrose on 82
light saturation of 81, 84
Photosynthetically active radiation 80–81
Phytochrome 166, 204
Polyamine oxidase 205
Polyamines 202–209
    effects on rooting 205–207
    effects on root growth 206
    effects on in vitro rooting 207
    metabolism of 203–205
    modes of action 207–208
Polyethylene 249, 250, 251, 252, 253, 254, 255, 258, 259, 260, 261
Polyethylene glycol 113, 225–226
Polyphenol oxidase 24, 41, 42, 92–93
Polyphenols 42
Potassium 62, 63, 64, 65, 66, 67
Protein synthesis 114
Pruning 16
Purines 93, 94, 95, 96
Putrescine 202–209
Psychometric studies 103–106
Pyrimidines 93, 94, 95, 96

Radioimmunoassay 279–280
Ramet 48, 53
Rejuvenation 16–18, 291–293
Relative humidity 81, 83, 236, 238
Research, future directions 303–309
Respiration 80, 95, 218, 220
Respiratory enzymes 89
Ri plasmid 285
RNase 96
Root cuttings 192–193
Root growth 194–196
Rooting cofactors 20, 24, 41–42, 84, 92, 114, 175, 188, 244, 275
Rooting compounds 132–146
Rooting inhibitors 21–22, 221, 222, 244
Rooting media 260, 266–268, 270

S-adenosyl methionine 204–208
Salvage pathway 96
Sanitation 269–270
Sclerenchyma 221
Sclerification 37–38, 39
Seasonal effects on rooting 179, 215, 216, 221–222
Seaweed concentrates 192, 196
Senescence 82, 188

Shade curtains 252
Shading 33–34, 103, 104, 105, 106, 112, 249, 250, 251–252, 256–257, 259, 261, 263, 264, 265
Shoot culture 290
Shoot cuttings 191–192
Shoot growth 83–84, 167–168, 174–182
Silver ion 155
Silver nitrate 155, 157
Silver thiosulfate 155
Sink activity 81
Somatic embryogenesis 49–51
Spermidine 202–209
Spermine 202–209
Sphaeroblasts 17
Stages of rooting 61, 71, 73–74, 89, 90, 187, 188, 189
Starch 71, 73, 74, 93
Stem blockage 110–111
Sterols 177
Stock plant 11–25, 29–43, 61, 62–63, 65, 72, 73, 140–141, 166, 167, 179, 214–229, 296
    carbon dioxide enrichment of 226–228
    lighting of 214–224, 226
    mineral nutrition of 228–229
    temperature 225
    treatments to overcome maturation 16–18
    water relations 113, 225
Stomata 80, 81, 83, 108–109, 111, 113, 175
Storage of cuttings 235–244, 269
    conditions for 236–238
    controlled atmosphere for 237–238
    duration of 238–241
    hypobaric conditions for 237–238
    physiological changes during 242–244
    pretreatment of cuttings for 241–242
    reasons for 235–236
Stress 175, 176
Suberization 38–39
Succinic dehydrogenase 89
Sucrose 73, 74, 75, 82, 94–95, 223
Sugars 74–75, 76, 82, 83, 94–95
Supplementary lighting 223–224

Temperature 33, 40, 71, 72, 74, 82, 85, 112–113, 222, 225, 236, 237, 243–244, 248–270
Thermo-blankets 249, 250
Tissue culture 13–14, 17–18, 19, 20, 24, 47, 48, 49–51, 163, 166, 168, 207, 262, 289–299, 307
    as a research tool 296–298
    establishment 290–294
Translocation 84, 114
Transpiration 40, 111, 236
Triiodobenzoic acid 181
Tryptophan 63, 118–119
Turgor pressure 104–106, 109, 253, 257

Uncouplers of oxidative phosphorylation 89, 95

Vapor pressure gradient 249, 253, 254, 255, 256, 261
Ventilation 263–264
Vitamin D_2 90
Vitrification 293

Water content 105, 106–108
Waterlogging 106, 108, 225
Water potential 103–106, 236
Water relations 34, 80, 81, 82, 84, 102–115, 225, 226, 228, 236, 241, 242, 248–270, 295
    roles in rooting 113–114
    stock plant 113
Water uptake 109–111
    through leaves 109–110
    through stem base 110
Wet tents 254–255
Wounding 110, 242

XE-1019 177, 179

Zinc 63, 66